EPIGENETIC REGULATION IN THE NERVOUS SYSTEM

ELSEVIER science & technology books

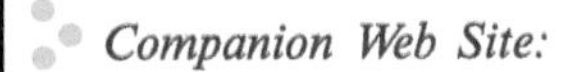

Companion Web Site:

http://booksite.elsevier.com/9780123914941/

Epigenetic Regulation in the Nervous System

J. David Sweatt, Michael J. Meaney, Eric J. Nestler, and Schahram Akbarian

Resources available:

- All figures from the volume available

EPIGENETIC REGULATION IN THE NERVOUS SYSTEM

Basic Mechanisms and Clinical Impact

Edited by

J. David Sweatt
McKnight Brain Institute
Department of Neurobiology
University of Alabama at Birmingham
Birmingham, Alabama, USA

Michael J. Meaney
Departments of Psychiatry, Neurology, and Neurosurgery
Douglas Institute
McGill University
Montreal, Quebec, Canada

Eric J. Nestler
Fishberg Department of Neuroscience
Friedman Brain Institute
Mount Sinai School of Medicine
New York, New York, USA

Schahram Akbarian
Department of Psychiatry
Friedman Brain Institute
Mount Sinai School of Medicine
New York, New York, USA

AMSTERDAM • BOSTON • HEIDELBERG • LONDON
NEW YORK • OXFORD • PARIS • SAN DIEGO
SAN FRANCISCO • SINGAPORE • SYDNEY • TOKYO
Academic Press is an imprint of Elsevier

Academic Press is an imprint of Elsevier
32 Jamestown Road, London NW1 7BY, UK
225 Wyman Street, Waltham, MA 02451, USA
525 B Street, Suite 1800, San Diego, CA 92101-4495, USA

Front Cover: ***"DNA with Methyl-Cytosine and Hydroxy-Methyl-Cytosine"***
J. David Sweatt, acrylic on wood panel (24 × 48), 2012

Back Cover: ***"DNA with Methyl-Cytosine and Hydroxy-Methyl-Cytosine – Abstraction"***
J. David Sweatt, acrylic on wood panel (24 × 48), 2012

British Library Cataloguing-in-Publication Data
A catalogue record for this book is available from the British Library

Library of Congress Cataloging-in-Publication Data
A catalog record for this book is available from the Library of Congress

ISBN: 978-0-12-391494-1

For information on all Academic Press publications
visit our website at www.store.elsevier.com

Typeset by MPS Limited, Chennai, India
www.adi-mps.com

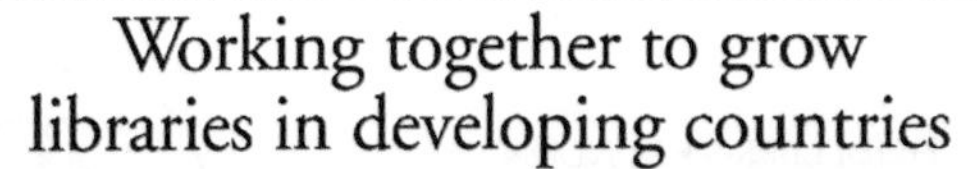

Contents

9. miRNAs and Neurodevelopmental Disorders 251

QUAN LIN AND YI E. SUN

10. Imprinting in the CNS and Neurodevelopmental Disorders 267

WESTON T. POWELL AND JANINE M. LASALLE

11. Neuronal Genomic and Epigenetic Diversity 281

MICHAEL J. McCONNELL AND FRED H. GAGE

12. Adult Neurogenesis 301

JENNY HSIEH AND HONGJUN SONG

13. Transgenerational Inheritance in Mammals 323

ISABELLE M. MANSUY, RAHIA MASHOODH AND FRANCES A. CHAMPAGNE

Preface

In a living cell, the functional definition of the human genome cannot be captured solely by its linear sequence of 3 (or 6 when diploid) billion base pairs. It is the *epigenome*, with highly regulated modifications of DNA cytosine and more than 100 site- and amino acid residue specific histone modifications, histone variants and other types of epigenetic markings which, in concert, define localized chromatin structures and functions, provide a molecular bridge between genes and "the environment", and orchestrate the expression of tens of thousands of transcriptional units, condensed chromatin clusters and many other features that distinguish between various cell types and development- or disease-states sharing the same nascent genome within the same subject.

Epigenetics in the nervous system, in particular, has made breath-taking advances over the course of the last 10–15 years. Initially, there were only a handful of studies, mainly focused on a single mark, DNA cytosine methylation, in the context of brain aging and development. Fast forward to the present, and the database grew to hundreds of studies, collectively indicating that epigenetic landscapes in brain maintain their highly dynamic and bi-directional regulation throughout the lifespan, and play a critical role in the mechanisms of learning, memory and, more generally, neuronal plasticity. Furthermore, a rapidly expanding repertoire of chromatin modifying drugs has been shown to exhibit an unexpectedly broad therapeutic potential for a wide range of degenerative and functional disorders of the nervous system and, furthermore, epigenetic dysregulation at selected loci, or even genome-wide, is thought to play a key role for the molecular pathology of major psychiatric disorders (including some cases diagnosed with autism, schizophrenia and depression) or maladaptive mechanisms associated with addiction and substance dependence and abuse.

Given these recent advances, there is clearly a need for a book that addresses molecular, cellular, behavioral and clinical roles for epigenetic mechanisms in the nervous system. It appears that the time is ripe to introduce a foundational book that will be broadly relevant to a wide variety of emerging research programs beginning to investigate the role of epigenetics in neural and CNS function and dysfunction. We hope that this book will capture and communicate to the interested reader some of the excitement that has gripped the neuroepigenetics field for the last several years, both from the basic and clinical science perspective.

Finally, the Editors would like to express their gratitude towards the various other authors and co-authors of these book chapters, without whom the compilation of these various chapters would not have been possible. We are also indebted to the most valuable support of the Elsevier editorial staff, including Kristi Anderson, and the anonymous reviewers whose valuable comments helped greatly to improve the quality of this book.

J. David Sweatt
Michael J. Meaney
Eric J. Nestler
Schahram Akbarian

List of Contributors

Ted Abel Department of Biology, School of Arts and Sciences, University of Pennsylvania, Philadelphia, Pennsylvania, USA

Schahram Akbarian Department of Psychiatry, Departments of Psychiatry and Neuroscience, Friedman Brain Institute, Mount Sinai School of Medicine, New York, New York and Brudnick Neuropsychiatric Research Institute, University of Massachusetts Medical School, Worcester, Massachusetts, USA

Rahul Bharadwaj Brudnick Neuropsychiatric Research Institute, University of Massachusetts Medical School, Worcester, Massachusetts, USA

Morgan Bridi Department of Neuroscience, Perelman School of Medicine, University of Pennsylvania, Philadelphia, Pennsylvania, USA

Christian Caldji Sackler Program for Epigenetics Psychobiology and Departments of Psychiatry and Neurology & Neurosurgery, McGill University, Montreal, Canada

Frances A. Champagne Columbia University, Department of Psychology, New York, USA

J. David Sweatt McKnight Brain Institute, Department of Neurobiology, University of Alabama at Birmingham, Birmingham, Alabama, USA

Jeremy J. Day Department of Neurobiology and Evelyn F. McKnight Brain Institute, University of Alabama at Birmingham, Birmingham, Alabama, USA

Josie C. Diorio Sackler Program for Epigenetics Psychobiology and Departments of Psychiatry and Neurology & Neurosurgery, McGill University, Montreal, Canada

Sabine Dhir Sackler Program for Epigenetics Psychobiology and Departments of Psychiatry and Neurology & Neurosurgery, McGill University, Montreal, Canada

Daniel M. Fass Broad Institute of Harvard and MIT, Cambridge, Massachusetts, USA; Center for Human Genetic Research, Massachusetts General Hospital, Harvard Medical School, Boston, Massachusetts, USA

Jian Feng Fishberg Department of Neuroscience and Friedman Brain Institute, Mount Sinai School of Medicine, New York, New York, USA

Fred H. Gage Laboratory of Genetics, Salk Institute for Biological Studies, La Jolla, California, USA

Junjie U. Guo Institute for Cell Engineering, Department of Neurology, The Solomon H. Snyder Department of Neuroscience, Johns Hopkins University School of Medicine, Baltimore, Maryland, USA

Stephen J. Haggarty Center for Human Genetic Research, Massachusetts General Hospital, Departments of Neurology and Psychiatry, Harvard Medical School, Boston, Massachusetts; Broad Institute of Harvard and MIT, Cambridge, Massachusetts, USA

Jenny Hsieh Department of Molecular Biology, University of Texas Southwestern Medical Center, Dallas, Texas, USA

Janine M. LaSalle Medical Microbiology and Immunology, Genome Center, Medical Institute of Neurodevelopmental Disorders, University of California, Davis School of Medicine, Davis, California, USA

Quan Lin Department of Psychiatry and Behavioral Sciences; Intellectual Development and Disabilities Research Center, David Geffen School of Medicine, University of California Los Angeles, Los Angeles, USA

Isabelle M. Mansuy Medical Faculty of the University of Zurich and Department of

Health Science and Technology, Swiss Federal Institute of Technology, Zurich, Switzerland

Rahia Mashoodh Columbia University, Department of Psychology, New York, USA

Michael J. McConnell Department of Biochemistry and Molecular Genetics, Center for Brain Immunology and Glia, School of Medicine, University of Virginia, Charlottesville, Virginia, USA

Michael J. Meaney Departments of Psychiatry, Neurology, and Neurosurgery, Douglas Institute, McGill University, Montreal, Quebec, Canada

Guo-li Ming Institute for Cell Engineering, Department of Neurology, The Solomon H. Snyder Department of Neuroscience, Johns Hopkins University School of Medicine, Baltimore, Maryland, USA

Eric J. Nestler Department of Neuroscience, Friedman Brain Institute, Mount Sinai School of Medicine, New York, New York USA

Alexi Nott Picower Institute for Learning and Memory, Massachusetts Institute of Technology, Cambridge, Massachusetts, USA

Cyril J. Peter Brudnick Neuropsychiatric Research Institute, University of Massachusetts Medical School, Worcester, Massachusetts, USA

Weston T. Powell Medical Microbiology and Immunology, Genome Center, Medical Institute of Neurodevelopmental Disorders, University of California, Davis School of Medicine, Davis, California, USA

Alfred J. Robison Fishberg Department of Neuroscience and Friedman Brain Institute, Mount Sinai School of Medicine, New York, USA

Hongjun Song Institute for Cell Engineering, Department of Neurology, The Solomon H. Snyder Department of Neuroscience, Johns Hopkins University School of Medicine, Baltimore, Maryland, USA

Yi E. Sun Department of Psychiatry and Behavioral Sciences; Intellectual Development and Disabilities Research Center, David Geffen School of Medicine; Department of Molecular and Medical Pharmacology, University of California, Los Angeles, California, USA; Stem Cell Translational Research Center, Shanghai Tongji Hospital, Department of Regenerative Medicine, Tongji University School of Medicine, Shanghai, China

Gustavo Turecki Sackler Program for Epigenetics Psychobiology and Departments of Psychiatry and Neurology & Neurosurgery, McGill University, Montreal, Quebec, Canada

Li-Huei Tsai Picower Institute for Learning and Memory, Massachusetts Institute of Technology, Howard Hughes Medical Institute, Cambridge, Massachusetts; Broad Institute of Harvard and MIT, Cambridge, Massachusetts, USA

Tie Yuan Zhang Sackler Program for Epigenetics Psychobiology and Departments of Psychiatry and Neurology & Neurosurgery, McGill University, Montreal, Quebec, Canada

"Histone Subunit Exchange"

J. David Sweatt, acrylic on canvas (40 x 30), 2012

CHAPTER

1

An Overview of the Molecular Basis of Epigenetics

J. David Sweatt,[1] Eric J. Nestler,[2] Michael J. Meaney[3] and Schahram Akbarian[4]

[1]McKnight Brain Institute, Department of Neurobiology, University of Alabama at Birmingham, Birmingham, Alabama, USA

[2]Fishberg Department of Neuroscience, Friedman Brain Institute, Mount Sinai School of Medicine, New York, New York, USA

[3]Departments of Psychiatry, Neurology, and Neurosurgery, Douglas Institute McGill University, Montreal, Quebec, Canada

[4]Department of Psychiatry, Friedman Brain Institute, Mount Sinai School of Medicine, New York, New York, USA

INTRODUCTION

The role of epigenetic molecular mechanisms in regulation of CNS function is one of the most exciting areas of contemporary molecular neuroscience. This emerging field, variously referred to by neologisms such as *Behavioral Epigenetics* or *Neuroepigenetics*,[1,2] is being driven by shifts in our understanding of several of the fundamental concepts of traditional epigenetics and cognitive neurobiology. These changes in viewpoint can be categorized in a broad fashion into two domains: first, how does neuroepigenetics differ from traditionally defined developmental epigenetics; and second, what is the impact of epigenetics on the historical debate of "Nature versus Nurture"?

After a brief introduction to the basics of epigenetics at the molecular level in this chapter, this book overall will describe the current understanding of the roles of epigenetic processes at the molecular and cellular level, their impact on neural development and behavior, and the potential roles of these mechanisms in neurological and psychiatric disorders. Our goal is for the book to be the first unified synthesis of information concerning the role of epigenetic

Epigenetic Regulation in the Nervous System
DOI: http://dx.doi.org/10.1016/B978-0-12-391494-1.00001-X

mechanisms in nervous system function. This chapter is an introduction to the overall contents of the book, which spans the range of topics including molecular epigenetics, development, cellular physiology and biochemistry, synaptic and neural plasticity, and behavioral models, and also incorporates chapters on epigenetically based disorders of the CNS.

One objective of the book is to begin to embrace the complexity of epigenetic mechanisms in the context of behavioral change. This book represents a critical first step toward synthesizing the complex puzzle of the molecular basis of behavioral plasticity and neural epigenetics.

What is Epigenetics?

Epigenetics and its associated terminology have several different connotations, and specific terms need to be defined before we can discuss them in detail. We will start by defining the *genome* as DNA and the nucleotide sequence that it encodes. In contrast, the *epigenome* is the sum of both histone-associated chromatin assembly and the pattern of DNA methylation, thereby defining the moldings and three-dimensional structure of the genomic material inside the cell nucleus and providing a "molecular bridge" between genes and the environment. Despite these precise structural definitions for genome and epigenome, three definitions for the term "epigenetic" are currently in use in the literature.

The broadest definition includes the transmission and perpetuation of information that is not based on the sequence of DNA, for example, perpetuation of cellular phenotype through meiosis or mitosis. This process is not restricted to DNA-based transmission and can also be protein-based. This definition is broadly used in the yeast literature, as one example, wherein phenotypes that can be inherited by daughter cells are perpetuated past cell division using protein-based (e.g. prion-like) mechanisms.[3–5] Whether such mechanisms operate in mammalian neurons is a subject of current investigation.

Developmental biologists and cancer researchers tend to utilize a second definition for epigenetic: meiotically and mitotically heritable changes in gene expression that are not coded in the DNA sequence itself. The altered patterns of gene expression can occur through the impact on gene transcription of several mechanisms that are based on DNA, RNA, or proteins[6] (see below). The principal criterion for this definition of epigenetic is heritability. It is worth noting that the issue of heritability is fundamental to developmental biology where a major issue is the fidelity of cellular phenotype across proliferation that is critical for tissue differentiation.

A third definition posits that epigenetics is the mechanism for stable maintenance of gene expression changes that involves physically "marking" DNA or its associated proteins, which allow genotypically identical cells (such as all cells in an individual human) to be phenotypically distinct (e.g. a neuron is phenotypically distinct from a liver cell). The molecular basis for this type of change in DNA or chromatin structure in the nervous system is the focus of this chapter.[7–9] By this definition, the regulation of chromatin structure and attendant DNA chemical modification is equivalent to epigenetic regulation.

The common theme that is shared across all of the definitions is that epigenetics is a mechanism for storing and perpetuating a "memory" at the cellular level. The catalyzing phenomenon that has focused attention on these mechanisms is cell division. It is clear from

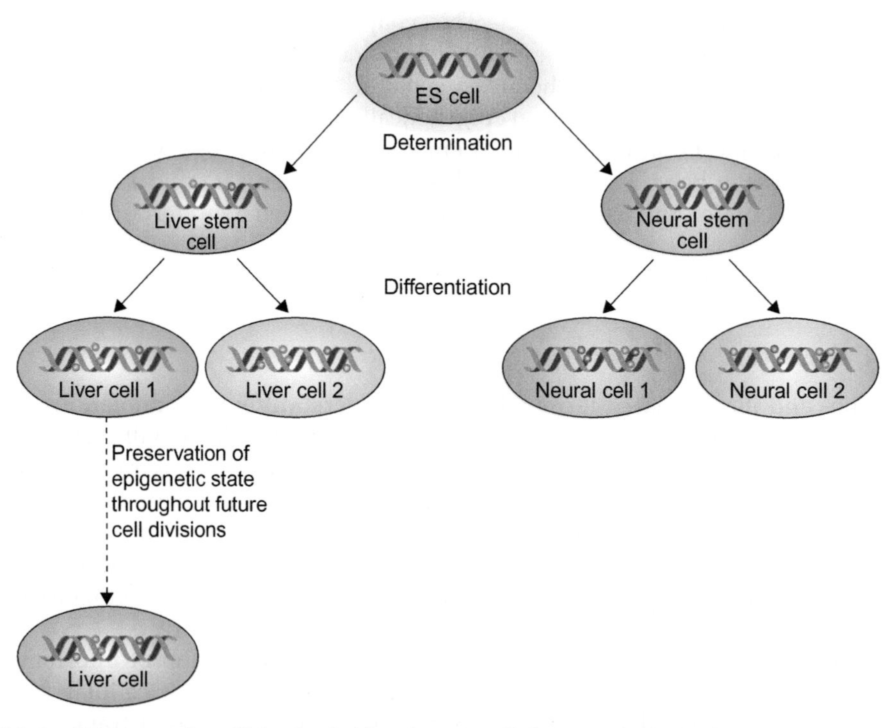

FIGURE 1.1 **Memory at the cellular level.** All embryonic cells begin with identical genotypes and phenotypes. External signals trigger developmental events that lead to the differentiation of cells. Mature cells become phenotypically distinct, but remain genotypically identical. The differences in gene expression persist in the face of numerous cell divisions, which indicates that they are self-sustaining. These developmentally induced changes in gene expression in mature cells are mediated by epigenetic regulation of gene expression. *Adapted from Levenson and Sweatt, 2005.*[7]

developmental studies that a mechanism is necessary for transferring information that concerns the differentiated state of the cell from mother cell to daughter cell; the phenotype must be perpetuated through many subsequent cell divisions which dilute any non self-perpetuating chemical marks. The mechanism for cellular memory does not rely on alterations in the sequence of DNA. This point is remarkable: a neuron and a liver cell, for example, differentiate from the same primordial embryonic stem cell and have the same DNA sequence within an individual (Figure 1.1). Therefore, the self-perpetuating mechanism for differentiation, which originates when the common parent cell divides, cannot be a change in DNA sequence. The distinct phenotypes of each cell are maintained by epigenetic mechanisms that can be detected in the pattern of expression of mRNA and protein in each cell.

How *Neuroepigenetics* Differs from Traditional Epigenetics

The term *epigenetics* is derived from the theoretical and experimental work of Waddington.[10] Waddington coined the term to describe a conceptual solution to a conundrum that arises as a fundamental consideration of developmental biology. All of the different cells in the body of one individual have exactly the same genome, that is, exactly the same DNA nucleotide sequence, with only a few exceptions in the reproductive, immune and nervous systems. Thus, in the vast majority of instances, one's liver cells have exactly the same DNA as the neurons. However, those two types of cells clearly are vastly different in terms of the gene products that they produce. How can two cells have exactly the same DNA but be so different? Especially when what *makes* them different is that they produce different gene transcripts that are read directly from the identical DNA. Waddington coined the term *epigenesis* to describe the conceptual solution to this problem. Some level of mechanism must exist, he reasoned, that was "above" the level of the genes encoded by the DNA sequence, that controlled the DNA readout. The mechanisms that allow this to happen are what we now refer to as *epigenetic* mechanisms. These epigenetic mechanisms specify in a neuron that genes A, C, D, L… are turned into functional products and, in a liver cell, that genes A, B, C, E… are turned into functional products. Epigenetic marks are put in place (or remodeled) during cell fate determination and serve as a cellular information storage system perpetuating cellular phenotype over the lifespan (see Figure 1.1).

A central tenet in the epigenetics field historically has been that epigenetic marks, once laid down as part of development, are immutable within a single cell and are subsequently inheritable across cell divisions. This concept has served developmental biologists well, and explains the permanence of, for example, cellular phenotype over the lifespan of an animal. Indeed, as will be described in more detail in Chapters 3 and 10, epigenetically driven programming of cells in the *nervous system* is a critical mechanism for cell fate determination and perpetuation of cellular phenotype. Moreover, disruption of these processes in humans contributes to a wide variety of neurodevelopmental disorders of behavior and cognition.

However, while stability of cell type is critical for any multicell organism, so too is the ability of cells to adapt phenotype to circumstance at various phases of the life cycle. Thus, recent studies of the CNS have indicated that while the permanence of epigenetic marks is a good general rule, there are some exceptions to that generalization in play in the nervous system and potentially other tissues as well. Thus, in some instances, epigenetic molecular mechanisms appear to be recruited to help drive acquired experience-dependent modifications in cognition and behavior. Numerous examples of these emerging discoveries, and the resulting change in viewpoint of the epigenome as being actively regulated as opposed to static, will be the topics of most of the chapters of this book.

When thinking about how to define epigenetics, consideration of the non-dividing nature of mature neurons in the nervous system leads immediately to a violation for these cells of the traditional defining aspect of epigenetics: heritability. By definition anything that happens in a non-dividing neuron cannot be *epigenetic*, if epigenetic connotes heritability. Nevertheless, it is clear that epigenetic *molecular mechanisms* are in action in non-dividing neurons in the nervous system, driven by organismal experience and cellular signaling. It seems that neurons have co-opted the classic epigenetic machinery that underlies cell-type specification to establish mechanisms for more subtle phenotypic variation. For this reason, the term epigenetic is undergoing a redefinition to accommodate the fact that epigenetic molecular changes can occur in cells but not necessarily be heritable in the traditional sense. Thus, our increased understanding of the role of epigenetic molecular mechanisms in experience-dependent transcriptional regulation in the nervous system is driving a re-formulation of how we should define epigenetics.[11]

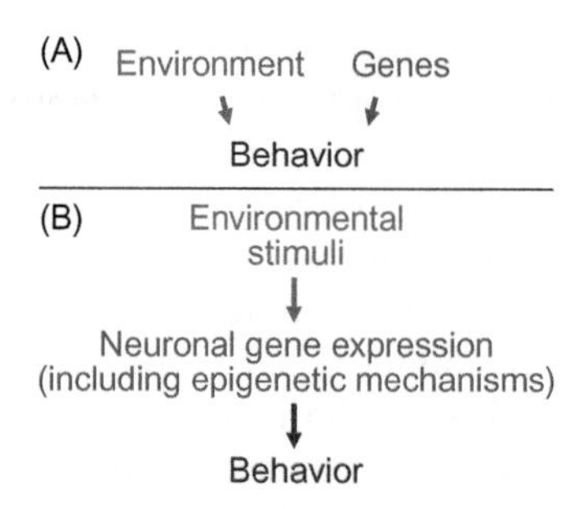

FIGURE 1.2 **The historical model with separate and distinct influences of genes and environment on behavior** (A) is now known to be incorrect. The historical model with separate and distinct influences of genes and environment on behavior has long been held to be biologically implausible as has been discussed by authors such as Lewontin, Gottelieb, Hebb, and Lehrman. Instead, contemporary studies have illustrated that environment and experience act in part through altering gene readout in the CNS in order to achieve their effects on behavior (B). One component of the processes by which the environment and experience alter individual behavior includes epigenetic molecular mechanisms such as regulation of chromatin structure and DNA methylation. The historical dichotomy between "nature" (genes) and "nurture" (environment and experience) is a false one – genes and experience are mechanistically intertwined. Epigenetic molecular mechanisms contribute to this intertwining.

Epigenetics and the Historical Debate of Nature vs Nurture

In the broader context of cognitive neurobiology, the emerging field referred to as *behavioral epigenetics* has deep implications for the historical debate of the role of genetics versus environment in controlling behavior, a debate colloquially referred to as "Nature vs Nurture". The historical model with separate and distinct influences of genes and environment on behavior (Figure 1.2A) is now known to be incorrect. Instead, contemporary studies have illustrated that environment and experience act in part through altering gene readout in the CNS in order to achieve their effects on behavior (Figure 1.2B). A major component of the processes by which the environment and experience alter individual behavior includes epigenetic molecular mechanisms such as regulation of chromatin structure and DNA methylation. The historical dichotomy between "nature" (genes) and "nurture" (environment and experience) is a false one – genes and experience are mechanistically intertwined. The emerging discovery is that epigenetic molecular mechanisms contribute importantly to this intertwining. Epigenetic molecular mechanisms represent a previously hidden mechanistic layer that sits at the interface of genes and environmental experience. In a literal sense, epigenetic mechanisms in the nervous system are the site where experience modifies the genome.

Thus, a new aspect of epigenetic control of gene expression is now emerging from recent studies of epigenetic molecular mechanisms in the nervous system. Convincing evidence has accumulated that epigenetic mechanisms do not just contribute to phenotypic hardwiring at the cellular level. Rather, in the nervous system, with its abundance of terminally differentiated, non-dividing cells, epigenetic mechanisms also play a role in acute regulation of gene expression in response to a wide range of environmental signals, such as behavioral experience, stress, drugs of abuse, and many others. In addition, epigenetic mechanisms appear to contribute to both psychiatric and neurological disorders. In retrospect, these roles for epigenetic molecular mechanisms are not surprising. Even in their role in development, epigenetic mechanisms sit at the interface of the environment and the genome. However, discoveries of active regulation of the epigenome in the adult CNS illustrate the unified, as opposed to dichotomous, relationship of genes and environment.

What are the Epigenetic Marks and What do they do?

There are two basic *molecular* epigenetic mechanisms that are widely studied at present – regulation of chromatin structure through histone post-translational modifications, and covalent modification of DNA principally through DNA methylation. These two mechanisms will be discussed in the next sections of this chapter. Other epigenetic molecular mechanisms such as regulation of gene expression through non-coding RNAs, and recombination of non-genic DNA, are also known to exist and will be briefly discussed. Finally, for the last part of this chapter we will highlight a number of emerging *functional* roles for epigenetic mechanisms in the nervous system as an introduction to the rest of this book.

DNA MODIFICATIONS

Covalent Modification of DNA – Cytosine Methylation

A major mechanism whereby the genome can be epigenetically marked is DNA methylation. Methylation of DNA is a *direct chemical modification* of a cytosine C5 side-chain that adds a -CH_3 group through a covalent bond (Figure 1.3). Methylation of DNA is catalyzed by a class of enzymes known as *DNA methyltransferases* (DNMTs).[12] DNMTs transfer methyl groups to cytosine nucleotides within a continuous stretch of DNA, specifically at the 5-position of the pyrimidine ring.[13,14] Not all cytosines can be methylated; usually (but

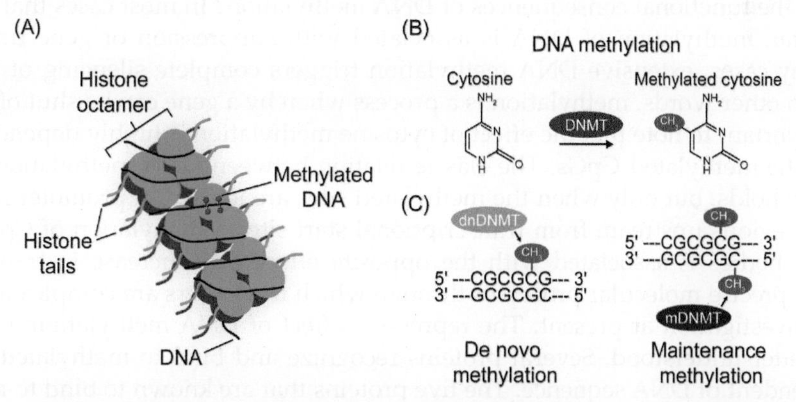

FIGURE 1.3 **DNA methylation and DNMTs.** DNA methylation. (A) Inside a cell nucleus, DNA is wrapped tightly around an octamer of highly basic histone proteins to form chromatin. Epigenetic modifications can occur at histone tails or directly at DNA via DNA methylation. (B) DNA methylation occurs at cytosine bases when a methyl group is added at the 5′ position on the pyrimidine ring by a DNMT. (C) Two types of DNMTs initiate DNA methylation. De novo DNMTs methylate previously non-methylated cytosines, whereas maintenance DNMTs methylate hemi-methylated DNA at the complementary strand. *Adapted from Day and Sweatt, 2010.*[2]

not always) cytosines must be immediately followed by a guanine in order to be methylated.[15,16] These "*CpG*" dinucleotide sequences are highly underrepresented in the genome relative to what would be predicted by random chance; however, about 70% of the CpG dinucleotides that are present are methylated.[17] The rest of the normally unmethylated CpG dinucleotides occur in small clusters, known as "*CpG islands*" that can occur both near gene transcription start sites and intragenically.[18,19] Among the methylated cytosines, only a minute portion (<3%) are located at the 5′ end of genes, with the remaining 97% of methylated cytosines found in intra- and intergenic sequences and within DNA repeats.[20] Thus, CpGs in regions of the genome that actively regulate gene transcription, such as promoters, are largely unmethylated.

There are two variants of DNMTs: *maintenance* DNMTs and *de novo* DNMTs. DNMT1 is the maintenance DNMT, DNMTs 3a and 3b are the de novo DNMT isoforms. Both maintenance and de novo DNMTs are expressed in most cells in the body including brain, although DNMT3b expression tends to be low in the adult CNS.[21] The two variants of DNMTs differ in one important respect, related to the conditions under which they will methylate DNA. De novo DNMTs methylate previously unmethylated CpG sites in DNA – sites which have no methyl-cytosine on either DNA strand. The maintenance DNMT isoform methylates *hemi-methylated DNA* – DNA which has a methylated CpG already present on one strand but no methyl-cytosine on the complementary strand. These two different isoforms thereby serve distinct roles in the cell (see Figure 1.3). De novo DNMTs place new methylation marks on DNA, for example, when specific genes are first silenced as part of cell fate determination. Maintenance DNMTs perpetuate methylation marks after cell division. They regenerate the methyl-cytosine marks on the newly synthesized complementary DNA strand that arises from DNA replication. Thus, in summary: DNMT1, the maintenance DNMT, propagates epigenetic marks through cell generations in dividing cells, while DNMTs 3a and 3b, the de novo DNMTs, are responsible for laying down the initial patterns of DNA methylation when cell fate is determined.

What are the functional consequences of DNA methylation? In most cases that have been studied so far, methylation of DNA is associated with suppression of gene transcription and, in many cases, extensive DNA methylation triggers complete silencing of the associated gene. In other words, methylation is a process whereby a gene can be shut off functionally. It is important to note that the effect of cytosine methylation is highly dependent on the location of the methylated CpGs. The classic relation between DNA methylation and gene transcription holds, but only when the methylated sites are located in promoter regions (i.e. non-coding regions upstream from transcriptional start sites). Methylation of CpGs located within gene bodies is associated with the opposite effect – an increase in transcriptional activity. The precise molecular processes through which this occurs are complex and an area of intense investigation at present. The repressive effect of DNA methylation in gene promoters is better understood. Several proteins recognize and bind to methylated CpG residues independent of DNA sequence. The five proteins that are known to bind to methylated CpGs are MeCP2, MBD1, MBD2, MBD4 and Kaiso.[22,23]

One simplified model for how methyl-DNA binding proteins might suppress transcription is shown in Figure 1.4. In essence, the concept is that methylation of cytosines at CpG dinucleotides recruits methyl-DNA binding proteins at specific sites in the genome. Proteins that bind to methylated DNA have both a *methyl-DNA binding domain* (MBD) and a

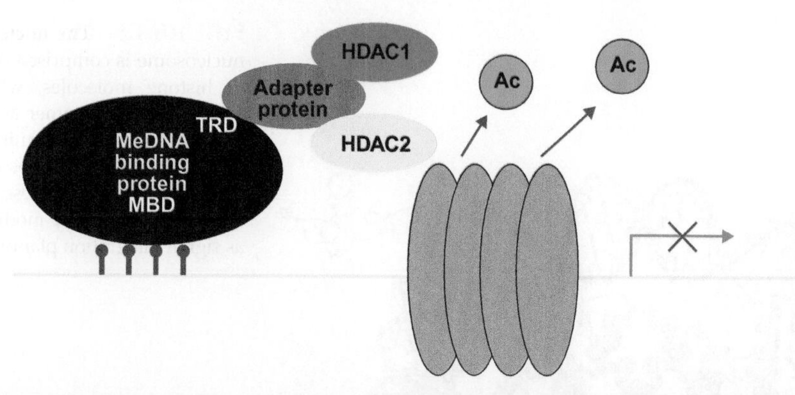

FIGURE 1.4 **A simplified scheme for DNA methylation-dependent gene silencing.** Methylation of cytosines at CpG dinucleotides (red lollipops) recruits methyl-DNA binding proteins, of which MeCP2 is a specific example. All proteins that bind to methylated DNA have both a methyl-DNA binding domain (MBD) and a transcription-regulatory domain (TRD). The TRD recruits adapter proteins such as Sin3A which, in turn, recruit histone deacetylases (HDACs). The HDACs alter chromatin structure locally through removing acetyl groups (Ac) from histone core proteins (gray spheres), leading to compaction of chromatin and transcriptional suppression. *Reproduced with permission from Elsevier.*

transcription-regulatory domain (TRD). The TRD recruits adapter/scaffolding proteins which, in turn, recruit histone deacetylases (HDACs) to the site. HDACs appear critical as they are the major common component to the two MBD repressive complexes, the Sin3a and NurF complexes. The HDACs alter *chromatin structure* locally – "chromatin" is the term describing nuclear DNA/protein complexes (Figure 1.5).[24,25] HDACs alter chromatin structure through removing acetyl groups from histone core proteins, leading to compaction of chromatin and transcriptional suppression. Thus, through this complex and highly regulated biochemical machinery, methylation of DNA triggers localized regulation of the three-dimensional structure of DNA and its associated histone proteins, resulting in a higher-affinity interaction between DNA and the histone core, and transcriptional repression by allosteric means. Consideration of this mechanism thus leads us to the second major category of epigenetic marks, histone post-translational modifications. However, before proceeding to histones, we will address a few additional aspects of regulation of DNA methylation that warrant our attention.

Methylation Regulates Transcription through Multiple Mechanisms

It is important to note that while DNA methylation is usually (and historically) associated with transcriptional suppression, recent studies have indicated that DNA methylation can also be associated with transcriptional activation, by mechanisms that have not yet been determined.[26,27] Also, while DNA methylation leads to marked changes in the structure of chromatin that ultimately result in significant downregulation of transcription, it also can directly interfere with the ability of transcription factors to bind to DNA regulatory elements at specific nucleotide sequences. For example, the transcription factor Ets-1 and the boundary element CCCTC binding factor (CTCF) can efficiently bind to non-methylated, but not methylated, DNA.[28]

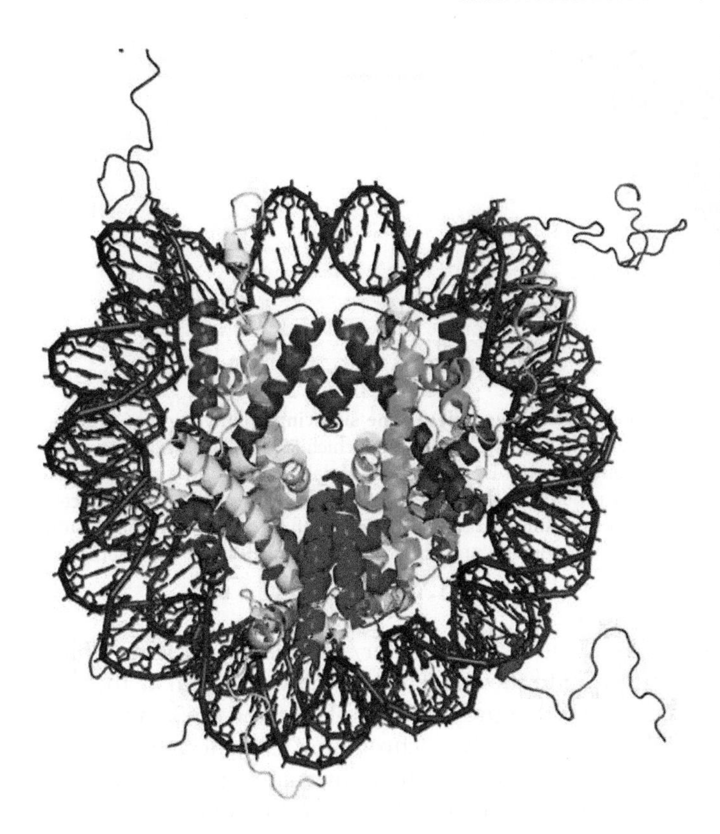

FIGURE 1.5 **The nucleosome.** Each nucleosome is comprised of an octamer of histone molecules, which consists of an $H3_2$-$H4_2$ tetramer and two H2A-H2B dimers. The N-termini of histones project out of the nucleosome core and interact with DNA. These histone tails can be epigenetically modified, and act as signal integration platforms.

Active Regulation of DNA Demethylation

The idea of the occurrence of *active* DNA demethylation has been contentious.[29,30] Traditional epigenetic studies have posited only passive DNA demethylation as a result of cell division and failure to replicate DNA methylation marks when DNA daughter strands are synthesized post-mitotically. However, active demethylation through direct chemical removal of methyl groups on cytosines (or methylcytosines themselves) has been proposed by several groups including those of Szyf and Meaney and Sweatt and colleagues based on their early findings in this area.[31–36] Moreover, replication-independent active demethylation is a defining feature of embryonic development since the DNA methylation of the parental genomes is erased in early development followed by a *re*methylation in later fetal development.[37]

Thus, several pieces of recent information motivate investigating a potential role for active DNA demethylation in non-dividing cells in the mature CNS. First, indirect evidence exists for active DNA demethylation in the adult CNS in response to DNMT inhibitor application or behavioral training (fear conditioning) based on non-quantitative methods, such as PCR-based methods and methylation-dependent immunoprecipitation.[2,31,33,38] Second, two recent publications[39,40] have demonstrated rapid DNA demethylation and remethylation, referred to as "cycling" of methyl-cytosine (hereafter abbreviated mC) in cultured cells.

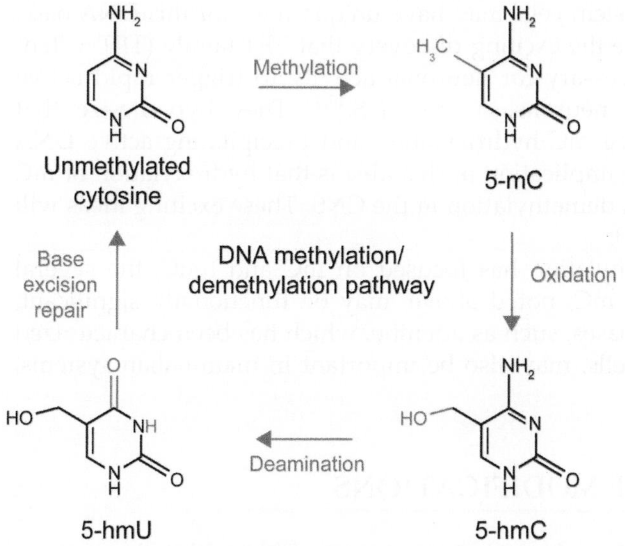

FIGURE 1.6 Proposed role for hydroxymethylcytosine in active DNA cytosine demethylation. 5-mC is 5-methyl-cytosine; 5-hmC is 5-hydroxymethyl-cytosine, and 5-hmU is 5-hydroxymethyl-uracil.

This demethylation occurs too rapidly to be explained by passive demethylation through cell division and must therefore be due to an active demethylation process.

This has led to the hypothesis of the existence of rapid, active DNA demethylation in the adult CNS.[41–44] Investigators working in this area have proposed a specific demethylation mechanism: C-to-T conversion of mC, followed by base-excision repair of the resulting nucleotide mismatch. Most recently, exciting work from Hongjun Song and colleagues supported this idea[45–48] – these investigators demonstrated that DNA repair mechanisms are utilized to demethylate DNA in non-dividing neurons, specifically through base-excision repair mechanisms controlled by the Growth Arrest and DNA Damage 45 (GADD45-beta) regulatory system. This finding demonstrates that demethylation can occur independent of DNA replication, and in a terminally differentiated neuron.

This now substantial body of evidence supports the idea that active DNA demethylation can occur in non-dividing neurons, findings which make viable the idea that active control of DNA methylation may play a role in activity-dependent processes in the CNS throughout the life cycle.

Other Forms of DNA Methylation

Two new studies[49,50] have shown that a novel DNA base, 5-hydroxymethyl-cytosine (hmC), may uniquely occur in the CNS and may serve as a precursor nucleoside for active demethylation. According to this idea, hmC is converted into 5-formyl-cytosine (5fC) and 5-carboxyl-cytosine (5caC) and, finally, restored to cytosine. The existence of this new sixth base (Figure 1.6) was only recently demonstrated convincingly.[49,50] HmC is most abundant in two categories of cells: the totipotent fertilized zygote and the CNS neuron. This observation is highly suggestive, as these are the two main cell types in which active DNA demethylation has been most convincingly demonstrated. One intriguing possibility is that hyperplastic cells such as neurons and stem cells may have unique roles for this DNA base. Using a seizure model, Song's lab made the exciting discovery that TET-family (TET = Ten-Eleven Translocation) oxidases are necessary for neuronal activity to trigger rapid active DNA demethylation in non-dividing neurons in the CNS.[45,46] They hypothesize that these effects are due to TET1 increasing mC hydroxylation and precipitating active DNA demethylation by this mechanism. The implication of this idea is that hydroxylation of mC is a gateway for regulating active DNA demethylation in the CNS. These exciting ideas will be discussed in more detail in Chapter 3.

While most attention on DNA methylation has focused on mC and hmC, the several intermediate steps between hmC and mC, noted above, may be functionally significant. Moreover, methylation of other DNA bases, such as adenine, which has been characterized in prokaryotic and lower eukaryotic cells, may also be important in mammalian systems. This remains an area of active research.

HISTONE MODIFICATIONS

Histones are highly basic proteins whose function is to organize DNA within the nucleus. As mentioned above, in the nucleus, DNA is tightly packaged into chromatin, a DNA-protein complex that consists of DNA in a double helix, histone proteins, and many associated regulatory proteins. Modification of histones is a crucial mechanism for epigenetic tagging of the genome.[25,51] Histone modification can occur as a consequence of DNA methylation, or can be mediated by mechanisms that are independent of DNA methylation and controlled by intracellular signaling.

The basic unit of chromatin is the *nucleosome*, which is composed of an octomer of histone proteins (containing two copies each of histones 2A, 2B, 3, and 4) around which is wrapped – like a rope on a windlass – the DNA double helix. The degree to which nucleosomes are condensed or packed is a critical determinant of the transcriptional activity of the associated DNA and this is mediated in part by chemical modifications of the N-terminal tails of histone proteins (Figure 1.7). Structural studies indicate that these N-terminal tails protrude from the nucleosome and are extensively modified post-translationally.[24]

Currently, four distinct post-translational modifications of histone tails have been well characterized: acetylation, methylation, ubiquitination and phosphorylation. All of these modifications serve as epigenetic tags or marks.[25] We will discuss each of these briefly in the following four sections; a more extensive discussion is included in Chapter 2.

Acetylation

Acetylation of histones occurs at lysine residues, specifically on their side-chain amino group, which effectively neutralizes their positive charge. Histone acetyltransferases (HATs) catalyze the direct transfer of an acetyl group from acetyl-CoA to the ε-NH^+ group of the lysine residues within a histone.[52,53] Histone acetylation is a reversible process, and the enzymes that catalyze the reversal of histone acetylation are known as HDACs.

Classical isoforms of HDACs catalyze the removal of acetyl groups from lysine residues through a Zn^{2+}-dependent charge-relay system.[54,55] By way of background, there are a total

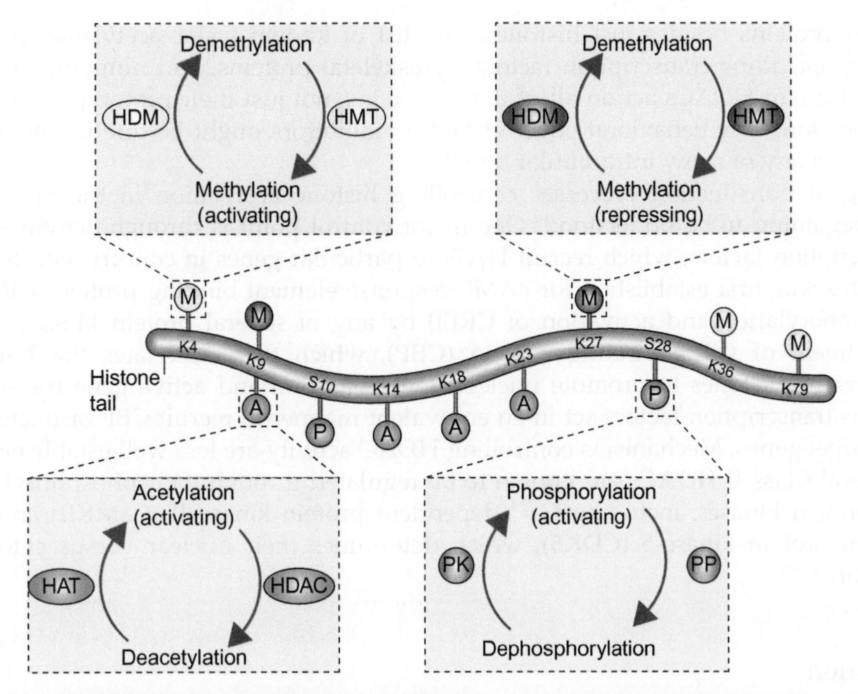

FIGURE 1.7 Post-translational modifications of histones. The first 30 amino acids in the N-terminus of the human histone H3 are illustrated. Many sites in the N-terminus can be targets for epigenetic tagging such as acetylation (A), phosphorylation (P) and methylation (M). Regulation of each site is independent, and the integration of epigenetic tags elicits a finely tuned transcriptional response. The integration of signaling at the level of epigenetics is commonly referred to as the histone code. *Adapted from Levenson and Sweatt, 2005[7] and Tsankova et al., 2007.[9]*

of eleven HDAC isoforms broadly divided into two classes. HDACs 1, 2, 3, and 8 are Class I HDACs, while Class II encompasses HDAC isoforms 4, 5, 6, 7, 9, 10, and 11. The newly characterized SIR2 family of HDACs (the "Sirtuins"), termed Class III, operate through an NAD^+-dependent mechanism, but are not discussed further here.[56]

As will be discussed in Chapter 8, HDAC inhibitors are undergoing a period of rapid development in the pharmaceutical industry because of their potential applicability in cancer treatment and the emerging possibility of their utility in neurological and psychiatric disorders. HDAC inhibitors are the principal way to manipulate the epigenome pharmacologically at present. Trichostatin A (TsA) inhibits HDACs broadly across both Class I and Class II, while sodium butyrate, suberoylanylide hydroxamic acid (SAHA, aka Vorinostat or Zolinza), and MS275 (aka Entinostat) are more selective for Class I HDACs. Tubacin selectively inhibits HDAC6, for which tubulin is an established substrate. Valproate is also an HDAC inhibitor, but this drug also has several additional targets and the role of HDAC inhibition in valproate's clinical efficacy is unclear at this time.

The principal caveat to interpreting all studies utilizing HDAC inhibitors is the fact that "histone deacetylase" is actually a misnomer. HDACs should be more accurately described as "lysine deacetylases". Lysine amino-acid side chains are acetylated in a wide variety of cellular proteins besides just histones. The list of known lysine-acetylated proteins is quite long, including transcription factors, cytoskeletal proteins, and numerous metabolic enzymes. Certain HDACs act on all of these proteins, not just their prototype histone substrate. Therefore, any behavioral effect of HDAC inhibitors might be due to alterations in acetylation of any of many intracellular targets.

The signal transduction processes controlling histone acetylation in the mature CNS are just beginning to be understood. One major control point is through activity-dependent transcription factors, which recruit HATs to particular genes in concert with their activation. This was first established for cAMP-response element binding protein (CREB).[57–61] The phosphorylation and activation of CREB by any of several protein kinases leads to its recruitment of CREB binding protein (CBP), which then acetylates the N-terminal tails of nearby histones to promote nucleosome separation and active gene transcription. Numerous transcription factors act in an equivalent manner to recruit CBP or related HATs to their target genes. Mechanisms controlling HDAC activity are less well established, however, several Class II HDACs are known to be regulated through their phosphorylation by several protein kinases, including Ca^{2+}-dependent protein kinase II (CaMKII) and cyclin-dependent protein kinase-5 (CDK5), which determines their nuclear versus cytoplasmic localization.[62–66]

Methylation

Histone methylation is another major histone-directed epigenetic tag.[67] Similar to acetylation, methylation of histones occurs on ε-NH^{+} groups of lysine residues, and is mediated by lysine methyltransferases (KMTs). Unlike acetylation, methylation of lysines preserves their positive charge. In addition, lysines can accept up to three methyl groups and thus exist in mono-, di- or trimethylated states. The effect of histone lysine methylation on gene regulation is highly complex, with the various valences of methylation on several distinct lysine residues mediating either transcription repression, activation, or elongation (Table 1.1). Arginine residues within histones can also be mono- or dimethylated on their guanidine nitrogen. This reaction is catalyzed by protein arginine methyltransferases (PRMTs). An overview of the major histone methylating and demethylating enzymes is given in Table 1.1.

Ubiquitination

Ubiquitination of histones was identified 29 years ago[68] but has only recently begun to be characterized in detail. Ubiquitin, a protein with 76 amino acids that is named for its ubiquitous distribution in all cell types and high degree of conservation across species, is usually, but not always, attached to proteins as a signal for degradation by the proteasome.[69] Like other proteins, histones are ubiquitinated through attachment of a ubiquitin to the ε-NH^{+} group of a lysine.[70] Ubiquitination of histones H2A, H2B, H3 and H1 has been observed.[71–73] Most histones appear to be mono-ubiquitinated, although there is evidence for poly-ubiquitination. The role of histone ubiquitination in the control of gene transcription in the nervous system remains poorly understood.

TABLE 1.1 Examples of Mechanisms for Regulation of Histone Acetylation and Methylation

Modification	Function	Writers	Erasers
Acetylation	Activates (H3, H4)	CBP, p300	HDACs 1–11
Methylation	Activates (H3K4)	MLL1, SetD1a	JARID/SMCX
	Elongates (H3K36)	Set2	JHDM1
	Represses		
	(H3K9)	G9a, SUV39H1	JMJD2a, LSD1
	(H3K27)	EZH2	JMJD3, UTX
	(H4K20)	SetD8	PHF8

Even more complicated, because each methylated residue can be mono-, di- or trimethylated, with distinct functional consequences.
JMJD2A also is a H3K4me3 eraser; and LSD1 is also active at K3me sites.

Phosphorylation

Phosphorylation of histones H1 and H3 was first observed in the context of chromosome condensation during mitosis.[74,75] H3 was the first histone whose phosphorylation was characterized in response to activation of mitogenic signaling pathways.[76] Phosphorylation of serine 10 on H3 is mediated by at least three kinases in the CNS: ribosomal S6 kinase-2 (Rsk2), which is downstream of extracellular signal-regulated kinase (ERK); mitogen- and stress-activated kinase-1 (Msk1), which is downstream of both ERK and p38 mitogen-activated kinases (MAPK); and the aurora kinase family member Ipl1.[77–80] Evidence also implicates aurora kinases in the phosphorylation of serine 28 in histone H3.[81] Phosphatases remove phosphate groups from histones.[82] To date, the phosphatases PP1 and PP2A, and the PP1 inhibitor DARPP32, have been shown to regulate H3 phosphorylation in the CNS.[78,83,84] In most cases, phosphorylation of histones is associated with gene activation, although much further work is needed to define the precise mechanisms involved.

Histone Subunit Exchange

Besides direct chemical post-translational modification of histones, an additional mechanism for altering the function of the histone components of the nucleosome is exchange of histone isotypic variants into and out of the histone octamer (Figure 1.8). This energy-dependent *chromatin remodeling* process is one mechanism for persistently altering the transcriptional efficacy of a given chromatin particle. Certain histone isomers, such as histone H2A.Z are associated with the absence of DNA methylation and transcriptional activation. The individual histone variants appear to differ in the capacity to support specific modifications, with some likely more associated with increased epigenetic "plasticity" and others more closely allied to epigenetically stable genomic regions. Determining whether active histone subunit exchange of this sort is actively regulated in the CNS is an area of contemporary investigation.

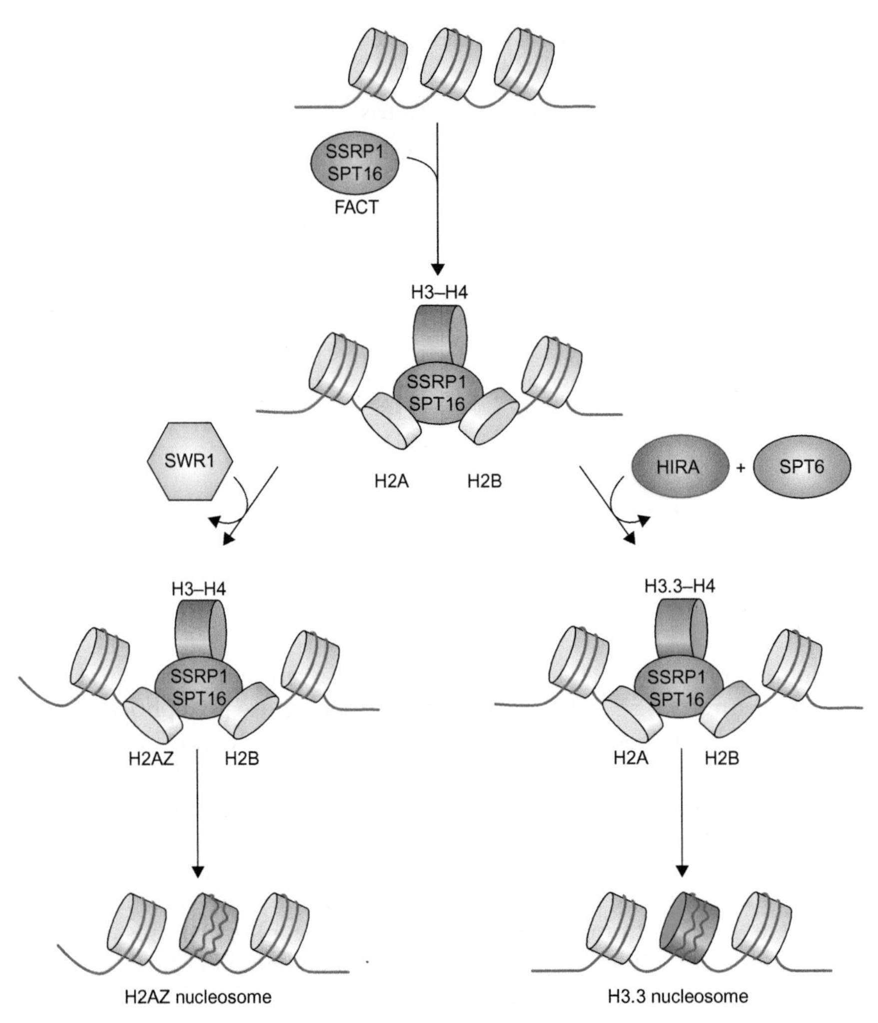

FIGURE 1.8 **Histone subunit exchange.** Under certain circumstances, histone subunits are replaced within existing nucleosomes, in concert with gene regulation. Examples of histone variants are: Histone H3: H3.1, H3.2, H3.3; Histone H2: H2 A.Z, MacroH2A. SSRP1, SPT16, SWR1, HIRA, and SPT6 are chromatin remodeling enzymes involved in subunit exchange. *Reproduced with permission from Nature Publishing Group.*

A Histone Code for Regulating CNS Function?

The plethora of functional histone modifications capable of affecting gene transcription has led to the proposal that one purpose for the complex biochemical signaling at this locus is that a *histone code* might be involved in transcriptional regulation. The basic concept is that specific patterns of histone post-translational modifications might help encode the salience of cell-surface signals and their contingencies. This general hypothesis that there is an epigenetic "histone code" for regulating CNS function is new and still quite speculative (see Chapter 2), and derives from an earlier idea proposed by David Allis and colleagues.[25] The overall concept is that multiple post-translational histone modifications may be integrated together, combinatorially driving neuronal gene expression patterns by recruiting signaling complexes and thereby remodeling the structure of chromatin. As different histone modifications may be driven by different upstream signaling pathways, multiple signals might thus converge on the nucleus, controlling gene readout through regulating chromatin structure. This would result in a mapping of multiple histone alteration states onto subsets of genes that are transcribed as a result of those changes. In principle, these specific patterns of histone (or more broadly speaking, chromatin) modifications might help encode a transient or lasting set of signals reflecting specific experience-dependent patterns of neuronal activity. This fascinating idea of a combinatorial histone/epigenetic code operating in CNS function is the topic of the following chapter in this book.

Other Mechanisms of Epigenetic Tagging in the CNS

We have so far focused on epigenetic mechanisms that are DNA-centric, which result in modification of either the DNA itself or associated histone proteins. According to the broadest definition of epigenetics – which includes any non-DNA-sequence-based system for the perpetuation of information across time and across cell replication – any protein- or RNA-based system for storage of cellular memory is also epigenetic. Our approach is to focus on the actual function of epigenetic mechanisms, that of transcriptional regulation, as opposed to an emphasis on the assumed physiochemical nature of the modification or heritability. This more inclusive approach provides neurobiologists with a richer and more integrated set of candidate mechanisms. In the following sections, we will briefly comment on several additional molecular modalities for epigenetic perpetuation of cellular phenotype that operate in the CNS.

NON-CODING RNAs

An additional set of epigenetic mechanisms operating in the CNS derives from the activity of several types of non-coding RNA molecules. The best characterized are small RNAs, which include microRNAs, small interfering RNAs (siRNAs), and small nuclear RNA (snRNAs). RNA interference (RNAi) is a mechanism whereby the expression of cognate genes is disrupted through the action of double-stranded RNA molecules.[85] Pioneering studies suggested that the RNAi machinery is used in the nucleus and is involved in the formation of heterochromatin and epigenetic tagging of histones in yeast. Genetic disruption of RNAi pathways leads to relaxation of heterochromatin around centromeres, which causes erroneous expression of normally silent genic regions and a decrease in the repressive methylation of histone H3.[86,87] Small RNAs that are produced by a specialized ribonuclease can associate with DNA and direct the formation of a protein complex that promotes the formation of heterochromatin.[88] Small RNAs have multiple functions within a cell, including activation, repression, or interference with gene expression, and have been implicated in a number of cognitive disorders. For example, microRNA binds to 3′ untranslated regions of messenger RNA and thereby either promotes the degradation of the messenger RNA or suppresses its expression through translational mechanisms. MicroRNAs may thereby control expression of the majority of genes within the genome, and represent a critical component of normal physiology and function in the developing and adult nervous system. Moreover, microRNAs are often an integral force within the complexes that are attracted to the genome by specific DNA or histone marks, and thus link epigenetic signatures to transcriptional activity.

Less well studied is a series of long non-coding RNAs. These are typically longer than 200 nucleotides and can be spliced like messenger RNAs to form active biological molecules, including small RNAs. Long non-coding RNAs have been shown to regulate the recruitment of chromatin remodeling complexes to particular genes in simpler systems and likely play an important role in the controlling gene transcription.[89] The concept that activity-dependent regulation of neuronal gene expression can be mediated by small and large non-coding RNAs is discussed further in Chapter 5.

NON-GENIC DNA

Most investigations to date concerning the role of epigenetic mechanisms in the nervous system have focused on promoter regulation, as we have been discussing thus far. However, recent work from the research groups of Adrian Bird, Rusty Gage, Eric Nestler, and Michael Meaney has begun to explore dynamic DNA/histone changes in association with the presence of DNA repeat sequences in CNS neurons.[90]

One example of these new studies involves L1 retrotransposition. L1 elements belong to the long interspersed element (LINE) class of repeat sequences, which are an active class of non-LTR (long terminal repeat) retro-elements in the human genome. Full-length, functional L1 elements are autonomous because, once expressed, they encode proteins (e.g. reverse transcriptase, endonucleases) necessary for their own retrotransposition (i.e. reincorporation into the genome). L1 elements are retrotransposons that insert extra copies of themselves throughout the genome using a copy-and-paste mechanism, and are thus able to influence chromosome integrity and gene expression upon reinsertion.

A particularly intriguing current hypothesis is that L1 elements are active and "jumping" during neuronal differentiation, potentially allowing L1 insertions to generate genomic plasticity in neurons by altering the transcriptome of individual cells. Among a number of mechanisms, L1 element insertion could alter neuronal gene transcription by affecting promoter location and efficacy, altering splice sites within genes, or triggering aberrant activation of a gene by local insertion of binding sites for transcription factors. In a mechanism specifically involving epigenetic mechanisms, an L1 element that has inserted upstream of another gene might become methylated, silencing transcription. By these mechanisms L1-induced variation could affect neuronal plasticity and behavior, broadening the spectrum of behavioral phenotypes that can originate from any single genome – a topic that is discussed in more detail in Chapter 11.

PRION-BASED EPIGENETIC INHERITANCE

Prions are proteins encoded in the DNA of most eukaryotes that are capable of a conformationally dependent self-perpetuating biochemical reaction. The basis of this reaction is that prion proteins are synthesized in an inactive form that, when triggered by an exogenous signal, convert into an active form that can alter a cellular phenotype. Moreover, the active form, once generated, can act upon other inactive prion molecules and render them activated. By this means, once activated in a cell, the prion proteins are able to establish a self-perpetuating biochemical reaction that is both persistent across time and might be heritable across cell division.

Thus, prions represent a viable, protein-based system for epigenetic memory. Once a protein has been converted into its prion form, that protein promotes the transition of other cognate proteins into the prion form. This epigenetic mechanism is broadly used in yeast, as noted above, wherein phenotypes that can be inherited by daughter cells are perpetuated past cell division using protein-based mechanisms.[3–5]

Recently, a provocative series of studies has suggested that, in *Aplysia*, the cytoplasmic polyadenylation element binding protein (ApCPEB) assumes a prion-like conformation after synapses are strengthened.[5] By assuming a prion-like conformation, it is hypothesized that ApCPEB can maintain a stable synaptic state in the face of protein turnover. This hypothesis, which requires further investigation, will be discussed in more detail in Chapter 5.

EPIGENOME ORGANIZATION AND HIGHER ORDER CHROMATIN STRUCTURES

While wrapping of genomic DNA into nucleosomal structures results in a several-fold increase in packaging density, as compared to naked DNA, the actual level of compaction in the vertebrate nucleus is about three orders of magnitude higher.[91] These chromosomal arrangements in the interphase nucleus are not random and loci with active transcription are more likely to be clustered together and positioned towards a central position within the nucleus, while heterochromatin and silenced loci tend to locate towards the nuclear periphery.[92,93] Chromatin loopings, in particular, are among the most highly regulated "supranucleosomal" structures and pivotal for the orderly process of gene expression, by enabling distal regulatory enhancer or silencer elements positioned a few, or many hundred kilobases apart from a gene, to interact directly with that gene's promoter.[94,95] There is a growing realization of the importance of these and other higher order chromatin structures for transcriptional regulation, but very little is known about their role in the context of epigenetic regulation in the nervous system. To date, less than a handful of studies have explored loop formations in brain tissue.[96–98] Clearly, functional explorations of higher order chromatin structures in the context of learning and plasticity, or various brain disorders, will provide important new insights into this layer of regulation.

ROLES FOR EPIGENETIC MECHANISMS IN THE NERVOUS SYSTEM

In the previous sections, we have presented a brief overview of the basic molecular and biochemical mechanisms governing epigenetic regulation. In the following sections, we explore the functional significance of epigenetics in several aspects of *neural* systems.

EPIGENETIC MECHANISMS IN NERVOUS SYSTEM DEVELOPMENT

The role of epigenetic mechanisms in nervous system development is a focus of several chapters of this book. However, by way of brief introduction here, we will highlight one of the prototype examples of a role for epigenetic mechanisms in neural development, a core process which contributes to cell fate determination in neurons and impinges upon epigenetic molecular mechanisms for its readout.

Neurons express a complement of proteins that are important for their function, but would be detrimental to other cell types. These include proteins that are involved in excitability, transmitter release and the maintenance of transmembrane potential. Many genes that are to be expressed in neurons, but not in other cell types, have a *neuron-restrictive silencer element* (NRSE) in their promoters.[99–101] This regulatory element, which is approximately 21–24 base pairs long, can completely silence a gene in non-neuronal cells.[101]

The first step toward understanding how NRSEs confer tissue-specific regulation of gene expression was identification of the transcription factors that bind to this regulatory element (Figure 1.9). The *RE1-silencing transcription factor* (REST or NRSF for neuron-restrictive

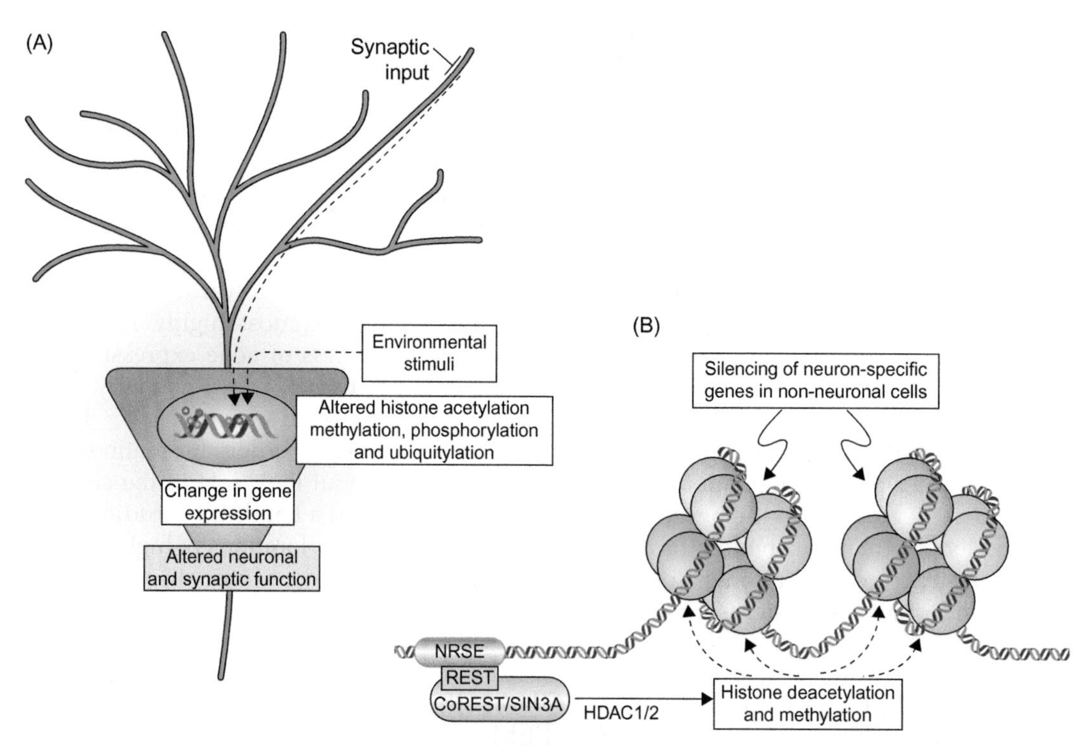

FIGURE 1.9 **Epigenetics in nervous system development.** The RE1-silencing transcription factor (REST)/REST co-repressor (CoREST) system. The NRSE upstream of genes to be silenced in non-neuronal cells recruits REST as a mediator of transcriptional repression. Sin3A, CoREST and REST, acting in concert with additional factors such as HDAC1 and 2, leads to chromatin condensation and gene silencing. *Adapted from Levenson and Sweatt, 2005.*[7]

silencer factor) was the first transcription factor that was shown to bind to NRSEs and repress gene expression.[102] The REST protein is ubiquitously expressed in cells outside the nervous system, where it acts to repress the expression of neuronal genes.[102] Deletion of the REST gene or functional inhibition of the protein in non-neuronal tissues leads to erroneous expression of neuronal genes and embryonic lethality, whereas ectopic expression of REST in the nervous system inhibits expression of neuronal genes, and results in developmental dysfunction.[102–104] Therefore, REST is important in determining whether a cell has a neuronal phenotype.

REST-dependent gene silencing requires the action of transcriptional co-repressors, two of which have been identified as the REST-binding proteins *Sin3A* and *CoREST*.[105–107] The cellular expression pattern of Sin3A is nearly identical to that of REST, which indicates that most REST-dependent gene repression might be co-mediated by Sin3A.[108] The expression of CoREST is more restricted, which indicates that it might be important in mediating specific gene expression patterns in subtypes of cells.

REST-mediated gene silencing requires modulation of chromatin structure. REST/Sin3A repressor complexes are associated with HDAC1, whereas REST/CoREST complexes with HDAC2.[20,106,107,109] Thus, REST-dependent gene silencing with either co-repressor seems to involve decreases in histone acetylation. CoREST has also been shown to associate with members of the hSWI–SNF complex, which is an ATP-dependent chromatin remodeling complex.[110] Interestingly, REST/CoREST-dependent chromatin remodeling, including decreases in histone acetylation and increases in DNA methylation, does not seem to be restricted to the immediate region around an NRSE silencer sequence; rather, the formation of heterochromatin extends across several genes that flank an NRSE.[111] These observations indicate that REST-dependent gene silencing, and thus cellular differentiation, involves the action of several proteins which, through decreases in histone acetylation and/or increases in DNA methylation, ultimately mark DNA epigenetically for repression. Interestingly, recent work has implicated REST in controlling dynamic, activity-dependent changes in gene expression within fully differentiated adult neurons.[112] This highlights a general theme that proteins that serve a particular function in development often play very different roles in adult tissues.

NEUROGENESIS IN THE ADULT CNS

Not too long ago, the widely held dogma was that there is no new generation of neurons in the adult CNS. However, fascinating results have shown that neurogenesis does indeed continue into the adult in a small number of brain regions, including the hippocampal dentate gyrus, a phenomenon shown by Fred Gage and his collaborators to occur in the adult *human* hippocampus as well.[113] How was this established? Cancer patients sometimes receive treatment with the drug bromo-deoxy uridine (BrdU). It selectively affects dividing cells by being incorporated into their DNA upon de novo DNA synthesis. An ancillary aspect of this is that BrdU selectively labels freshly divided cells. Post-mortem analysis of the brains of cancer patients who had received BrdU as a chemotherapeutic treatment revealed that indeed new dentate granule cells are produced in an ongoing fashion in the adult human brain.

Investigation of how epigenetic mechanisms regulate this process is one of the most exciting areas of contemporary study concerning the roles of epigenetic molecular mechanisms in the functioning of the adult CNS. This topic is covered in detail in Chapter 12.

CIRCADIAN RHYTHMS

The physiology of most organisms is modulated by time of day. These daily rhythms persist in the absence of external environmental cues, have a period of approximately 24 hours, and are commonly referred to as circadian rhythms. Circadian rhythms are generated endogenously by a biological timekeeping mechanism known as the circadian clock, which comprises intricate feedback loops of transcription and translation.[114,115] In addition, the mechanisms that are responsible for entrainment of the circadian clock to the environment (such as light) rely on signaling pathways that induce changes in transcription within particular brain regions. In mammals, the master circadian clock – which entrains the body's rhythm to light – resides in the suprachiasmatic nucleus (SCN), which is situated in the anterior hypothalamus.[116,117] However, most other brain regions and most peripheral tissues have been shown to have endogenous circadian clocks which presumably control rhythms in diverse physiological functions.[114,117,118]

The heart of any circadian clock lies in the transcription–translation feedback loop, which is known to be modulated by epigenetic mechanisms. Thus, the genome likely undergoes daily changes in its epigenetic state. The acetylation of histones H3 and H4 associated with the promoters of genes that form part of the core molecular clock mechanism are differentially regulated during a circadian cycle.[114,119] Moreover, infusion of the HDAC inhibitor trichostatin A into the SCN increases the expression of the clock genes mPer1 and mPer2, which indicates that epigenetic states directly affect the expression of the molecular components of the circadian clock.

Adjusting the phase of the circadian clock also requires transcription. The most salient phase-resetting environmental stimulus is light, and pulses of light induce changes in the transcription of several genes that comprise the molecular clock.[115,120] Epigenetic mechanisms seem to be associated with this regulation, as discrete pulses of light induce increases in acetylation of histones H3 and H4 associated with the promoters of mPer1 and mPer2.[119] Moreover, discrete light pulses induce significant increases in the phosphorylation of histone H3 in the SCN in vivo.[121] These observations indicate that regulation of the epigenetic state of the nucleus is a core molecular mechanism of the circadian clock, which is used to generate rhythmic gene expression and to establish a stable phase relationship between gene expression, an animal's behavior and physiology, and the environment.

PERSISTING EFFECTS OF LIFE EXPERIENCE: NURTURING AND TRANSGENERATIONAL EFFECTS

Mother rats that exhibit strong nurturing behaviors toward their pups, for example, by frequently licking and grooming their offspring, produce lasting alterations in the patterns of DNA methylation in the central nervous systems (CNS) of their pups, which persist throughout adulthood. Studies by Meaney and colleagues have presented evidence that these changes in DNA structure result in decreased anxiety-like behavior and a strong maternal nurturing instinct in the adult offspring as compared with offspring of mothers that show lower levels of grooming behavior.[34] These observations will be described in more detail in Chapter 4.

There are several interesting implications of these types of studies, which demonstrate persisting epigenetic marks and altered adult behavior in response to life experiences. First, this work indicates that experientially acquired alterations in DNA methylation affect behaviors in the adult. Second, the persistence of neonatally acquired patterns of DNA methylation in the mature CNS is consistent with the hypothesis that epigenetic mechanisms contribute to lasting cellular effects, that is, cellular memory in the CNS. Finally, and perhaps most importantly, studies of this sort suggest a specific epigenetic mechanism in the CNS for behaviorally perpetuating an acquired behavioral characteristic across generations – a particularly robust example of behavioral memory that is potentially subserved by epigenetics. The idea of transgenerational perpetuation of acquired epigenetic marks and the data supporting their existence will be described in Chapter 13.

EPIGENETIC MECHANISMS AND CELLULAR INFORMATION STORAGE

As already alluded to previously, there are numerous examples that illustrate the importance of epigenetic mechanisms in information storage at the cellular level. They indicate that epigenetic mechanisms are widely used for the formation and storage of cellular information in response to transient environmental signals. Storage of information at the cellular level is in some ways analogous to behavioral memory storage in the adult nervous system. Moreover, the lasting cellular changes are triggered by a transient signal in each case, which is also a commonality between cellular memory and the formation of behavioral memory in the CNS.

A prototype example of the analogy between developmental memory and behavioral memory is mammalian cellular differentiation. Once an embryonic precursor cell is triggered to differentiate into a particular cell type (e.g. a liver cell), that cell and its subsequent daughter cells might be required to undergo thousands of cell divisions over the lifetime of the animal. How does a liver cell remember that it is a liver cell when, over the course of cell division, it must replicate de novo its entire genome? The information clearly cannot be contained in the DNA sequence itself. As mentioned above, the answer to this question involves epigenetic mechanisms, which allow the cell's identity to be manifest as the subset of genomic DNA that it expresses. The DNA is marked by, for example, DNA methylation at specific sites that are acquired as part of the differentiation process but are self-perpetuating during DNA replication and cell division. A role for non-coding RNAs in this process is also likely. Thus, a liver cell perpetuates its specific acquired pattern of gene expression across cellular generations and over time through these epigenetic marks – an example of memory at the cellular level.

The formation of epigenetic memory is not limited to mammalian cells. Plants are induced to flower by a process called vernalization that also involves epigenetic mechanisms. For example, a biennial plant must experience a period of cold weather between its first and second years of existence for its flowering to be triggered. Exposure to cold in biennial plants results in activation of epigenetic mechanisms that involve methylation of DNA-binding proteins and acetylation of histones, and these processes trigger mitotically stable changes in the pattern of gene expression. In this way, plant cells "remember" their exposure to the winter cold and are prepared to allow the plant to flower during the next spring.

Another example involves T cells of the mammalian immune system. The commitment of T-lymphocyte precursors to a wide variety of differentiated states with different patterns of gene expression is triggered by numerous epigenetic mechanisms that involve DNA methylation and histone modifications. These processes are important in the formation of long-lasting immunological memory in response to a transient signal from the environment.

However, epigenetic mechanisms are also extant and operable in non-dividing, terminally differentiated neurons in the adult CNS. Adult neurons no longer have to deal with the problem of heritability, but the basic epigenetic mechanisms important for information storage during development are also important for storing memory that manifests itself behaviorally in the adult. Chapter 5 will discuss the idea that these mechanisms are conserved in the adult nervous system, where they have been co-opted to serve the formation of behavioral memories. Thus, current hypotheses posit that epigenetic mechanisms subserve changes in neuronal function in the adult that are components of memory at the behavioral level. Epigenetic processes may constitute a unified set of molecular mechanisms that allow information storage in systems as diverse as yeast, plants, and cellular differentiation and memory storage in the mammalian CNS.

HUMAN COGNITION AND COGNITIVE DISORDERS

As a final comment we would be remiss if we did not highlight the fact that there is a considerable body of evidence, albeit indirect, implicating disruption of epigenetic mechanisms as a causal basis for *human* cognitive dysfunction. Here we will briefly describe several instances wherein derangements in molecular components of the epigenetic apparatus have been implicated in human cognitive disorders. These issues are discussed in much greater detail in Chapters 6–10. In broad overview, in interpreting these findings of a role for epigenetic molecular mechanisms in human behavior, an important caveat applies. When considering these cases it is important to distinguish between a developmental need for epigenetic mechanisms, to allow formation of a normal nervous system, versus an ongoing need for these mechanisms as part of cognitive processing per se in the adult. The majority of the attention to date has justifiably focused on developmental roles for epigenetics in establishing the capacity for cognitive function in the adult. However, the experimental results outlined above and in a number of chapters in this book implicate an ongoing and active role for epigenetic mechanisms in cognition and behavior in the adult. Thus, we believe it is timely and worthwhile to consider a possible component of cognitive disruption in those disorders outlined below to be due to a loss of active utilization of epigenetic mechanisms, necessary for normal cognition, in the mature post-developmental CNS.

TABLE 1.2 Disorders of Human Cognition Partly Attributed to Dysfunction in the Mechanisms that Underlie Epigenetic Marking of the Genome

Disease	Gene	Function	Epigenetic Affect	References
Rubinstein–Taybi syndrome	CREB-binding protein (CBP)	CBP is a histone acetyltransferase	↑ histone acetylation	8,109
Rett syndrome	MecP2	MeCP2 binds to CpG dinucleotides and recruits HDACs	↓ histone acetylation	14,43,79,88, 129,136
Fragile X mental retardation	Trinucleotide expansions in FMR1 and FMR2 genes	Expansion of CGG or CCG repeats results in aberrant DNA methylation around FMR1 and FMR2 genes	↑ DNA methylation. ↓ histone acetylation	4,5,90,133
Alzheimer's disease	Amyloid precursor protein	APP intracellular domain acts as a Notch-like transcription factor. Associated with the HAT Tip60	↑ histone acetylation	25,50,53, 60,84,126
Schizophrenia	Reelin	Reelin is an extracellular matrix protein, involved in synapse development	↑ DNA methylation around reelin gene	9,80

In this vein, several disorders of human cognition can be at least partly attributed to dysfunction in the mechanisms that underlie epigenetic marking of the genome (Table 1.2). Rubinstein–Taybi syndrome (RTS), an inherited autosomal dominant disease, is due to mutation of the gene encoding CBP, the transcriptional co-activator and HAT discussed earlier.[122,123] Several studies using animal models to investigate the molecular basis of RTS indicate that deficiency in CBP has severe consequences for long-term memory formation. Rett syndrome (RS) is an inherited, X-linked disease that appears to be due, in most cases, to loss-of-function mutations in the gene encoding MeCP2, the methyl-DNA binding protein.[124–126] Using genetic animal models, it was discovered that overexpression of MeCP2 enhanced long-term memory formation and the induction of hippocampal long-term potentiation (LTP), indicating that MeCP2 modulates memory formation and induction of synaptic plasticity.[127] Fragile X syndrome, the most commonly inherited form of mental retardation, is brought about by an abnormal expansion of repeated trinucleotide sequences within one of two different Fragile X genes: FMR1 and FMR2.[128,129] Both FMR1 and FMR2 contain a polymorphic trinucleotide repeat, CGG and CCG respectively, in their 5′ untranslated regions responsible for the loss of gene expression.[130,131] Expansion of these repeats results in hypermethylation of these regions and flanking CpG islands, leading to transcriptional silencing of the FMR and surrounding genes. The most widespread of senile dementias, Alzheimer's disease, appears to be due, in part, to an increase in soluble β-amyloid peptides in the brain.[132] These peptides are created by endo-proteolytic cleavage of the transmembrane amyloid precursor protein (APP) by β- and γ-secretases.[133] Interestingly, cleavage of APP results not only in production of an extracellular β-amyloid fragment, but also an intracellular fragment, the APP intracellular domain (AICD), that regulates transcription through recruitment of the adapter protein Fe65 and the HAT Tip60, suggesting that some of the pathology of Alzheimer's disease might be due to dysregulation of histone acetylation.[134–137] Finally, schizophrenia is a serious disorder of cognition, rendering sufferers unable to function normally in social situations and in performing everyday cognitive tasks. An emerging body of evidence suggests that deficiencies in the extracellular matrix protein reelin may contribute to the pathophysiology of schizophrenia, at least in a subset of patients.[138] The promoter of reelin contains several sites for DNA methylation, and inhibitors of HDAC and DNMT activity increase expression of reelin, indicating that epigenetic mechanisms govern reelin expression.[139]

All of these observations indicate that dysfunction of the normal epigenetic status of the genome can have dramatic consequences on normal cognitive function.[7] These studies also suggest that drugs which target the epigenome might represent viable therapies in treating various diseases affecting cognition, as will be discussed in more detail in Chapters 6–8.

SUMMARY – ACTIVE REGULATION OF EPIGENETIC MARKS IN THE NERVOUS SYSTEM

In this brief overview, we have presented an emerging view of the epigenome and its role in the adult CNS. New studies are being published at a rapid pace demonstrating that epigenetic mechanisms are involved in mediating diverse experience-driven changes in the CNS. These experience-driven changes in the adult CNS are manifest at the molecular, cellular, circuit, and behavioral levels. Overall, these diverse observations demonstrate that the epigenome resides at the interface of the environment and the genome. Furthermore, it is now becoming clear that epigenetic mechanisms exert a powerful influence over behavior. Future studies geared toward understanding the role of the epigenome in experience-dependent behavioral modification will clearly be important for, and relevant to, not only the memory field but studies of diverse types of psychiatric and neurological disorders as well.

Chromatin is a dynamic structure that integrates potentially hundreds of signals from the cell surface and effects a coordinated and appropriate transcriptional response. It is increasingly clear that epigenetic marking of chromatin and DNA itself is an important component of the signal integration that is performed by the genome as a whole. Moreover, changes in the epigenetic state of chromatin can have lasting effects on behavior. We hypothesize that the CNS has co-opted mechanisms of epigenetic tagging of the genome for use in the formation of long-term memory and many other forms of long-lasting neural plasticity seen in both health and disease. In our estimation, understanding the epigenetic regulation of neural and glial function will be vital for fully understanding the molecular processes that govern normal brain function as well as the range of brain abnormalities that underlie diverse disease states.

References

1. Lester BM, Tronick E, Nestler E, et al. Behavioral epigenetics. *Ann N Y Acad Sci*. 2011;1226:14–33.
2. Day JJ, Sweatt JD. DNA methylation and memory formation. *Nat Neurosci*. 2010;13(11):1319–1323.

3. Pray L. Epigenetics: genome, meet your environment. *The Scientist*. 2004:18.
4. Si K, Giustetto M, Etkin A, et al. A neuronal isoform of CPEB regulates local protein synthesis and stabilizes synapse-specific long-term facilitation in aplysia. *Cell*. 2003;115(7):893–904.
5. Si K, Lindquist S, Kandel ER. A neuronal isoform of the aplysia CPEB has prion-like properties. *Cell*. 2003;115:879–891.
6. Rakyan VK, Preis J, Morgan HD, Whitelaw E. The marks, mechanisms and memory of epigenetic states in mammals. *Biochem J*. 2001;356:1–10.
7. Levenson JM, Sweatt JD. Epigenetic mechanisms in memory formation. *Nat Rev Neurosci*. 2005;6(2):108–118.
8. Robison AJ, Nestler EJ. Transcriptional and epigenetic mechanisms of addiction. *Nat Rev Neurosci*. 2011;12(11):623–637.
9. Tsankova N, Renthal W, Kumar A, Nestler EJ. Epigenetic regulation in psychiatric disorders. *Nat Rev Neurosci*. 2007;8(5):355–367.
10. Waddington CH. *The Strategy of the Genes*. New York: MacMillan; 1957.
11. Bird AP. Perceptions of epigenetics. *Nature*. 2007;447:396–398.
12. Okano M, Xie S, Li E. Cloning and characterization of a family of novel mammalian DNA (cytosine-5) methyltransferases. *Nat Genet*. 1998;19:219–220.
13. Chen L, MacMillan AM, Chang W, Ezaz-Nikpay K, Lane WS, Verdine GL. Direct identification of the active-site nucleophile in a DNA (cytosine-5)-methyltransferase. *Biochemistry*. 1991;30(46):11018–11025.
14. Santi DV, Garrett CE, Barr PJ. On the mechanism of inhibition of DNA-cytosine methyltransferases by cytosine analogs. *Cell*. 1983;33:9–10.
15. Bird AP. Use of restriction enzymes to study eukaryotic DNA methylation: II. The symmetry of methylated sites supports semi-conservative copying of the methylation pattern. *J Mol Biol*. 1978;118:49–60.
16. Cedar H, Solage A, Glaser G, Razin A. Direct detection of methylated cytosine in DNA by use of the restriction enzyme MspI. *Nucleic Acids Res*. 1979;6:2125–2132.
17. Cooper DN, Krawczak M. Cytosine methylation and the fate of CpG dinucleotides in vertebrate genomes. *Hum Genet*. 1989;83:181–188.
18. Bird AP. CpG-rich islands and the function of DNA methylation. *Nature*. 1986;321:209–213.
19. Gardiner-Garden M, Frommer M. CpG islands in vertebrate genomes. *J Mol Biol*. 1987;196:261–282.
20. Ballas N, Battaglioli E, Atouf F, et al. Regulation of neuronal traits by a novel transcriptional complex. *Neuron*. 2001;31:353–365.
21. Feng J, Zhou Y, Campbell SL, et al. Dnmt1 and Dnmt3a maintain DNA methylation and regulate synaptic function in adult forebrain neurons. *Nat Neurosci*. 2010;13(4):423–430.
22. Hendrich B, Bird A. Identification and characterization of a family of mammalian methyl-CpG binding proteins. *Mol Cell Biol*. 1998;18:6538–6547.
23. Prokhortchouk A, Hendrich B, Jørgensen H, et al. The p120 catenin partner Kaiso is a DNA methylation-dependent transcriptional repressor. *Genes Dev*. 2001;15(13):1613–1618.
24. Luger K, Mader AW, Richmond RK, Sargent DF, Richmond TJ. Crystal structure of the nucleosome core particle at 2.8 A resolution. *Nature*. 1997;389:251–260.
25. Strahl BD, Allis CD. The language of covalent histone modifications. *Nature*. 2000;403:41–45.
26. Chahrour M, Jung SY, Shaw C, et al. MeCP2, a key contributor to neurological disease, activates and represses transcription. *Science*. 2008;320:1224–1229.
27. Cohen S, Zhou Z, Greenberg ME. Activating a repressor. *Science*. 2008;320:1172–1173.
28. Bell AC, Felsenfeld G. Methylation of a CTCF-dependent boundary controls imprinted expression of the Igf2 gene. *Nature*. 2000;405:482–485.
29. Barrès R, Yan J, Egan B, et al. Acute exercise remodels promoter methylation in human skeletal muscle. *Cell Metab*. 2012;15(3):405–411.
30. Ooi SK, Bestor TH. The colorful history of active DNA demethylation. *Cell*. 2008;133:1145–1148.
31. Levenson JM, Roth TL, Lubin FD, et al. Evidence that DNA (cytosine-5) methyltransferase regulates synaptic plasticity in the hippocampus. *J Biol Chem*. 2006;281:15763–15773.
32. Meaney MJ, Szyf M, Seckl JR. Epigenetic mechanisms of perinatal programming of hypothalamic-pituitary-adrenal function and health. *Trends Mol Med*. 2007;13(7):269–277.
33. Miller CA, Sweatt JD. Covalent modification of DNA regulates memory formation. *Neuron*. 2007;53(6): 857–869.

34. Weaver IC, Champagne FA, Brown SE, et al. Reversal of maternal programming of stress responses in adult offspring through methyl supplementation: altering epigenetic marking later in life. *J Neurosci.* 2005;25(47):11045–11054.
35. Weaver IC, Meaney MJ, Szyf M. Maternal care effects on the hippocampal transcriptome and anxiety-mediated behaviors in the offspring that are reversible in adulthood. *Proc Natl Acad Sci USA.* 2006;103(9):3480–3485.
36. Weaver IC, Cervoni N, Champagne FA, et al. Epigenetic programming by maternal behavior. *Nat Neurosci.* 2004;7(8):847–854.
37. Reik W, Dean W, Walter J. Epigenetic reprogramming in mammalian development. *Science.* 2001;293(5532):1089–1093.
38. Lubin FD, Roth TL, Sweatt JD. Epigenetic regulation of BDNF gene transcription in the consolidation of fear memory. *J Neurosci.* 2008;28(42):10576–10586.
39. Volpe TA, Kidner C, Hall IM, Teng G, Grewal SI, Martienssen RA. Regulation of heterochromatic silencing and histone H3 lysine-9 methylation by RNAi. *Science.* 2002;297:1833–1837.
40. Métivier R, Gallais R, Tiffoche C, et al. Cyclical DNA methylation of a transcriptionally active promoter. *Nature.* 2008;452(7183):45–50.
41. Gehring M, Reik W, Henikoff S. DNA demethylation by DNA repair. *Trends Genet.* 2009;25(2):82–90.
42. Niehrs C. Active DNA demethylation and DNA repair. *Differentiation.* 2009;77(1):1–11.
43. Schmitz KM, Schmitt N, Hoffmann-Rohrer U, Schäfer A, Grummt I, Mayer C. TAF12 recruits Gadd45a and the nucleotide excision repair complex to the promoter of rRNA genes leading to active DNA demethylation. *Mol Cell.* 2009;33(3):344–353.
44. Wu H, Sun YE. Reversing DNA methylation: new insights from neuronal activity-induced Gadd45b in adult neurogenesis. *Sci Signal.* 2009;2(64):pe17.
45. Guo JU, Ma DK, Mo H, et al. Neuronal activity modifies the DNA methylation landscape in the adult brain. *Nat Neurosci.* 2011;14(10):1345–1351.
46. Guo JU, Su Y, Zhong C, Ming GL, Song H. Hydroxylation of 5-methylcytosine by TET1 promotes active DNA demethylation in the adult brain. *Cell.* 2011;145:423–434.
47. Ma DK, Guo JU, Ming GL, Song H. DNA excision repair proteins and Gadd45 as molecular players for active DNA demethylation. *Cell Cycle.* 2009;8(10):1526–1531.
48. Ma DK, Jang MH, Guo JU, et al. Neuronal activity-induced Gadd45b promotes epigenetic DNA demethylation and adult neurogenesis. *Science.* 2009;323(5917):1074–1077.
49. Kriaucionis S, Heintz N. The nuclear DNA base 5-hydroxymethylcytosine is present in Purkinje neurons and the brain. *Science.* 2009;324(5929):929–930.
50. Tahiliani M, Koh KP, Shen Y, et al. Conversion of 5-methylcytosine to 5-hydroxymethylcytosine in mammalian DNA by MLL partner TET1. *Science.* 2009;324(5929):930–935.
51. Kim SY, Levenson JM, Korsmeyer S, Sweatt JD, Schumacher A. Developmental regulation of Eed complex composition governs a switch in global histone modification in brain. *J Biol Chem.* 2007;282:9962–9972.
52. Lau OD, Courtney AD, Vassilev A, et al. p300/CBP-associated factor histone acetyltransferase processing of a peptide substrate. Kinetic analysis of the catalytic mechanism. *J Biol Chem.* 2000;275(29):21953–21959.
53. Tanner KG, Langer MR, Denu JM. Kinetic mechanism of human histone acetyltransferase P/CAF. *Biochemistry.* 2000;39:11961–11969.
54. Buggy JJ, Sideris ML, Mak P, Lorimer DD, McIntosh B, Clark JM. Cloning and characterization of a novel human histone deacetylase, HDAC8. *Biochem J.* 2000;350(Pt 1):199–205.
55. Finnin MS, Donigian JR, Cohen A, et al. Structures of a histone deacetylase homologue bound to the TSA and SAHA inhibitors. *Nature.* 1999;401(6749):188–193.
56. Buck SW, Gallo CM, Smith JS. Diversity in the Sir2 family of protein deacetylases. *J Leukoc Biol.* 2004;75:939–950.
57. Brami-Cherrier K, Valjent E, Hervé D, et al. Parsing molecular and behavioral effects of cocaine in mitogen- and stress-activated protein kinase-1-deficient mice. *J Neurosci.* 2005;25:11444–11454.
58. Chwang WB, Arthur JS, Schumacher A, Sweatt JD. The nuclear kinase mitogen- and stress-activated protein kinase 1 regulates hippocampal chromatin remodeling in memory formation. *J Neurosci.* 2007;27:12732–12742.
59. Chwang WB, O'Riordan KJ, Levenson JM, Sweatt JD. ERK/MAPK regulates hippocampal histone phosphorylation following contextual fear conditioning. *Learn Mem.* 2006;13:322–328.

60. Swank MW, Sweatt JD. Increased histone acetyltransferase and lysine acetyltransferase activity and biphasic activation of the ERK/RSK cascade in insular cortex during novel taste learning. *J Neurosci*. 2001;21:3383–3391.
61. Vo N, Goodman RH. CREB-binding protein and p300 in transcriptional regulation. *J Biol Chem*. 2001;276(17):13505–13508.
62. Chandramohan Y, Droste SK, Reul JM. Novelty stress induces phospho-acetylation of histone H3 in rat dentate gyrus granule neurons through coincident signalling via the N-methyl-D-aspartate receptor and the glucocorticoid receptor: relevance for c-fos induction. *J Neurochem*. 2007;101:815–828.
63. Haberland M, Montgomery RL, Olson EN. The many roles of histone deacetylases in development and physiology: implications for disease and therapy. *Nat Rev Genet*. 2009;10(1):32–42.
64. Lubin FD, Sweatt JD. The IkappaB kinase regulates chromatin structure during reconsolidation of conditioned fear memories. *Neuron*. 2007;55:942–957.
65. Oliveira AM, Wood MA, McDonough CB, Abel T. Transgenic mice expressing an inhibitory truncated form of p300 exhibit long-term memory deficits. *Learn Mem*. 2007;14:564–572.
66. Yeh SH, Lin CH, Gean PW. Acetylation of nuclear factor-kappaB in rat amygdala improves long-term but not short-term retention of fear memory. *Mol Pharmacol*. 2004;65:1286–1292.
67. Murray K. The occurrence of epsilon-N-methyl lysine in histones. *Biochemistry*. 1964;127:10–15.
68. Goldknopf IL, Taylor CW, Baum RM, et al. Isolation and characterization of protein A24, a "histone-like" non-histone chromosomal protein. *J Biol Chem*. 1975;250:7182–7187.
69. Pickart CM. Mechanisms underlying ubiquitination. *Annu Rev Biochem*. 2001;70:503–533.
70. Nickel BE, Davie JR. Structure of polyubiquitinated histone H2A. *Biochemistry*. 1989;28:964–968.
71. Chen HY, Sun JM, Zhang Y, Davie JR, Meistrich ML. Ubiquitination of histone H3 in elongating spermatids of rat testes. *J Biol Chem*. 1998;273:13165–13169.
72. Pham AD, Sauer F. Ubiquitin-activating/conjugating activity of TAFII250, a mediator of activation of gene expression in Drosophila. *Science*. 2000;289:2357–2360.
73. West MH, Bonner WM. Histone 2B can be modified by the attachment of ubiquitin. *Nucleic Acids Res*. 1980;8:4671–4680.
74. Bradbury EM, Inglis RJ, Matthews HR, Sarner N. Phosphorylation of very-lysine-rich histone in Physarum polycephalum. Correlation with chromosome condensation. *Eur J Biochem*. 1973;33:131–139.
75. Gurley LR, Walters RA, Tobey RA. Cell cycle-specific changes in histone phosphorylation associated with cell proliferation and chromosome condensation. *J Cell Biol*. 1974;60:356–364.
76. Mahadevan LC, Willis AC, Barratt MJ. Rapid histone H3 phosphorylation in response to growth factors, phorbol esters, okadaic acid, and protein synthesis inhibitors. *Cell*. 1991;65:775–783.
77. Di Agostino S, Rossi P, Geremia R, Sette C. The MAPK pathway triggers activation of Nek2 during chromosome condensation in mouse spermatocytes. *Development*. 2002;129:1715–1727.
78. Hsu JY, Sun ZW, Li X, et al. Mitotic phosphorylation of histone H3 is governed by Ipl1/aurora kinase and Glc7/PP1 phosphatase in budding yeast and nematodes. *Cell*. 2000;102:279–291.
79. Sassone-Corsi P, Mizzen CA, Cheung P, et al. Requirement of Rsk-2 for epidermal growth factor-activated phosphorylation of histone H3. *Science*. 1999;285:886–891.
80. Thomson S, Clayton AL, Hazzalin CA, Rose S, Barratt MJ, Mahadevan LC. The nucleosomal response associated with immediate-early gene induction is mediated via alternative MAP kinase cascades: MSK1 as a potential histone H3/HMG-14 kinase. *Embo J*. 1999;18:4779–4793.
81. Goto H, Yasui Y, Nigg EA, Inagaki M. Aurora-B phosphorylates histone H3 at serine28 with regard to the mitotic chromosome condensation. *Genes Cells*. 2002;7:11–17.
82. Ajiro K, Yoda K, Utsumi K, Nishikawa Y. Alteration of cell cycle-dependent histone phosphorylations by okadaic acid. Induction of mitosis-specific H3 phosphorylation and chromatin condensation in mammalian interphase cells. *J Biol Chem*. 1996;271:13197–13201.
83. Nowak SJ, Pai CY, Corces VG. Protein phosphatase 2A activity affects histone H3 phosphorylation and transcription in Drosophila melanogaster. *Mol Cell Biol*. 2003;23:6129–6138.
84. Stipanovich A, Valjent E, Matamales M, et al. A phosphatase cascade by which rewarding stimuli control nucleosomal response. *Nature*. 2008;453(7197):879–884.
85. Montgomery MK, Xu S, Fire A. RNA as a target of double-stranded RNA-mediated genetic interference in Caenorhabditis elegans. *Proc Natl Acad Sci USA*. 1998;95:15502–15507.

86. Hall IM, Shankaranarayana GD, Noma K, Ayoub N, Cohen A, Grewal SI. Establishment and maintenance of a heterochromatin domain. *Science*. 2002;297:2232–2237.
87. Kangaspeska S, Stride B, Métivier R, et al. Transient cyclical methylation of promoter DNA. *Nature*. 2008;452(7183):112–115.
88. Verdel A, Jia S, Gerber S, et al. RNAi-mediated targeting of heterochromatin by the RITS complex. *Science*. 2004;303:672–676.
89. Wang KC, Chang HY. Molecular mechanisms of long noncoding RNAs. *Mol Cell*. 2011;43(6):904–914.
90. Singer T, McConnell MJ, Marchetto MC, Coufal NG, Gage FH. LINE-1 retrotransposons: mediators of somatic variation in neuronal genomes? *Trends Neurosci*. 2010;33:345–354.
91. Belmont AS. Mitotic chromosome structure and condensation. *Curr Opin Cell Biol*. 2006;18(6):632–638.
92. Cremer T, Cremer C. Chromosome territories, nuclear architecture and gene regulation in mammalian cells. *Nat Rev Genet*. 2001;2(4):292–301.
93. Duan Z, Andronescu M, Schutz K, et al. A three-dimensional model of the yeast genome. *Nature*. 2010;465(7296):363–367.
94. Gaszner M, Felsenfeld G. Insulators: exploiting transcriptional and epigenetic mechanisms. *Nat Rev Genet*. 2006;7(9):703–713.
95. Wood AJ, Severson AF, Meyer BJ. Condensin and cohesin complexity: the expanding repertoire of functions. *Nat Rev Genet*. 2010;11(6):391–404.
96. Dhar SS, Ongwijitwat S, Wong-Riley MT. Chromosome conformation capture of all 13 genomic Loci in the transcriptional regulation of the multisubunit bigenomic cytochrome C oxidase in neurons. *J Biol Chem*. 2009;284(28):18644–18650.
97. Horike S, Cai S, Miyano M, Cheng JF, Kohwi-Shigematsu T. Loss of silent-chromatin looping and impaired imprinting of DLX5 in Rett syndrome. *Nat Genet*. 2005;37(1):31–40.
98. Jiang Y, Jakovcevski M, Bharadwaj R, et al. Setdb1 histone methyltransferase regulates mood-related behaviors and expression of the NMDA receptor subunit NR2B. *J Neurosci*. 2010;30(21):7152–7167.
99. Li L, Suzuki T, Mori N, Greengard P. Identification of a functional silencer element involved in neuron-specific expression of the synapsin I gene. *Proc Natl Acad Sci USA*. 1993;90:1460–1464.
100. Maue RA, Kraner SD, Goodman RH, Mandel G. Neuron-specific expression of the rat brain type II sodium channel gene is directed by upstream regulatory elements. *Neuron*. 1990;4:223–231.
101. Mori N, Schoenherr C, Vandenbergh DJ, Anderson DJ. A common silencer element in the SCG10 and type II Na+ channel genes binds a factor present in nonneuronal cells but not in neuronal cells. *Neuron*. 1992;9:45–54.
102. Chong JA, Tapia-Ramírez J, Kim S, et al. REST: a mammalian silencer protein that restricts sodium channel gene expression to neurons. *Cell*. 1995;80:949–957.
103. Chen ZF, Paquette AJ, Anderson DJ. NRSF/REST is required in vivo for repression of multiple neuronal target genes during embryogenesis. *Nat Genet*. 1998;20:136–142.
104. Paquette AJ, Perez SE, Anderson DJ. Constitutive expression of the neuron-restrictive silencer factor (NRSF)/REST in differentiating neurons disrupts neuronal gene expression and causes axon pathfinding errors in vivo. *Proc Natl Acad Sci USA*. 2000;97:12318–12323.
105. Andres ME, Burger C, Peral-Rubio MJ, et al. CoREST: a functional corepressor required for regulation of neural-specific gene expression. *Proc Natl Acad Sci USA*. 1999;96:9873–9878.
106. Huang Y, Myers SJ, Dingledine R. Transcriptional repression by REST: recruitment of Sin3A and histone deacetylase to neuronal genes. *Nat Neurosci*. 1999;2:867–872.
107. Naruse Y, Oh-hashi K, Iijima N, Naruse M, Yoshioka H, Tanaka M. Circadian and light-induced transcription of clock gene Per1 depends on histone acetylation and deacetylation. *Mol Cell Biol*. 2004;24:6278–6287.
108. Grimes JA, Nielsen SJ, Battaglioli E, et al. The co-repressor mSin3A is a functional component of the REST-CoREST repressor complex. *J Biol Chem*. 2000;275:9461–9467.
109. Roopra A, Sharling L, Wood IC, et al. Transcriptional repression by neuron-restrictive silencer factor is mediated via the Sin3-histone deacetylase complex. *Mol Cell Biol*. 2000;20:2147–2157.
110. Battaglioli E, Andrés ME, Rose DW, et al. REST repression of neuronal genes requires components of the hSWI.SNF complex. *J Biol Chem*. 2002;277:41038–41045.
111. Lunyak VV, Burgess R, Prefontaine GG, et al. Corepressor-dependent silencing of chromosomal regions encoding neuronal genes. *Science*. 2002;298:1747–1752.

112. Noh KM, Hwang JY, Follenzi A, et al. Repressor element-1 silencing transcription factor (REST)-dependent epigenetic remodeling is critical to ischemia-induced neuronal death. *Proc Natl Acad Sci USA*. 2012;109(16):E962–E971.
113. Eriksson PS, Perfilieva E, Bjork-Eriksson T, et al. Neurogenesis in the adult human hippocampus. *Nat Med*. 1998;4:1313–1317.
114. Bellet MM, Sassone-Corsi P. Mammalian circadian clock and metabolism – the epigenetic link. *J Cell Sci*. 2010;123(Pt 22):3837–3848.
115. Reppert SM, Weaver DR. Coordination of circadian timing in mammals. *Nature*. 2002;418:935–941.
116. Klein DC, Moore RY, Reppert SM. *Suprachiasmatic Nucleus: The Mind's Clock*. New York: Oxford University Press; 1991.
117. Zylka MJ, Shearman LP, Weaver DR, Reppert SM. Three period homologs in mammals: differential light responses in the suprachiasmatic circadian clock and oscillating transcripts outside of brain. *Neuron*. 1998;20:1103–1110.
118. Balsalobre A, Damiola F, Schibler U. A serum shock induces circadian gene expression in mammalian tissue culture cells. *Cell*. 1998;93:929–937.
119. Naruse Y, Aoki T, Kojima T, Mori N. Neural restrictive silencer factor recruits mSin3 and histone deacetylase complex to repress neuron-specific target genes. *Proc Natl Acad Sci USA*. 1999;96:13691–13696.
120. Glossop NR, Hardin PE. Central and peripheral circadian oscillator mechanisms in flies and mammals. *J Cell Sci*. 2002;115:3369–3377.
121. Crosio C, Cermakian N, Allis CD, Sassone-Corsi P. Light induces chromatin modification in cells of the mammalian circadian clock. *Nat Neurosci*. 2000;3:1241–1247.
122. Blough RI, Petrij F, Dauwerse JG, et al. Variation in microdeletions of the cyclic AMP-responsive element-binding protein gene at chromosome band 16p13.3 in the Rubinstein-Taybi syndrome. *Am J Med Genet*. 2000;90:29–34.
123. Petrij F, Petrij F, Giles RH, et al. Rubinstein-Taybi syndrome caused by mutations in the transcriptional co-activator CBP. *Nature*. 1995;376:348–351.
124. Amir RE, Van den Veyver IB, Wan M, Tran CQ, Francke U, Zoghbi HY. Rett syndrome is caused by mutations in X-linked MECP2, encoding methyl- CpG-binding protein 2. *Nat Genet*. 1999;23:185–188.
125. Ellaway C, Christodoulou J. Rett syndrome: clinical characteristics and recent genetic advances. *Disabil Rehabil*. 2001;23:98–106.
126. Sirianni N, Naidu S, Pereira J, Pillotto RF, Hoffman EP. Rett syndrome: confirmation of X-linked dominant inheritance, and localization of the gene to Xq28. *Am J Hum Genet*. 1998;63:1552–1558.
127. Collins AL, Levenson JM, Vilaythong AP, et al. Mild overexpression of MeCP2 causes a progressive neurological disorder in mice. *Hum Mol Genet*. 2004;13(21):2679–2689.
128. Ashley CT, Sutcliffe JS, Kunst CB, et al. Human and murine FMR-1: alternative splicing and translational initiation downstream of the CGG-repeat. *Nat Genet*. 1993;4:244–251.
129. Turner G, Webb T, Wake S, Robinson H. Prevalence of fragile X syndrome. *Am J Med Genet*. 1996;64:196–197.
130. Gecz J, Gedeon AK, Sutherland GR, Mulley JC. Identification of the gene FMR2, associated with FRAXE mental retardation. *Nat Genet*. 1996;13:105–108.
131. Gu Y, Shen Y, Gibbs RA, Nelson DL. Identification of FMR2, a novel gene associated with the FRAXE CCG repeat and CpG island. *Nat Genet*. 1996;13:109–113.
132. Kuo YM, Emmerling MR, Vigo-Pelfrey C, et al. Water-soluble Abeta (N-40, N-42) oligomers in normal and Alzheimer disease brains. *J Biol Chem*. 1996;271:4077–4081.
133. Selkoe DJ. The cell biology of beta-amyloid precursor protein and presenilin in Alzheimer's disease. *Trends Cell Biol*. 1998;8:447–453.
134. Cao X, Sudhof TC. A transcriptionally [correction of transcriptively] active complex of APP with Fe65 and histone acetyltransferase Tip60. *Science*. 2001;293:115–120.
135. Kimberly WT, Zheng JB, Guenette SY, Selkoe DJ. The intracellular domain of the beta-amyloid precursor protein is stabilized by Fe65 and translocates to the nucleus in a notch-like manner. *J Biol Chem*. 2001;276:40288–40292.
136. Sastre M, Steiner H, Fuchs K, et al. Presenilin-dependent gamma-secretase processing of beta-amyloid precursor protein at a site corresponding to the S3 cleavage of Notch. *EMBO Rep*. 2001;2:835–841.

137. Von Rotz RC, Kohli BM, Bosset J, et al. The APP intracellular domain forms nuclear multiprotein complexes and regulates the transcription of its own precursor. *J Cell Sci*. 2004;117:4435–4448.
138. Costa E, Chen Y, Davis J, et al. Reelin and schizophrenia: a disease at the interface of the genome and the epigenome. *Mol Intervent*. 2002;2:47–57.
139. Chen Y, Sharma RP, Costa RH, Costa E, Grayson DR. On the epigenetic regulation of the human reelin promoter. *Nucleic Acids Res*. 2002;30:2930–2939.

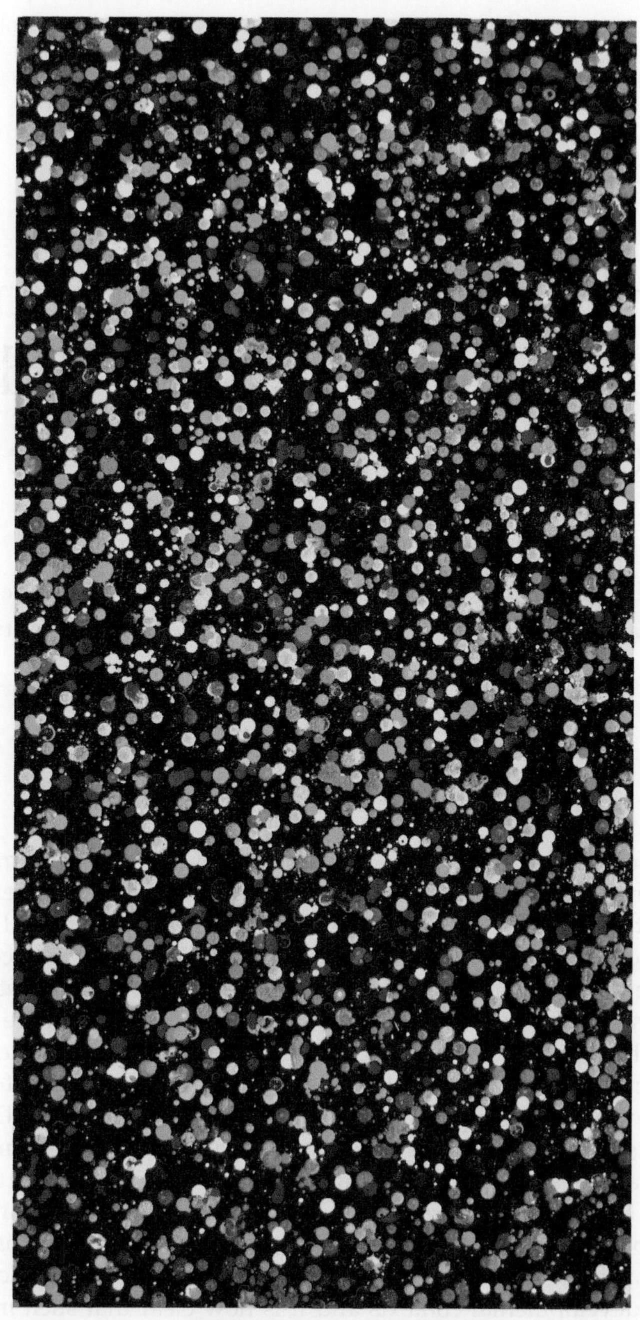

"A Histone Code for CNS Function"

J. David Sweatt, acrylic on wood panel (24 x 48), 2012

CHAPTER

2

Histone Modifications in the Nervous System and Neuropsychiatric Disorders

Morgan Bridi[1] *and Ted Abel*[2]

[1]Department of Neuroscience, Perelman School of Medicine

[2]Department of Biology, School of Arts and Sciences, University of Pennsylvania, Philadelphia, Pennsylvania, USA

INTRODUCTION

The previous chapter dealt with the "nuts-and-bolts" issues of chromatin biology and the histone code: the ways in which post-translational modification of the core histone proteins can contribute to higher-order chromatin structure and differential patterns of gene expression, a concept referred to broadly as *epigenetics*. This term was coined by Conrad Waddington, a developmental biologist, who referred to an "epigenetic landscape" to explain the observation that cells sharing the same origin and genotype display different phenotypes over the course of development.[1,2] This concept was updated in the decades to follow as our understanding of genetics grew, and epigenetics came to refer to "potentially heritable changes in gene expression that do not involve changes in DNA sequence".[3,4] Epigenetics was first conceived as a way to explain developmental phenomena and, indeed, epigenetic modifications are heritable through cell division, influencing the transcriptome and phenotype of subsequent generations of cells. Epigenetic changes are also mitotically stable enough to be heritable transgenerationally and passed on from parent to offspring and influence behavioral and genetic phenotypes.[5,6] However, epigenetic modifications are not limited to developmental contexts and it is now clear that such mechanisms are at work even in terminally differentiated cells such as neurons. The modern understanding of epigenetics reflects this, and we now define epigenetics as "the sum of the alterations to the chromatin template that collectively establish and propagate different patterns of gene expression and silencing from the same genome".[7]

Epigenetic Regulation in the Nervous System
DOI: http://dx.doi.org/10.1016/B978-0-12-391494-1.00002-1

Changes to the epigenome can be stable and long lasting. They are also dynamic and can be recruited in response to external stimuli in non-dividing cells, such as neurons.[8] Research into the various epigenetic systems at work in the central nervous system (CNS) is a burgeoning field, and investigators have begun to reveal essential roles for post-translational histone modifications in a diverse array of neuronal systems. In this chapter, we explore some of the implications of histone modifications in central nervous system function, as well as the relationship between histone modifications and neuropsychiatric and neurological disorders.

THE HISTONE CODE: COMPLEX AND COMBINATORIAL

Chromatin is the tightly packaged complex of DNA and proteins that facilitates the proper organization, storage, and transcription of the genome within the nucleus. The basic unit of chromatin is the nucleosome core particle, 147 base pairs of DNA wrapped around an octamer composed of two copies each of the canonical core histone proteins H2A, H2B, H3, and H4.[9] A number of post-translational modifications can be made to the histones, altering the structure of nucleosomes and chromatin and thereby potentially altering patterns of gene expression.[10,11] The *histone code* hypothesis was first proposed by David Allis and colleagues over a decade ago.[12] At its most basic, this idea was that histone modifications act in combination to effect downstream changes in gene expression in response to external stimuli. The histone code has served the field well as a general hypothesis and framework for experimental design. As our knowledge of chromatin biology has expanded and matured, so too has the formulation of the histone code hypothesis changed to keep pace with current research.[8,13,14] A more current view of the histone code is that of an "epigenetic index"[8] in which the different combinations of histone modifications correspond to particular transcriptional and epigenomic states. This idea is taken further with the concept of a *"histone language"*,[14] in which the downstream effects of histone modifications are context-sensitive and cross-talk between modifications influences the addition, removal and, ultimately, the readout of epigenetic histone marks.

There are at least 14 distinct modifications that have been identified in the literature, occurring at over 100 sites on the N-terminal tails and the globular bodies of histone proteins (Figure 2.1).[15,16] Histone modifications regulate various processes including activation or inactivation of transcription, DNA repair, replication, and chromatin condensation.[11,17] Histone acetylation occurs at lysine residues on all of the core histone proteins, and is usually associated with active transcription.[18] Phosphorylation at serine, threonine, and tyrosine residues is also correlated with actively-transcribed genes, and it is frequently found to co-occur with histone acetylation, although some uncertainty surrounds the nature of this relationship.[19,20] Histone methylation is a more complicated modification. Methyl groups are added to both lysine and arginine residues. Lysines can be mono-, di-, or trimethylated, while arginine residues can only have one or two methyl groups added. Histone methylation is an uncharged modification and is associated with both activation and inactivation of transcription, depending on the site modified and the number of methyl groups added.[22]

The other histone modifications are generally not as well understood, especially as they relate to brain function. Ubiquitination is a very large modification, associated with

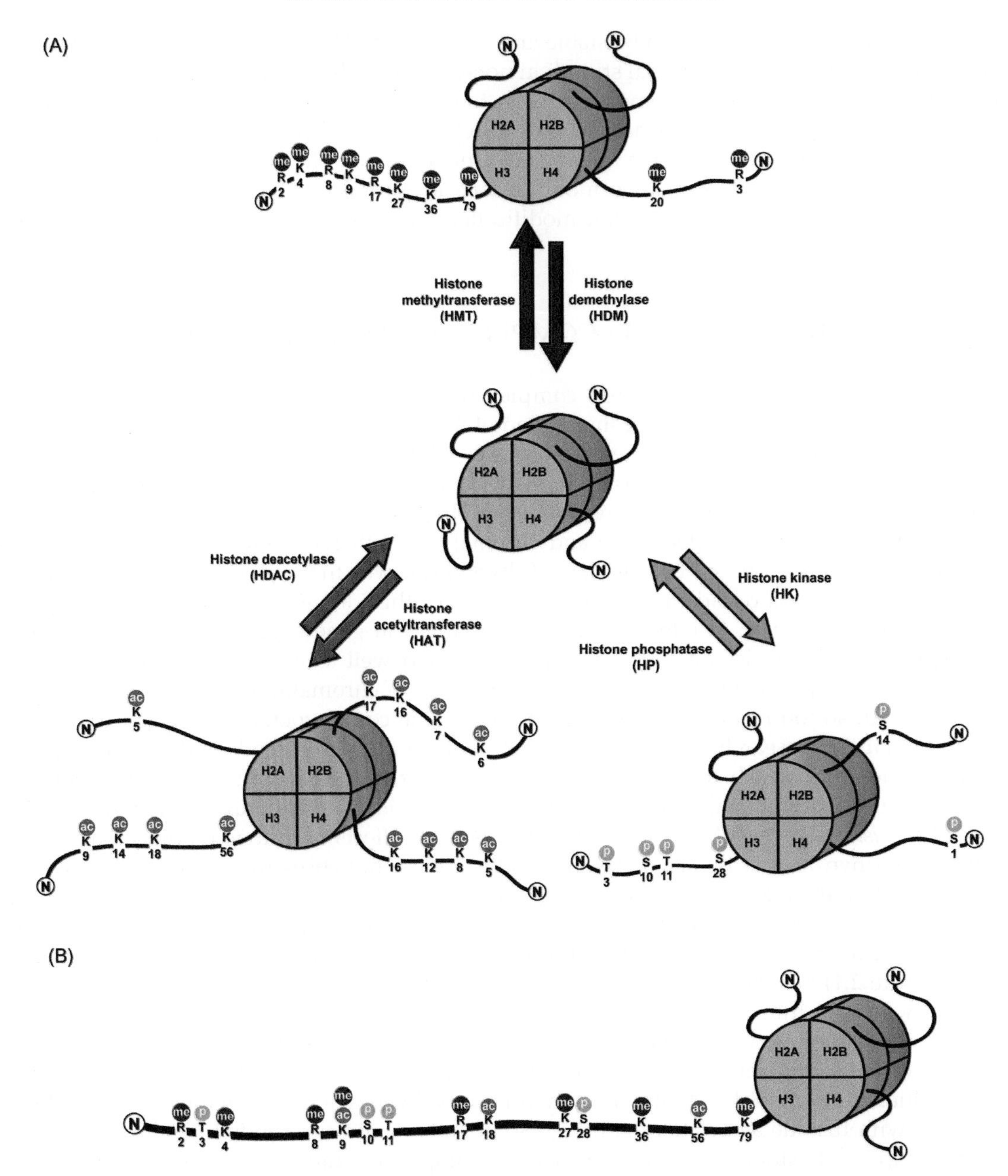

FIGURE 2.1 **Post-translational histone modifications.** (A) Histone marks are added and removed by particular classes of enzymes. Histone methylation occurs on lysine and arginine residues. Methyl groups are added by histone methyltransferase enzymes, and are removed by histone demethylase enzymes. Acetyl groups are added to lysine residues by the histone acetyltransferase enzymes; histone deacetylase enzymes remove them. Histone kinases phosphorylate serine, tyrosine, and threonine residues; histone phosphatase enzymes erase these marks.[10,11,17,19] (B) Multiple histone modifications of different types can occur together on the same histone protein.[21]

repression, and it may interfere with histone acetylation. Sumoylation is related to ubiquitination, though it is smaller and antagonizes both acetylation and ubiquitination.[11,17] Deimination, or citrullination, is the conversion of an arginine residue to citrulline, which possibly interferes with transcriptional activation by arginine methylation,[23,24] although this relationship may also go the other way.[25] There are still other histone modifications: proprionylation, formylation, butyrylation, ADP ribosylation, and proline isomerization have all been reported, but it is not yet known what role they play in CNS function. Novel histone modifications and modification sites are still being discovered. The addition of beta-N-acetylglucosamine (O-GlcNAc) to serine and threonine on histones H2A, H2B, and H4 was first described in 2010. Histone O-GlcNAcylation was found to be responsive to heat shock and mitotic cell division, and may be involved in chromatin remodeling.[26] Crotonylation is a recently described modification, discovered in 2011.[27] Histone crotonylation was observed at lysine residues and is associated with the promoters of actively transcribed genes in male germ cells. The same study that identified crotonylation also found another novel, as-yet-uncharacterized mark, histone tyrosine hydroxylation.[27] Clearly, future research is needed to define these histone modifications and what roles they play in neurons.

A rough, back-of-the-envelope calculation can give us a feel for the complexity of the histone code by estimating how many potential epigenetic states histone modification could establish at the promoter of any particular gene. To begin, consider the number of modifications that are possible on the nucleosome core proteins: approximately 100 modifications between histones H3 and H4 and another 10 between histones H2A and H2B. If all the different combinations are taken into account, this gives a possible 10^6 states. Some redundancy may be involved, because not all modifications make unique contributions and not all possible combinations are observed in vivo. In that case, we can trim the number down to a more conservative 10^3 possible combinations per gene. Multiply that figure by the approximately 30000 genes identified in the human genome, and we could estimate that there are 30000000 possible histone modification states that could be established in the nucleus.

The potential number of epigenetic states created by histone modifications is large and the cross-talk between histone modifications is complex, but the histone code seems to follow particular rules that could make understanding it more tractable. Genome-wide studies in yeast, *Drosophila,* and mammalian cells have mapped histone modifications and found that not all of the possible histone modification patterns actually occur in vivo.[30–32] Smaller-scale experiments have also found some of the relationships and rules that dictate the histone code. For example, not all histone modifications make unique contributions, and the addition of some histone modifications is dependent on others. In yeast, mutagenesis studies of the lysine residues on the tail of histone H4 revealed two apparently independent histone acetylation mechanisms.[33] Acetylation on three of the four H4 lysine residues (H4K5, H4K8, and H4K12) acted as a cumulative mark; the number of acetylated lysines, but not the identity, led to higher levels of transcription of one set of genes. On the other hand, H4K16 acetylation functioned separately, regulating an entirely different transcriptional program than the other three acetyl marks.

Biochemical evidence indicates that causal relationships exist between histone modifications, and that they are not made independently. Some evidence from yeast[34] and *Drosophila*[32] indicates that only a limited number of histone modification combinations may occur. Many histone-modifying enzymes occur together in large regulatory complexes, which may facilitate the co-occurrence of some histone marks. The interdependence of histone modifications could also result in them being "written" in a particular order or as part of a group, leading to a characteristic pattern of modifications and transcriptional outcomes. Epigenomic data from *Saccharomyces cerevisiae* has been scrutinized for evidence of such relationships, and multiple causal combinations of histone acetylation and histone methylation have been reported,[35] specifically associated with both high and low levels of transcription. Interplay between histone modifications has also been implicated in memory formation.[36] Investigators have observed that contextual fear conditioning induces concurrent phosphorylation (at S10) and acetylation (K14) of histone H3,[37] a combination of modifications associated with transcriptional activation. There is some evidence that histone acetylation and phosphorylation are coupled,[38] with H3S10 phosphorylation acting to recruit an enzyme that is able to acetylate H3K14.[39,40] It is hypothesized that the combined phospho-acetylation of histone H3 could be part of a histone modification state that regulates the gene expression required for memory formation.[36,37] It is very likely that more of these causal histone modification interactions are waiting to be discovered, both within and between nucleosomes. These results should not be interpreted as a truly "simple" histone code, however. The ultimate effects of a histone modification are dependent on time and context, as well as the identity of the writers, erasers, and readers involved. Future studies must account for these variables to understand fully the sometimes contradictory effects of histone cross-talk.[14]

OTHER EPIGENETIC MECHANISMS

Epigenetic modifications extend beyond histone modifications to include DNA methylation, chromatin remodeling, and deposition of variant histone proteins, RNA-mediated regulation, and even protein conformational changes (Figure 2.2). All of these processes contribute to epigenetic regulation in mature eukaryotic cells, operating in concert with post-translational histone modifications. There are well-documented links between histone modification and DNA methylation, an epigenetic modification associated with transcriptional silencing. Many of genes related to learning and plasticity have transcription start sites that are located within CpG islands, the substrate for DNA methylation.[41] Memory formation is accompanied by bi-directional changes in DNA methylation.[49] Inhibition of DNA methylation impairs memory, but these deficits can be counteracted by inhibition of histone deacetylase enzymes, leading to histone hyperacetylation.[50] Conversely, DNA methyltransferase inhibition in hippocampal slices blocks histone acetylation following the induction of long-term potentiation.[50] Work from the lab of Michael Meaney at McGill University has demonstrated that maternal care influences response to stress in offspring, and that such maternal behaviors are transmitted across generations in female pups.[51] The transmission of these behaviors and manifestation of altered adult stress behavior have been attributed to epigenetic programming due to the style of maternal care, and differences in DNA methylation and histone acetylation have been observed at relevant genes in the hippocampus. The epigenetic changes are stable enough to be passed to future generations, but still labile

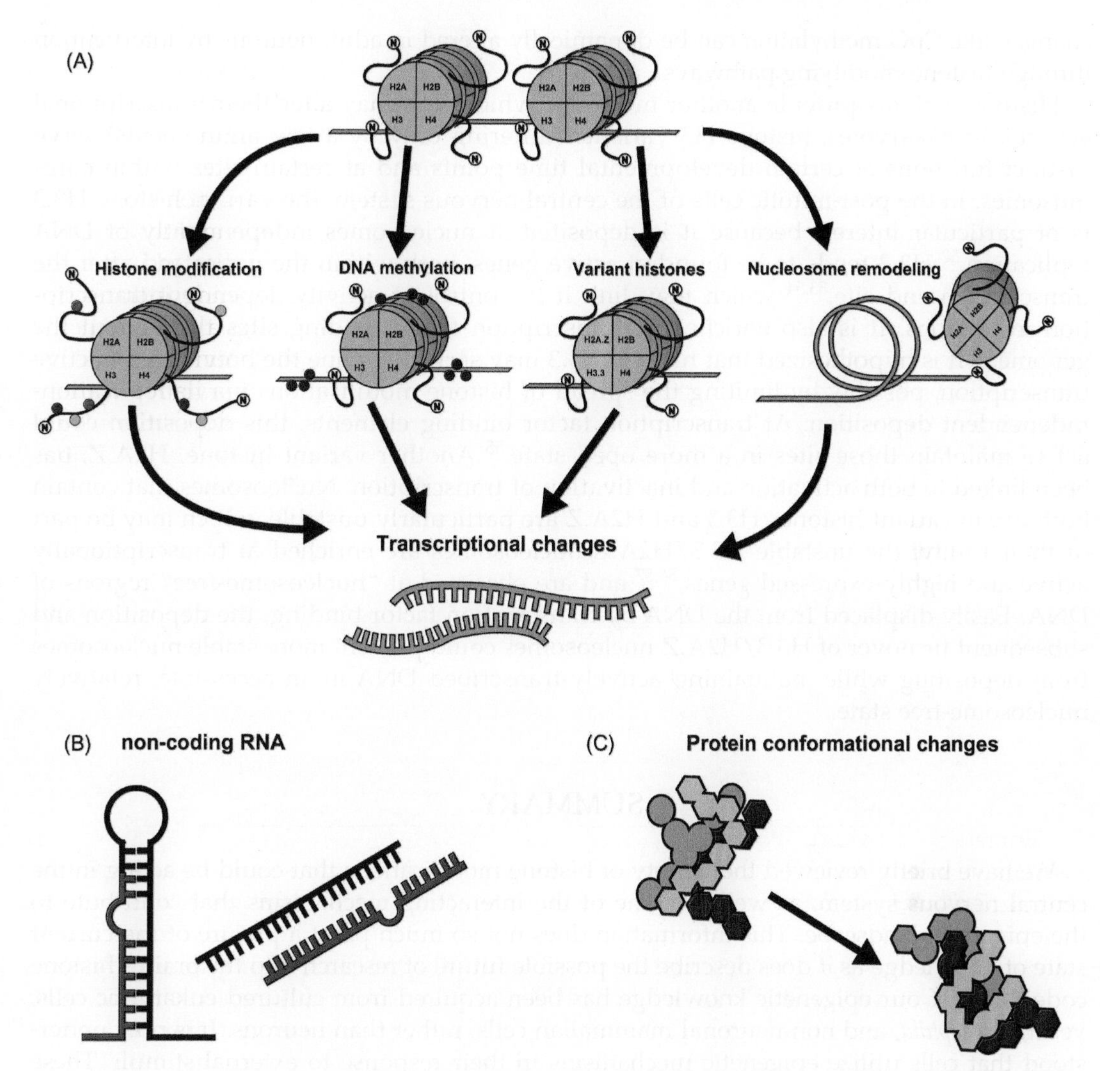

FIGURE 2.2 **The variety of epigenetic mechanisms at work in the central nervous system.** (A) Epigenetic modifications that directly affect DNA and chromatin include post-translational modification of histones,[8] DNA modification by the addition of methyl groups,[41] deposition of variant histone proteins,[42] and ATP-dependent chromatin remodeling.[43] (B) Non-coding RNA and microRNA have been implicated in the regulation of memory formation and neural plasticity. Non-coding RNA mediates post-transcriptional gene expression silencing by regulating the degradation, stability, editing, and translation of target mRNA in an activity-dependent manner.[44,45] (C) In some systems, the maintenance of long-term changes in synaptic strength associated with memory is supported by self-propagating conformational changes in proteins like neuronal CPEB, which form amyloid-like oligomers at the synapse.[46–48]

and responsive to pharmacological manipulation in adulthood, as the differences in DNA methylation and stress response can be reversed in by treatment with a drug that alters histone acetylation.[5,6] These experiments provide evidence of the close relationships between histone modification and other epigenetic mechanisms, and show that even extremely stable changes like CpG methylation can be dynamically altered in adult neurons by intervention through histone-modifying pathways.

Histone variants provide another means by which cells may alter their transcriptional activity. In eukaryotes, histone H3 variants (differing only by a few amino acids) serve distinct functions at certain developmental time points and at certain sites within chromosomes. In the post-mitotic cells of the central nervous system, the variant histone H3.3 is of particular interest because it is deposited at nucleosomes independently of DNA replication.[52] H3.3 tends to be found at active genes, both within the genes and after the transcription end site,[53,54] which may link it to control of activity-dependent transcription in neurons. It is also enriched at transcription factor binding sites throughout the genome.[55] It is hypothesized that histone H3.3 may serve to define the boundary of active transcription, possibly by limiting the spread of histone modifications via its replication-independent deposition. At transcription factor binding elements, this deposition could act to maintain those sites in a more open state.[55] Another variant histone, H2A.Z, has been linked to both activation and inactivation of transcription. Nucleosomes that contain both of the variant histones H3.3 and H2A.Z are particularly unstable, which may be part of their utility; the unstable H3.3/H2A.Z nucleosomes are enriched at transcriptionally active and highly-expressed genes,[56,57] and are observed at "nucleosome-free" regions of DNA. Easily displaced from the DNA by transcription factor binding, the deposition and subsequent turnover of H3.3/H2A.Z nucleosomes could prevent more stable nucleosomes from depositing while maintaining actively-transcribed DNA in an accessible, relatively nucleosome-free state.

SUMMARY

We have briefly reviewed the variety of histone modifications that could be acting in the central nervous system, as well as some of the interacting mechanisms that contribute to the epigenetic landscape. This information does not so much paint a picture of the current state of knowledge as it does describe the possible future of research into the brain's histone code. Most of our epigenetic knowledge has been acquired from cultured eukaryotic cells, yeast, *Drosophila*, and non-neuronal mammalian cells, rather than neurons. It is now understood that cells utilize epigenetic mechanisms in their response to external stimuli. These mechanisms, especially histone modification, can be rapidly made and are stable enough to mediate long-term changes in the transcriptome. However, the histone language is complicated and dynamic, making it well suited to the needs of cells like neurons that must receive thousands of inputs. The complex cross-talk of histone modifications at plasticity-related genes offers a system that can integrate such input and produce the appropriate pattern of gene expression in response. Research into the neural histone code has focused primarily on the best-characterized histone modifications (acetylation, phosphorylation, and methylation) and, accordingly, this chapter will provide an overview of many of the central nervous system functions that are influenced, controlled, or regulated by this epigenetic language. These topics – learning and memory, disease, aging, addiction, neuropsychiatric illness – will receive more detailed discussion in their own chapters, but they are briefly discussed below.

THE HISTONE CODE IN LEARNING, MEMORY AND SYNAPTIC PLASTICITY

Gene transcription and de novo protein synthesis are required for the formation of long-lasting memories.[58,59] These requirements are shared by long-lasting forms of long-term potentiation, an activity-dependent change in synaptic strength that is viewed as a cellular and molecular correlate of learning.[60,61] The histone code, posited to be an epigenetic network responsible for much of the transcriptional flexibility demonstrated by our cells, sits poised to regulate both the long- and short-term changes in chromatin environment and gene expression patterns that support memory consolidation. Over the past decade, epigenetics research has shown us that chromatin modifications are required for memory acquisition and synaptic plasticity in many areas of the brain including the cortex, hippocampus, striatum, and amygdala (Figure 2.3). Only a few of these histone marks have been studied in the context of memory formation; most of the work to date has focused on histone tail acetylation, phosphorylation, and methylation. In the next section, these modifications and their involvement in memory and plasticity are our focus.

Histone Acetylation

Histone acetylation is the best studied of the chromatin modifications that have been linked learning and memory.[62] This makes sense, as histone acetylation is associated with active transcription,[18] and the active transcription of a subset of genes is necessary for both memory formation and long-term changes in synaptic plasticity.[63,64] Schmitt and Matthies[65] were the first to demonstrate changes in acetylation in the brain after learning. Using radiolabeled acetate, they showed that histone acetylation after behavioral training was upregulated in the hippocampus, but downregulated in other brain regions. It took a little over two decades for histone acetylation to get another look, this time in the classic invertebrate model of learning *Aplysia californica*[66] where long-term facilitation of synaptic strength in response to administration of 5-HT increased the acetylation of histone H3 and histone H4.

Over the past decade, histone acetylation has been repeatedly connected to learning and plasticity in both invertebrate and mammalian model systems, and it is now known to be a vital mechanism underlying both memory consolidation[28,67–70] and long-term potentiation (LTP).[69,70] The changes in histone acetylation that accompany learning occur rapidly, peaking at approximately 60 minutes post-training. However, these global changes in histone acetylation are relatively transient and return to baseline levels by 24 hours post-training.[69] The time course of these changes in histone acetylation dovetails very well with their proposed role in transcription of memory-related genes, occurring on the same timescale as the initial wave of transcription that facilitates memory consolidation and LTP.[64,71–74]

What are the specific histone acetylation marks that interact with the processes of memory consolidation and plasticity? In *Aplysia*, long-term facilitation is associated with acetylation of histone H3K14 and histone H4K8.[66] The findings in rodent models are more varied. For example, contextual fear conditioning increases histone H3, but not histone H4, acetylation in hippocampal area CA1, while latent inhibition training has the opposite outcome.[69] Fear conditioning also increases the expression of *homer1* mRNA in the hippocampus, which is accompanied by increased histone H3 acetylation at the *homer1* promoter with no change

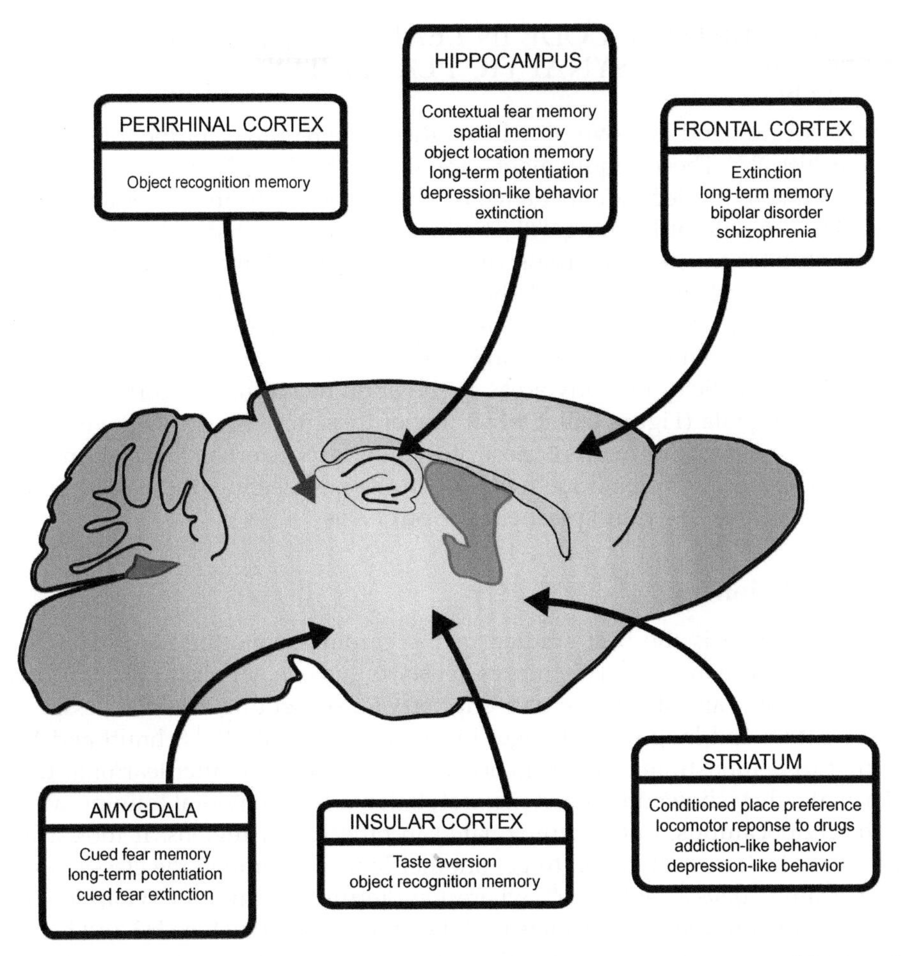

FIGURE 2.3 **Central nervous system functions known to be associated with histone modifications in different brain areas.** The epigenetic regulation of histone modification occurs in a number of different brain regions, and is involved in a wide array of different central nervous system functions. Some of the known roles of histone modification are shown, along with the brain areas where they have been observed. *Adapted with permission from Sultan and Day, 2011.*[41]

in acetylation of histone H4.[75] A study on aging and memory found that, in young mice, fear conditioning increased acetylation at H3 (K9 and K14) and H4 (K5, K8, and K12), while old mice had memory impairments and were deficient in H4K12 acetylation.[76] Morris Water Maze training increases acetylation of H3 (K9 and K14), H4 (K12), and H2B, but the acetylation of H3 was also found in a visual-platform version of this task, indicating that H3 acetylation could be involved in more passive spatial learning[77]; another study found that water maze training increases H3K18 acetylation.[78] Mice with a deletion of the histone deacetylase enzyme HDAC2 had enhanced memory and synaptic plasticity, and increased H4K5 and H4K12 acetylation.[79] In hippocampal slices, the induction of LTP by forskolin stimulation increased H3K14 acetylation.[80] It is clear that different training paradigms have revealed that specific patterns of histone acetylation are associated with specific forms of memory and plasticity, but acetylation of histone H3 or histone H4 lysine residues are the most commonly reported modifications.

Acetyl groups are added to histone tail lysine residues by the histone acetyltransferase (HAT) enzymes. CREB-binding protein, CBP, is a transcriptional co-activator with intrinsic HAT activity. CBP binds CREB, a transcription factor that is a well-known mediator of transcription of memory-related genes.[81,82] Many groups have used genetic tools to dissect out the role of CBP in memory, and we now understand that both the HAT activity[68] and CREB-binding activity[83] of CBP are important for long-term memory formation and late-phase LTP. HDAC inhibition can rescue memory deficits in some strains of CBP mutant mice,[28,68] but the enhancement of memory and plasticity by HDAC inhibition requires CREB and the interaction of CBP with CREB,[70,84] which implicates CREB-mediated genes as targets of regulation by histone modification in memory. The closely related HATs p300 and PCAF have also been implicated in learning and memory, though their precise roles and the targets of their acetyltransferase activity seem to differ from those of CBP.[85–88]

The histone deacetylase (HDAC) enzymes are responsible for the removal of acetyl groups from histone tails, acting in opposition to the HATs. There are currently 18 identified HDAC enzymes, divided into four different classes based on sequence and structure. In the mammalian CNS, the class I enzymes (HDAC1, HDAC2, HDAC3, and HDAC8) are the most active in regulating memory formation, whereas others are involved in different plastic processes such as stress, depression, and addiction.[89] The supporting evidence is both pharmacological and genetic. Overexpression of HDAC2 results in impaired long-term memory and hippocampal LTP, while knocking it out has the inverse effect. Overexpression of HDAC1 does not produce any differences in learning or plasticity, indicating that HDAC2, but not HDAC1, is particularly important for memory.[79] In mice, conditional deletion of HDAC3 in the dorsal hippocampus enhances long-term contextual fear memory,[90] an effect which is phenocopied by specific pharmacological inhibition of HDAC3. Intrahippocampal infusion of the class I-specific HDAC inhibitor MS-275 enhances long-term memory for object location; MS-275 preferentially inhibits HDAC1 and HDAC2 over HDAC3 and HDAC8, which supports the importance of HDAC2 in memory.[91] The HDAC enzymes do not bind to DNA directly, but function as parts of large, multisubunit repressor complexes.[92] HDAC1 and HDAC2 can be found as part of the Sin3a, NurD, and Co-REST repressor complexes, while HDAC3 is found with the co-repressors N-CoR and SMRT. Additionally, the Sin3a co-repressor complex interacts with both SMRT and N-CoR, so HDAC2 and HDAC3 may be found at promoters together or separately, adding another layer of complexity and specificity to the regulation of histone modification.[93] It still remains to be seen where and when the various co-repressor complexes are localized, and exactly what recruits them. A large number of transcription-regulating factors have been identified that could interact with and recruit repressive deacetylase activity, making the puzzle of these upstream regulators of histone acetylation even more interesting and complicated.

Small-molecule histone deacetylase inhibitors are now widely available, and have been used in many experimental systems. These drugs belong to a variety of chemical families and boast various degrees of specificity in which HDAC enzymes they target.[94] The experiments performed in *Aplysia* found that treatment with trichostatin-A (TSA) enhanced long-term facilitation.[66] In rodents, TSA and another HDAC inhibitor, sodium butyrate (NaB), enhance LTP in ex vivo hippocampal slices[69,70] and memory formation in vivo following systemic administration.[69] Infusion of TSA directly into the hippocampus also enhances contextual fear memory, but not short-term memory or hippocampal-independent cued fear.[70] The memory-enhancing effects of HDAC inhibitors are not limited to contextual and spatial memory, which tells us that HDAC inhibition can facilitate histone hyperacetylation in plastic processes outside of the hippocampus. For example, the infusion of an HDAC inhibitor directly into the lateral amygdala enhances the memory for the tone-footshock association of cued fear conditioning, and LTP in the amygdala is also enhanced by HDAC inhibition.[95]

Memory processes besides acquisition and consolidation are also influenced by histone acetylation. Reconsolidation, the phenomenon of a consolidated memory becoming temporarily labile when it is recalled, involves many of the same molecular mechanisms as the initial consolidation, and extinction, which is the weakening of a previously-learned association, share many behavioral and biochemical mechanisms with memory acquisition.[96–100] The frontal cortex plays a major part in extinction processes, and it has been reported that systemic treatment with the HDAC inhibitor valproic acid (VPA) enhances extinction of conditioned fear by increasing acetylation of *bdnf* promoter regions P1 and P4 (and thus increasing expression of BDNF exon IV) in the prefrontal cortex.[101] Researchers also found that VPA administration facilitates both the reconsolidation and the extinction of a conditioned fear memory, depending upon the conditions of memory retrieval.[102] The hippocampus also plays a role in extinction learning, demonstrated by the enhanced extinction of fear memory that results from intrahippocampal injection of the HDAC inhibitor TSA before memory retrieval.[103] The extinction of cued fear memory, which is amygdala-dependent, is also enhanced by a single treatment with sodium butyrate.[104]

Although highly specific agents are currently under development and have seen some limited experimental use,[78,90,91,105] most of the commonly-used HDAC inhibitors (TSA, VPA, SAHA, NaB) affect a broad range of HDAC enzymes.[94] It cannot always be certain if the effects of HDAC inhibition are attributable to altered acetylation of histone or non-histone proteins[106]; these drugs are powerful tools in the investigator's arsenal, but they tend to be sledgehammers rather than fine scalpels. The use of HDAC inhibitors has provided powerful evidence that histone acetylation plays a pivotal role in normal processes of memory and plasticity. They have also shown us that HDAC enzymes are able to negatively regulate memory, as learning and plasticity are both enhanced by their inhibition.

The multifactor, multilayered regulation of histone acetylation in molecular memory processes does present difficulties for research, and great effort will be required to dissect the mechanisms that enable the precise, opposing activities of HAT and HDAC complexes. This complication does present a large number of potential targets for pharmacological and genetic manipulation waiting to be exploited by experimenters. Other histone-modifying enzymes, involved in phosphorylation and methylation and implicated in memory, are being found to interact with the HAT- and HDAC-containing complexes. Histone acetylation is important but, when it comes to memory, it is far from the only game in town. Investigation of the details of histone acetylation, together with the combined action and cross-talk with complexed histone methyltransferases and histone kinases, will provide avenues of investigation for years to come.

Histone Phosphorylation

Histone phosphorylation plays a role in the transcriptional programs triggered by external stimuli[107,108] in many cellular systems, including the central nervous system. Like histone acetylation, it has been associated with active gene transcription, but histone phosphorylation is not yet as well characterized. The enzymes responsible for writing this mark are the histone kinases, which add phosphate groups to serine, threonine, and tyrosine residues on histone tails, mostly on histone H3. Only eight of these marks have been fully characterized in *S. cerevisiae*.[19] While there are many kinases that are capable of modifying histones, few of them have been linked to histone phosphorylation in neurons.[19,109,110] Cell culture experiments provided the initial clues that would connect histone phosphorylation with memory; in vitro, the ERK/MAPK intracellular signaling cascade, which is essential for the formation of associative memories,[111] has been placed upstream of histone H3 phosphorylation through the phosphorylation and activation of the kinases MSK1 and MSK2.[112]

Investigators were interested in determining if histone phosphorylation, with its links to the MAPK signaling pathway and transcriptional activation, is important for memory and synaptic plasticity. Both in vivo and ex vivo experiments have found correlations between histone phosphorylation and memory processes. In hippocampal slice preparations, pharmacological activation of the ERK pathway through either of the kinases PKC or PKA significantly increases both H3S10 phosphorylation and H3K14 acetylation, as well as concurrent H3 phospho-acetylation – a reminder of the combinatorial nature of the histone code.[37] Moving in vivo, the same histone modifications, along with levels of phosphorylated, active ERK, peaked at 60 minutes after contextual fear conditioning. This observation corresponds to the observed peaks in histone acetylation and the initial wave of transcription that follows neuronal activation or memory induction.

The ability to add and remove phosphate groups to histone proteins in neurons has so far only been demonstrated for two enzymes, the kinase MSK1 (mitogen- and stress-activated kinase 1) and the phosphatase PP1 (protein phosphatase 1). In mice, knockout of MSK1 impairs memory for contextual fear and attenuates both the H3 acetylation and phosphorylation that are usually found after learning.[73] MSK1 and histone phosphorylation are thus connected to memory, but it should be kept in mind that MSK1, like many other histone-modifying enzymes, also has non-histone targets in the nucleus such as CREB and NFκB.[113] PP1 is known to dephosphorylate histone H3, and it is also known to interact with HDAC enzymes and to affect their activity.[114] Overexpression of PP1 attenuates H3K14 and H4K5 acetylation, whereas inhibition of nuclear PP1 also decreases HDAC activity, increases H4K5 and H2B acetylation, and enhances object recognition memory.[115]

Although there is evidence for the co-dependence of acetylation and phosphorylation of histone H3,[37,38,40] some conflicting reports provide evidence that acetylation and phosphorylation are targeted to histone H3 through independent mechanisms,[116] and the observed H3 phospho-acetylation could be a matter of coincidence rather than coordinated action. Even with independent targeting and writing mechanisms in some systems, phosphorylation may still be involved in cross-talk with histone acetyl marks in the brain.[36,117] Whether on its own or in combination with histone acetylation, histone phosphorylation merits further inquiry as part of the histone code of the central nervous system.

Histone Methylation

Histone methylation is the third major histone modification implicated in memory and CNS function, and it is potentially a much more complicated phenomenon due to its diversity. Methylation of histone tails occurs at both lysine and arginine residues, and in three different possible variations – monomethylated, dimethylated, and trimethylated. These combinations can have different effects on chromatin structure and transcriptional activity,[10,11,22] and histone methylation is known to play both repressive and facilitative roles in transcription.[10] Methylation was once considered to be a stable, maybe irreversible modification, and the rate of turnover for histone methyl marks is much slower than that of histone acetylation.[22] However, the discovery of several enzymes that are able to reverse arginine and lysine methylation has revealed that histone methylation is in fact dynamically regulated.[17,22]

Gupta and colleagues[118] were the first to investigate the role of histone methylation in hippocampus-dependent memory. They found that contextual fear conditioning increased trimethylation of H3K4, a chromatin modification associated with active transcription. This increase in permissive histone methylation was observed at the promoter regions of the immediate-early gene *zif268* and the neuronal growth factor *bdnf*, both of which have established roles in memory formation.[63] Much like the acetylation of histone H3, which can occur in response to exposure to a novel environment,[77] behavioral training is not required for changes in histone methylation to occur. When Gupta et al placed mice in a novel context, H3K9 dimethylation was increased 60 minutes later independent of footshock delivery, indicating that this change in methylation is involved in spatial, not associative, learning.[118] Interestingly, 24 hours after context exposure, when histone acetylation induced by learning has been reduced to baseline levels, the dimethylation of H3K9 dipped to levels below those seen at baseline, the first report of a memory-related change in histone modification persisting beyond the initial consolidation phase.[118]

Several of the enzymes that mediate histone methylation in the central nervous system have been identified. Trimethylation of H3K4, an activating mark, is specifically handled by the histone methyltransferase (HMT) enzyme MLL, and deletion of the *Mll* gene leads to deficits in long-term memory for contextual fear.[118] The HMT enzyme Setdb1 (also known as Eset) catalyzes the dimethylation of H3K9, and represses expression of the NMDA receptor subunit NR2B.[119] The related enzymes GLP and G9a, which form a complex together and catalyze the repressive di- and trimethylation of H3K9, have also been linked to behavior and cognition.[120] These HMTs are involved in repression of lineage-specific genes in the central nervous system, and postnatal knock-down of GLP in the forebrain results in a number of behavioral abnormalities including memory deficits for cued and contextual fear.[121] Pharmacological inhibition of GLP/G9a in the entorhinal cortex enhances both the consolidation and the extinction of fear memory, while the same manipulation of GLP/G9a in hippocampal area CA1 impairs long-term contextual fear memory.[122] Corresponding effects on synaptic plasticity were observed ex vivo in brain slices. Long-term LTP at the CA3-CA1 Schaffer collaterals was blocked by GLP/G9a inhibition but, at the synapses of the temporoammonic pathway (from the entorhinal cortex to hippocampal CA1), GLP/G9a inhibition had no effect on LTP.[122]

For some time, histone methylation was thought of as a more-or-less permanent histone modification. That view has changed with the discovery of multiple histone demethylase

(HDM) enzymes in the past decade. HDM enzymes capable of targeting methylated lysine (KDM1/LSD1, JMJD2) and arginine (JMJD6) at all methylation states have been identified,[17,22] but little is known of their role in the central nervous system. Some interesting evidence has come from human genetic studies, however, where mutation in the H3K4 demethylase KDM5 has been linked to an intellectual disability phenotype.[123]

Some connections have been made between the roles of histone acetylation and methylation in memory. Systemic administration of the HDAC inhibitor NaB before fear conditioning enhances long-term memory and increases histone acetylation, and is accompanied by a decrease in repressive H3K9 dimethylation.[118] Furthermore, memory enhancement induced by GLP/G9a inhibition is not only accompanied by H3 methylation changes, but also by increased H3K9 acetylation.[122] The activation and repression complexes that regulate histone acetylation in plastic processes are also linked with histone methylation enzymes, though these links have not been explored in memory-formation experiments. The histone methyltransferase MLL associates with the histone acetyltransferase CBP as well as HDAC1 and HDAC2,[124] while Setdb1 can interact with the co-repressor complex mediated by Sin3a.[125] The cross-talk and interdependence between different histone modifications remains largely unexplored, and investigating these relationships will prove crucial to our understanding of neuronal plasticity.

THE HISTONE CODE IN NEUROLOGICAL AND NEUROPSYCHIATRIC DISORDERS

The previous sections of this chapter illustrated the role of histone modifications in the processes that underlie our ability to learn and adapt to our environments. Epigenetic mechanisms enable neurons tightly to regulate gene expression, allowing the brain to respond to external events, forming new associations and memories. These findings have prompted speculation that we could enhance our cognitive abilities through careful pharmacological manipulation of the writers, erasers, and readers of histone modifications. Conversely, dysregulation of histone modifications contributes to many neurodevelopmental and psychiatric disorders (see Figure 2.3). Developmental disorders that result from mutation in the histone-modifying machinery are rare, but when they do occur they can lead to severe intellectual disability. In other cases, changes in histone modifications may be connected with widespread alterations in gene expression patterns that underlie serious, chronic behavioral and mood disorders. In such instances, the pharmacological interventions described above may provide therapeutic opportunities to ameliorate these conditions.

NEURODEVELOPMENTAL DISORDERS AND INTELLECTUAL DISABILITY

Rubinstein–Taybi Syndrome

Rubinstein–Taybi syndrome, or RTS, is often referenced in the histone acetylation literature. Named for the authors of the 1963 paper that first described this syndrome,[126] Rubinstein–Taybi is a developmental disorder that is characterized by skeletal abnormalities, impaired growth, and intellectual disability (which, though usually profound, can vary widely in its severity). Approximately 1 out of every 125000 births is affected by RTS. The genetic underpinnings of RTS are generally understood, based on evidence gathered from human patients and studies of genetically-modified mouse models. Most instances of RTS can be traced to dysfunction of the transcriptional co-activator and histone acetyltransferase CBP,[127,128] attributable to a number of mutations in the gene encoding the CBP protein (*CREBBP* in humans, *cbp* in rodents). More recent data have also implicated similar mutations in the gene encoding p300 (a HAT closely related to CBP) as a cause of Rubinstein–Taybi syndrome.[129–131]

Most of our knowledge of the molecular and genetic basis of RTS comes from research performed in mouse models of the disease, and they tend to follow one of two distinct approaches in studying the syndrome. The first is the haploinsufficiency model, in which two functional copies of *CREBBP* are necessary for normal development and function.[128] These studies have used null mutations of one copy of the *cbp* allele, either globally or in a restricted fashion, to reduce the expression of CBP protein.[28,132] The other model typically proposes a dominant-negative effect of altered CBP protein produced by *CREBBP* mutation.[133,134] Experimenters have also used a number of *cbp* mutant lines expressing truncated CBP protein[67,133] or proteins with point mutations disrupting transcription-factor binding or enzymatic activity[68,83] to produce a dominant-negative CBP protein.

The first attempts at modeling RTS used global manipulation of *cbp*. Mice without a functioning *cbp* allele died early during development.[135] However, heterozygous *CBP+/−* mutants with just one functional allele of *cbp* were viable. These mice expressed CBP protein at levels half that of wild-type animals and displayed physiological characteristics that resembled RTS in humans.[132] Behavioral and electrophysiological phenotyping of these CBP+/− mice did not occur until years later, when Alarcón and colleagues[28] found that the reduction in CBP protein was accompanied by impaired long-term memory for contextual fear and novel object recognition, but not spatial navigation memory, as well as impairments in the late phase of hippocampal LTP.

Truncated and modified forms of CBP also produce deficits in long-term memory and synaptic plasticity.[67,68,83,133] A different line of CBP mutant mice, with one wild-type *cbp* allele and one producing a truncated CBP protein lacking the HAT domain and C-terminus, displayed long-term memory deficits in cued fear conditioning and a step-through passive avoidance task.[133] The memory impairment caused by truncated CBP was not limited to fear memory, as later experiments with the same truncated CBP mutants also found long-term memory deficits in novel object recognition.[67] Transgenic expression of a similar truncated CBP protein in forebrain neurons impaired LTP induced by pairing tetanic stimulation with D1 dopamine receptor activation. Expression of the CBP transgene also impaired spatial memory and contextual, but not cued, fear memory.[136]

These experiments revealed an important role for CBP in memory formation, but did not demonstrate which of its functions was responsible for the memory deficits and RTS-like symptoms. To investigate specifically the role of the HAT activity of CBP, experimenters generated a line of mice expressing a dominant-negative CBP transgene carrying two point mutations in the HAT domain that blocked the acetyltransferase activity of CBP but did not affect any of its protein–protein interaction domains.[68] Disruption of the HAT activity of CBP impaired long-term object and spatial memory, but not short-term memory.[68] In addition to its HAT activity, CBP acts as co-activator of the transcription factor CREB, binding to the KID domain of phosphorylated CREB via its KIX domain.[134] Mutation of the CBP KIX domain disrupts CREB binding but leaves the HAT activity of CBP intact,[137] and this mutation impairs memory for contextual fear and novel object recognition without affecting cued fear memory.[83] Thus, there are at least two distinct roles for CBP in memory, one depending on the CREB/CBP interaction and the other on the enzymatic HAT activity of CBP (Figure 2.4B).

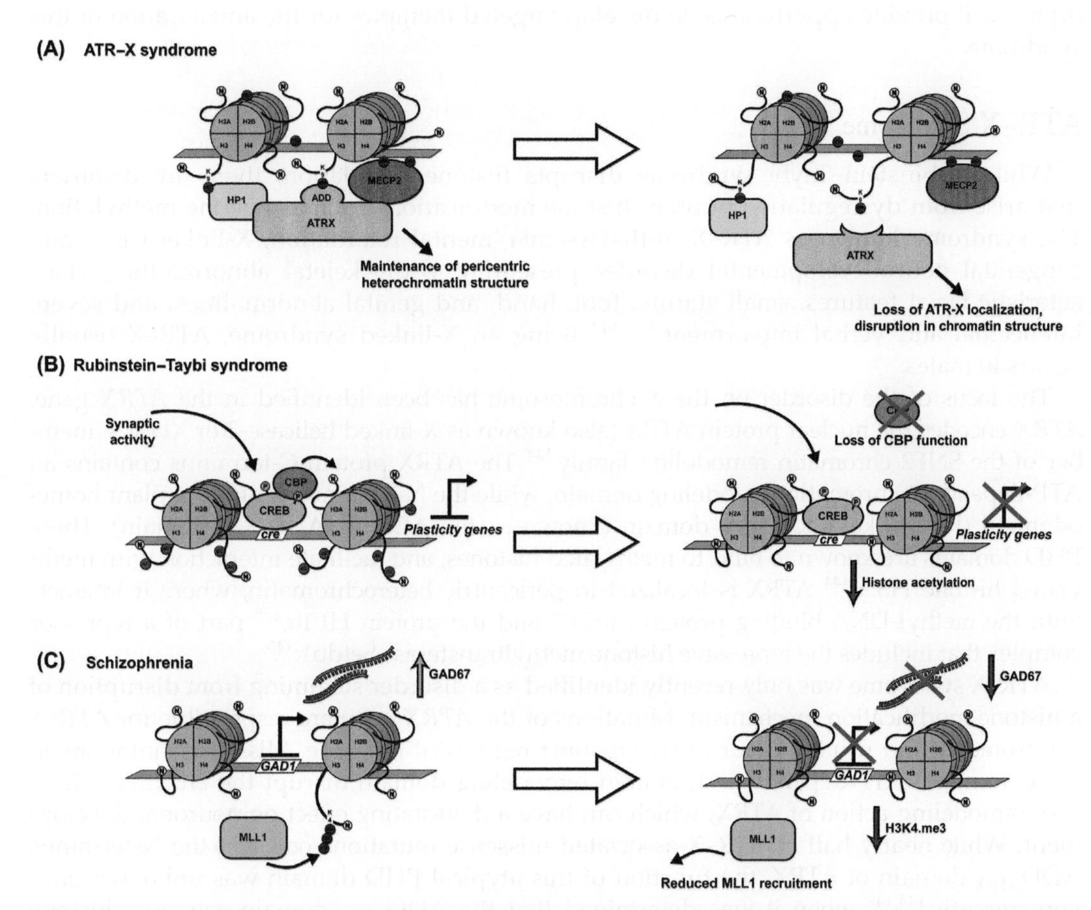

FIGURE 2.4 **Disorders and dysregulation of histone modification in the brain.** (A) A model of a potential cause of ATR–X syndrome. ATRX is recruited to heterochromatin by binding to trimethylated H3K9, where it associates with repressive proteins like HP1α and MeCP2. Mutation of the ADD_{ATRX} domain prevents this localization of ATRX, and could lead to destabilization of heterochromatin structure.[149] (B) Rubinstein–Taybi syndrome (RTS) is thought to be caused by the mutation and loss of function of the transcriptional co-activator and histone acetyltransferase CBP. CBP binds phosphorylated CREB, and the loss of CBP activity and histone acetylation at *cre*-mediated genes leads to impairment in memory and plasticity.[28,68,83,133,136,138] (C) Reduced H3K4 methylation and *GAD1* expression have been observed in patients with schizophrenia, and the loss of the histone methyltransferase MLL1 from the *GAD1* promoter may be involved in these changes.[29,168]

RTS-like symptoms can be linked with deficits in histone acetylation. Null-allele CBP+/− mice have significantly diminished histone H2B acetylation levels.[28] More recent studies have further characterized the acetylation deficits in CBP mutants, with reductions in H3K14, H4K8, and H2BK12 observed following focal knockout of CBP in the hippocampus.[138] These particular H3 and H4 acetyl marks have established links to active transcription[139] and memory,[76,77] and evidence supports a similar role for acetylation of H2B.[140] HDAC inhibitors can rescue RTS-like phenotypes in some *cbp* mutant animals. This finding is encouraging for the development of new pharmacological interventions.[28,68] Further understanding of how dysregulation of histone acetylation leads to Rubinstein–Taybi syndrome will provide opportunities to develop targeted therapies for the amelioration of this syndrome.

ATR–X Syndrome

While Rubinstein–Taybi syndrome disrupts histone acetylation, there are disorders that arise from dysregulation of other histone modifications such as histone methylation. The syndrome known as ATR–X (α-thalassemia/mental retardation, X-linked) is a rare congenital neurodevelopmental disorder presenting with skeletal abnormalities, characteristic facial features, small stature, foot, hand, and genital abnormalities, and severe intellectual and verbal impairment.[141–143] Being an X-linked syndrome, ATR–X usually occurs in males.

The locus of the disorder on the X chromosome has been identified in the *ATRX* gene. *ATRX* encodes the nuclear protein ATRX (also known as X-linked helicase-2 or XH2), a member of the SNF2 chromatin remodeling family.[144] The ATRX protein C-terminus contains an ATP-dependent chromatin-remodeling domain, while the N-terminus contains a plant homeodomain (PHD)-like zinc-finger domain (known as the ADD or ADD_{ATRX} domain). These PHD domains are known to bind to methylated histones, and facilitate interaction with methylated histone H3.[22,144] ATRX is localized to pericentric heterochromatin, where it interacts with the methyl-DNA binding protein MeCP2 and the protein HP1α,[145] part of a repressor complex that includes the repressive histone methyltransferase Setdb1.[146]

ATR–X syndrome was only recently identified as a disorder stemming from disruption of a histone modification mechanism. Mutations of the *ATRX* gene are responsible for ATR–X syndrome,[147] and usually occur in two distinct regions of the gene. Missense mutations in the C-terminal ATP-dependent chromatin remodeling domain disrupt the enzymatic helicase remodeling action of ATRX, which can have a devastating effect on neuronal development. While nearly half of ATR–X-associated missense mutations occur in the N-terminus ADD_{ATRX} domain of ATRX, the function of this atypical PHD domain was unknown until very recently,[144,148] when it was determined that the ADD_{ATRX} domain acts as a histone H3-binding module and contains a structurally unique H3K9 trimethyl binding pocket.[149] Binding of ADD_{ATRX} to histone H3 is promoted by trimethylation of H3K9, a repressive chromatin mark, but inhibited by the simultaneous presence of di- or trimethylated H3K4, a histone modification characteristic of active transcription.[148,149] This revealed that ATRX acts as a reader of the combined methylation states of K4 and K9 on histone H3, preferentially binding to the repressive, trimethylated H3K9 characteristic of heterochromatin in the absence of methylated H3K4.

The implications of these studies are important, because the cases of ATR–X syndrome caused by mutation of the ADD domain have been reported to be more severe than those attributed to missense mutation of the C-terminus helicase portion of the protein.[144,148,150] A proposed model of ATR–X syndrome focuses on the role of the ADD_{ATRX} domain in reading the methylation state of histone H3.[149] Researchers found that mutating the ADD_{ATRX} domain disrupts H3K9 binding, prevents localization of ATRX to pericentric heterochromatin, and reduces ATRX protein stability.[148,149] The loss of ATRX at pericentric chromatin could result in reduced disruption in heterochromatin structure and defects in chromosome segregation, leading to the death of neural progenitor cells (see Figure 2.4A).[149]

Addiction

Drug addiction – as opposed to drug use – is a chronic behavioral disorder in which individuals exhibit pathological drug-seeking and drug-taking behavior despite the severe negative outcomes of those behaviors.[151,152] It is thought that there is some element of genetic predisposition in the development of addiction, but environmental factors probably account for a large part of the susceptibility to addiction.[153] The search to explain the biological mechanisms that underlie these factors is a major endeavor in the biomedical research community.

The behavioral patterns that develop from drug addiction are very persistent and resistant to extinction. These changes are thus hypothesized to be supported by drug-induced, long-term alterations in gene expression, and epigenetic mechanisms may be involved in such stable changes in transcription. The brain regions involved in addiction include the reward pathways of the nucleus accumbens, the limbic system, the hippocampus, and the prefrontal cortex. Drug exposure induces neuronal plasticity, through many of the same signaling pathways that are important for learning and memory in the hippocampus and amygdala, and activating many of the very same transcription factors and immediate early genes that are known to be regulated by histone modification.

Acute and chronic drug exposure have different effects on gene expression and histone modification. Acute exposure to drugs of abuse tends to result in short, transient changes in chromatin modification and gene expression. The acute administration of cocaine (the most commonly used drug in animal models of addiction, though substances like alcohol and amphetamines are also well-studied) briefly increases histone H4 acetylation in the nucleus accumbens at the promoters of the plasticity-related immediate-early genes *cFos* and *FosB*.[154] Chronic drug exposure results in correspondingly long-lasting changes in gene expression and chromatin modification at a much larger scale, especially in the nucleus accumbens. Chronic exposure to cocaine leads to persistent upregulation of a set of plasticity-related genes including b*dnf* and *cdk5*,[155,156] and hyperacetylation of histone H3 at the promoters of these genes persists in the nucleus accumbens for as long as one week after repeated cocaine exposure.[154] Taken together, the evidence indicates that cocaine exposure facilitates plasticity via histone hyperacetylation in the nucleus accumbens, possibly mediated by a different set of modifications depending on the duration and frequency of drug exposure.[153]

Several HAT and HDAC enzymes are known to have roles in drug-induced histone modification. CBP mutant mice with just one functional *CBP* allele and reduced CBP protein levels are not as sensitive to chronic cocaine exposure as their wild-type counterparts, with reduced locomotor response and *fosb* expression after cocaine injection.[157] Focal deletion of CBP in the nucleus accumbens achieves similar results. This manipulation leads to decreased H3K14 and H2BK12 acetylation in response to acute and chronic cocaine administration. It also reduces the locomotor response to cocaine and blocks the formation of a preference for a cocaine-paired context.[158] While the class I HDAC enzymes were discussed earlier in the chapter in relation to memory, the class II HDACs seem to be involved in drug response. Viral overexpression of HDAC4[154] or HDAC5[159] decreases the rewarding effects of cocaine, as measured by conditioned place preference. Conversely, knockout of HDAC5 does not alter naïve cocaine response, but prior drug exposure sensitizes mutant animals to cocaine; they exhibit enhanced conditioned place preference compared to wild-type mice after such treatment.[159] Chronic (but not acute) exposure to cocaine actually induces phosphorylation of HDAC5 and its subsequent export from the nucleus,[159] and this loss of repression by HDAC5 may contribute to the increased acetylation, gene expression, and behavioral sensitization that accompany chronic drug administration.[153,160]

In addition to acetylation, histone phosphorylation and methylation have been associated with addiction and the response to drug exposure. In the striatum, histone H3 phosphorylation and phospho-acetylation are induced rapidly and transiently by cocaine administration.[154,161] Co-administration of cocaine and the HDAC inhibitor NaB significantly increases striatal H3 phospho-acetylation, and also potentiates the locomotor response to cocaine, linking histone phosphorylation to the behavioral response to drug exposure.[154] Cocaine-induced histone H3 phosphorylation can be been attributed to the histone kinase MSK1, as the cocaine-induced phosphorylation of histone H3S10 was not observed in MSK1 knockout mice.[161] The loss of MSK1 also prevented the drug-induced expression of *c-fos* and blocked the locomotor sensitization that occurs with chronic cocaine administration.[161] Genome-wide promoter analysis of the nucleus accumbens after chronic cocaine administration found altered patterns of histone H3K9 dimethylation.[162] Maze and colleagues[163] found that 24 hours after repeated exposure to cocaine, the level of H3K9 dimethylation was downregulated in the nucleus accumbens, as was the expression of the histone methyltransferase enzymes G9a and GLP. G9a protein levels were also reduced, and the investigators linked the cocaine-induced repression of G9a to sensitized transcriptional response to cocaine in the nucleus accumbens due to the loss of repressive methylation by the enzyme.[163]

Administration of the HDAC inhibitors TSA, NaB, and SAHA can promote behavioral sensitivity to drug exposure, determined by an increase in conditioned place preference following cocaine exposure.[154,159] Although HDAC inhibiting agents can facilitate addiction by enhancing neuronal plasticity, they may also offer a way to enhance the extinction of drug-seeking and other addiction-related behavior. Consistent with the enhancements in extinction that have been observed in hippocampus-dependent memory,[103] HDAC inhibitor treatment also facilitates extinction of cocaine-induced conditioned place preference.[158] These findings suggest that pharmacological manipulation of histone proteins may provide an opportunity for therapies to break the addiction cycle.

Schizophrenia

Schizophrenia is a very prevalent disorder, affecting an estimated 1.1% of the adult population of the USA.[164] The disorder is characterized by a number of classic symptoms, usually divided into three categories (positive, negative, and cognitive) based on their manifestation. Positive symptoms include auditory and visual hallucinations, delusions, and movement disorders, and are the signs most usually associated with schizophrenia. Negative symptoms are more difficult to identify, and present very similarly to the symptoms of depression. The cognitive symptoms of schizophrenia, such as deficits in attention and working memory, can make the day-to-day functioning of schizophrenia patients very difficult.[165]

The root causes of schizophrenia are still not well understood, but it is thought that the disorder is the result of a combination of genetic and environmental factors. Investigation of epigenetic mechanisms could be vital to our understanding of the changes in gene expression that are found in schizophrenia and how they are influenced by the environment and experience. Analysis of post-mortem human brain tissue has found changes in both facilitative and repressive histone methyl modifications. Researchers have found significantly increased methylation of histone H3R17 in post-mortem cortical tissue samples from a subgroup of schizophrenic patients.[166] The increase they observed in histone H3R17 methylation was associated with decreased expression of four metabolic genes in the identified subgroup.[166] While these changes in gene expression do not seem to be representative of schizophrenia in general, this study was the first to demonstrate gene regulation by histone modifications in schizophrenia.

In schizophrenia, cortical and hippocampal dysfunction is hypothesized to involve the disruption of GABAergic neurotransmission.[167] GAD67, encoded by the gene *GAD1*, is an isoform of glutamic acid decarboxylase, critical for the synthesis of the neurotransmitter γ-aminobutyric acid (GABA). Over the course of normal development in humans, expression of *GAD1* increases, a change that is accompanied by increasing H3K4 methylation at the *GAD1* transcription start site.[168] Furthermore, the histone methyltransferase MLL1, which catalyzes H3K4 methylation,[118] is expressed in GABAergic cortical neurons in both humans and mice, and occupies the *GAD1* promoter in mice. Genetically modified mice with lower *Mll1* levels have decreased H3K4 methylation at the promoters of *GAD1*, *Sst*, and *Npy*, all GABAergic genes, demonstrating a mechanistic link between MLL1 and GABA production proteins.

Decreases in the level of cortical GAD67 mRNA are commonly reported in patients with schizophrenia, and dysregulation of GAD67 expression has been attributed to multiple causes.[167] Recently, changes in histone methylation have been linked to the decreased expression of GAD67 in schizophrenic patients. The antipsychotic drug clozapine, used to treat schizophrenia, increases MLL1 promoter occupancy and H3K4 methylation at *Gad1*. In the human tissue samples, the researchers found that female schizophrenic patients exhibited decreased *GAD1* expression and H3K4 trimethylation. Further, a subset of schizophrenic patients with polymorphisms in *GAD1* had lower GAD1 expression, and a shift from H3K4 trimethylation to the repressive H3K27 trimethyl mark.[168] While the connection in human patients between H3K4 methylation deficits and changes in MLL1 activity is not firm, this research demonstrates a link between *GAD1* expression and histone modification mediated by MLL1. It provides an interesting model of *GAD1* dysregulation in schizophrenia; decreased H3K4 methylation and *GAD1* expression could result from the loss of MLL1 at the *GAD1* promoter during development[29,168] (see Figure 2.4C).

Histone acetylation may also be involved in schizophrenia. An analysis of microarray data from the National Brain Databank compared the expression levels of different histone deacetylase enzymes in the prefrontal cortices of schizophrenic, bipolar, and control subjects.[169] Expression of *GAD1* was found to be negatively correlated with expression of HDAC1, HDAC3, and HDAC4, all of which were more highly expressed in schizophrenic patients than in matched controls. Perhaps in part because of these findings, treatment of schizophrenia with HDAC inhibitors has received some renewed consideration. However, the results of such studies have been controversial and at times conflicting. In animal models of schizophrenia, HDAC inhibition seems to have the potential for relief. Researchers employing a methionine-induced mouse model of schizophrenia have used valproate (valproic acid, VPA) to rescue the physiological (promoter hypermethylation and downregulation of *reelin*) and behavioral (prepulse inhibition and social interaction deficits) schizophrenia-like phenotypes of the mice, indicating that HDAC inhibitors could be useful in treating at least some of the known symptoms of schizophrenia in human patients.[170]

In the clinical realm, the HDAC inhibitor valproic acid has seen use as a supplemental medication, along with a regimen of more established antipsychotic medications, in severely affected patients,[171] and may serve to accelerate the antipsychotic actions of established schizophrenia-treatment drugs.[172] However, treatment with VPA has not been as successful in controlled, randomized trials,[171,172] and it may be the case that HDAC inhibitor treatment is best suited for individuals with very severe cases of schizophrenia who do not respond to traditional medications.[173]

Histone methyltransferase inhibitors and histone demethylase inhibitors have only been described in the literature within the past decade, but they hold great potential because of the reports linking histone methylation changes and downregulation of GABA synthesis genes.[167] The effects of HMT inhibitors in living animals are mostly unknown, but inhibition of the repressive HMT enzymes GLP and G9a has been found to enhance or to impair memory, in a region-specific fashion.[122] Inhibition of HDM enzymes is able to block the demethylation of trimethyl H3K4, a mark associated with open chromatin,[174] which could be useful in facilitating the expression of *GAD1*. Monoamine oxidase inhibitors administered as antidepressants also have inhibitory effects on HDM enzymes, and combined treatment with particular monoamine oxidase inhibitors and traditional antipsychotic medications can relieve negative schizophrenic symptoms.[175] While some of these treatments may be promising, we still know very little about differential histone modification in schizophrenia and how histone modifications may affect the onset and progression of the disease.

MOOD DISORDERS

Depression

Major depressive disorder, also known as major depression or unipolar depression, is a debilitating mental disorder in which extreme feelings of sadness, anger, and frustration persist chronically and interfere with the ability to lead one's life. Depression is often predicated by chronic stress in susceptible individuals. The incidence of major depression is significant, with an estimated 16% of the population suffering from depression at least once at some point in their lives and 6% of the population dealing with it every year, according to the NIMH. People with depression are at elevated risk for suicide and often suffer disproportionately from other illnesses and physical health problems, making it a serious issue in public health.

The effects of depression in the central nervous system are widespread, implicating the nucleus accumbens, the limbic system and the hippocampus, and the frontal cortex. It is hypothesized that depression involves coordinated changes in the regulation of a number of genes, leading to large-scale alterations in gene expression patterns. The *bdnf* gene encoding brain-derived neurotrophic factor (BDNF) is of particular interest. Chronic stress, a precursor of depression and a common paradigm used in animal models of depression, downregulates BDNF levels, but this change is prevented by the administration of antidepressant drugs.[176] Hippocampal BDNF infusion can also relieve depression-like behaviors,[176] and animals lacking the *bdnf* gene are not responsive to treatment with antidepressants.[177] These results provide an important potential link between depression and histone modification, as the promoter regions of *bdnf* are known to be regulated by histone acetylation and expression of the various BDNF isoforms is sensitive to HDAC inhibitor treatment. Initial work on the role of histone modifications in depression focused on the regulation of hippocampal *bdnf*[178] in an animal model of depression, the chronic social defeat stress paradigm. The behavioral effects that resulted from chronic stress were accompanied by downregulation of *bdnf* transcription from the p3 and p4 promoters of the gene, while treatment with the antidepressant drug imipramine rescued these behavioral effects and reversed the *bdnf* expression deficits. Chronically stressed animals also had large increases in repressive H3K27 dimethylation at the p3 and p4 promoters of *bdnf*, but imipramine treatment did not exert its effects by reversing these repressive marks. Instead, it induced hyperacetylation of histone H3 at these same promoters and simultaneously reduced expression of HDAC5, a class II histone deacetylase enzyme, apparently counteracting the repression of *bdnf* transcription attributed to H3K27 histone methylation.

HDAC inhibitors are being considered as drugs for the treatment of depression because of this promising research, but some doubts about their efficacy linger. These agents may not be effective when administered alone. For example, in an experiment that found significant effects of NaB on reducing immobility in the tail-suspension test, the effects of HDAC inhibition were not found to extend to the other aspects of the behavioral despair battery employed by the researchers.[179] Some investigators have shown that increasing histone acetylation on its own may not be enough to relieve depression-like behaviors, as assayed by the forced-swim test and novelty-induced hypophagia.[180] Neither chronic nor acute NaB treatment reduced immobility in the forced swim test and, in acute doses, HDAC inhibition tended to lead to more severe symptoms, perhaps attributable to the neuronal plasticity induced by NaB administration. Instead, HDAC inhibitors may need to accompany other, more traditional antidepressant medications in order to reveal their effects.[180]

Chronic social defeat, another model of depression, produced time-dependent alterations in acetylation of H3K14 in the nucleus accumbens: decreased at 60 minutes after the final defeat, but elevated at 24 hours and even as far out as 10 days. This unusual finding was bolstered by the discovery that H3K14 acetylation was highly enriched in post-mortem nuclceus accumbens tissue samples from depressed patients compared to matched control individuals. It is a different story in the hippocampus, where decreased H3K14 acetylation was observed at 10 days after the final defeat, indicating that different histone modification mechanisms are involved in mediating the hippocampal and amygdalar responses to stress.[181]

Some of these issues have been addressed by the use of more potent and specific HDAC inhibitors. Recently, investigators have reported some success with the use of the class I-specific HDAC inhibitor MS-275, finding it to reduce depression-like behaviors when administered in both the nucleus accumbens[182] and the hippocampus.[181] HDAC2 protein levels are decreased both in the nucleus accumbens of chronically-defeated mice and in post-mortem tissue from depressed patients,[182] an observation which could account for the long-term increase in H3K14 acetylation induced by chronic-defeat stress. Chronic infusion of the HDAC inhibitors SAHA or MS-275 has antidepressant-like effects in socially-defeated mice, and it even further increases H3K14 acetylation to levels higher than those found in untreated, stressed control mice. It is still unclear how this hyperacetylation of H3K14 relieves depression-like symptoms, but it is possible that it could be related to a reversal in gene expression profiles that were altered by the experience of chronic social defeat.[182]

The role of histone modifications in stress and depression is complicated, but the research is promising. Although reports have provided some conflicting data on efficacy of certain manipulations, it is very clear that histone modification operates on a large scale in the symptomology of depression. Techniques to manipulate these histone modifications could provide great leaps in the therapeutic treatment of the underlying mechanisms of depression.

Bipolar Disorder

Bipolar disorder is one of the major mood disorders that threatens public health and mental well-being, along with major depressive disorder. According to the NIMH, it is estimated that 2.6% of the adult US population has bipolar disorder. Patients suffering from bipolar disorder experience alternating swings between periods of mania and depression, with differing degrees of severity in the intensity of these mood changes. Like schizophrenia, the onset of bipolar disorder most often occurs in late adolescence or early adulthood, but it affects both sexes equally often. Bipolar disorder is difficult to diagnose early, and those who have bipolar disorder may suffer for years before receiving an appropriate diagnosis and treatment.

The triggers and risk factors for bipolar disorder remain mostly unknown, but there may be some genetic contribution as those diagnosed with bipolar disorder are more likely to have a relative with the same diagnosis. Like schizophrenia, bipolar disorder has been associated with patterns of altered gene expression in the prefrontal cortex and the hippocampus that may be connected to the psychosis that is frequently present in both disorders. Expression of the genes *RELN* and *GAD1* is downregulated in the prefrontal cortex in both schizophrenia and bipolar disorder patients,[183] while decreased expression of GAD67 and GAD65 is also found in the hippocampus of bipolar patients.[184] It is unclear what changes in histone modification, if any, are associated with these gene expression changes in bipolar disorder. However, similar gene expression patterns in schizophrenia have been connected to altered histone methylation.[168,185] Given the similarities described here, it is plausible that some of the same mechanisms proposed to operate in schizophrenia are at work in GABAergic downregulation in bipolar disorder (see Figure 2.4C). Microarray analysis of frontal cortex samples from bipolar patients has also shown a trend (although not a significant one) towards increased HDAC1 expression.[169] The analysis of peripheral white blood cells from bipolar disorder patients found that, compared to control subjects, the expression of HDAC4 was upregulated when patients were in a remissive state, while HDAC6 and HDAC8 expression was lower in both depressive and remissive states.[186]

This is possibly related to the successful use of mood-stabilizing drugs to treat bipolar disorder, as one of the principal medications employed is valproate,[187] or valproic acid, a broad-spectrum HDAC inhibitor.[94] While the success of an HDAC inhibitor in the treatment of bipolar disorder provides another potential link to histone modification,[175] it is important to consider an alternative interpretation that HDAC inhibitors exert these effects via acetylation of non-histone proteins.[106]

Aging

A general decline in memory and cognitive function accompanies aging.[188–190] Memory recall is especially susceptible to age-related deterioration, and this loss of cognitive ability has been linked to altered gene expression in the hippocampus and the prefrontal cortex,[189,191] structures highly involved in the formation, storage, and recollection of declarative memories. Earlier sections of this chapter described the robust evidence that links histone modification to gene expression in normal processes of memory and synaptic plasticity. It is quite plausible that some of the epigenetic mechanisms that control learning and memory are also involved in aging-related cognitive decline.

Because of the connections that have been established between histone modifications and memory, Peleg and colleagues[76] hypothesized that disruption in these mechanisms may have some part in cognitive aging. The researchers found that aged (16 month-old) mice had significant impairment in hippocampus-dependent memory, in both contextual fear conditioning and the Morris Water Maze task, compared to young mice. The observed memory deficits could not be attributed to basal changes in HAT or HDAC levels. Instead, the differences were caused by dysregulation in activity-dependent histone H4 acetylation at K12 in the coding regions of memory-associated genes and an associated failure to initiate hippocampal gene transcription programs associated with memory consolidation.[76] Intrahippocampal infusion of the HDAC inhibitor SAHA rescued memory and the learning-induced acetylation of H4K12. Other forms of memory are impaired by aging; deficits in long-term memory for novel object recognition were observed in 24-month old rats compared to 3-month old controls, and this memory impairment was rescued by systemic administration of NaB immediately after training.[192]

Aging also impairs hippocampal synaptic plasticity.[193] In rats, significant impairment in the late maintenance phase of LTP was found in hippocampal slices from old animals compared to young controls, and this deficit in plasticity was rescued by pretreating the slices with the HDAC inhibitor TSA. Unlike the aged mice examined by Peleg and colleagues,[76] old rats did exhibit some differences in their histone acetylation enzymes. The expression of HDAC2 was increased and the expression of CBP was decreased in the hippocampi of aged rats, indicating that a major shift in the balance of histone acetylation could occur in the aged brain. Reduced acetylation of H3K9 and H4K12 was observed at all nine promoters of the *bdnf* gene in the aged rats, with a corresponding decrease in the level of BDNF protein.[193] The downregulation of BDNF was accompanied by reduced levels of phosphorylated, active CaMKII and ERK, two kinases activated by BDNF signaling and important for memory formation.[194] HDAC inhibitor treatment, which had rescued the LTP deficits in the old rats, also increased histone H3 and H4 acetylation at *bdnf* promoters and reversed the deficits in CaMKII and ERK phosphorylation, evidence that aging-related histone acetylation deficits alter hippocampal BDNF signaling, possibly impairing synaptic plasticity and memory.

While our understanding of the changes in histone modification that underlie aging-related cognitive decline is still mostly incomplete, the work that has been done so far is promising. Histone hypoacetylation has been implicated,[76,193] and may relate to dysregulation of transcriptional elongation.[76,195] Histone modifications regulating the neurotrophic factor *bdnf* may also be involved, which could lead to reduced BDNF signaling and disruption in memory-related transcriptional programs.[196,197] As in many of the other disorders discussed in this chapter, treatment with HDAC inhibitors has shown some potential for relieving age-related memory deficits.[76,192,193] However, there are other histone modifications involved in memory and plasticity that have not yet been investigated in the context of the aging brain, and future research into the histone language could have major ramifications for maintaining the cognitive health of an aging population.

SUMMARY

An organism's survival depends on its ability to perceive and interact with a constantly changing environment, and the central nervous system mediates these processes. The brain must form and store associations and act on them accordingly; it must update these connections as new information becomes available and conditions change. Within circuits of neurons of the central nervous system, changes in patterns of gene expression ultimately enable the storage and expression of memories and behaviors. The epigenetic control of neuronal gene expression is in large part controlled by a combinatorial histone code, a stable but reversible and highly modifiable language that enables the fine control of transcription.

In this chapter, we covered the best-characterized mechanisms and the most prominent roles of histone modification in central nervous system function: learning and memory, aging, addiction, and neuronal plasticity. We also reviewed some the manifestations of histone code dysfunction: cognitive impairment, schizophrenia, depression, and bipolar disorder. What should be clear is the utility of the histone code in providing both plasticity and stability to the transcriptional mechanisms that allow neurons to function and respond to changing inputs. The same chromatin modifications may be utilized in different nuclei or circuits, under the control of a different set of signaling pathways and histone-modifying enzymes.

Continued research into the role of histone modification in the brain is invaluable. As we begin to explore more fully the interrelationship and cooperativity between the different histone modifying marks and learn how the molecular machinery that both regulates them and exerts their downstream effects in neurons, we will gain a greater understanding of the function of how the central nervous system functions, both in health and disease.

Acknowledgments

We would like to thank Shane Poplawski, Lucia Peixoto, Marcel Estévez, and Michelle Dumoulin for contributing their discussion and comments. This publication was made possible by the support of predoctoral NRSA training grant NS079019 (to M. Bridi) and R01-MH087463 (to T. Abel).

References

1. Waddington CH, ed. *The Strategy of the Genes: A Discussion of Some Aspects of Theoretical Biology*. London: Allen & Unwin; 1957.
2. Waddington CH. Canalization of development and the inheritance of acquired characters. *Nature*. 1942;150:563–565.
3. Holliday R, Pugh JE. DNA modification mechanisms and gene activity during development. *Science*. 1975;187(4173):226–232.
4. Jaenisch R, Bird A. Epigenetic regulation of gene expression: how the genome integrates intrinsic and environmental signals. *Nat Genet*. 2003;33(suppl):245–254.
5. Weaver ICG, Cervoni N, Champagne FA, et al. Epigenetic programming by maternal behavior. *Nat Neurosci*. 2004;7(8):847–854.
6. Weaver ICG, Meaney MJ, Szyf M. Maternal care effects on the hippocampal transcriptome and anxiety-mediated behaviors in the offspring that are reversible in adulthood. *Proc Natl Acad Sci USA*. 2006;103(9):3480–3485.
7. Allis CD, Jenuwein T, Reinberg D, Caparros M-L, eds. *Epigenetics*. Cold Spring Harbor, NY: Cold Spring Harbor Laboratory Press; 2007.
8. Borrelli E, Nestler EJ, Allis CD, Sassone-Corsi P. Decoding the epigenetic language of neuronal plasticity. *Neuron*. 2008;60(6):961–974.
9. Richmond RK, Sargent DF, Richmond TJ, Luger K, Ma AW. Crystal structure of the nucleosome core particle at 2.8 A resolution. *Nature*. 1997;389(6648):251–260.
10. Berger SL. The complex language of chromatin regulation during transcription. *Nature*. 2007;447(7143): 407–412.
11. Kouzarides T. Chromatin modifications and their function. *Cell*. 2007;128(4):693–705.
12. Strahl BD, Allis CD. The language of covalent histone modifications. *Nature*. 2000;403(6765):41–45.
13. Jenuwein T, Allis CD. Translating the histone code. *Science*. 2001;293(5532):1074–1080.
14. Lee J-S, Smith E, Shilatifard A. The language of histone crosstalk. *Cell*. 2010;142(5):682–685.
15. Martin C, Zhang Y. Mechanisms of epigenetic inheritance. *Curr Opin Cell Biol*. 2007;19(3):266–272.
16. Ruthenburg AJ, Li H, Patel DJ, Allis CD. Multivalent engagement of chromatin modifications by linked binding modules. *Nat Rev Mol Cell Biol*. 2007;8(12):983–994.
17. Bannister AJ, Kouzarides T. Regulation of chromatin by histone modifications. *Cell Res*. 2011;21(3):381–395.
18. Hebbes TR, Thorne AW. A direct link between core histone acetylation and transcriptionally active chromatin. *EMBO J*. 1988;7(5):1395–1402.
19. Berger SL. Cell signaling and transcriptional regulation via histone phosphorylation. *Cold Spring Harb Symp Quant Biol*. 2010;75:23–26.
20. Zippo A, Serafini R, Rocchigiani M, Pennacchini S, Krepelova A, Oliviero S. Histone crosstalk between H3S10ph and H4K16ac generates a histone code that mediates transcription elongation. *Cell*. 2009;138(6):1122–1136.
21. Gardner KE, Allis CD, Strahl BD. Operating on chromatin, a colorful language where context matters. *J Mol Biol*. 2011;409(1):36–46.
22. Ng SS, Yue WW, Oppermann U, Klose RJ. Dynamic protein methylation in chromatin biology. *Cell Mol Life Sci*. 2009;66(3):407–422.
23. Cuthbert GL, Daujat S, Snowden AW, et al. Histone deimination antagonizes arginine methylation. *Cell*. 2004;118(5):545–553.
24. Wang Y, Wysocka J, Sayegh J, et al. Human PAD4 regulates histone arginine methylation levels via demethylimination. *Science*. 2004;306(5694):279–283.

25. Raijmakers R, Zendman AJW, Egberts WV, et al. Methylation of arginine residues interferes with citrullination by peptidylarginine deiminases in vitro. *J Mol Biol*. 2007;367(4):1118–1129.
26. Sakabe K, Wang Z, Hart GW. Beta-N-acetylglucosamine (O-GlcNAc) is part of the histone code. *Proc Natl Acad Sci USA*. 2010;107(46):19915–19920.
27. Tan M, Luo H, Lee S, et al. Identification of 67 histone marks and histone lysine crotonylation as a new type of histone modification. *Cell*. 2011;146(6):1016–1028.
28. Alarcón JM, Malleret G, Touzani K, et al. Chromatin acetylation, memory, and LTP are impaired in CBP+/− mice: a model for the cognitive deficit in Rubinstein-Taybi syndrome and its amelioration. *Neuron*. 2004;42(6):947–959.
29. Akbarian S, Huang H-S. Epigenetic regulation in human brain-focus on histone lysine methylation. *Biol Psychiatry*. 2009;65(3):198–203.
30. Bernstein BE, Kamal M, Lindblad-Toh K, et al. Genomic maps and comparative analysis of histone modifications in human and mouse. *Cell*. 2005;120(2):169–181.
31. Kurdistani SK, Tavazoie S, Grunstein M. Mapping global histone acetylation patterns to gene expression. *Cell*. 2004;117(6):721–733.
32. Schübeler D, Macalpine DM, Scalzo D, et al. The histone modification pattern of active genes revealed through genome-wide chromatin analysis of a higher eukaryote. *Genes Dev*. 2004;18(11):1263–1271.
33. Dion MF, Altschuler SJ, Wu LF, Rando OJ. Genomic characterization reveals a simple histone H4 acetylation code. *Proc Natl Acad Sci USA*. 2005;102(15):5501–5506.
34. Liu CL, Kaplan T, Kim M, et al. Single-nucleosome mapping of histone modifications in S. cerevisiae. *PLoS Biol*. 2005;3(10):e328.
35. Cui X-J, Li H, Liu G-Q, Wiley J. Combinatorial patterns of histone modifications in saccharomyces cerevisiae. *Yeast*. 2011;28(9):683–691.
36. Wood MA, Hawk JD, Abel T. Combinatorial chromatin modifications and memory storage: a code for memory? *Learn Mem*. 2006;13(3):241–244.
37. Chwang WB, O'Riordan KJ, Levenson JM, Sweatt JD. ERK/MAPK regulates hippocampal histone phosphorylation following contextual fear conditioning. *Learn Mem*. 2006;13(3):322–328.
38. Cheung P, Tanner KG, Cheung WL, Sassone-Corsi P, Denu JM, Allis CD. Synergistic coupling of histone H3 phosphorylation and acetylation in response to epidermal growth factor stimulation. *Mol Cell*. 2000;5(6):905–915.
39. Brownell JE, Zhou J, Ranalli T, et al. Tetrahymena histone acetyltransferase A: a homolog to yeast Gcn5p linking histone acetylation to gene activation. *Cell*. 1996;84(6):843–851.
40. Lo W-S, Trievel RC, Rojas JR, et al. Phosphorylation of serine 10 in histone H3 is functionally linked in vitro and in vivo to Gcn5-mediated acetylation at lysine 14. *Mol Cell*. 2000;5(6):917–926.
41. Sultan FA, Day JJ. Epigenetic mechanisms in memory and synaptic function. *Epigenomics*. 2011;3(2):157–181.
42. Kamakaka RT, Biggins S. Histone variants: deviants? *Genes Dev*. 2005;19(3):295–310.
43. Kundu S, Peterson CL. Role of chromatin states in transcriptional memory. *Biochim Biophys Acta*. 2009;1790(6):445–455.
44. Bredy TW, Lin Q, Wei W, Baker-Andresen D, Mattick JS. MicroRNA regulation of neural plasticity and memory. *Neurobiol Learn Mem*. 2011;96(1):89–94.
45. Mercer TR, Dinger ME, Mariani J, Kosik KS, Mehler MF, Mattick JS. Noncoding RNAs in long-term memory formation. *Neuroscientist*. 2008;14(5):434–445.
46. Majumdar A, Cesario WC, White-Grindley E, et al. Critical role of amyloid-like oligomers of drosophila orb2 in the persistence of memory. *Cell*. 2012;148(3):515–529.
47. Si K, Choi Y-B, White-Grindley E, Majumdar A, Kandel ER. Aplysia CPEB can form prion-like multimers in sensory neurons that contribute to long-term facilitation. *Cell*. 2010;140(3):421–435.
48. Si K, Lindquist S, Kandel ER. A neuronal isoform of the aplysia CPEB has prion-like properties. *Cell*. 2003;115(7):879–891.
49. Miller CA, Sweatt JD. Covalent modification of DNA regulates memory formation. *Neuron*. 2007;53(6):857–869.
50. Miller CA, Campbell SL, Sweatt JD. DNA methylation and histone acetylation work in concert to regulate memory formation and synaptic plasticity. *Neurobiol Learn Mem*. 2008;89(4):599–603.
51. Francis D, Diorio J, Liu D, Meaney MJ. Nongenomic transmission across generations of maternal behavior and stress responses in the rat. *Science*. 1999;286(5442):1155–1158.

52. Ahmad K, Henikoff S. The histone variant H3.3 marks active chromatin by replication-independent nucleosome assembly. *Mol Cell*. 2002;9(6):1191–1200.
53. Goldberg AD, Banaszynski LA, Noh K-M, et al. Distinct factors control histone variant H3.3 localization at specific genomic regions. *Cell*. 2010;140(5):678–691.
54. Henikoff S. Labile H3.3 + H2A.Z nucleosomes mark "nucleosome-free regions". *Nat Genet*. 2009; 41(8): 865–866.
55. Henikoff S. Nucleosome destabilization in the epigenetic regulation of gene expression. *Nat Rev Genet*. 2008;9(1):15–26.
56. Jin C, Felsenfeld G. Nucleosome stability mediated by histone variants H3.3 and H2A.Z. *Genes Dev*. 2007;21(12):1519–1529.
57. Jin C, Zang C, Wei G, et al. H3.3/H2A.Z double variant-containing nucleosomes mark "nucleosome-free regions" of active promoters and other regulatory regions. *Nat Genet*. 2009;41(8):941–945.
58. Agranoff BW, Davis RE, Casola L, Lim R. Actinomycin D blocks formation of memory of shock-avoidance in goldfish. *Science*. 1967;158(3808):1600–1601.
59. Flood JF, Rosenzweig MR, Bennett EL, Orme AE. The influence of duration of protein synthesis inhibition on memory. *Physiol Behav*. 1973;10(3):555–562.
60. Huang YY, Nguyen PV, Abel T, Kandel ER. Long-lasting forms of synaptic potentiation in the mammalian hippocampus. *Learn Mem*. 1996;3(2–3):74–85.
61. Nguyen PV, Abel T, Kandel ER. Requirement of a critical period of transcription for induction of a late phase of LTP. *Science*. 1994;265(5175):1104–1107.
62. Peixoto L, Abel T. The role of histone acetylation in memory formation and cognitive impairments. *Neuropsychopharmacology*. 2012:1–15.
63. Loebrich S, Nedivi E. The function of activity-regulated genes in the nervous system. *Physiol Rev*. 2009;89(210):1079–1103.
64. Tischmeyer W, Grimm R. Activation of immediate early genes and memory formation. *Cell Mol Life Sci*. 1999;55(4):564–574.
65. Schmitt M, Matthies H. [Biochemical studies on histones of the central nervous system. III. Incorporation of [14C]-acetate into the histones of different rat brain regions during a learning experiment]. *Acta Biol Med Ger*. 1979;38(4):683–689.
66. Guan Z, Giustetto M, Lomvardas S, et al. Integration of long-term-memory-related synaptic plasticity involves bidirectional regulation of gene expression and chromatin structure. *Cell*. 2002;111(4):483–493.
67. Bourtchouladze R, Lidge R, Catapano R, et al. A mouse model of Rubinstein-Taybi syndrome: defective long-term memory is ameliorated by inhibitors of phosphodiesterase 4. *Proc Natl Acad Sci USA*. 2003;100(18):10518–10522.
68. Korzus E, Rosenfeld MG, Mayford M. CBP histone acetyltransferase activity is a critical component of memory consolidation. *Neuron*. 2004;42(6):961–972.
69. Levenson JM, O'Riordan KJ, Brown KD, Trinh MA, Molfese DL, Sweatt JD. Regulation of histone acetylation during memory formation in the hippocampus. *J Biol Chem*. 2004;279(39):40545–40559.
70. Vecsey CG, Hawk JD, Lattal KM, et al. Histone deacetylase inhibitors enhance memory and synaptic plasticity via CREB:CBP-dependent transcriptional activation. *J Neurosci*. 2007;27(23):6128–6140.
71. Duvarci S, Nader K, LeDoux JE. De novo mRNA synthesis is required for both consolidation and reconsolidation of fear memories in the amygdala. *Learn Mem*. 2008;15(10):747–755.
72. Igaz LM, Vianna MRM, Medina JH, Izquierdo I. Two time periods of hippocampal mRNA synthesis are required for memory consolidation of fear-motivated learning. *J Neurosci*. 2002;22(15):6781–6789.
73. Katche C, Bekinschtein P, Slipczuk L, et al. Delayed wave of c-Fos expression in the dorsal hippocampus involved specifically in persistence of long-term memory storage. *Proc Natl Acad Sci USA*. 2010;107(1): 349–354.
74. Squire LR, Barondes SH. Actinomycin-D: effects on memory at different times after training. *Nature*. 1970;225(5233):649–650.
75. Mahan AL, Mou L, Shah N, Hu J-H, Worley PF, Ressler KJ. Epigenetic modulation of Homer1a transcription regulation in amygdala and hippocampus with Pavlovian fear conditioning. *J Neurosci*. 2012;32(13):4651–4659.
76. Peleg S, Sananbenesi F, Zovoilis A, et al. Altered histone acetylation is associated with age-dependent memory impairment in mice. *Science*. 2010;328(5979):753–756.

77. Bousiges O, Vasconcelos APD, Neidl R, et al. Spatial memory consolidation is associated with induction of several lysine-acetyltransferase (histone acetyltransferase) expression levels and H2B/H4 acetylation-dependent transcriptional events in the rat hippocampus. *Neuropsychopharmacology*. 2010;35(13):2521–2537.
78. Engmann O, Hortobágyi T, Pidsley R, et al. Schizophrenia is associated with dysregulation of a Cdk5 activator that regulates synaptic protein expression and cognition. *Brain*. 2011;134(8):2408–2421.
79. Guan J-S, Haggarty SJ, Giacometti E, et al. HDAC2 negatively regulates memory formation and synaptic plasticity. *Nature*. 2009;459(7243):55–60.
80. Chwang WB, Arthur JS, Schumacher A, Sweatt JD. The nuclear kinase mitogen- and stress-activated protein kinase 1 regulates hippocampal chromatin remodeling in memory formation. *J Neurosci*. 2007;27(46):12732–12742.
81. Pittenger C, Huang YY, Paletzki RF, et al. Reversible inhibition of CREB/ATF transcription factors in region CA1 of the dorsal hippocampus disrupts hippocampus-dependent spatial memory. *Neuron*. 2002;34(3):447–462.
82. Sakamoto K, Karelina K, Obrietan K. CREB: a multifaceted regulator of neuronal plasticity and protection. *J Neurochem*. 2011;116(1):1–9.
83. Wood MA, Attner MA, Oliveira AMM, Brindle PK, Abel T. A transcription factor-binding domain of the coactivator CBP is essential for long-term memory and the expression of specific target genes. *Learn Mem*. 2006;13(5):609–617.
84. Haettig J, Stefanko DP, Multani ML, Figueroa DX, McQuown SC, Wood MA. HDAC inhibition modulates hippocampus-dependent long-term memory for object location in a CBP-dependent manner. *Learn Mem*. 2011;18(2):71–79.
85. Duclot F, Jacquet C, Gongora C, Maurice T. Alteration of working memory but not in anxiety or stress response in p300/CBP associated factor (PCAF) histone acetylase knockout mice bred on a C57BL/6 background. *Neurosci Lett*. 2010;475(3):179–183.
86. Maurice T, Duclot F, Meunier J, et al. Altered memory capacities and response to stress in p300/CBP-associated factor (PCAF) histone acetylase knockout mice. *Neuropsychopharmacology*. 2008;33(7):1584–1602.
87. Oliveira AMM, Estévez MA, Hawk JD, Grimes S, Brindle PK, Abel T. Subregion-specific p300 conditional knock-out mice exhibit long-term memory impairments. *Learn Mem*. 2011;18(3):161–169.
88. Oliveira AMM, Wood MA, McDonough CB, Abel T. Transgenic mice expressing an inhibitory truncated form of p300 exhibit long-term memory deficits. *Learn Mem*. 2007;14(9):564–572.
89. Ferland CL, Schrader LA. Regulation of histone acetylation in the hippocampus of chronically stressed rats: a potential role of sirtuins. *Neuroscience*. 2011;174:104–114.
90. McQuown SC, Barrett RM, Matheos DP, et al. HDAC3 is a critical negative regulator of long-term memory formation. *J Neurosci*. 2011;31(2):764–774.
91. Hawk JD, Florian C, Abel T. Post-training intrahippocampal inhibition of class I histone deacetylases enhances long-term object-location memory. *Learn Mem*. 2011;18(6):367–370.
92. Sengupta N, Seto E. Regulation of histone deacetylase activities. *J Cell Biochem*. 2004;93(1):57–67.
93. Jones PL, Sachs LM, Rouse N, Wade PA, Shi YB. Multiple N-CoR complexes contain distinct histone deacetylases. *J Biol Chem*. 2001;276(12):8807–8811.
94. Bolden JE, Peart MJ, Johnstone RW. Anticancer activities of histone deacetylase inhibitors. *Nat Rev Drug Discov*. 2006;5(9):769–784.
95. Monsey MS, Ota KT, Akingbade IF, Hong ES, Schafe GE. Epigenetic alterations are critical for fear memory consolidation and synaptic plasticity in the lateral amygdala. *PLoS One*. 2011;6(5):e19958.
96. Alberini CM. The role of protein synthesis during the labile phases of memory: revisiting the skepticism. *Neurobiol Learn Mem*. 2008;89(3):234–246.
97. Debiec J, LeDoux JE, Nader K. Cellular and systems reconsolidation in the hippocampus. *Neuron*. 2002;36(3):527–538.
98. Kaplan GB, Moore KA. The use of cognitive enhancers in animal models of fear extinction. *Pharmacol Biochem Behav*. 2011;99(2):217–228.
99. Lattal KM, Radulovic J, Lukowiak K. Extinction: does it or doesn't it? the requirement of altered gene activity and new protein synthesis. *Biol Psychiatry*. 2006;60(4):344–351.

100. Stafford JM, Lattal KM. Is an epigenetic switch the key to persistent extinction? *Neurobiol Learn Mem.* 2011;96(1):35–40.
101. Bredy TW, Wu H, Crego C, Zellhoefer J, Sun YE, Barad M. Histone modifications around individual BDNF gene promoters in prefrontal cortex are associated with extinction of conditioned fear. *Learn Mem.* 2007;14(4):268–276.
102. Bredy TW, Barad M. The histone deacetylase inhibitor valproic acid enhances acquisition, extinction, and reconsolidation of conditioned fear. *Learn Mem.* 2008;15(1):39–45.
103. Lattal KM, Barrett RM, Wood MA. Systemic or intrahippocampal delivery of histone deacetylase inhibitors facilitates fear extinction. *Behav Neurosci.* 2007;121(5):1125–1131.
104. Itzhak Y, Anderson KL, Kelley JB, Petkov M. Histone acetylation rescues contextual fear conditioning in nNOS KO mice and accelerates extinction of cued fear conditioning in wild type mice. *Neurobiol Learn Mem.* 2012;97(4):409–417.
105. Simonini MV, Camargo LM, Dong E, et al. The benzamide MS-275 is a potent, long-lasting brain region-selective inhibitor of histone deacetylases. *Proc Natl Acad Sci USA.* 2006;103(5):1587–1592.
106. Yeh S-H, Lin C-H, Gean P-W. Acetylation of nuclear factor-kappaB in rat amygdala improves long-term but not short-term retention of fear memory. *Mol Pharmacol.* 2004;65(5):1286–1292.
107. Crosio C, Cermakian N, Allis CD, Sassone-Corsi P. Light induces chromatin modification in cells of the mammalian circadian clock. *Nat Neurosci.* 2000;3(12):1241–1247.
108. Crosio C, Heitz E, Allis CD, Borrelli E, Sassone-Corsi P. Chromatin remodeling and neuronal response: multiple signaling pathways induce specific histone H3 modifications and early gene expression in hippocampal neurons. *J Cell Sci.* 2003;116(24):4905–4914.
109. Bode AM, Dong Z. Inducible covalent posttranslational modification of histone H3. *Sci STKE.* 2005;2005(281):re4.
110. Zippo A, De Robertis A, Serafini R, Oliviero S. PIM1-dependent phosphorylation of histone H3 at serine 10 is required for MYC-dependent transcriptional activation and oncogenic transformation. *Nat Cell Biol.* 2007;9(8):932–944.
111. Atkins CM, Selcher JC, Petraitis JJ, Trzaskos JM, Sweatt JD. The MAPK cascade is required for mammalian associative learning. *Nat Neurosci.* 1998;1(7):602–609.
112. Soloaga A, Thomson S, Wiggin GR, et al. MSK2 and MSK1 mediate the mitogen- and stress-induced phosphorylation of histone H3 and HMG-14. *EMBO J.* 2003;22(11):2788–2797.
113. Arthur JSC. MSK activation and physiological roles. *Front Biosci.* 2008;13:5866–5879.
114. Brush MH, Guardiola A, Connor JH, Yao T-P, Shenolikar S. Deactylase inhibitors disrupt cellular complexes containing protein phosphatases and deacetylases. *J Biol Chem.* 2004;279(9):7685–7691.
115. Koshibu K, Gräff J, Beullens M, et al. Protein phosphatase 1 regulates the histone code for long-term memory. *J Neurosci.* 2009;29(41):13079–13089.
116. Thomson S, Clayton AL, Mahadevan LC. Independent dynamic regulation of histone phosphorylation and acetylation during immediate-early gene induction. *Mol Cell.* 2001;8(6):1231–1241.
117. Banerjee T, Chakravarti D. A peek into the complex realm of histone phosphorylation. *Mol Cell Biol.* 2011;31(24):4858–4873.
118. Gupta S, Kim SY, Artis S, et al. Histone methylation regulates memory formation. *J Neurosci.* 2010;30(10):3589–3599.
119. Jiang Y, Jakovcevski M, Bharadwaj R, et al. Setdb1 histone methyltransferase regulates mood-related behaviors and expression of the NMDA receptor subunit NR2B. *J Neurosci.* 2010;30(21):7152–7167.
120. Ding N, Zhou H, Esteve P-O, et al. Mediator links epigenetic silencing of neuronal gene expression with X-linked mental retardation. *Mol Cell.* 2008;31(3):347–359.
121. Schaefer A, Sampath SC, Intrator A, et al. Control of cognition and adaptive behavior by the GLP/G9a epigenetic suppressor complex. *Neuron.* 2009;64(5):678–691.
122. Gupta-Agarwal S, Franklin AV, DeRamus T, et al. G9a/GLP histone lysine dimethyltransferase complex activity in the hippocampus and the entorhinal cortex is required for gene activation and silencing during memory consolidation. *J Neurosci.* 2012;32(16):5440–5453.
123. Santos-Rebouças CB, Fintelman-Rodrigues N, Jensen LR, et al. A novel nonsense mutation in KDM5C/JARID1C gene causing intellectual disability, short stature and speech delay. *Neurosci Lett.* 2011;498(1):67–71.
124. Xia Z-B, Anderson M, Diaz MO, Zeleznik-Le NJ. MLL repression domain interacts with histone deacetylases, the polycomb group proteins HPC2 and BMI-1, and the corepressor C-terminal-binding protein. *Proc Natl Acad Sci USA.* 2003;100(14):8342–8347.

125. Yang L, Mei Q, Zielinska-Kwiatkowska A, et al. An ERG (ets-related gene)-associated histone methyltransferase interacts with histone deacetylases 1/2 and transcription co-repressors mSin3A/B. *Biochem J*. 2003;369:651–657.
126. Rubinstein JH, Taybi H. Broad thumbs and toes and facial abnormalities. A possible mental retardation syndrome. *Am J Dis Child*. 1963;105:588–608.
127. Blough RI, Petrij F, Dauwerse JG, et al. Variation in microdeletions of the cyclic AMP-responsive element-binding protein gene at chromosome band 16p13.3 in the Rubinstein-Taybi syndrome. *Am J Med Genet*. 2000;90(1):29–34.
128. Petrij F, Giles RH, Dauwerse HG, et al. Rubinstein-Taybi syndrome is caused by mutations in the transcriptional co-activataor CBP. *Nature*. 1995;376:348–351.
129. Roelfsema JH, White SJ, Ariyu Y, et al. Genetic heterogeneity in Rubinstein-Taybi syndrome: mutations in both the CBP and EP300 genes cause disease. *Am J Hum Genet*. 2005;76(4):572–580.
130. Tsai AC-H, Dossett CJ, Walton CS, et al. Exon deletions of the EP300 and CREBBP genes in two children with Rubinstein-Taybi syndrome detected by aCGH. *Eur J Hum Genet*. 2011;19(1):43–49.
131. Viosca J, Lopez-Atalaya JP, Olivares R, Eckner R, Barco A. Syndromic features and mild cognitive impairment in mice with genetic reduction on p300 activity: differential contribution of p300 and CBP to Rubinstein-Taybi syndrome etiology. *Neurobiol Dis*. 2010;37(1):186–194.
132. Tanaka Y, Naruse I, Maekawa T, Masuya H, Shiroishi T, Ishii S. Abnormal skeletal patterning in embryos lacking a single Cbp allele: a partial similarity with Rubinstein-Taybi syndrome. *Proc Natl Acad Sci USA*. 1997;94(19):10215–10220.
133. Oike Y, Hata A, Mamiya T, et al. Truncated CBP protein leads to classical Rubinstein-Taybi syndrome phenotypes in mice: implications for a dominant-negative mechanism. *Hum Mol Genet*. 1999;8(3):387–396.
134. Parker D, Ferreri K, Nakajima T, et al. Phosphorylation of CREB at Ser-133 induces complex formation with CREB-binding protein via a direct mechanism. *Mol Cell Biol*. 1996;16(2):694–703.
135. Tanaka Y, Naruse I, Hongo T, et al. Extensive brain hemorrhage and embryonic lethality in a mouse null mutant of CREB-binding protein. *Mech Dev*. 2000;95(1–2):133–145.
136. Wood MA, Kaplan MP, Park A, et al. Transgenic mice expressing a truncated form of CREB-binding protein (CBP) exhibit deficits in hippocampal synaptic plasticity and memory storage. *Learn Mem*. 2005;12(2):111–119.
137. Kasper LH, Boussouar F, Ney PA, et al. A transcription-factor-binding surface of coactivator p300 is required for haematopoiesis. *Nature*. 2002;419(6908):738–743.
138. Barrett RM, Malvaez M, Kramar E, et al. Hippocampal focal knockout of CBP affects specific histone modifications, long-term potentiation, and long-term memory. *Neuropsychopharmacology*. 2011;36(8):1545–1556.
139. Pokholok DK, Harbison CT, Levine S, et al. Genome-wide map of nucleosome acetylation and methylation in yeast. *Cell*. 2005;122(4):517–527.
140. Myers FA, Chong W, Evans DR, Thorne AW, Crane-Robinson C. Acetylation of histone H2B mirrors that of H4 and H3 at the chicken beta-globin locus but not at housekeeping genes. *J Biol Chem*. 2003;278(38):36315–36322.
141. Ausió J, Levin DB, De Amorim GV, Bakker S, Macleod PM. Syndromes of disordered chromatin remodeling. *Clin Genet*. 2003;64(2):83–95.
142. Gibbons RJ, Picketts DJ, Villard L, Higgs DR. Mutations in a putative global transcriptional regulator cause X-linked mental retardation with alpha-thalassemia (ATR-X syndrome). *Cell*. 1995;80(6):837–845.
143. McPherson EW, Clemens MM, Gibbons RJ, Higgs DR. X-linked alpha-thalassemia/mental retardation (ATR-X) syndrome: a new kindred with severe genital anomalies and mild hematologic expression. *Am J Med Genet*. 1995;55(3):302–306.
144. Gibbons RJ, Wada ÃT, Fisher CA, et al. Mutations in the chromatin-associated protein ATRX. *Hum Mutat*. 2008;29(6):796–802.
145. Lechner MS, Schultz DC, Negorev D, Maul GG, Rauscher FJ. The mammalian heterochromatin protein 1 binds diverse nuclear proteins through a common motif that targets the chromoshadow domain. *Biochem Biophys Res Commun*. 2005;331(4):929–937.
146. Loyola A, Tagami H, Bonaldi T, et al. The HP1alpha-CAF1-SetDB1-containing complex provides H3K9me1 for Suv39-mediated K9me3 in pericentric heterochromatin. *EMBO Rep*. 2009;10(7):769–775.
147. Gibbons R. Alpha thalassaemia-mental retardation, X linked. *Orphanet J Rare Dis*. 2006;1:15.
148. Dhayalan A, Tamas R, Bock I, et al. The ATRX-ADD domain binds to H3 tail peptides and reads the combined methylation state of K4 and K9. *Hum Mol Genet*. 2011;20(11):2195–2203.

149. Iwase S, Xiang B, Ghosh S, et al. ATRX ADD domain links an atypical histone methylation recognition mechanism to human mental-retardation syndrome. *Nat Struct Mol Biol*. 2011;18(7):769–776.
150. Badens C, Lacoste C, Philip N, et al. Mutations in PHD-like domain of the ATRX gene correlate with severe psychomotor impairment and severe urogenital abnormalities in patients with ATRX syndrome. *Clin Genet*. 2006;70(1):57–62.
151. Hyman SE, Malenka RC, Nestler EJ. Neural mechanisms of addiction: the role of reward-related learning and memory. *Annu Rev Neurosci*. 2006;29:565–598.
152. Koob G, Kreek MJ. Stress, dysregulation of drug reward pathways, and the transition to drug dependence. *Am J Psychiatry*. 2007;164(8):1149–1159.
153. Maze I, Nestler EJ. The epigenetic landscape of addiction. *Ann NY Acad Sci*. 2011;1216:99–113.
154. Kumar A, Choi K-H, Renthal W, et al. Chromatin remodeling is a key mechanism underlying cocaine-induced plasticity in striatum. *Neuron*. 2005;48(2):303–314.
155. Bibb JA, Chen J, Taylor JR, et al. Effects of chronic exposure to cocaine are regulated by the neuronal protein Cdk5. *Nature*. 2001;410(6826):376–380.
156. Grimm JW, Lu L, Hayashi T, Hope BT, Su T-P, Shaham Y. Time-dependent increases in brain-derived neurotrophic factor protein levels within the mesolimbic dopamine system after withdrawal from cocaine: implications for incubation of cocaine craving. *J Neurosci*. 2003;23(3):742–747.
157. Levine AA, Guan Z, Barco A, Xu S, Kandel ER, Schwartz JH. CREB-binding protein controls response to cocaine by acetylating histones at the fosB promoter in the mouse striatum. *Proc Natl Acad Sci USA*. 2005;102(52):19186–19191.
158. Malvaez M, Mhillaj E, Matheos DP, Palmery M, Wood MA. CBP in the nucleus accumbens regulates cocaine-induced histone acetylation and is critical for cocaine-associated behaviors. *J Neurosci*. 2011;31(47):16941–16948.
159. Renthal W, Maze I, Krishnan V, et al. Histone deacetylase 5 epigenetically controls behavioral adaptations to chronic emotional stimuli. *Neuron*. 2007;56(3):517–529.
160. Renthal W, Nestler EJ. Chromatin regulation in drug addiction and depression. *Dialogues Clin Neurosci*. 2009;11(3):257–268.
161. Brami-Cherrier K, Valjent E, Hervé D, et al. Parsing molecular and behavioral effects of cocaine in mitogen- and stress-activated protein kinase-1-deficient mice. *J Neurosci*. 2005;25(49):11444–11454.
162. Renthal W, Kumar A, Xiao G, et al. Genome-wide analysis of chromatin regulation by cocaine reveals a role for sirtuins. *Neuron*. 2009;62(3):335–348.
163. Maze I, Covington HE, Dietz DM, et al. Essential role of the histone methyltransferase G9a in cocaine-induced plasticity. *Science*. 2010;327(5962):213–216.
164. Sawa A, Snyder SH. Schizophrenia: diverse approaches to a complex disease. *Science*. 2002;296(5568):692–695.
165. Roth TL, Lubin FD, Sodhi M, Kleinman JE. Epigenetic mechanisms in schizophrenia. *Biochim Biophys Acta*. 2009;1790(9):869–877.
166. Akbarian S, Ruehl MG, Bliven E, et al. Chromatin alterations associated with down-regulated metabolic gene expression in the prefrontal cortex of subjects with schizophrenia. *Arch Gen Psychiatry*. 2005;62(8):829–840.
167. Akbarian S, Huang H-S. Molecular and cellular mechanisms of altered GAD1/GAD67 expression in schizophrenia and related disorders. *Brain Res Rev*. 2006;52(2):293–304.
168. Huang H-S, Matevossian A, Whittle C, et al. Prefrontal dysfunction in schizophrenia involves mixed-lineage leukemia 1-regulated histone methylation at GABAergic gene promoters. *J Neurosci*. 2007;27(42):11254–11262.
169. Sharma RP, Grayson DR, Gavin DP. Histone deactylase 1 expression is increased in the prefrontal cortex of schizophrenia subjects: analysis of the national brain databank microarray collection. *Schizophr Res*. 2008;98(1–3):111–117.
170. Tremolizzo L, Doueiri M-S, Dong E, et al. Valproate corrects the schizophrenia-like epigenetic behavioral modifications induced by methionine in mice. *Biol Psychiatry*. 2005;57(5):500–509.
171. Citrome L. Adjunctive lithium and anticonvulsants for the treatment of schizophrenia: what is the evidence? *Expert Rev Neurotherapeutics*. 2009;9(1):55–71.
172. Casey DE, Daniel DG, Wassef AA, Tracy KA, Wozniak P, Sommerville KW. Effect of divalproex combined with olanzapine or risperidone in patients with an acute exacerbation of schizophrenia. *Neuropsychopharmacology*. 2003;28(1):182–192.
173. Grayson DR, Kundakovic M, Sharma RP. Is there a future for histone deacetylase inhibitors in the pharmacotherapy of psychiatric disorders? *Mol Pharmacol*. 2010;77(2):126–135.

174. Huang Y, Greene E, Murray Stewart T, et al. Inhibition of lysine-specific demethylase 1 by polyamine analogues results in reexpression of aberrantly silenced genes. *Proc Natl Acad Sci USA*. 2007;104(19):8023–8028.
175. Gavin DP, Akbarian S. Epigenetic and post-transcriptional dysregulation of gene expression in schizophrenia and related disease. *Neurobiol Dis*. 2012;46(2):255–262.
176. Shirayama Y, Chen AC-H, Nakagawa S, Russell DS, Duman RS. Brain-derived neurotrophic factor produces antidepressant effects in behavioral models of depression. *J Neurosci*. 2002;22(8):3251–3261.
177. Monteggia LM, Barrot M, Powell CM, et al. Essential role of brain-derived neurotrophic factor in adult hippocampal function. *Proc Natl Acad Sci USA*. 2004;101(29):10827–10832.
178. Tsankova NM, Berton O, Renthal W, Kumar A, Neve RL, Nestler EJ. Sustained hippocampal chromatin regulation in a mouse model of depression and antidepressant action. *Nat Neurosci*. 2006;9(4):519–525.
179. Schroeder FA, Lin CL, Crusio WE, Akbarian S. Antidepressant-like effects of the histone deacetylase inhibitor, sodium butyrate, in the mouse. *Biol Psychiatry*. 2007;62(1):55–64.
180. Gundersen BB, Blendy JA. Effects of the histone deacetylase inhibitor sodium butyrate in models of depression and anxiety. *Neuropharmacology*. 2009;57(1):67–74.
181. Covington HE, Vialou VF, LaPlant Q, Ohnishi YN, Nestler EJ. Hippocampal-dependent antidepressant-like activity of histone deacetylase inhibition. *Neurosci Lett*. 2011;493(3):122–126.
182. Covington HE, Maze I, LaPlant QC, et al. Antidepressant actions of histone deacetylase inhibitors. *J Neurosci*. 2009;29(37):11451–11460.
183. Guidotti A, Auta J, Davis JM, et al. Decrease in reelin and glutamic acid decarboxylase 67 (GAD 67) expression in schizophrenia and bipolar disorder: a postmortem brain study. *Arch Gen Psychiatry*. 2000;57(11):1061–1069.
184. Heckers S, Stone D, Walsh J, Shick J, Koul P, Benes FM. Differential hippocampal expression of glutamic acid decarboxylase 65 and 67 messenger RNA in bipolar disorder and schizophrenia. *Arch Gen Psychiatry*. 2002;59(6):521–529.
185. Huang H-S, Akbarian S. GAD1 mRNA expression and DNA methylation in prefrontal cortex of subjects with schizophrenia. *PLoS One*. 2007;2(8):e809.
186. Hobara T, Uchida S, Otsuki K, et al. Altered gene expression of histone deacetylases in mood disorder patients. *J Psych Res*. 2010;44(5):263–270.
187. Phiel CJ, Zhang F, Huang EY, Guenther MG, Lazar MA, Klein PS. Histone deacetylase is a direct target of valproic acid, a potent anticonvulsant, mood stabilizer, and teratogen. *J Biol*. 2001;276(39):36734–36741.
188. Bach ME, Barad M, Son H, et al. Age-related defects in spatial memory are correlated with defects in the late phase of hippocampal long-term potentiation in vitro and are attenuated by drugs that enhance the cAMP signaling pathway. *Proc Natl Acad Sci USA*. 1999;96(9):5280–5285.
189. Burke SN, Barnes CA. Neural plasticity in the ageing brain. *Nat Rev Neurosci*. 2006;7(1):30–40.
190. Gallagher M, Rapp PR. The use of animal models to study the effects of aging on cognition. *Annu Rev Psychol*. 1997;48:339–370.
191. Schimanski LA, Barnes CA. Neural protein synthesis during aging: effects on plasticity and memory. *Front Aging Neurosci*. 2010;2(26):1–16.
192. Reolon GK, Maurmann N, Werenicz A, et al. Posttraining systemic administration of the histone deacetylase inhibitor sodium butyrate ameliorates aging-related memory decline in rats. *Beh Brain Res*. 2011;221(1):329–332.
193. Zeng Y, Tan M, Kohyama J, et al. Epigenetic enhancement of BDNF signaling rescues synaptic plasticity in aging. *J Neurosci*. 2011;31(49):17800–17810.
194. Minichiello L. TrkB signalling pathways in LTP and learning. *Nat Rev Neurosci*. 2009;10(12):850–860.
195. Stilling RM, Fischer A. The role of histone acetylation in age-associated memory impairment and Alzheimer's disease. *Neurobiol Learn Mem*. 2011;96(1):19–26.
196. Alonso M, Vianna MRM, Izquierdo I, Medina JH. Signaling mechanisms mediating BDNF modulation of memory formation in vivo in the hippocampus. *Cell Mol Neurobiol*. 2002;22(5–6):663–674.
197. Tyler WJ, Alonso M, Bramham CR, Pozzo-Miller LD. From acquisition to consolidation: on the role of brain-derived neurotrophic factor signaling in hippocampal-dependent learning. *Learn Mem*. 2002;9(5):224–237.

"Active DNA Demethylation"

J. David Sweatt, acrylic on canvas (detail, 36 x 48), 2011

CHAPTER

3

Active DNA Demethylation and 5-Hydroxymethylcytosine

Junjie U. Guo, Guo-li Ming and Hongjun Song

Institute for Cell Engineering, Department of Neurology, The Solomon H. Snyder Department of Neuroscience, Johns Hopkins University School of Medicine, Baltimore, Maryland, USA

INTRODUCTION

C5-methylation of the cytosine bases is thus far the only covalent modification of the metazoan genomic DNA that has clear biological functions.[1–3] DNA methylation is the most extensively studied epigenetic modification and is generally linked to transcriptional repression, although new evidence suggests that the regulatory role of DNA methylation is highly context-dependent.[4–8] Due to its stability, DNA methylation has been considered one of the possible vehicles of cellular transcriptional memory for many years.[9,10] The mechanism, initially proposed by Pugh and Holliday, in which two classes of DNA methyltransferases (DNMTs) coordinate to form and maintain transcriptional memories,[9] has been nicely confirmed by the identification and the detailed characterization of the de novo methyltransferases (DNMT3A/B in mammals), the maintenance methyltransferase DNMT1, and their associated factors.[11]

The reversal of DNA methylation, which may erase and reset DNA methylation-based transcriptional memories, has been much more puzzling. This is in sharp contrast to various histone modifications, for which many classes of reversal enzymes (e.g. deacetylases, demethylases, phosphatases, etc.) have been described.[12,13] The belief that an active DNA demethylation mechanism, that is, DNA demethylation that does not involve DNA replication, should exist is supported by a number of observations. The most dramatic example is the observation that, in the zygotes of several animal species, cytosine C5-methylation (5mC) immunostaining signals from the paternal pronucleus rapidly disappear soon after fertilization, whereas the maternal pronucleus seems to be resistant to this wave of demethylation.[14,15] Such a global loss of the paternal 5mC signals is followed by multiple

Epigenetic Regulation in the Nervous System
DOI: http://dx.doi.org/10.1016/B978-0-12-391494-1.00003-3

rounds of cell division with the passive dilution of DNA demethylation of both copies of genomes,[16] which leads to a pre-implantation embryo with a low global level of DNA methylation. Similar global DNA demethylation has also been described in primordial germ cell development, in which parent-of-origin-specific genomic imprints need to be erased, and sex-specific imprints need to be established.[14,17] In contrast, active DNA demethylation in somatic tissues is highly locus-specific. For example, rapid demethylation of the promoter of Il2 gene has been shown to activate Il2 expression in B lymphocytes.[18] Emerging evidence supports that active DNA demethylation also occurs in the nervous system. In this chapter, we first describe the discovery of active DNA demethylation and its function in gene regulation in the nervous systems, followed by a review of underlying molecular mechanisms, and conclude with a discussion of the potential role of active DNA demethylation in neurological disorders.

REGULATION OF NEURONAL GENE EXPRESSION BY ACTIVE DNA DEMETHYLATION

The long-lived nature of post-mitotic neurons poses particular challenges to the regulation of their genomes. A high level of genome stability has to be maintained to ensure the functional integrity of neurons for decades. On the other hand, the neuronal transcriptional landscape also exhibits remarkable plasticity and responsiveness to external stimuli mediated through synaptic activation of neurons.

Neuronal depolarization induces activation/repression of a large number of genes, many of which play important roles in regulating synaptic plasticity and intrinsic excitability among other neuronal properties.[19] Many molecular messengers, including calcium, protein kinases, and several transcription factors [e.g. cAMP-responsive element (CRE) binding factor (CREB),[20] serum-responsive element binding factor (SRF)[21]], have been shown to mediate signal transduction from the synapses to the nucleus. However, how these transient signals are detected and, in some cases, perpetuated in the neuronal genome to influence gene expression in a long lasting manner is much less well understood.[22] In cultured cortical neurons, it has been shown that potassium chloride (KCl)-induced depolarization could lead to promoter DNA demethylation of the brain-derived neurotrophic factor (*Bdnf*) gene,[20] which is a well-established paradigm for activity-dependent gene regulation.[19] DNA demethylation was accompanied by the release of a methylated DNA binding protein MeCP2 and its associated co-repressors, therefore upregulating *Bdnf* transcription[20] (Figure 3.1A). DNA demethylation was later shown to occur in vivo by Sweatt and colleagues when they monitored DNA methylation status of several learning-regulated genes during context fear conditioning training.[25–27] They found that neuronal genes could undergo dynamic and bidirectional changes (both de novo methylation and demethylation) at different stages of memory formation in the adult rat brain as discussed in Chapter 5.

In a survey of epigenetic modifiers that are dynamically regulated by neuronal activity of the mature dentate granule neurons in vivo, a member of the Gadd45 (growth arrest and DNA damage-induced 45) gene family *Gadd45b* was identified as a neuronal activity-induced immediate early gene.[28] Electroconvulsive stimulation (ECS) rapidly induced *Gadd45b* expression in dentate gyrus granule neurons of the adult mouse hippocampus by over 30 folds. In addition, behavior stimuli, such as exploration in a novel environment, could also induce *Gadd45b* expression in Arc (activity-regulated cytoskeleton-associated protein)-expressing neurons in the adult dentate gyrus.[28] Gadd45 proteins have been shown to promote DNA demethylation in multiple studies, including several unbiased overexpression screens.[29,30] Detailed analysis of the DNA methylation patterns of neuronal gene promoters showed that Gadd45b induction did not lead to a global decrease in DNA methylation, but rather caused DNA demethylation of the promoters of *Fgf1B* (fibroblast growth factor 1, isoform B) and *Bdnf IX* (*Bdnf*, isoform IX),[28] two brain-specific transcripts that encoded important neurogenic niche factors. In the *Gadd45b*-null mice, both ECS-induced DNA demethylation and the ECS-induced expression of both *Fgf1B* and *Bdnf IX* were significantly attenuated, leading to a block to ECS-promoted neurogenesis in the dentate gyrus[28,31] (see Figure 3.1B).

A comprehensive analysis of the dentate gyrus neuronal DNA methylome before and after ECS further revealed thousands of CpG sites where neuronal activity induced bi-directional changes in methylation levels.[32] Interestingly, these changes do not always associate with changes in the transcript levels of the local genes, implying that other layers of regulation may be required to achieve transcriptional regulation. An alternative hypothesis is that dynamic DNA methylation changes may not only regulate the overall transcription levels of their associated genes, but also other co-transcriptional events such as co-transcriptional pre-mRNA processing.[33] Recent studies have clearly shown that epigenetic modifications

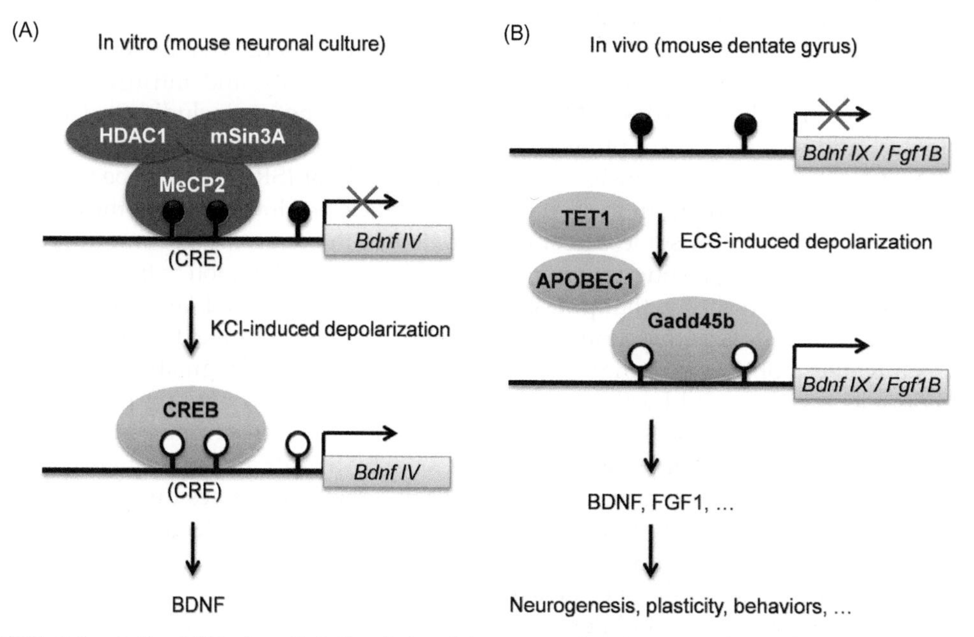

FIGURE 3.1 **Active DNA demethylation induced by neuronal activation.** (A) In cultured mouse neurons, KCl-induced depolarization induces demethylation of specific CpGs located in the *Bdnf IV* promoter, which facilitates CREB binding and leads to the dissociation of MeCP2 and associated co-repressors from the promoter. *Modified from Martinowich et al., 2003.*[20] (B) ECS induces *Gadd45b* expression and *Gadd45b*-dependent demethylation of *Bdnf IX* and *Fgf1B* promoters,[23] which also requires *TET1* and *APOBEC1*.[24]

including DNA methylation have major roles in regulating alternative splicing of many transcripts.[34,35] Given the extraordinary complexity of the transcriptome in the mammalian brain, it is safe to predict that further exploration of the potential role of DNA methylation/demethylation in fine-tuning transcript splicing will generate new insights into the biological functions of these dynamic changes in the neuronal DNA methylome.

ACTIVE DNA DEMETHYLATION MEDIATED BY DNA REPAIR

Despite the accumulating evidence for both global and locus-specific DNA demethylation in animal cells, the underlying mechanism is still unclear and is hotly debated (Figure 3.2).[36,37] Both unbiased and targeted searches have made significant contributions to our understanding of this process. The role of Gadd45 proteins was first suggested by overexpression screening for proteins that could reactivate in vitro methylated plasmid reporters.[29,30] Biochemical characterization further suggested that Gadd45 proteins might promote DNA demethylation by interacting with various components in several DNA repair pathways.[38] Notably, members of the Gadd45 protein family have been shown to respond differentially to a variety of cellular signals,[28,39] suggesting that dynamic changes of DNA methylation may have extensive roles in mediating signal-induced gene regulation across different cell types. Other proteins [e.g. elongation protein 3 (ELP3),[40] RING finger protein 4 (RNF4)[30]] have also been linked to active

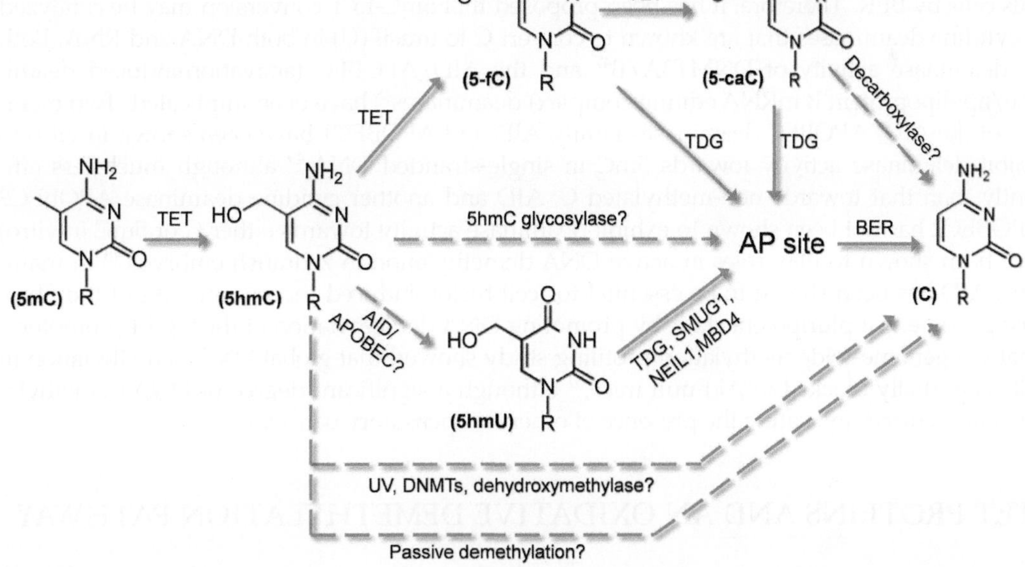

FIGURE 3.2 **Molecular pathways of active DNA demethylation.** 5hmC may mediate DNA demethylation in many possible ways, including BER-dependent and BER-independent pathways. Individual steps that have been proposed but have not been experimentally proven are indicated by dashed arrows. *Modified from Guo et al., 2011.*[23]

DNA demethylation, although how they can be incorporated into the current framework of knowledge is still unclear.

The first well-established DNA demethylases were described in plants, in which DNA demethylation involves the base-excision repair (BER) pathway.[41] To initiate BER, the Demeter family of 5mC DNA glycosylases, including DME, ROS1, and DML2/3, specifically recognize and excise the 5mC base and generate an apurinic/apurymidinic (AP) site on the double-stranded DNA. Strand excision was further promoted by the AP endonuclease, and followed by the refill of mononucleotides and ligation. Although new regulatory components are likely still to be discovered,[42,43] the core enzymatic mechanism is relatively clear.

The search for metazoan orthologs of plant 5mC DNA glycosylases has been largely unsuccessful. Although several mammalian DNA glycosylases, including thymine DNA glycosylase (TDG)[44] and methylated-DNA binding domain-containing protein (MBD4),[45] have been shown to possess low levels of 5mC DNA glycosylase activity in vitro, no experimental evidence has been found to support their roles as 5mC DNA glycosylases in vivo. It should be noted that it is by no means a trivial task to separate this activity from their other possible roles (see discussion below). Nevertheless, pharmacological and genetic inhibition of key BER enzymes led to a partial block of global DNA demethylation in the mouse germ line,[46] strongly supporting an evolutionarily conserved role for BER in mediating active DNA demethylation in animal cells.

One prevalent hypothesis is that BER may be initiated in an indirect manner, which means that 5mC needs to be converted to a secondary form before being recognized by a DNA glycosylase and fed into the BER pathway. Compared to C, 5mC has a higher tendency to undergo deamination and becomes thymine (T), which is part of the basis for the significant depletion of CpG dinucletides in animal genomes. T:G mismatches can be efficiently repaired in most animals cells by BER. Therefore, it has been proposed that 5mC-to-T conversion may be catalyzed by cytidine deaminases that are known to convert C to uracil (U) in both DNA and RNA. Both the deaminase activity of DNMT3A/B[47] and the AID/APOBEC (activation-induced deaminase/apolipoprotein B mRNA editing complex) deaminases[48] have been implicated. Two members of the AID/APOBEC deaminases family, AID and APOBEC1 have been shown in vitro to exhibit deaminase activity towards 5mC in single-stranded DNA,[49] although much less efficiently than that towards non-methylated C. AID and another cytidine deaminase APOBEC2 (APOBEC2 has not been shown to exhibit deaminase activity towards either C or 5mC in vitro) have been shown to play roles in active DNA demethylation in zebrafish embryos.[48] In mammals, AID has been shown to be essential for cell-fusion-induced reprogramming of the fibroblast genome to a pluripotent state, by promoting DNA demethylation of the OCT4 promoter.[50] Finally, a genome-wide methylation profiling study showed that global DNA demethylation in PGC is partially blocked in Aid-null mice,[51] although a significant degree of DNA demethylation still occurred, indicating the presence of other compensatory pathways.

TET PROTEINS AND AN OXIDATIVE DEMETHYLATION PATHWAY

Demethylation of lysine methylation in histones involves an oxidative mechanism.[12,13,52] Some types of aberrant methylation of DNA (e.g. 1-methyladenosine and 3-methylcytosine) can also be repaired by AlkB dioxygenase and its mammalian homologs.[53,54] Therefore, it has been a tempting hypothesis that demethylation of 5mC can also be achieved by an oxidative mechanism.[24,55] This notion received strong support in 2009, when two groups independently (re-)discovered the presence of 5hmC,[56,57] an oxidative product of 5mC first described in the 1970s,[58] in the mammalian genomic DNA. It was further shown that hydroxylation of 5mC was catalyzed by the dioxygenase encoded by ten-eleven-translocated gene 1 (*TET1*). Later, it was reported that all three members of the TET proteins family exhibit the 5mC hydroxylase activity.[59] Like AlkB homologs, TET1-catalyzed 5mC-to-5hmC conversion required Fe(II) and 2-oxoglutarate as co-factors.

Not surprisingly, TET proteins and 5hmC have received much attention regarding their possible roles in DNA demethylation. Early studies have mostly investigated the effects of gain and/or loss of function of TET proteins on genomic DNA methylation. When overexpressing TET1 in HEK293 cells, Rao and colleagues observed a small yet significant increase in the global abundance of non-methylated cytosines in bulk DNA.[57] Knock-down of endogenous Tet1 expression with RNAi in mouse embryonic stem cells (ESC) led to hypermethylation of the promoter of pluripotency gene Nanog,[59] which correlated with the downregulation of Nanog expression and defects in ESC self-renewal. Later, it was further shown that Tet1 knock-down led to global hypermethylation at Tet1 binding sites (mostly CpG-rich regions) in ESC,[60,61] supporting a role of Tet1 in maintaining CpG islands at a non-methylated state.

Support for the oxidative demethylation hypothesis has also emerged from an interesting intersection between TET protein biology and cancer biology. Around the same time as the identification of TET proteins as 5mC hydroxylases, *TET2* was shown to be frequently mutated in multiple types of hematopoietic malignancies.[62] It was later shown that the oncogenic metabolite 2-hydroxyglutarate (2HG) generated by the mutated isocitrate dehydrogenases IDH1/2 was an inhibitor for 2-oxoglutarate-dependent dioxygenases including both histone demethylases[63] and TET proteins.[64] Inhibition of TET2 led to hypermethylation of a large number of genes,[64] as well as a block to hematopoietic differentiation, which may underlie the pathogenesis of these subtypes of cancers. However, not all the early data directly pointed to a clear role for TET proteins in DNA demethylation. For example, Tet1/2 knock-down in ESCs resulted in a complex DNA methylation phenotype at specific loci.[65] Tet2-deficient, low-5hmC-abundance cancer tissues showed predominantly hypomethylation at certain loci, compared to their high-5hmC-abundance counterparts.[66] Some of these results may reflect both direct and indirect consequences of TET protein deficiency. In addition, TET proteins and 5hmC may have DNA demethylation-independent functions (see discussion below). Notably, TET1 has been shown to direct the Polycomb repressive complex 2 (PRC2) to a large number of genes in ESC,[60] indicating a high rank of TET proteins in the hierarchy of chromatin modifications. These confounding factors can make it difficult to interpret conclusively the effects of TET protein changes on genomic DNA methylation.

To assess directly the roles of TET proteins and 5hmC in active DNA demethylation, the experimental system needs to satisfy several criteria: (1) the DNA substrate should have a well-defined, preferably uniform methylation pattern; (2) the DNA substrate should not undergo replication so that any observed demethylation cannot be attributed to the passive dilution of DNA methylation; and (3) the system should be able to distinguish the 5mC hydroxylase-dependent functions of TET proteins from the hydroxylase-independent ones. Early attempts revolved around using replication-deficient DNA plasmids that are methylated in vitro at all the CpG sites by the bacterial methyltransferase M.SssI.[23,67] As mentioned above, such plasmids have facilitated the identification of other components of active DNA demethylation in previous functional screening studies. Two independent groups both reported the capability of TET1 in reactivating the reporter gene expression from the in-vitro methylated plasmid,[23,67] which was likely linked to active DNA demethylation. In addition, the mutations that disrupted the co-factor-binding sites in the catalytic domain, and thus rendered TET1 catalytically dead, abolished the ability of TET1 to reactivate the in-vitro methylated plasmid, indicating that the 5mC hydroxylase activity is required in this process.

To ascertain directly the fate of 5hmC in DNA, linear DNA probes (a green fluorescent protein expression cassette) were generated in which every single cytosine is replaced by 5hmC (5hmC-GFP).[23,24] This was achieved by replacing non-methylated 2′-deoxycytidine 5′-triphosphate (dCTP) with synthetically modified 5-hydroxymethyl-2′-deoxycytidine 5′-triphosphate (5hmdCTP) in the PCR reaction (Figure 3.3A). If 5hmC-to-C conversion indeed occurs, the presence of non-methylated cytosines can be detected by methylation-sensitive restriction enzymes that are similarly blocked by either 5mC or 5hmC in their restriction sites, or by bisulfite sequencing where any 5hmC-to-C conversion will be indicated by C-to-T transitions in the Sanger sequencing result. Two days after transfection into HEK293 cells, 5hmC-GFP probes were retrieved by extracting total DNA from the culture. Both sensitivity to HpaII (a methylation-sensitive restriction enzyme that cuts non-methylated CCGG sequences) and C-to-T transitions in bisulfite sequencing were observed on the transfected 5hmC-GFP (Figure 3.3B),[23] indicating that some of the 5hmCs in the original 5hmC-GFP were replaced

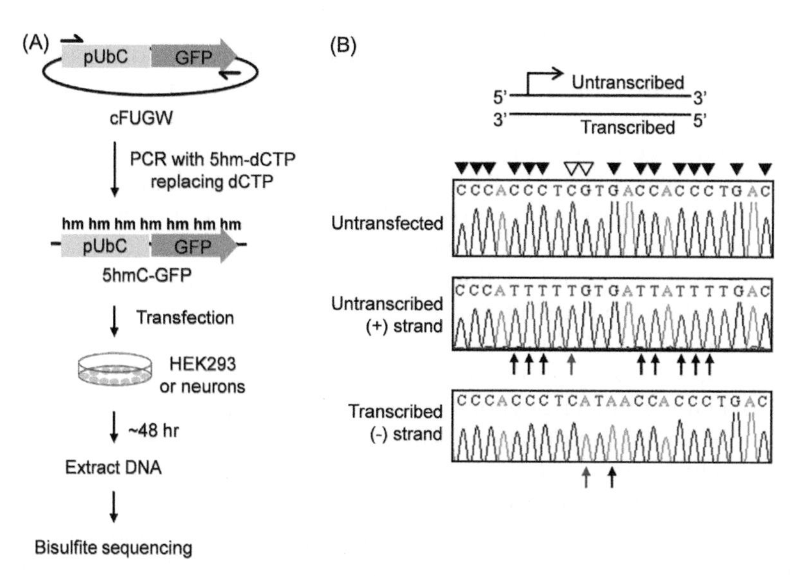

FIGURE 3.3 **The 5hmC demethylase footprint assay.** (A) A schematic illustration of the 5hmC demethylase footprint assay. (B) Sample sequencing traces from the 5hmC demethylase footprint assay. C-to-T or G-to-A transitions can distinguish the two DNA strands-of-origin. Open and filled arrowheads indicate cytosines in CpG and non-CpG sequence contexts, respectively. Red and black arrows indicate demethylated 5hmCs in CpG and non-CpG contexts, respectively. Note that demethylated 5hmCs were often found in clusters. *From Guo et al., 2011.*[23]

with non-methylated cytosines. Intriguingly, similar demethylation events did not spontaneously occur to a significant degree in 5mC-GFP, which was generated with a similar method to 5hmC-GFP using synthetically modified 5-methyl-2′-deoxycytidine 5′-triphosphate (5m-dCTP).[23] These results suggest that 5mC-to-5hmC conversion is a critical and rate-limiting step in active DNA demethylation. Once 5mC is oxidized, the downstream 5hmC-demethylating machinery is ubiquitous and effective in mammalian cells.

The simplicity of this cell-based "demethylase footprint" assay allowed pharmacological and genetic manipulation to interrogate the biochemical identity of the underlying 5hmC demethylase activity. Consistent with previous observations in mouse germline DNA demethylation,[46] pharmacological inhibition of key BER enzymes, including the AP endonuclease and poly (ADP-ribose) polymerase, could significantly attenuate the 5hmC demethylase activity,[23] indicating an involvement of BER in the 5hmC-to-C conversion. If BER is directly initiated by a 5hmC-specific DNA glycosylase like the plant 5mC-specific DNA glycosylases, overexpression of this DNA glycosylase may efficiently remove the 5hmC in the genomic DNA, and/or enhance the 5hmC demethylase activity towards the 5hmC-GFP probe. 5hmC DNA glycosylase activity was reported more than two decades ago,[68] although the identity of this glycosylase is still unknown. Intriguingly, overexpression of none of the 12 known human DNA glycosylases exhibited these effects,[23] suggesting that BER was initiated in an indirect manner, and was possibly preceded by the conversion of 5hmC to a tertiary (or even quaternary) intermediate product, which was in turn recognized by a DNA glycosylase (see Figure 3.2).

What could this tertiary intermediate be? One hypothesis is that AID/APOBEC enzymes could intersect with the 5hmC-mediated DNA demethylation pathway by deaminating 5hmC, and generating 5-hydroxymethyluracil (5hmU). Interestingly, in the demethylase footprint assay, overexpression of AID could increase the 5hmC-to-C conversion in the 5hmC-GFP probe.[23] Moreover, overexpression of AID could also decrease the 5hmC immunoreactive signals both in vitro (exogenously introduced 5hmC by TET1 overexpression in HEK293 cells) and in vivo (endogenous genomic 5hmC in the dentate gyrus in the adult mouse brain). Finally, the single-base, single-DNA-strand resolution of the demethylase footprint assay allowed detailed characterization of the 5hmC demethylase activity (see Figure 3.3B), which turned out to be (1) highly processive, (2) sequence-selective, (3) transcription-dependent, and (4) strand-biased.[23] All these properties coincided with the known catalytic properties of AID,[69,70] rendering these enzymes as promising candidates. Notably, mammalian cells possess relatively high repair capability for 5hmU.[71] Several DNA glycosylases have been shown to exhibit 5hmU DNA glycosylase activity, including single-strand-selective monofunctional uracil DNA glycosylase 1 (SMUG1),[72] Nei endonuclease VIII-like 1 (NEIL1),[73] Nth endonuclease III-like 1 (NTHL1),[73] TDG,[74] and MBD4.[75] It will be very important for future studies to investigate (1) whether AID/APOBEC enzymes exhibit deaminase activity towards 5hmC in vitro, (2) whether endogenous 5hmC levels are altered in AID/APOBEC knockout mice, and (3) whether the deamination product 5hmU can be detected endogenously.

In the thymidine salvage pathway, thymine-7-hydroxylase (THase) can iteratively oxidize the 7-methyl group of thymine to a formyl and then a carboxyl group and generate iso-orotate, which is finally decarboxylated by the iso-orate decarboxylase to generate uracil.[76] It is therefore a tempting hypothesis that 5hmC may also be the first step of iterative oxidation, possibly all carried out by TET proteins.[55] This hypothesis received strong support by in vitro studies that showed TET proteins could indeed further oxidize 5hmC to 5-formylcytosine (5fC) and/or 5-carboxylcytosine (5caC), although the processivity varies in different studies.[77,78] In addition, endogenous 5fC and 5caC were also detected in the ESC genomic DNA by ultrasensitive mass spectrometry-based methods.[77,78] Although there has not been direct evidence showing that 5fC or 5caC could be converted to non-methylated C in vivo, potential repair mechanisms for 5fC and 5caC have been proposed. Consistent with a BER-based mechanism, both 5fC and 5caC have been shown to be efficiently excised by TDG.[77,79,80] It is also possible that 5caC could be directly converted to non-methylated C by an unknown decarboxylase,[78] eliminating the need for the relatively energy-expensive BER. Also independent of BER is the surprising demonstration that bacterial DNMTs can directly dehydroxymethylate 5hmC in vitro,[81] although no evidence has been shown that mammalian DNMTs can perform this function in vivo.

In post-mitotic cells such as neurons, DNA demethylation is by definition active in its nature. On the other hand, 5hmC has also been proposed to induce passive DNA demethylation in dividing cells. The first hint of this mode of action was the observation that the maintenance methyltransferase DNMT1 does not recognize hemi-hydroxylmethylated CpG as much as hemi-methylated CpG.[82] However, it was later shown that the SRA (SET and RING-associated) domain in the DNMT1-associated factor UHRF1 (ubiquitin-like with PHD and RING finger domain 1, a critical factor in DNA methylation maintenance) could bind to both hemi-methylated and hemi-hydroxymethylated CpGs with similar affinity.[83] Therefore, it remains to be determined whether hydroxylation of a methylated CpG has any major impact on the maintenance of DNA methylation in dividing cells.

Related to this question is the global DNA demethylation in pre-implantation embryos. As discussed above, the 5mC immunoreactive signal of the paternal pronucleus was extensively lost soon after fertilization before DNA replication occurs. Soon after the identification of 5hmC and TET proteins, it was found that the loss of 5mC signal in the paternal pronucleus was largely due to 5mC-to-5hmC conversion,[84,85] most likely catalyzed by the least studied TET protein family member Tet3. By immunostaining the mitotic chromosome spread samples prepared at different stages of the mouse pre-implantation embryos, Inoue and Zhang nicely showed that passive DNA demethylation occurred on a global scale, resulting in an embryo with an extremely low global level of DNA methylation.[16] However, the passive DNA demethylation did not only occur in the 5hmC-enriched paternal chromosomes, but also in the 5mC-enriched maternal chromosomes, which suggested that passive demethylation might not be a specific consequence of 5mC-to-5hmC conversion. Nevertheless, the maternal Tet3-deficiency in mice caused a failure in demethylation and transcriptional activation of some key developmental genes, and led to severe developmental deficits,[15] indicating that Tet3 played a critical role in epigenetic reprogramming in embryonic development.

FUNCTIONS OF TET PROTEINS AND 5HMC BEYOND DNA DEMETHYLATION

TET proteins are generally large in size (e.g. TET1 has a predicted molecular weight of 235 kD), containing multiple conserved functional domains. The 5mC hydroxylase catalytic function has been mapped to their C-termini, whereas the biological functions of their N-termini are largely unknown. This points to two possibilities: (1) TET proteins may be regulated by multiple protein–protein and/or protein–DNA interaction mediated by their N-termini. For example, the CXXC domain in the TET1 N-terminus may play an important role in the targeting of TET1 to the specific genomic loci. (2) The N-termini of TET proteins may harbor 5mC hydroxylase-independent functions. This possibility was first confirmed by Helin and colleagues when they examined the effect of TET1 loss of function on gene expression profiles in both wild-type and triple DNMT knockout mouse ESC, in which DNA methylation was completely eliminated.[86] Surprisingly, the influences of TET1 loss of function on gene expression were largely similar in the two genetic backgrounds, indicating that the regulatory function of TET1 on these targets is 5mC hydroxylase activity-independent. They further showed both physical and functional interaction between TET1 and the well-studied SIN3A co-repressor complex, which may mediate the hydroxylase-independent functions observed for TET1. As mentioned above, TET1 has also been shown to be required for the targeting of PRC2 and H3K27me3 to a large number of genes in ESC.[60,87] Whether TET1 has an instructive or permissive role for PRC2 targeting, and whether this role is through direct interaction between TET1 and PRC2, or mediated by 5hmC or other associated factors, are still widely open questions.

In the genome, 5mC is recognized by a family of methylated DNA binding proteins (MBPs),[88] which in turn recruits downstream effector complexes to regulate gene expression. 5mC by itself may also directly influence chromatin structure and/or protein-DNA interaction (e.g. CpG methylation in the center of the binding motif of CREB can prevent its binding[89]). In the nervous system, 5hmC accounts for as much as 20–40% of all methylated cytosines.[56,90–92] It is therefore reasonable to hypothesize that 5hmC may also influence transcription either on its own or by recruiting specific protein readers without triggering active DNA demethylation. Furthermore, it has been shown that most, if not all, MBPs do not recognize 5hmC, or bind to 5hmC to a much lesser extent when compared to 5mC.[75,93] Therefore, 5mC-to-5hmC conversion may dislodge MBP from its targets. Protein readers that specifically recognize 5hmC have yet to be discovered.

ROLE OF ACTIVE DNA DEMETHYLATION IN NEUROLOGICAL DISORDERS?

Given the central roles of chromatin modifications in orchestrating the transcriptional programs, it is not too surprising that abnormalities in many epigenetic mechanisms, including DNA methylation, have been linked to a large number of neurological and mental diseases[94,95] (see also Chapters 7, 8 and 10). For example, genetic alternations of *MECP2*, which encodes a methylated DNA binding protein, are known causes of a monogenic neurodevelopmental disorder, Rett syndrome.[96,97] Another putative methylated DNA binding protein gene *MBD5* has been recently linked to autism spectrum disorders and other neurological symptoms.[98,99] With the ever increasing number of human genomes being sequenced, it is safe to predict that more epigenetic modifiers will be implicated. Both genome-wide and locus-specific analyses have linked changes in DNA methylation to aging,[100,101] neurodegeneration,[102,103] and mental diseases,[104,105] some of which may be due to dysregulation of the active DNA demethylation mechanisms. Expression of several molecular components in active DNA demethylation has also been shown to be altered in psychosis patient samples.[106] More comprehensive genetic association and better animal models will allow researchers to examine further whether a causal link exists between defects in active DNA demethylation and brain disorders.

CONCLUDING REMARKS

Accumulating evidence has strongly supported important roles of epigenetic mechanisms in orchestrating both the basal and the activity-dependent transcriptional programs in the nervous system. In particular, neurobiological studies on active DNA demethylation have not only revealed a novel molecular mechanism underlying neuronal plasticity, but have also contributed significantly to the more general and fundamental understanding of genome regulation. Expansion of the DNA modification "code" demands the development of new experimental techniques that can simultaneously differentiate all known types of cytosine modifications,[20] which may also reveal new types of DNA modifications. Deciphering the molecular details of how DNA methylation is dynamically regulated in the neuronal genome will become an integral part of understanding the plastic nature of the nervous system.

References

1. Bird A. DNA methylation patterns and epigenetic memory. *Genes Dev*. 2002;16(1):6–21.
2. Law JA, Jacobsen SE. Establishing, maintaining and modifying DNA methylation patterns in plants and animals. *Nat Rev Genet*. 2010;11(3):204–220.
3. Cedar H, Bergman Y. Linking DNA methylation and histone modification: patterns and paradigms. *Nat Rev Genet*. 2009;10(5):295–304.
4. Suzuki MM, Bird A. DNA methylation landscapes: provocative insights from epigenomics. *Nat Rev Genet*. 2008;9(6):465–476.
5. Hellman A, Chess A. Gene body-specific methylation on the active X chromosome. *Science*. 2007;315(5815):1141–1143.
6. Wu H, Coskun V, Tao J, et al. Dnmt3a-dependent nonpromoter DNA methylation facilitates transcription of neurogenic genes. *Science*. 2010;329(5990):444–448.
7. Deng J, Shoemaker R, Xie B, et al. Targeted bisulfite sequencing reveals changes in DNA methylation associated with nuclear reprogramming. *Nat Biotechnol*. 2009;27(4):353–360.
8. Lister R, Pelizzola M, Dowen RH, et al. Human DNA methylomes at base resolution show widespread epigenomic differences. *Nature*. 2009;462(7271):315–322.
9. Holliday R, Pugh JE. DNA modification mechanisms and gene activity during development. *Science*. 1975;187(4173):226–232.
10. Wigler MH. The inheritance of methylation patterns in vertebrates. *Cell*. 1981;24(2):285–286.
11. Goll MG, Bestor TH. Eukaryotic cytosine methyltransferases. *Annu Rev Biochem*. 2005;74:481–514.
12. Shi Y, Lan F, Matson C, et al. Histone demethylation mediated by the nuclear amine oxidase homolog LSD1. *Cell*. 2004;119(7):941–953.
13. Klose RJ, Zhang Y. Regulation of histone methylation by demethylimination and demethylation. *Nat Rev Mol Cell Biol*. 2007;8(4):307–318.
14. Surani MA, Hajkova P. Epigenetic reprogramming of mouse germ cells toward totipotency. *Cold Spring Harb Symp Quant Biol*. 2010;75:211–218.
15. Gu TP, Guo F, Yang H, et al. The role of Tet3 DNA dioxygenase in epigenetic reprogramming by oocytes. *Nature*. 2011;477(7366):606–610.

16. Inoue A, Zhang Y. Replication-dependent loss of 5-hydroxymethylcytosine in mouse preimplantation embryos. *Science*. 2011;334(6053):194.
17. Hajkova P, Ancelin K, Waldmann T, et al. Chromatin dynamics during epigenetic reprogramming in the mouse germ line. *Nature*. 2008;452(7189):877–881.
18. Bruniquel D, Schwartz RH. Selective, stable demethylation of the interleukin-2 gene enhances transcription by an active process. *Nat Immunol*. 2003;4(3):235–240.
19. West AE, Greenberg ME. Neuronal activity-regulated gene transcription in synapse development and cognitive function. *Cold Spring Harb Perspect Biol*. 2011;3(6):1–21.
20. Martinowich K, Hattori D, Wu H, et al. DNA methylation-related chromatin remodeling in activity-dependent BDNF gene regulation. *Science*. 2003;302(5646):890–893.
21. Ramanan N, Shen Y, Sarsfield S, et al. SRF mediates activity-induced gene expression and synaptic plasticity but not neuronal viability. *Nat Neurosci*. 2005;8(6):759–767.
22. Meaney MJ, Ferguson-Smith AC. Epigenetic regulation of the neural transcriptome: the meaning of the marks. *Nat Neurosci*. 2010;13(11):1313–1318.
23. Guo JU, Su Y, Zhong C, Ming GL, Song H. Hydroxylation of 5-methylcytosine by TET1 promotes active DNA demethylation in the adult brain. *Cell*. 2011;145(3):423–434.
24. Guo JU, Su Y, Zhong C, Ming GL, Song H. Emerging roles of TET proteins and 5-hydroxymethylcytosines in active DNA demethylation and beyond. *Cell Cycle*. 2011;10(16):2662–2668.
25. Day JJ, Sweatt JD. DNA methylation and memory formation. *Nat Neurosci*. 2010;13(11):1319–1323.
26. Miller CA, Sweatt JD. Covalent modification of DNA regulates memory formation. *Neuron*. 2007;53(6): 857–869.
27. Lubin FD, Roth TL, Sweatt JD. Epigenetic regulation of BDNF gene transcription in the consolidation of fear memory. *J Neurosci*. 2008;28(42):10576–10586.
28. Ma DK, Jang MH, Guo JU, et al. Neuronal activity-induced Gadd45b promotes epigenetic DNA demethylation and adult neurogenesis. *Science*. 2009;323(5917):1074–1077.
29. Barreto G, Schafer A, Marhold J, et al. Gadd45a promotes epigenetic gene activation by repair-mediated DNA demethylation. *Nature*. 2007;445(7128):671–675.
30. Hu XV, Rodrigues TM, Tao H, et al. Identification of RING finger protein 4 (RNF4) as a modulator of DNA demethylation through a functional genomics screen. *Proc Natl Acad Sci USA*. 2010;107(34):15087–15092.
31. Ma DK, Marchetto MC, Guo JU, Ming GL, Gage FH, Song H. Epigenetic choreographers of neurogenesis in the adult mammalian brain. *Nat Neurosci*. 2010;13(11):1338–1344.
32. Guo JU, Ma DK, Mo H, et al. Neuronal activity modifies the DNA methylation landscape in the adult brain. *Nat Neurosci*. 2011;14(10):1345–1351.
33. Luco RF, Allo M, Schor IE, Kornblihtt AR, Misteli T. Epigenetics in alternative pre-mRNA splicing. *Cell*. 2011;144(1):16–26.
34. Luco RF, Pan Q, Tominaga K, Blencowe BJ, Pereira-Smith OM, Misteli T. Regulation of alternative splicing by histone modifications. *Science*. 2010;327(5968):996–1000.
35. Shukla S, Kavak E, Gregory M, et al. CTCF-promoted RNA polymerase II pausing links DNA methylation to splicing. *Nature*. 2011;479(7371):74–79.
36. Bhutani N, Burns DM, Blau HM. DNA demethylation dynamics. *Cell*. 2011;146(6):866–872.
37. Ooi SK, Bestor TH. The colorful history of active DNA demethylation. *Cell*. 2008;133(7):1145–1148.
38. Schmitz KM, Schmitt N, Hoffmann-Rohrer U, Schafer A, Grummt I, Mayer C. TAF12 recruits Gadd45a and the nucleotide excision repair complex to the promoter of rRNA genes leading to active DNA demethylation. *Mol Cell*. 2009;33(3):344–353.
39. Liebermann DA, Tront JS, Sha X, Mukherjee K, Mohamed-Hadley A, Hoffman B. Gadd45 stress sensors in malignancy and leukemia. *Crit Rev Oncog*. 2011;16(1–2):129–140.
40. Okada Y, Yamagata K, Hong K, Wakayama T, Zhang Y. A role for the elongator complex in zygotic paternal genome demethylation. *Nature*. 2010;463(7280):554–558.
41. Zhu JK. Active DNA demethylation mediated by DNA glycosylases. *Annu Rev Genet*. 2009;43:143–166.
42. Martinez-Macias MI, Qian W, Miki D, et al. A DNA 3′ phosphatase functions in active DNA demethylation in Arabidopsis. *Mol Cell*. 2012;45(3):357–370.
43. Zheng X, Pontes O, Zhu J, et al. ROS3 is an RNA-binding protein required for DNA demethylation in Arabidopsis. *Nature*. 2008;455(7217):1259–1262.

44. Zhu B, Zheng Y, Hess D, et al. 5-methylcytosine-DNA glycosylase activity is present in a cloned G/T mismatch DNA glycosylase associated with the chicken embryo DNA demethylation complex. *Proc Natl Acad Sci USA*. 2000;97(10):5135–5139.
45. Zhu B, Zheng Y, Angliker H, et al. 5-Methylcytosine DNA glycosylase activity is also present in the human MBD4 (G/T mismatch glycosylase) and in a related avian sequence. *Nucleic Acids Res*. 2000;28(21):4157–4165.
46. Hajkova P, Jeffries SJ, Lee C, Miller N, Jackson SP, Surani MA. Genome-wide reprogramming in the mouse germ line entails the base excision repair pathway. *Science*. 2010;329(5987):78–82.
47. Metivier R, Gallais R, Tiffoche C, et al. Cyclical DNA methylation of a transcriptionally active promoter. *Nature*. 2008;452(7183):45–50.
48. Rai K, Huggins IJ, James SR, Karpf AR, Jones DA, Cairns BR. DNA demethylation in zebrafish involves the coupling of a deaminase, a glycosylase, and gadd45. *Cell*. 2008;135(7):1201–1212.
49. Morgan HD, Dean W, Coker HA, Reik W, Petersen-Mahrt SK. Activation-induced cytidine deaminase deaminates 5-methylcytosine in DNA and is expressed in pluripotent tissues: implications for epigenetic reprogramming. *J Biol Chem*. 2004;279(50):52353–52360.
50. Bhutani N, Brady JJ, Damian M, Sacco A, Corbel SY, Blau HM. Reprogramming towards pluripotency requires AID-dependent DNA demethylation. *Nature*. 2010;463(7284):1042–1047.
51. Popp C, Dean W, Feng S, et al. Genome-wide erasure of DNA methylation in mouse primordial germ cells is affected by AID deficiency. *Nature*. 2010;463(7284):1101–1105.
52. Tsukada Y, Fang J, Erdjument-Bromage H, et al. Histone demethylation by a family of JmjC domain-containing proteins. *Nature*. 2006;439(7078):811–816.
53. Aas PA, Otterlei M, Falnes PO, et al. Human and bacterial oxidative demethylases repair alkylation damage in both RNA and DNA. *Nature*. 2003;421(6925):859–863.
54. Falnes PO, Johansen RF, Seeberg E. AlkB-mediated oxidative demethylation reverses DNA damage in Escherichia coli. *Nature*. 2002;419(6903):178–182.
55. Wu SC, Zhang Y. Active DNA demethylation: many roads lead to Rome. *Nat Rev Mol Cell Biol*. 2010;11(9):607–620.
56. Kriaucionis S, Heintz N. The nuclear DNA base 5-hydroxymethylcytosine is present in Purkinje neurons and the brain. *Science*. 2009;324(5929):929–930.
57. Tahiliani M, Koh KP, Shen Y, et al. Conversion of 5-methylcytosine to 5-hydroxymethylcytosine in mammalian DNA by MLL partner TET1. *Science*. 2009;324(5929):930–935.
58. Penn NW, Suwalski R, O'Riley C, Bojanowski K, Yura R. The presence of 5-hydroxymethylcytosine in animal deoxyribonucleic acid. *Biochem J*. 1972;126(4):781–790.
59. Ito S, D'Alessio AC, Taranova OV, Hong K, Sowers LC, Zhang Y. Role of Tet proteins in 5mC to 5hmC conversion, ES-cell self-renewal and inner cell mass specification. *Nature*. 2010;466(7310):1129–1133.
60. Wu H, D'Alessio AC, Ito S, et al. Dual functions of Tet1 in transcriptional regulation in mouse embryonic stem cells. *Nature*. 2011;473(7347):389–393.
61. Xu Y, Wu F, Tan L, et al. Genome-wide regulation of 5hmC, 5mC, and gene expression by tet1 hydroxylase in mouse embryonic stem cells. *Mol Cell*. 2011;42(4):451–464.
62. Bejar R, Levine R, Ebert BL. Unraveling the molecular pathophysiology of myelodysplastic syndromes. *J Clin Oncol*. 2011;29(5):504–515.
63. Xu W, Yang H, Liu Y, et al. Oncometabolite 2-hydroxyglutarate is a competitive inhibitor of alpha-ketoglutarate-dependent dioxygenases. *Cancer Cell*. 2011;19(1):17–30.
64. Figueroa ME, Abdel-Wahab O, Lu C, et al. Leukemic IDH1 and IDH2 mutations result in a hypermethylation phenotype, disrupt TET2 function, and impair hematopoietic differentiation. *Cancer Cell*. 2010;18(6):553–567.
65. Koh KP, Yabuuchi A, Rao S, et al. Tet1 and Tet2 regulate 5-hydroxymethylcytosine production and cell lineage specification in mouse embryonic stem cells. *Cell Stem Cell*. 2011;8(2):200–213.
66. Ko M, Huang Y, Jankowska AM, et al. Impaired hydroxylation of 5-methylcytosine in myeloid cancers with mutant TET2. *Nature*. 2010;468(7325):839–843.
67. Zhang H, Zhang X, Clark E, Mulcahey M, Huang S, Shi YG. TET1 is a DNA-binding protein that modulates DNA methylation and gene transcription via hydroxylation of 5-methylcytosine. *Cell Res*. 2010;20(12):1390–1393.
68. Cannon SV, Cummings A, Teebor GW. 5-Hydroxymethylcytosine DNA glycosylase activity in mammalian tissue. *Biochem Biophys Res Commun*. 1988;151(3):1173–1179.

69. Chaudhuri J, Tian M, Khuong C, Chua K, Pinaud E, Alt FW. Transcription-targeted DNA deamination by the AID antibody diversification enzyme. *Nature*. 2003;422(6933):726–730.
70. Pham P, Bransteitter R, Petruska J, Goodman MF. Processive AID-catalysed cytosine deamination on single-stranded DNA simulates somatic hypermutation. *Nature*. 2003;424(6944):103–107.
71. Rusmintratip V, Sowers LC. An unexpectedly high excision capacity for mispaired 5-hydroxymethyluracil in human cell extracts. *Proc Natl Acad Sci USA*. 2000;97(26):14183–14187.
72. Boorstein RJ, Cummings Jr A, Marenstein DR, et al. Definitive identification of mammalian 5-hydroxymethyluracil DNA N-glycosylase activity as SMUG1. *J Biol Chem*. 2001;276(45):41991–41997.
73. Zhang QM, Yonekura S, Takao M, Yasui A, Sugiyama H, Yonei S. DNA glycosylase activities for thymine residues oxidized in the methyl group are functions of the hNEIL1 and hNTH1 enzymes in human cells. *DNA Repair (Amst)*. 2005;4(1):71–79.
74. Cortellino S, Xu J, Sannai M, et al. Thymine DNA glycosylase is essential for active DNA demethylation by linked deamination-base excision repair. *Cell*. 2011;146(1):67–79.
75. Hashimoto H, Liu Y, Upadhyay AK, et al. Recognition and potential mechanisms for replication and erasure of cytosine hydroxymethylation. *Nucleic Acids Res*. 2012;40(11):4841–4849.
76. Warn-Cramer BJ, Macrander LA, Abbott MT. Markedly different ascorbate dependencies of the sequential alpha-ketoglutarate dioxygenase reactions catalyzed by an essentially homogeneous thymine 7-hydroxylase from Rhodotorula glutinis. *J Biol Chem*. 1983;258(17):10551–10557.
77. He YF, Li BZ, Li Z, et al. Tet-mediated formation of 5-carboxylcytosine and its excision by TDG in mammalian DNA. *Science*. 2011;333(6047):1303–1307.
78. Ito S, Shen L, Dai Q, et al. Tet proteins can convert 5-methylcytosine to 5-formylcytosine and 5-carboxylcytosine. *Science*. 2011;333(6047):1300–1303.
79. Zhang L, Lu X, Lu J, et al. Thymine DNA glycosylase specifically recognizes 5-carboxylcytosine-modified DNA. *Nat Chem Biol*. 2012;8(4):328–330.
80. Maiti A, Drohat AC. Thymine DNA glycosylase can rapidly excise 5-formylcytosine and 5-carboxylcytosine: potential implications for active demethylation of CpG sites. *J Biol Chem*. 2011;286(41):35334–35338.
81. Liutkeviciute Z, Lukinavicius G, Masevicius V, Daujotyte D, Klimasauskas S. Cytosine-5-methyltransferases add aldehydes to DNA. *Nat Chem Biol*. 2009;5(6):400–402.
82. Valinluck V, Sowers LC. Endogenous cytosine damage products alter the site selectivity of human DNA maintenance methyltransferase DNMT1. *Cancer Res*. 2007;67(3):946–950.
83. Frauer C, Hoffmann T, Bultmann S, et al. Recognition of 5-hydroxymethylcytosine by the Uhrf1 SRA domain. *PLoS One*. 2011;6(6):e21306.
84. Iqbal K, Jin SG, Pfeifer GP, Szabo PE. Reprogramming of the paternal genome upon fertilization involves genome-wide oxidation of 5-methylcytosine. *Proc Natl Acad Sci USA*. 2011;108(9):3642–3647.
85. Wossidlo M, Nakamura T, Lepikhov K, et al. 5-Hydroxymethylcytosine in the mammalian zygote is linked with epigenetic reprogramming. *Nat Commun*. 2011;2:241.
86. Williams K, Christensen J, Pedersen MT, et al. TET1 and hydroxymethylcytosine in transcription and DNA methylation fidelity. *Nature*. 2011;473(7347):343–348.
87. Pastor WA, Pape UJ, Huang Y, et al. Genome-wide mapping of 5-hydroxymethylcytosine in embryonic stem cells. *Nature*. 2011;473(7347):394–397.
88. Klose RJ, Bird AP. Genomic DNA methylation: the mark and its mediators. *Trends Biochem Sci*. 2006;31(2):89–97.
89. Zhang X, Odom DT, Koo SH, et al. Genome-wide analysis of cAMP-response element binding protein occupancy, phosphorylation, and target gene activation in human tissues. *Proc Natl Acad Sci USA*. 2005;102(12):4459–4464.
90. Globisch D, Munzel M, Muller M, et al. Tissue distribution of 5-hydroxymethylcytosine and search for active demethylation intermediates. *PLoS One*. 2010;5(12):e15367.
91. Kinney SM, Chin HG, Vaisvila R, et al. Tissue specific distribution and dynamic changes of 5-hydroxymethylcytosine in mammalian genome. *J Biol Chem*. 2011;286:24685–24693.
92. Munzel M, Globisch D, Bruckl T, et al. Quantification of the sixth DNA base hydroxymethylcytosine in the brain. *Angew Chem Int Ed Engl*. 2010;49(31):5375–5377.
93. Valinluck V, Tsai HH, Rogstad DK, Burdzy A, Bird A, Sowers LC. Oxidative damage to methyl-CpG sequences inhibits the binding of the methyl-CpG binding domain (MBD) of methyl-CpG binding protein 2 (MeCP2). *Nucleic Acids Res*. 2004;32(14):4100–4108.

94. Ding N, Zhou H, Esteve PO, et al. Mediator links epigenetic silencing of neuronal gene expression with x-linked mental retardation. *Mol Cell*. 2008;31(3):347–359.
95. Urdinguio RG, Sanchez-Mut JV, Esteller M. Epigenetic mechanisms in neurological diseases: genes, syndromes, and therapies. *Lancet Neurol*. 2009;8(11):1056–1072.
96. Guy J, Cheval H, Selfridge J, Bird A. The role of MeCP2 in the brain. *Annu Rev Cell Dev Biol*. 2011;27:631–652.
97. Moretti P, Zoghbi HY. MeCP2 dysfunction in Rett syndrome and related disorders. *Curr Opin Genet Dev*. 2006;16(3):276–281.
98. Talkowski ME, Mullegama SV, Rosenfeld JA, et al. Assessment of 2q23.1 microdeletion syndrome implicates MBD5 as a single causal locus of intellectual disability, epilepsy, and autism spectrum disorder. *Am J Hum Genet*. 2011;89(4):551–563.
99. O'Roak BJ, Vives L, Girirajan S, et al. Sporadic autism exomes reveal a highly interconnected protein network of de novo mutations. *Nature*. 2012;485(7397):246-250.
100. Szulwach KE, Li X, Li Y, et al. 5-hmC-mediated epigenetic dynamics during postnatal neurodevelopment and aging. *Nat Neurosci*. 2011;14(12):1607–1616.
101. Numata S, Ye T, Hyde TM, et al. DNA methylation signatures in development and aging of the human prefrontal cortex. *Am J Hum Genet*. 2012;90(2):260–272.
102. Mill J. Toward an integrated genetic and epigenetic approach to Alzheimer's disease. *Neurobiol Aging*. 2011;32(7):1188–1191.
103. Laffita-Mesa JM, Bauer PO, Kouri V, et al. Epigenetics DNA methylation in the core ataxin-2 gene promoter: novel physiological and pathological implications. *Hum Genet*. 2012;131(4):625–638.
104. Sabunciyan S, Aryee MJ, Irizarry RA, et al. Genome-wide DNA methylation scan in major depressive disorder. *PLoS One*. 2012;7(4):e34451.
105. Melas PA, Rogdaki M, Osby U, Schalling M, Lavebratt C, Ekstrom TJ. Epigenetic aberrations in leukocytes of patients with schizophrenia: association of global DNA methylation with antipsychotic drug treatment and disease onset. *Faseb J*. 2012;26:2712–2718.
106. Gavin DP, Sharma RP, Chase KA, Matrisciano F, Dong E, Guidotti A. Growth arrest and DNA-damage-inducible, beta (GADD45b)-mediated DNA demethylation in major psychosis. *Neuropsychopharmacology*. 2012;37(2):531–542.

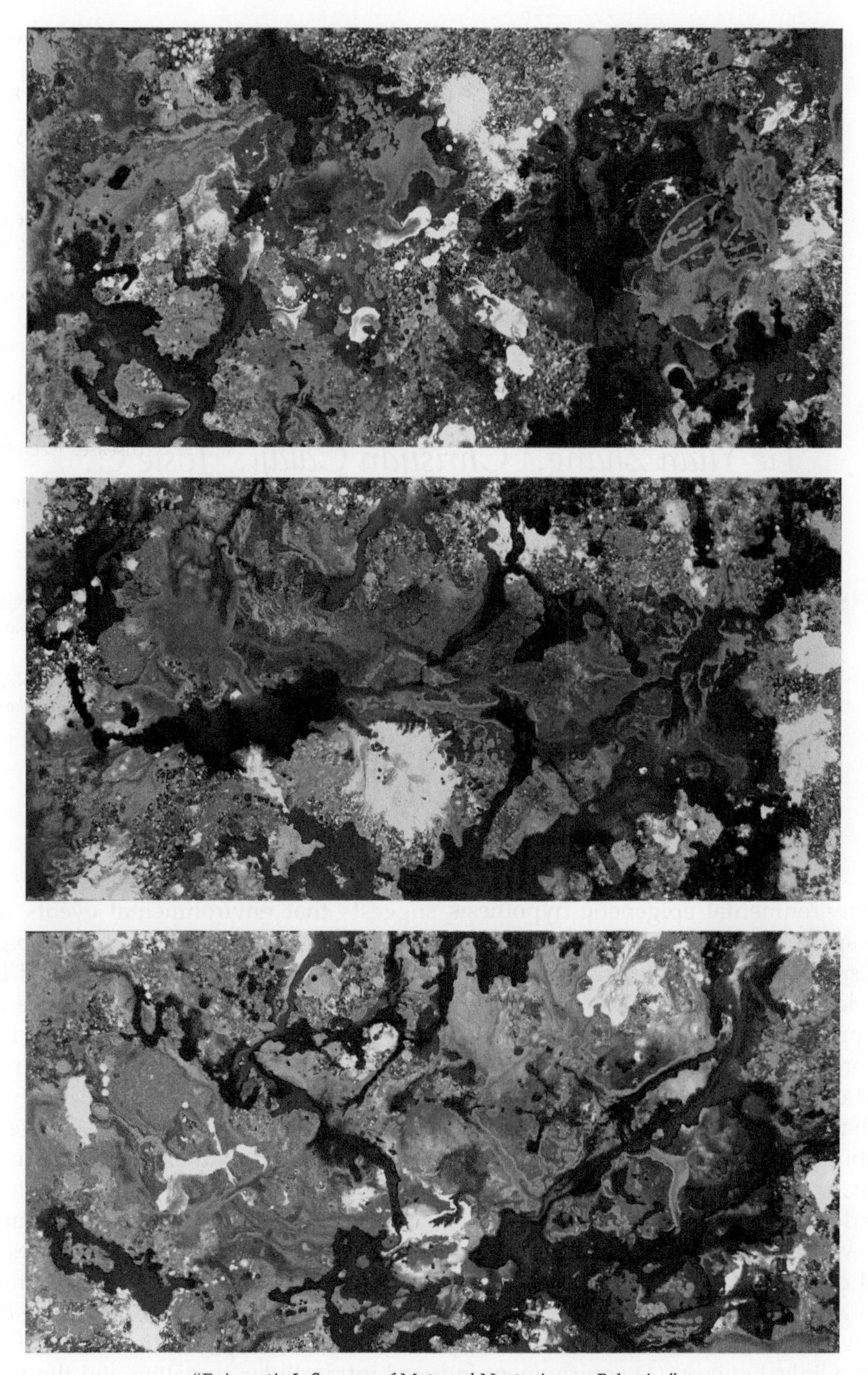

"Epigenetic Influences of Maternal Nurturing on Behavior"

J. David Sweatt, acrylic on wood panel (triptych of three 48 x 24 panels), 2012

CHAPTER

4

The Epigenetics of Parental Effects

Tie Yuan Zhang,[1] Christian Caldji,[1] Josie C. Diorio,[1] Sabine Dhir,[1] Gustavo Turecki[1] and Michael J. Meaney[1,2]

[1]Sackler Program for Epigenetics Psychobiology and Departments of Psychiatry and Neurology & Neurosurgery, McGill University, Montreal, Canada

[2]Departments of Psychiatry, Neurology, and Neurosurgery, Douglas Institute McGill University, Montreal, Quebec, Canada

OVERVIEW

The environmental epigenetic hypothesis suggests that environmental events reliably activate specific intracellular signaling pathways that initiate the remodeling of epigenetic state of specific genomic regions, leading to stable alterations in transcription, and thus in broader levels of phenotypic variation. We review studies that focus on variations in parent–offspring interactions as the relevant environmental condition, and examine the evidence for the role of epigenetic mechanisms in mediating the relation between parental care and offspring development. We begin this chapter by placing these studies of molecular mechanism within the context of the developmental origins of individual differences in vulnerability/resistance for psychopathology focusing on individual differences in stress reactivity, which appears to mediate the effects of childhood experience on the risk for mental disorders (Figure 4.1). Importantly, these studies suggest that variations in parental care operate as a determinant of phenotypic variation across all members of the species. Thus, parental care is relevant not only as a profound influence on health outcomes in humans, but also as a major source of phenotypic variation for evolutionary diversity. The chapter then provides a review of the evidence for the environmental epigenetic hypothesis focusing on the links between (1) environment-induced intracellular signaling and the resulting

Epigenetic Regulation in the Nervous System
DOI: http://dx.doi.org/10.1016/B978-0-12-391494-1.00004-5

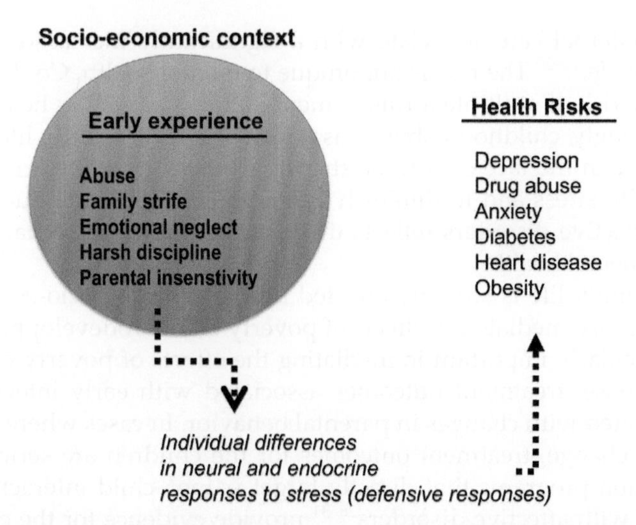

FIGURE 4.1 **An overview of the context and rationale for the studies described within this review.** Existing studies strongly link the quality of parent–child interactions as determinants of multiple health outcomes, actually extending beyond the list provided to the right of this schema. Likewise studies in human and non-human models suggest that this relation is mediated by influences on stress reactivity. It is important to appreciate that familial influences occur within a broader socio-economic context and there is strong evidence revealing the influence of conditions such as poverty on the quality of parent–child interactions. This chapter focuses on studies suggesting that the relation between parenting and stress reactivity in the offspring is mediated by epigenetic influences on the regulation of the neural transcriptome in selected brain regions.

epigenetic modifications and (2) epigenetic modification and phenotypic outcome. We conclude with a summary of the challenges facing future studies of the environmental regulation of epigenetic states.

PARENT–OFFSPRING INTERACTIONS AND THE MENTAL HEALTH OF THE OFFSPRING

The quality of family life influences the development of individual differences in vulnerability for multiple forms of mental illness, including affective illnesses and addictions. As adults, victims of childhood physical, sexual or emotional abuse, or of parental neglect are at considerably greater risk for affective disorders.[1–7] These findings were confirmed in a prospective, longitudinal study that established a direct link between abuse/neglect and depression.[8] Moreover, childhood maltreatment associates with an increased severity of illness, reduced treatment responsivity and increased co-morbidity.[8] Broader forms of familial dysfunction including persistent emotional neglect, family conflict, and harsh, inconsistent discipline compromise cognitive and emotional development[9–12] and increase the risk for depression and anxiety disorders[13–15] to a level comparable to that for abuse. Low scores on measures of parental bonding, reflecting cold, distant parent–child relationships, particularly low maternal care, associate with a significantly increased risk of depression and anxiety in later life.[16–20] The risk is not unique to mental health. Cold and detached parenting increases the risk of multiple forms of mental illness as well as heart disease and diabetes.[21] Not surprisingly, childhood abuse has a similar effect.[4] Family life can also serve as a source of resilience in the face of chronic stress.[22] Thus, warm, nurturing families tend to promote resistance to stress and to diminish vulnerability to stress-induced illness.[11,23] The epidemiology of affective disorders reflects the profound influence of family life on neural development and mental health.

The quality of family life is strongly affected by the broader socio-economic context[11,24] and parental factors also mediate the effects of poverty on neurodevelopment.[25–27] Parenting appears to be particularly important in mediating the effects of poverty on socio-emotional development. Likewise, treatment outcomes associated with early intervention programs are routinely correlated with changes in parental behavior. In cases where parental behavior proves resistant to change, treatment outcomes for the children are seriously limited. The effects of intervention programs that directly target parent–child interactions on endophenotypes associated with affective disorders[28–31] provide evidence for the causal influence of parenting on mental health.

There is considerable evidence in favor of the idea that the quality of parent–offspring interactions reflects that of the prevailing environment.[11,32] Poor quality human environments associate with family dysfunction.[33–36] Environmental adversity, including economic hardship and marital strife, compromise the emotional well-being of the parent and thus influence the quality of parent–child relationships.[11] High levels of maternal stress are associated with increased parental anxiety, less sensitive childcare[37,38] and insecure parental attachment.[38,39] Parents in poverty or other environmental stressors, experience more negative emotions, irritable, depressed, and anxious moods, which associate with punitive parenting[32,40,41] and abuse.[42] Such findings likely explain the profound influence of childhood socio-economic status (SES), independent of that in adulthood, on depression[43–46] as well as on endophenotypes that associate with affective disorders such as neuroticism, emotional regulation and coping style.[6,47]

The greater the number of environmental stressors (e.g. lesser education of parents, low income, many children, being a single parent), the less supportive are mothers of their children; such mothers are more likely to threaten, push, or grab them, and display more controlling attitudes. Parental stress is increased within abusive families.[11,42] Fleming[48] reported the anxiety of the mother is the best predictor of the mother's attitude (also see[49]). While the results of such studies are generally correlational, there is compelling evidence for a direct effect of environmental adversity on parent–infant interactions among non-human primates.[50–52] Bonnet macaque mother–infant dyads placed under conditions of uncertain food availability show a progressive deterioration in the quality of interaction, with greater evidence of conflict, resulting in increased infant timidity and fearfulness by comparison to control dyads. Infants reared under the more challenging foraging conditions came to show signs of depression, commonly observed in maternally-separated macaque infants, even when the infants are in contact with their mothers. These findings brilliantly underscore the link between the quality of the parental environment, that of parent–offspring interactions and the expression of endophenotypes for affective illness in the offspring.

EVOLUTIONARY BIOLOGY OF PARENTAL SIGNALING

There are at least two ways to view the effects of parental care on the development of the offspring. The first, which is common within the biomedical and social sciences, views parental abuse and neglect as a "toxic" event that impairs neurodevelopment, rendering the individual vulnerable for disease over the lifespan. The second interpretation of such relations, one derived from evolutionary biology, considers parental care as a developmental signal that directs development along specific pathways.[53,54] The critical consideration in defining the nature of developmental influence is whether the resulting developmental outcome might be adaptive under specific environmental conditions. There are no doubt environmental exposures occurring during early life that are undeniably toxic. Few would argue that exposure to elevated levels of lead or alcohol during prenatal development results in any phenotypic modification that is adaptive under any conditions. However, parental effects on traits that associate with an increased risk for affective disorders, including an increased reactivity to stressful conditions, may be adaptive despite the cost associated with vulnerability. Such effects are viewed in evolutionary terms as trade-offs. It is important to bear in mind that within the context of this discussion the term "adaptive" refers to traits that enhance survival and the opportunity for reproduction irrespective of the impact on health, per se.

We suggest that a meaningful understanding of the relation between early experience and health in adulthood must consider the question of why adversity in early life alters phenotypic development in such a predictable and persistent manner and why parental investment in the offspring varies so greatly; if parental nurturance is so "beneficial" for neural development, then why do not all parents invest heavily in each individual offspring? The argument here is based on the idea that development is essentially a process of continuous adaptation[55,56] that permits the individual to function independently of parental support and best compete with co-specifics. Development occurring under conditions of adversity leads to increased sensitivity of defensive responses, as an adaptation to adversity, and thus an increased risk for multiple forms of chronic illness, which reflects the cost associated with an enhanced capacity for defensive responses. We suggest that: (1) the effect of environmental adversity on the development of defensive responses is mediated by parental signals in early development; and (2) increased sensitivity of the defensive responses is a natural, predictable and potentially adaptive byproduct of adversity imposed on the parent or the offspring during early development. There is strong support for these ideas from studies in evolutionary biology and ecology describing so-called "phenotypic plasticity" in defensive responses.[57–60] The "parental signaling" approach may resolve an apparent paradox in the literature on the developmental origins of psychopathology, whereby more subtle variations in the quality of parental care predict mental health outcomes to much the same degree as more overt and dramatic forms of adversity such as abuse and neglect (see above). The point here is not that abuse or neglect are not without consequence, but rather that cold/detached parenting appears to be almost of equal effect on health outcomes. However, if parenting acts as a developmental signal then it is perhaps not so surprising that developmental effects of such magnitude derive from variations in parental care that appear to lie within a normal range for the species.

The critical point is that, under appropriate circumstances, increased stress reactivity is of considerable advantage to the individual. Under conditions of increased risk for predation, more reactive guppies with shorter escape latencies show increased survival.[61] Under conditions of increased environmental demand, it is commonly in the animal's interest to enhance its behavioral (e.g. vigilance, fearfulness) and endocrine/autonomic (metabolic/cardiovascular) responses to stress. Such considerations explain why victims of childhood abuse are more sensitive to facial expressions that suggest anger.[62] These responses promote detection of potential threat, fear conditioning to stimuli associated with threat and avoidance learning. Adult rats reared by mothers that show a reduced frequency of pup licking/grooming show an increased capacity for contextual fear conditioning.[63] However, the molecular mechanisms associated with such enhanced learning under fearful conditions, also restrain synaptic plasticity and episodic learning under non-stressful conditions.

Likewise the endocrine correlates of enhanced stress reactivity can prove adaptive. The hormonal effectors of sympathoadrenal and hypothalamic–pituitary–adrenal (HPA) stress responses mobilize energy reserves through effects on the production and utilization of fats and glucose.[64] These effects are the hallmark of the shift to catabolism that occurs during periods of stress and are essential for animals exposed to famine. Indeed, the ability to survive sustained periods of nutrient deprivation depends upon the capacity to increase circulating levels of glucocorticoids and catecholamines. Impoverished environments are also commonly associated with multiple sources of infection. Under such conditions adrenal glucocorticoids are a potent defense against the increased immunological activity that could otherwise lead to septic shock. Increased HPA responses to agents such as bacterial endotoxins reduce the risk of sepsis. Adults exposed to a bacterial endotoxin during the first week of life exhibit enhanced HPA responses to stress, an effect that is mediated by increased hypothalamic expression of both corticotropin-releasing factor (CRF) and vasopressin.[65,66] These animals also show increased resistance to sepsis upon subsequent exposure to bacterial infection.[67] Conversely, postnatal handling, which increases the frequency of pup licking/grooming by rat mothers, reduces hypothalamic CRF expression, dampens HPA responses to stress, and is associated with increased vulnerability to endotoxin-induced sepsis.[65,68,69] These findings underscore the potentially adaptive value of increased HPA responses to threat. Impoverished environments, which are rife with the psychosocial conditions that promote increased HPA/sympathetic reactivity[70,71] are commonly characterized by violence, decreased nutrient availability and an increased density of pathogens. These are precisely the conditions under which increased HPA activity promotes survival.

There is also evidence for the potentially adaptive effects of increased stress reactivity that comes closer to the interests of child and adolescent psychiatry. Research by Farrington and colleagues[72] and Tremblay[73] on young males growing up in a low SES, high crime urban environments provides an excellent illustration of the potential advantages of increased emotional stress reactivity. In both studies, the males most successful in avoiding the pitfalls associated with such a "criminogenic" environment are shy and somewhat timid. Under such conditions, behavioral inhibition emerges as a protective factor.[73] Moreover, under such adverse conditions, a parental rearing style that favored the development of a greater level of stress reactivity to threat could be viewed as adaptive. The obvious conclusion is that there is no single ideal form of parenting. Different environments call for different forms of parental care. The importance of context is well illustrated in avian research. More fearful, less exploratory great tits can show either significantly increased or decreased survival rates by comparison to more aggressive, exploratory birds depending upon resource availability.[74] The differences in fearfulness/exploration are driven by maternally-regulated steroid levels in the yolk sac.[75,76]

This issue is also critical for studies of molecular mechanism. If indeed evolution has shaped the young to use environmental signals, such as variations in parental care, as predictors of the quality of the prevailing environment, then we would expect that a complex set of molecular mechanisms have evolved to serve such phenotypic plasticity. We suggest that these mechanisms include the epigenetic regulation of genome structure and function. This rationale underlies the *environmental epigenetics hypothesis*. Studies of environmental regulation of epigenetic states present a series of challenges. The first is that of identifying the specific environmental signal. Second, is that of identifying the relevant signaling pathway(s). Third is the imperative of characterizing the critical enzymatic events that catalyze the relevant epigenetic mark. Fourth is the challenge of directly linking the epigenetic mark to the relevant transcriptional event. We present here a review of studies using rodent models that examine the environmental epigenetics hypothesis in relation to variations in parental care.

THE BIOLOGY OF PARENT INFLUENCES

A critical question concerns the mechanisms that mediate the enduring parental influence on the health of offspring. The relationship between social influences over development and health in adulthood appears to be, in part, mediated by the development of individual differences in neural systems that underlie the expression of behavioral and endocrine responses to stress.[11,77–84] Thus, physical and sexual abuse in early life increases endocrine and autonomic responses to stress in adulthood.[70,85] Likewise, variations in parental care associate with individual differences in neuroendocrine and autonomic responses to stress in humans[71,86–90] as well as emotional reactivity[91] (and see[11]). Finally, there is considerable evidence in favor of the hypothesis that individual differences in stress reactivity associate with the risk for depression.[92,93] Thus, the influence of familial depressive illness is, in part at least, mediated by increased stress reactivity, enhancing the response of the individual to mild, regular stressors (i.e. hassles). A critical question is that of how parental signals operate during early development to influence stress reactivity and how such effects persist into adulthood.

We examine the influence of variations in maternal care in the rat on the development of individual differences in behavioral and endocrine responses to stress. We define variations in maternal care in lactating female Long-Evans rats (an out-bred strain of *Rattus norvegicus*) using a simple observational procedure[94] in which the behavior of the mother is scored every three minutes for 25 consecutive observation points. There are five such observational sessions each day over the first 6 days post-partum. The behavior showing the most consistent, individual differences across mother is that of the frequency of pup licking/grooming (LG[95]; Figure 4.2). Individual differences in the frequency of pup LG among adult female rats are reliable over time and across multiple litters, and thus a stable feature of the maternal phenotype.[95]

We use pup LG scores from cohorts of lactating rats to define mothers that show a frequency of pup LG that is 1SD above (high LG) or below (low LG) the mean for the cohort.

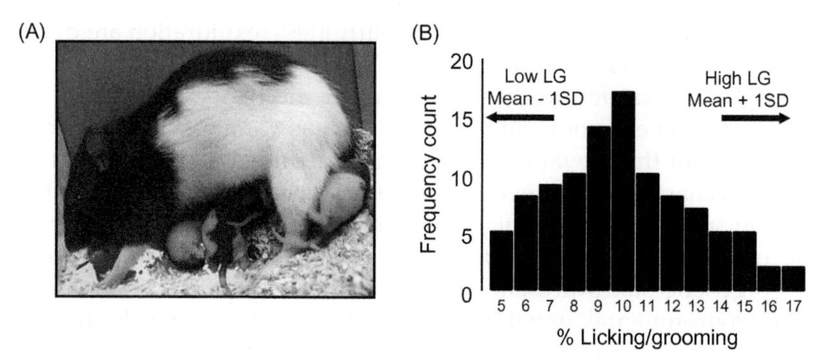

FIGURE 4.2 (A) Photo of pup licking/grooming (LG) during a nursing bout. (B) A frequency histogram of a representative cohort of pup LG showing a normal distribution and the selection of high and low LG mothers (i.e. females where the mean frequency of pup LG from standardized behavioral observations are 1 SD above [high LG mothers] or below [how LG mothers] the mean for the cohort).

Over 15 years of work with this model, the variation in mean pup LG scores across cohorts varies by less than 5%. Interestingly, the difference in the frequency of pup LG occurs not because of the frequency of individual bouts of pup LG, but because the duration of pup LG bouts are significantly longer in high compared to low LG mothers[95–97] (and see[98]).

The quality of parent–child interactions in humans is a reflection of the quality of the prevailing environment. This finding forms the basis for parental signaling. Importantly, there is evidence of a comparable environmental effect on mother–pup interactions in the rat. Chronic stress during gestation reduces the frequency of pup LG in previously characterized high LG mothers to levels comparable to those of low LG mothers[99] (also see[100]). Likewise impoverished environmental conditions during lactation also affect mother–pup interactions, reducing the frequency of pup LG.[98,101] For example, mothers consistently provided inadequate nesting materials, and thus obligated to rear their pups on mesh flooring, show a decreased frequency of pup LG. This effect was due to a shorter duration of pup LG bouts, typical of low LG mothers, than to a decrease in the frequency of pup LG bouts.[98] Likewise, stress decreased total contact time between the mother and her pups and increased some measures of maternal fearfulness.[98] Low LG mothers also show increased fearfulness,[102] and these findings are consistent with the idea that maternal stress establishes a maternal phenotype characteristic of the naturally-occurring low LG mothers. In contrast with the effects of stress, environmental enrichment, including larger caging, physical complexity and another mother and litter, increased pup LG. These suggest that, in the rat, as in humans, the quality of the prevailing environment directly affects maternal care.

Variations in pup LG over the first week of postnatal rat life affect the development of behavioral and HPA responses to stress in adulthood.[80,103–109] Behavioral responses to environmental stressors include a cessation of exploration or appetitive behavior,[80,104,109,110] as well as active attempts to escape from threat.[105] For example, in a novelty-induced suppression of feeding test in which food-deprived animals are provided food in a novel context, the adult offspring from high LG mothers show a shorter latency to begin eating and eat for a longer period of time[104] (O'Donnell & Meaney, unpublished observation). The offspring of low LG mothers also show increased vulnerability for stress-induced learned helplessness[111] and spend more time immobile in the forced swim test (O'Donnell & Meaney, unpublished observation).

Likewise there are differences in HPA responses to acute stress apparent in both circulating levels of pituitary adrenocorticotropin (ACTH) and adrenal corticosterone. As adults, the offspring of high LG mothers show more modest plasma ACTH and corticosterone responses to acute stress by comparison to animals reared by low LG mothers.[99,103,109,111,112] Circulating glucocorticoids act at glucocorticoid receptor (GR) sites in corticolimbic structures, such as the hippocampus, to regulate HPA activity. Such inhibitory feedback effects target CRF synthesis and release at the level of the paraventricular nucleus of the hypothalamus (PVNh). The offspring of high LG mothers show significantly increased hippocampal GR mRNA and protein expression, enhanced glucocorticoid negative feedback sensitivity and decreased hypothalamic CRF mRNA levels. Moreover, manipulations that enhance the frequency of pup LG by rat mothers associate with increased hippocampal GR expression and decreased CRF expression in the offspring.[80,113] Importantly, hippocampal infusion of a GR antagonist completely eliminates the maternal effect on HPA responses to stress, suggesting a direct relation between hippocampal GR expression and the magnitude of the HPA response to stress. These findings are consistent with the results of studies that have genetically modified forebrain GR expression. Thus, selective knock-down of GR expression in the corticolimbic system in rodents is associated with increased HPA activity under basal as well as stressful conditions.[114–116] Conversely, GR overexpression is associated with a dampened HPA response to acute stress.[117]

The effect of maternal care on GR expression is largely unique to the hippocampus (McGowan & Meaney, unpublished observation). We find no evidence of an alteration in GR expression in the PVNh, the amygdala or the prefrontal cortex is a function of variations in early postnatal maternal care. There is evidence for an increase in GR expression in the anterior pituitary in the adult offspring of low LG mothers. There are a number of relevant caveats to these data. Most important is the heterogeneity within the individual regions, particularly the amygdala and prefrontal cortex. We can exclude the possibility of sub-region-specific effects. Nevertheless, the findings do suggest regionally selective effects of maternal care on GR expression.

The effects of maternal care on gene expression and stress responses of the adult offspring are reversed with cross-fostering:[80,106,118] stress responses of adult animals born from low LG mothers and reared by high LG dams are comparable to normal offspring of high LG mothers (and vice versa). Moreover, variations in the frequency of pup LG towards individual pups of the same mother are significantly correlated with hippocampal GR expression in adulthood.[119] These findings, as well as those from studies that directly manipulate the frequency of pup LG by the dam reveal a direct relation between maternal care and the phenotypic development of the offspring.

MOLECULAR TRANSDUCTION OF MATERNAL SIGNALS

The results of the studies described above suggest that the influence of maternal care on HPA responses to stress are mediated by effects on hippocampal GR expression. In vivo studies with rat pups or in vitro studies using cultured hippocampal neurons reveal that maternal effects on hippocampal GR expression are mediated by increases in hippocampal serotonin (5-HT) turnover and a resulting increase in the expression of the nerve-growth factor-inducible factor-A (NGFI-A; aka egr-1, zif-268) transcription factor.[97,120–122] In vitro, 5-HT acts through a 5-HT_7 receptor to increase the activity of cAMP-dependent signaling pathways in hippocampal neurons, resulting in elevated expression of the transcription factor, NGFI-A (aka, egr-1, zyf-268). The effect of various 5-HT agonists on GR expression in hippocampal neurons is strongly correlated with the effect on cAMP formation.[123] Activation of these signaling cascades leads to an increased GR expression. The effect of 5-HT on GR expression in cultured hippocampal neurons is: (1) blocked by 5-HT_7 receptor antagonists or inhibitors of protein kinase A; (2) mimicked by 5-HT_7 receptor agonists or treatments with stable cAMP analogs: and (3) eliminated by antisense or siRNA knockdown of NGFI-A mRNA.[97,124] The effect of 5-HT treatment is mimicked by overexpression of NGFI-A.[97] In vivo, the effect on GR is blocked with 5-HT receptor antagonists.[122]

The in vivo increase in hippocampal 5-HT activity is associated with a maternally-regulated increase in the conversion of thyroxine to triidodithyronine (T3):[97] T3 administration in the neonatal period, which regulates the 5-HT systems activity, mimics the effects of increased pup LG on both NGFI-A expression and hippocampal GR programming.[97,122,125] Interestingly, the activation of ascending 5-HT systems during postnatal development also regulates the development of corticolimbic systems implicated in fear behavior[126,127] as well as hippocampal synaptic development.[128]

An issue concerns the identity of the genomic site for the regulation of GR transcription. The 5′ non-coding variable exon 1 region of the hippocampal GR gene (Figure 4.3) contains multiple alternate promoter sequences including a neuron-specific, exon 1_7 sequence.[129] Increased pup LG enhances hippocampal expression of GR mRNA splice variants

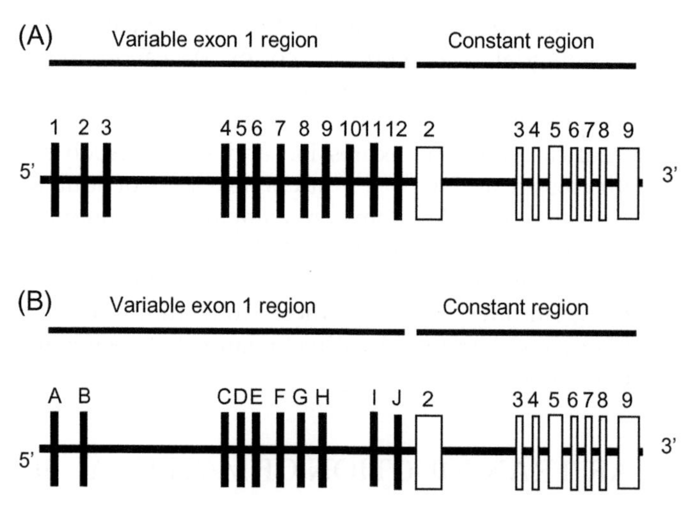

FIGURE 4.3 **Schema describing the organization of the rat and human glucocorticoid receptor gene, including the 9 exonic regions.** Exons 2–9 code for the glucocorticoid receptor protein. Exon 1 is comprised of multiple, tissue-specific promoter regions (rat is based on[129] and human on[130]). The rat exon 1_7 sequence shares ≈70% sequence homology with the human exon 1_F sequence, and both are highly expressed in hippocampus.

containing exon 1_7 sequence,[97,106,129] which contains an NGFI-A response element.[131] Since each of these exon 1 variants is associated with an individual promoter and is spliced onto a 5′ acceptor site, it seems reasonable to assume that the exon 1_7 promoter is more actively involved in the regulation of GR transcription in the adult hippocampus of animals reared by high compared to low LG mothers.

Pup LG from the mother increases hippocampal NGFI-A expression and binding to the exon 1_7 promoter in neonates;[97,106] and see below). Co-transfection of an NGFI-A vector and an exon 1_7–luciferase construct shows increased luciferase activity, reflecting NGFI-A-induced activation of transcription through the exon 1_7 promoter.[97,124] The effect of NGFI-A is eliminated by a site-directed mutation within the NGFI-A response element of the exon 1_7 promoter[124] that eliminates NGFI-A binding, revealing that it is the physical interaction of NGFI-A with its response element that triggers the increase in transcriptional activity. Moreover, infection of cultured hippocampal neurons with an NGFI-A expression plasmid increases both total GR mRNA and exon 1_7-containing GR mRNA.[97] Comparable effects on both total GR mRNA and exon 1_7-containing GR mRNA occur following treatment of cultured hippocampal neurons with 5-HT: these effects are blocked by an shRNA that targets NGFI-A.[97]

Interestingly, treatment of cultured hippocampal neurons with either 5-HT or with the NGFI-A expression vector also increased exon 1_{10}- and exon 1_{11}-containing GR mRNA despite the fact that these sequences do not contain an NGFI-A consensus sequence. These findings suggest the possibility of regulation in *cis* across the GR promoter regions.

A series of in vivo studies shows that the association of NGFI-A with the exon 1_7 promoter is actively regulated by pup LG.[97] Thus, chromatin-immunoprecipitation (ChIP) assays reveal increased binding of NGFI-A to the exon 1_7 promoter in pups of high compared to low LG mothers, but only in the period following a nursing bout with pup LG: hippocampal tissue samples obtained 20 min following a nursing bout, with no subsequent interaction between the mother and pup do not reveal the difference in NGFI-A association. Perhaps most convincingly, artificial tactile stimulation of pups (i.e. stroking) increases circulating T3 levels, hippocampal NGFI-A expression and NGFI-A binding to the exon 1_7 promoter. These findings directly link the tactile stimulation derived from pup LG with the activation of the intracellular signaling pathways that regulate hippocampal GR expression.

These findings are consistent with the idea that it is the tactile stimulation derived from maternal licking/grooming that appears to be the critical environmental signal for the regulation of hippocampal GR expression in the neonate.

Increased frequency of pup LG associates with an increase in the expression of multiple transcriptional regulators, including SP1 and the CREB-binding protein (CBP). Similar effects are seen following 5-HT treatment in cultured hippocampal neurons[97] (LaPlante & Meaney, unpublished observation). ChIP assays with hippocampal tissue obtained following either a period of active pup LG or no recent interaction show that the binding of both CBP and SP1 to the exon 1_7 promoter sequence, like that for NGFI-A, is actively regulated by maternal care.[97] Co-immunoprecipitation assays reveal that CBP and NGFI-A associate in nuclear fractions, suggesting they bind at common sites. Moreover, the same site-directed mutagenesis of the NGFI-A site in the exon 1_7 promoter that abolishes NGFI-A binding to the exon 1_7 promoter sequence, also eliminates CBP binding, suggesting that it is NGFI-A that targets CBP to this region. This is likely important since CBP is a known transcriptional co-factor that acts as a histone acetyltransferase.[132] Also of interest is the presence of an SP1 binding site that overlaps with that of the NGFI-A site in the exon 1_7 promoter sequence (see Figure 4.3) since SP1 has been implicated in the regulation of DNA methylation states at promoters[133,134] and see below).

There is a comparable effect of maternal care on other, NGFI-A regulated genes such as *GAD1*,[135] which encodes for glutamic acid decarboxylase, the rate-limiting enzyme for γ-aminobutyric acid (GABA) synthesis. The association of NGFI-A with the *GAD1* promoter in hippocampus is increased in the offspring of high compared to low LG mothers, but only following a nursing bout. Similarly, hippocampal neuronal cultures treated with 5-HT show an increase in *GAD1* expression and the effect is blocked by an siRNA targeting NGFI-A. These findings suggest that maternal care regulates the expression of a range of NGFI-A-sensitive genes.

These findings suggest that the tactile stimulation derived from pup LG increases hippocampal NGFI-A expression and NGFI-A binding to the exon 1_7 GR promoter, with a subsequent activation of GR gene transcription. This model explains the increased hippocampal GR expression in pups of high compared to low LG mothers.[124] However, the critical issue concerns the mechanism by which hippocampal GR expression remains elevated following weaning and separation from the mother. How is the maternal effect rendered stable? One possibility is that the increased NGFI-A–exon 1_7 interaction occurring within hippocampal neurons in the pups of high LG mothers might result in an epigenetic modification of the exon 1_7 sequence that alters NGFI-A binding and maintains the maternal effect into adulthood. We focused our initial studies on potential influences on DNA methylation with the assumption that this relatively stable covalent modification was a reasonable candidate mechanism for the enduring effects of maternal care on hippocampal gene expression in the rat.

THE EPIGENETICS OF PARENTAL EFFECTS

Preliminary studies using methylation-sensitive restriction enzymes and Southern blotting revealed greater methylation across the entire exon 1_7 GR promoter sequence in the hippocampus of adult offspring of low LG mothers. These findings suggested a maternal effect on DNA methylation patterns in the offspring. More focused approaches examined the methylation status of individual CpGs in the exon 1_7 sequence using sodium bisulfite mapping, which permits the characterization of the methylation state of individual CpG dinucleotides across a specific region of DNA. Sequencing reveals significant differences in methylation at the 5′ CpG dinucleotide of the NGFI-A consensus site on the exon 1_7 promoter. The site is hypermethylated in the offspring of low LG mothers, and hypomethylated in those of high LG dams. Cross-fostering reverses the differences in the methylation of the 5′ CpG site and suggests a direct relation between maternal care and DNA methylation of the exon 1_7 GR promoter.[106] The effect of maternal care is highly specific, with highly significant alterations in the methylation status of the 5′ CpG, and no effect at the 3′ site. Nevertheless, although less striking, there are differences in the frequency of methylation at other CpG sites on the exon 1_7 promoter.[112]

A series of studies suggests that the effect of maternal care on the DNA methylation status of the exon 1_7 GR promoter involves the same intracellular signaling pathways described above.[97,112] Thus, in cultured hippocampal neurons, the 5′ CpG site of the NGFI-A consensus sequence within the exon 1_7 promoter is hypomethylated following treatment with either 5-HT or a stable cAMP analog, 8-bromo-cAMP. These effects are blocked by an shRNA against NGFI-A. Overexpression of NGFI-A also results in the hypomethylation of the 5′ CpG site, and this effect is abolished by the site-directed mutation of the sequence that eliminates NGFI-A binding.[124] Importantly, ChIP assays confirm the increased in vivo association of both NGFI-A and CBP with the exon 1_7 promoter sequence in pups of high compared with low LG mothers.[97,124] The presence of CBP is interesting to consider. CBP is a histone acetyltransferase[132] and levels of histone 3 lysine 9 acetylation (H3K9ac) are increased in pups from high compared to low LG dams.[124,135] Histone deacetylase (HDAC) inhibitors, which also increase histone acetylation, can alter DNA methylation.[136,137]

These findings suggest that maternal care activates intracellular signaling pathways that initiate the remodeling of DNA methylation states. A major limitation to these studies is that the relevant catalytic agents have yet to be established. The difference in the methylation state of the exon 1_7 GR promoter emerges as a function of an apparent *de*methylation. Methylation levels of the 5′ CpG site are comparable at the time of birth in the offspring of high and low LG mothers, and the difference emerges over the first week of postnatal life with a significant decrease in methylation levels in the offspring of high and low LG mothers. The issue of demethylation remains.

5-hydroxymethylation

An alternative form of DNA methylation, 5-hydroxymethocytosine, has recently been re-discovered[138–140] (and see[141]), although its function is not fully understood. Interestingly, the ten-eleven translocation (TET) family of enzymes can convert 5-methylcytosine to 5-hydroxymethylcytosine.[139,140] 5-hydroxymethylcytosine is absent across the genome in cells depleted of all DNA methyltransferases (DNMTs), suggesting that the formation of 5-hydroxymethylcytosine is dependent upon existing 5-methylcytosine.[142] Interestingly both TET1 and 5-hydroxymethylcytosine signals are enriched at transcriptional start sites across the mouse genome, consistent with the enrichment of 5-hydroxymethylcytosine at regions intermediate and high in CpG density.[142] TET1- and 5-hydroxymethylcytosine-positive regions appear to be transcriptionally inactive.[142] 5-hydroxymethylcytosine is found widely distributed in embryonic stem (ES) cells and in neurons, suggesting a possible function in gene regulation in cells that express a high level of phenotypic plasticity. 5-hydroxymethylcytosine is also enriched in certain neuronal cells.[138]

Bisulfite sequencing or PCR-based approaches to the study of DNA methylation cannot distinguish between 5-methylcytosine and 5-hydroxymethylcytosine. Since maternal care has sustained effect on GR exon 1_7 methylation, we[143] analyzed levels of 5-hydroxymethylcytosine and 5-methylcytosine across the hippocampal GR exon 1_7 promoter in rats using antibody capture (i.e. 5-hydroxymethylcytosine- and 5-methylcytosine-dependent immunoprecipitation) of hippocampal DNA. The level of 5-hydroxymethylcytosine of the exon 1_7 GR promoter was three times higher in hippocampal samples from the offspring of low compared to high-LG mothers.[143] In contrast, 5-methylcytosine-dependent immunoprecipitation revealed no differences across the exon 1_7 GR promoter. While antibody capture approaches are substantially less sensitive than direct sequencing, these findings suggest that the differences in DNA methylation at this site reflect, in part at least, differences in 5-hydroxymethylcytosine. This conclusion is consistent with the finding that 5-hydroxymethylcytosine is enriched in regions surrounding transcriptional start sites, which are commonly devoid of 5-methylcytosine.[142,144] Moreover, the majority of 5-hydroxymethylcytosine-positive transcriptional start sites show little or no 5-methylcytosine.[142] The involvement of 5-hydroxymethylcytosine may also explain why our earlier studies with the exon 1_7 GR promoter had failed to reveal any increase in the binding of methylated-DNA binding proteins, e.g. MeCP-2 or MBD-2[124] (Zhang & Meaney, unpublished observation) in hippocampus from the offspring of low LG mothers, since 5-hydroxymethylcytosine does not attract these repressive mediators.[145] Nevertheless, in ES cells, most 5-hydroxymethylcytosine-positive genes are not expressed,[142,146] although this is less clear in neurons.

5-Hydroxymethylcytosine has been considered as a form of methylated cytosine that may lie along a pathway leading to demethylation.[141,147] Developmental time-course studies using bisulfite sequencing suggest that the differences in the methylation status of the exon 1_7 GR promoter emerges over the first week of life and involves a process of demethylation; on the day of birth, promoter methylation levels are comparable, and then selectively decreased in the offspring of high LG mothers.[112] It will be interesting to define the relative level of 5-hydroxymethylcytosine over this period. Such studies will determine whether the maternal effect involves a transient difference in 5-methylcytosine, or whether the differences in methylation status are uniquely in the form of 5-hydroxymethylation. Interestingly, there is no effect of maternal care on the level of 5-hydroxymethylation of the *GAD1* promoter (Zhang & Meaney, unpublished observation), suggesting that at this site the maternal influence is registered in the form of 5-methylcytosine.

Methylation and GR Transcription

Since DNA methylation favors a closed chromatin structure, the difference in methylation within the 5′ CpG dinucleotide of the NGFI-A response element suggests alteration of NGFI-A binding to the exon 1_7 sequence. This was verified using in vitro binding of purified recombinant NGFI-A protein to its response element using electrophoresis mobility shift assays to indicate that methylation of the 5′ CpG dinucleotide in the NGFI-A response element of the exon 1_7 GR promoter inhibits NGFI-A protein binding.[106] Transfection studies using HEK293 cells or cultured hippocampal neurons show that: (1) NGFI-A induces transcription through the exon 1_7 promoter; and (2) DNA methylation of a transfected exon 1_7 construct inhibits the ability of NGFI-A to bind and activate transcription.[97,124]

The recent findings of differences in 5-hydroxymethylation compromise the interpretation of the previous in vitro studies that focused on 5-methylcytosine. Nevertheless, such findings are consistent with the pattern of NGFI-A binding that is observed in vivo in the adult offspring of high and low LG mothers. There is a threefold higher level of NGFI-A binding to the exon 1_7 promoter in hippocampus from the adult offspring of high compared to low LG mothers. Such differences occur despite a comparable level of NGFI-A expression in adult hippocampus. Thus, the methylation status of the 5′ CpG site associates with an alteration in the "affinity" of the NGFI-A consensus sequence for its ligand, resulting in a decreased level of NGFI-A binding. Finally, the sequencing involved in these studies has yet to reveal any evidence for sequence variation in this region. Thus, to our knowledge, the individual differences in GR expression in this model associate with variation at the level of epigenetic state, and not in nucleotide sequence.

Likewise, ChIP assays indicate increased H3K9ac and a threefold greater binding of NGFI-A to the exon 1_7 GR promoter in hippocampal samples obtained from the adult offspring of high compared with low LG mothers.[106] This finding is important as H3K9ac associates with stably active transcription.[148,149] The electrostatic bonds formed between the positively-charged histone proteins and their negatively-charged DNA partners demands an active chromatin remodeling process for transcriptional activation.[150,151] Chromatin remodeling is achieved through biochemical modifications to the histone proteins that control chromatin structure and thus genome function. The post-translational modifications to the histones occur through a series of enzymes that bind to the histone tails and modify the local chemical properties of specific amino acids.[151–155] For example, histone acetylation neutralizes the positive charge on the histone tail, opening chromatin and increasing the access of transcription factors to their DNA binding sites. Acetylation commonly occurs at lysine residues, such as the H3K9, and is catalyzed by histone acetyltransferases and reversed by HDACs. HDACs remove acetyl groups from histone tails and prevent subsequent acetylation.[136,152] Cytosine methylation attracts repressor complexes comprised of HDACs such that DNA methylation and histone acetylation are usually inversely related. Thus, we find that maternal effects on the methylation status of the exon 1_7 promoter correspond to differences in H3K9ac and NGFI-A binding. H3K9ac associates with increased transcription and we found increased H3K9ac of the exon 1_7 GR promoter[106] genes in hippocampus from the adult offspring of high compared with low LG mothers. This pattern is similar to maternal effects on hippocampal *GAD1* or *Grm1* expression; in each case decreased DNA methylation within promoter regions associates with increases in both H3K9ac and gene transcription.[135,156] As noted above, H3K9ac tends to associate with stably transcribed regions of the genome, which is consistent with the idea of a persistent increase in hippocampal GR transcription in the adult offspring of high LG mothers.

Histone acetylation directly modifies chromatin structure through effects on the local physico-chemical environment that define the chromatin state.[150,151] Additional histone modifications, notably histone methylation, influence transcription through indirect pathways that involve a complex array of transcriptional mediators.[151–155,157–159] Multiple lysine and arginine residues on the histone tails are subject to methylation, which is catalyzed by distinct histone methyltransferases and reversed by histone *de*methylases. This process provides a signaling pathway that begins with the activation of the intracellular signals that activate the individual methylating or demethylating enzymes producing a specific epigenetic profile on the histone tails, thus linking specific intracellular signals to specific histone methylation marks. The methylation profile of the histone tails is highly variable. Methylation can occur at multiple sites along the histone tails and vary in the level of methylation (mono-, di-, or trimethylation). The resulting profile acts as a platform for various protein complexes that remodel chromatin and alter transcriptional activity; thus, the indirect influence of histone methylation on transcription.[151,154,155]

Certain histone modifications co-vary. An example of relevance here is that of H3K9ac and H3K4me. Both marks are generally present at actively transcribed regions of the genome.[148,160,161] Thus, we find increased H3K9ac and H3K4me3 at both regions of the exon 1_7 GR promoter and the levels of these individual marks are very highly correlated.[143] This same pattern is also apparent at the promoters for *GAD1* and *Grm1*, both of which show maternally-regulated differences in DNA methylation.[135,156]

H3K4me, whether in the mono-, di- or trimethylated state, appears to protect CpG islands against methylation.[162,163] Thus, genome-wide analyses reveal a negative correlation between H3K4me and CpG methylation; the relation to 5-hydroxymethylation is unknown. Interestingly, H3K4me3 appears actively to "repel" the binding of the DNA methyltransferase, DNMT3L, which is essential for de novo methylation and attracts complexes containing histone acetyltransferases that open chromatin and enhance transcription factor binding.[162] Indeed, the absence of H3K4me3 seems to be a prerequisite for the recruitment of de novo DNA methyltransferases and the acquisition of DNA methylation.[162–164] The same relation was apparent across the exon 1_7 GR, GAD1 and *GRM1* promoters, where the decreased level of DNA methylation was associated with an increased level of H3K4m3. H3K4me3 targets the chromatin remodeling factor (NURF) and the Yng1 protein in the NuA3 (nucleosomal acetyltransferase of histone H3) complex to genes increasing the level of histone acetylation and transcriptional activation. This process explains the tight correlation between the levels of H3K4me3 and H3K9ac, which are >0.90 across each of the promoters examined.

HDAC Inhibitors Reverse the Effect of Maternal Care

These findings suggest that variations in maternal care influence the epigenetic state of the exon 1_7 GR promoter in hippocampus, regulating NGFI-A binding, GR transcription and HPA stress responses. The effect of DNA methylation on gene expression is, in part, mediated by the recruitment of HDAC-containing repressor complexes.[150,165–170] HDAC inhibitors permit chromatin remodeling and transcription factor binding, and may thus liberate the expression of genes from methylation-induced repression. HDAC inhibition also reverses the maternal effects on hippocampal GR expression.[106] Chronic, central infusion of adult offspring of low LG mothers with the broad spectrum HDAC inhibitor, trichostatin A (TSA),[112] significantly increased H3K9 acetylation, NGFI-A binding to the GR-1_7 promoter, and GR expression to levels comparable to those observed in the offspring of high LG mothers. TSA infusion also eliminated the effect of maternal care on HPA responses to acute stress. These results suggest a direct relation between maternal care, histone acetylation of the GR-1_7 promoter, GR expression and HPA responses to stress.

Specificity is an issue with the use of HDAC inhibitors; might TSA influence simply liberate transcriptional repression across a wide range of the genome? Interestingly, this does not appear to be the case. Expression profiling revealed that <2% of the hippocampal transcriptome is affected by chronic TSA infusion.[110] The reason why only a small percentage of the transcriptome is affected by TSA is unknown, but such agents likely target only that fraction of the euchromatin that bears the epigenetic potential for transcriptional activation.[158,171]

An obvious concern is whether the effects of maternal care on hippocampal GR expression represent a more global process in which variations in parental signals affect the methylation status of broad regions of the genome. Other studies reveal that stress-induced variations in maternal care in rats, including the frequency of pup LG, alter the methylation state of the *bdnf* gene in the hippocampus.[172] In the mouse, prolonged periods of maternal separation alter the methylation state of the promoter for the arginine vasopressin gene (AVP), increasing hypothalamic AVP synthesis and HPA responses to stress.[173] Maternal separation in the rat associates with reduced $GABA_A$ receptor levels in the locus coeruleus (LC) and the nucleus tractus solitarius (NTS) as well as levels of the mRNA for the γ_2 subunit of the $GABA_A$ receptor complex, which confers high affinity benzodiazepine binding, in the amygdala as well as in the LC and NTS.[174] Both the amygdala and the ascending noradrenergic systems have been considered as critical sites for the anxiolytic effects of GABAergic inhibition. These findings suggest that parental influences might influence the epigenetic regulation of multiple regions of the genome in different brain regions to produce a coordinated effect on the stress response of the offspring.

The NGFI-A-regulated *GAD1* gene shows a similar increase in the level of promoter methylation in the hippocampus from the adult offspring of low LG mothers.[135] Moreover, a ChIP-chip study using high-density oligonucleotide microarrays tiling a contiguous 7 million base pair region of rat chromosome 18 containing the *NR3C1* gene at 100 bp spacing revealed coordinated alterations in H3K9ac, DNA methylation and gene expression across a number of areas in response to variations in maternal care, including a sub-region containing multiple protocadherin genes.[175] These results suggest a broad epigenomic response to variations in maternal care that associates with an extensive difference in gene expression.

DEVELOPMENTAL REGULATION OF HIPPOCAMPAL GR EXPRESSION IN HUMANS

A critical question is whether familial influences operating during early development in humans are linked to the stable epigenetic regulation of gene expression as in rodents. There are obvious constraints on tissue access for molecular studies of neural function in humans. These limitations are of considerable importance in the study of epigenetic mechanisms, which are potentially tissue- and even cell-type specific (see Chapter 7). We were able to establish a translational program focusing on human hippocampus by virtue of the resources of the Quebec Suicide Brain Bank (www.douglas.qc.ca/suicide). Approximately one-third of individuals who die by suicide have histories of childhood maltreatment, including childhood sexual and physical abuse, as well as parental neglect. We thus showed decreased hippocampal GR expression in samples from suicide completers with histories of childhood maltreatment compared with controls (sudden, involuntary fatalities).[176,177] The program is strengthened by a validated forensic interview that establishes developmental history and mental health status.[178–180] Regression analyses across the samples showed no significant correlations between psychopathology, notably depression and substance disorders, and hippocampal GR expression. Rather the decreased hippocampal GR expression associated with a history of childhood maltreatment. There were no differences in hippocampal GR expression in samples from suicides negative for a history of childhood maltreatment. Instead, the differences in hippocampal GR expression were unique to suicide completers with a history of childhood maltreatment.

Splice variant analysis, comparable to that performed in the rat, revealed decreased expression of non-coding exons 1_B, 1_C, 1_F and 1_H in suicides with a history of childhood maltreatment compared with both controls and suicides without a history of maltreatment. These expression differences correlated with differential DNA methylation patterns between groups in the corresponding exon 1 variant promoters. The exon 1_F sequence is of particular interest as it is the homolog of the rat exon 1_7, is highly expressed in brain and contains an NGFI-A response element[130,176] (see Figure 4.3). Moreover, the exon 1_F sequence shows increased DNA methylation and decreased NGFI-A binding in samples from suicide victims with a history of maltreatment. These findings bear considerable similarity to the maternal effect in the rat and are suggestive of early-environment regulation of the neural epigenome in humans. Of interest, recent studies in independent human samples investigating the effects of early-environmental adversity on exon 1_F methylation reported consistent results.[181,182]

Decreased expression levels of GR exon 1_B, 1_C, and 1_H transcripts were also associated with alterations in methylation of the respective sequences, with particular sites significantly correlated with expression levels. As expected on the basis of the expression data, the exon 1_B and 1_C regions showed increased methylation at predictive sites uniquely in samples from suicide/maltreatment subjects. However, analysis of the exon 1_H GR promoter yielded an interesting profile that contrasted starkly with that observed for the other exon 1 regions.[177] There was significantly increased DNA methylation of the exon 1_H promoter in hippocampal samples from both controls and suicide victims without a history of maltreatment by comparison to those positive for maltreatment. And the methylation of the exon 1_H promoter was *positively* correlated with hippocampal GR expression.

Most differentially methylated sites were found within putative transcription factor binding sites. Multiple transcription factors are predicted to bind promoters of GR non-coding exons[183] although, to date, only NGFI-A has been shown to activate transcription in the promoter of $GR1_F$.[97,124] Nevertheless, most of the CpG sites where the methylation states were investigated in $GR1_B$, $GR1_C$ and $GR1_H$ promoters are predicted to bind transcription factors such as Sp1 and Sp3. Sp1 and Sp3 regulate GR basal expression.[184] Interestingly, as discussed above, there is evidence that Sp1 binding can alter the underlying methylation state of the DNA.[133,134] Other known factors predicted to bind within the investigated promoter regions include NF-1, YY1 and members of the AP-1 family composed of Jun, Fos and ATF. Interestingly, when interacting with Sp1/Sp3, these transcription factors can activate or repress transcription by recruiting co-factors inducing the opening or the closing of chromatin.[185–193] Consequently, hypermethylation in $GR1_B$ and 1_C promoters represses the binding of these transcription factors reducing expression but, at the same time, the hypomethylated state in $GR1_H$ permits Sp1/Sp3 binding and the recruitment of HDACs closes the chromatin state. Such models are currently a matter of speculation, but underscore the importance of studies of the molecular mechanisms that link methylation at specific genomic regions with alterations in transcriptional activity. These findings also point to the potential for bi-directional relation between transcription factor binding and transcriptional activity and that of DNA methylation.[158,171]

Another important consideration is that of interpreting data from studies of DNA methylation in brain. DNA methylation is a digital signal; an allele is either methylated or unmethylated at a specific site in a given cell. The percentage of methylation represents the fraction of cells in which the allele is methylated. An increase in methylation levels indicates an increase in the number of cells that bear a methylated allele. As expected for functional promoters, many of which lie within CpG islands, methylation levels are commonly low. GR promoters are generally hypomethylated in the majority of neurons in the hippocampus,[176,183,194,195] suggesting that these regions are poised for transcriptional activation in the majority of neurons. Our results suggest that site-specific methylation in selected GR promoters, such as exons 1_B, 1_C, 1_F and 1_H varies in a fraction of cells in the hippocampus as a function of childhood maltreatment. We suggest that this difference results in the down-regulation of hippocampal GR expression.

Forebrain glucocorticoid receptor (GR) activation inhibits HPA activity through tonic negative-feedback inhibition of CRF expression.[196] Familial dysfunction in childhood associates with increased CRF activity[197] and enhanced HPA and autonomic stress reactivity.[70,71,85,198–200] Importantly, interventions that target parental care of high-risk children alter HPA activity.[201] In humans, decreased GR expression, altered corticosteroid feedback sensitivity and increased HPA activity are linked to major depressive disorder.[196,202] And there is evidence for decreased hippocampal GR expression in depression.[203] Polymorphisms in the *NR3C1* gene that encodes the GR result in corticosteroid resistance and enhance the risk for major depressive disorder.[204,205] While not all depressed patients show evidence for increased hypercortisolemia, psychotic and treatment-resistant forms of depression are commonly associated with increased HPA activity.[206,207] Interestingly, childhood maltreatment is associated with more severe, treatment-resistant forms of depression. Successful treatment of such populations with antidepressants may require a normalization of HPA activity.[207] The GR antagonist, mifepristone (RU486), which blocks the effects of elevated cortisol, has been successfully used as an adjunct in the treatment of psychotic depression.[208]

A recent genome-wide analysis in human populations suggests that the quality of the parental environment associates with the epigenetic state of the genome in the offspring.[198] DNA from epithelial cells obtained from buccal samples was subjected to methylation analysis using the Infinium assay that allows for the simultaneous measurement of the methylation status within ≈27 500 CpG sites over 13 500 genes, focusing on promoters and first exons of the coding region. Samples from adolescents showed systematic variation in methylation that correlated with the level of maternal, but not paternal stress in infancy with a considerably greater effect in daughters than sons. These findings map nicely onto known developmental influences at other levels of phenotype. Another recent study[209] showed that childhood socio-economic status was correlated with the DNA methylation state of individuals in their fifties.

Such studies bear very significant caveats,[210] including the weakness of DNA samples from blood or epithelial cells, rather than directly from brain. However, to some extent, the point of such studies transcends concerns of tissue specificity. The very fact that the social context during early life, or indeed that of the parent, might associate with the epigenome of the offspring is important. There are critical obstacles for translational research, but the findings are certainly consistent with the environmental epigenetics hypothesis.

REVERSIBILITY OF DNA METHYLATION

The issue of reversibility is critical for translational studies of the epigenomic consequences of early adversity. To our knowledge, the issue has yet to be directly addressed in humans, even in samples of non-neural origin. Nevertheless, there is considerable evidence that suggests a capacity for the remodeling of epigenetic marks over the lifespan, including DNA methylation. Across multiple tissues, including brain, the methylation levels at specific regions change with age,[211] reflecting the potential for dynamic variation. In moving forward, it will be important to appreciate that the capacity for environmentally-induced dynamic variation in methylation states, like any phenotype, will likely vary as a function of sequence-based variation across the genome.

The results of the TSA study described above suggest that DNA methylation patterns are dynamic and potentially reversible even in adult animals.[111] These findings are consistent with previous in vitro studies showing that increased histone acetylation associated with HDAC inhibitors can trigger demethylation.[136] Conversely, intra-hippocampal infusion of the methyl donor amino acid methionine[107] leads to a hypermethylation of the exon 1_7 GR promoter in the adult offspring of high LG animals. Thus, chronic central infusion of adult offspring of high or low LG mothers with methionine increases DNA methylation at the NGFI-A binding site and reduces NGFI-A binding to the exon 1_7 promoter sequence selectively in the offspring of high LG mothers. These effects eliminate group differences in both hippocampal GR expression and HPA responses to stress. Methionine increases the levels of S-adenosyl methionine (SAM) and DNA methylation.[212] SAM could increase DNA methylation through the activation of DNA methylation enzymes.[213] Likewise, studies of transcriptional regulation of *reelin* and *GAD1* reveal evidence for dynamic regulation of methylation states in mature cortical neurons through the disruption of repressor complexes and the inhibition of DNMT expression.[214–217] While the precise mechanisms for each of these effects is as yet unclear, these studies imply that mature brain cells express the enzymes necessary for both methylation and demethylation.

These findings are consistent with an emerging characterization of the potential for dynamic modifications in DNA methylation, first described in the brain in relation to activity-dependent changes in BDNF expression.[218] Perhaps the most compelling evidence for dynamic, experience-induced alterations in DNA methylation emerges from studies of contextual fear conditioning, a hippocampal-dependent learning paradigm whereby an animal associates a novel context with an aversive stimulus, is accompanied by broad increases H3K9ac[219–222] dependent upon activation of the ERK/MAPK signaling pathway and the CREB binding protein. Dynamic changes in DNA methylation at specific genomic sites appear crucial for learning and memory.[223] Adult neurons show high levels of expression for the de novo methylation enzymes, DNMT3a and 3b. Moreover, there is considerable regional specificity in DNMT expression in the adult rat brain, suggesting a specialized function in adulthood.[224] Increases in DNMT3a and 3b expression accompany contextual fear conditioning, and drugs that block DNMT activity impair conditioning.[221] DNMT-deficient mice show impaired contextual fear conditioning.[225] More recent studies identify specific genomic targets. Fear conditioning results in an increased methylation and transcriptional silencing of the gene for protein phosphatase 1 (PP1), which impairs learning.[222] The same training results in the demethylation of a proximal promoter and transcriptional activation of the synaptic plasticity gene *reelin*.

More recent studies used next-generation sequencing for a genome-wide analysis of CpG methylation of adult mouse dentate granule neurons in vivo before and after synchronous neuronal activation (electroconvulsive stimulation).[141] About 1.4% of the CpGs examined showed rapid active demethylation or de novo methylation, with some modifications remaining stable for at least 24h. These activity-modified CpGs showed a broad genomic distribution with significant enrichment in low-CpG density regions, and were associated with brain-specific genes related to neuronal plasticity. The low CpG density regions are of interest since the tightest correlations between DNA methylation and transcription are observed in such regions.[226]

Taken together these findings suggest considerable capacity for active remodeling of DNA methylation. These findings are consistent with the prominent expression of DNMTs in neurons over adulthood and the degree to which DNMT expression as well as that of candidate demethylating agents is dynamically regulated by activity-dependent, extracellular signals.[227,228] Since treatments that target histone acetylation, such as HDAC inhibitors, can influence DNA methylation,[136] it might be possible to affect changes in DNA methylation through more accessible targets such as the histone post-translational modifications that directly regulate chromatin structure. However, the pathways that lead to the remodeling of DNA methylation, especially those implicated in DNA *de*methylation, have yet to be fully identified. A related question concerns the variability across the genome in the capacity for epigenetic remodeling. We have yet to identify the factors that determine the sensitivity of genomic regions to active remodeling. One interesting possibility is that such variation, either across genomic regions or within the same genomic regions and across individuals, may be related to underlying sequence variation.[229] In moving forward, it will be important to appreciate that the capacity for environmentally-induced dynamic variation in methylation states, like any phenotype, will likely vary as a function of sequence-based variation across the genome. One recent exemplar shows that methylation of a *BDNF* exon is associated with the well-known rs6265 (val66met) single nucleotide polymorphism in the *BDNF* gene.[210] Interestingly, the same polymorphism interacts with early life adversity to influence hippocampal volume and the risk for depression.[230] Studies linking genomic sequence variants to differential sensitivity to intervention[231] beg the question of whether such individual differences suggest a variation in the capacity for epigenetic remodeling.

NON-GENOMIC TRANSMISSION OF TRAITS FROM PARENT TO OFFSPRING

Affective illnesses show familial transmission. Such findings inevitably reflect an influence of sequence-based genomic variation. However, parental depression profoundly affects parenting, increasing the frequency of forms of parent–child interactions that promote phenotypic outcomes associated with vulnerability for depression and anxiety disorders.[1,15,232–237] Since childhood maltreatment increases the risk for depression, these findings suggest a transgenerational pathway extending from parental childhood maltreatment, to emotional well-being and parental behavior and, thus, child development. We (Bouvette-Turcotte et al, submitted) found direct evidence for this pathway within a longitudinal birth cohort study. Maternal childhood maltreatment predicted both her own mood states, as well as negative emotionality/behavioral regulation in her 3-year child. Interestingly, the link between maternal childhood maltreatment and child emotional function was apparent even after controlling for maternal mood. The results suggest that the effect is mediated by variations in parenting, and this interpretation is consistent with the finding that, across multiple species, individual differences in parental behaviors are transmitted from parent to offspring.[80,233,238–241] Such findings include the rat.

Transgenerational Transmission of Individual Differences in Maternal Behavior

Individual differences in the maternal care of lactating rats are transmitted from mother to daughter.[80,242] The lactating female offspring of high LG mothers exhibit significantly higher levels of pup LG than do those of low LG mothers. Cross-fostering completely reverses the pattern of transmission revealing evidence for a direct effect of maternal care.[80] These findings suggest that maternal–infant interactions may program specific reproductive behavior in the female offspring.[60,243] Fleming and colleagues provided direct experimental support for this conclusion. As lactating mothers, female rats artificially-reared in isolation with no maternal care following the first day of life show significantly reduced responsiveness to pups and reduced pup LG.[244,245] The effects of artificial rearing on maternal behavior are greatly reduced by providing the female pups with stroking, which mimics the tactile stimulation associated with maternal LG, as well as with social contact with peers. These findings suggest a direct relation between the quality of maternal care received in early life and that expressed as an adult.[246] Moreover, the inheritance of individual differences in maternal care in the rat is, in part at least, non-genomic and associated with tactile stimulation derived from pup LG.

The transgenerational effects on maternal behavior involve a series of neuroendocrine systems closely associated with maternal behavior in the rat. The elaborate pattern of maternal care that is evident shortly after parturition in the rat is orchestrated through a complex pattern of endocrine events during gestation. Throughout most of pregnancy, progesterone levels are high and accompanied by moderate levels of estrogen. Progesterone levels fall prior to parturition and there occurs a surge in estrogen levels.[247] Both events are obligatory for the onset of maternal behavior and of particular importance are the effects of estrogen at the level of the medial preoptic area (MPOA). This brain region is critical for the expression of maternal behaviors in the rat[248] including pup LG.[246]

Ovarian hormones initiate the onset of maternal behavior in the rat through effects on central oxytocinergic systems.[249] Estrogen increases oxytocin synthesis in the parvocelluar neurons of the PVNh that project to the MPOA as well as other brain regions that regulate maternal behavior.[250] Estrogen also increases oxytocin receptor gene expression and receptor binding in the MPOA[95,242,251,252] and this effect appears to be mediated through the estrogen receptor alpha (ERα).[253] Intracereboventricular administration of oxytocin rapidly stimulates maternal behavior in virgin rats[252,254] and the MPOA appears to be a critical site. This effect is estrogen dependent. The effect of oxytocin is abolished by ovariectomy and reinstated with estrogen treatment. Moreover, treatment with oxytocin-antisera or receptor antagonists blocks the effects of ovarian steroid treatments on maternal behavior.[252,255] Among lactating females, there are significantly higher levels of oxytocin receptors in the MPOA, the bed n. of the stria terminalis and the lateral septum of all animals,[249] however, the lactation-induced increase in receptors levels is substantially greater in the high LG mothers.[102,242] Not surprisingly, the oxytocin receptor binding levels are highly correlated with the frequency of pup licking/grooming.[99,242] Importantly, central infusion of an oxytocin receptor antagonist on day 3 of lactation completely eliminates the differences in maternal licking/grooming between high and low LG mothers.[242]

Oxytocin neurons in the MPOA project directly to the ventral tegmental area (VTA)[250] the origin of the mesocorticolimbic dopamine system. Oxytocin facilitates dopamine release through effects on oxytocin receptors in VTA neurons[250] and enhances dopamine-mediated behaviors.[256] Not surprisingly, the magnitude of the dopamine signal in the nucleus accumbens accompanying pup LG is significantly greater in high LG mothers.[96] Moreover, central infusion of a dopamine transporter blocker in lactating high and low LG females completely eliminated the differences in both dopamine levels in the nucleus accumbens and pup licking/grooming.[96] These findings are consistent with earlier reports that lesions of the nucleus accumbens significantly reduce pup LG.[246]

Differences in estrogen sensitivity mediate the differential effects of lactation on the induction of oxytocin receptors in high and low LG females (Figure 4.4). Among ovariectomized females provided estrogen replacement, there is a significantly greater estrogen effect on oxytocin receptor levels in the MPOA in high compared with low LG animals[242] suggesting stable differences in estrogen sensitivity. Indeed, among either lactating high LG

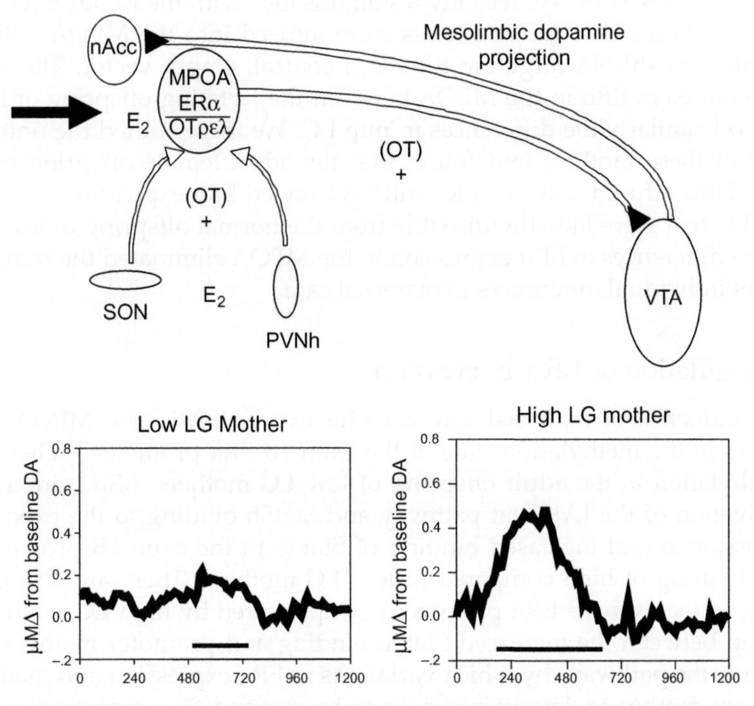

FIGURE 4.4 (Top) Working model for the neuroendocrine basis of individual differences in pup licking/grooming (LG) from studies of high and low LG mothers. Oxytocin projections from the paraventricular n. of the hypothalamus (PVNh) and the supraoptic n. (SON) activate estrogen-inducible oxytocin receptors in the medial preoptic area (MPOA), producing a feed-forward effect on oxytocin synthesis in the MPOA. Oxytocin projections from the MPOA to the ventral tegmental area (VTA) enhance activity in dopamine neurons increasing dopamine release at the level of the n. accumbens (nAcc). High LG females show increased estrogen receptor alpha expression in the MPOA, greater estrogen receptor sensitivity and greater oxytocin receptor binding in the MPOA. These effects are critical for the increased dopamine release in the nAcc[250] and thus for increased pup licking/grooming[96,242] and the differences in pup LG. (Bottom) Representative electrochemical recordings of the change (Δ) from baseline in the dopamine signal in the nAcc during pup LG (bar) in high and low LG mothers (from[96]).

mothers or in the virgin female offspring of high LG dams the expression of ERα, but not ERβ, is significantly increased in the medial preoptic area.[99,257] The effect is apparent at the level of mRNA and protein. The difference in ERα expression in the medial preoptic region is largely confined to the caudal regions of the medial and central subdivisions of the medial preoptic nucleus. In these regions, ERα mRNA expression is about twofold greater in the virgin offspring of high compared with low LG mothers. During gestation, the increase in circulating estrogen appears to enhance oxytocin receptor binding to a greater extent in the high LG mothers, leading to increased oxytocin activation of the ascending mesolimbic dopamine system and increased dopamine release in the nucleus accumbens.

ERα Mediates the Transgenerational Transmission of Maternal Care

According to this model, the critical feature for the transmission of the individual differences in maternal behavior from the mother to her female offspring are the differences in ERα expression in the MPOA. We recently tested this idea with the F1 adult female offspring of high and low LG mothers (F0). Animals were infused into the MPOA with a lentiviral vector containing an shRNA targeting ERα or a control, empty vector. The shRNA eliminated the differences in ERα in the MPOA between the lactating offspring of high and low LG mothers, and similarly the differences in pup LG. We then studied the untreated female offspring (F2) of these mothers and found that the adult female offspring of F1 mothers from high LG dams infused with the ERα shRNA showed ERα expression in the MPOA and levels of pup LG that were indistinguishable from the normal offspring of low LG mothers. Eliminating the differences in ERα expression in the MPOA eliminated the transgenerational transmission of individual differences in maternal care.

Epigenetic Regulation of ERα Expression

The stable influence of maternal care on ERα expression in the MPOA is associated with differences in the methylation state of the exon 1B ERα promoter.[99] There is increased promoter methylation in the adult offspring of low LG mothers. ERα transcription is regulated by activation of the JAK-Stat pathway and Stat5b binding to the exon 1B ERα promoter. ChIP assays reveal increased binding of Stat5b to the exon 1B promoter in MPOA of the adult offspring of high compared to low LG mothers. There are also differences in Stat5b binding to the exon 1B ERα promoter in pups reared by high LG mothers, however, the relationship between the increased Stat5b binding and promoter methylation is as yet unknown. While the pathways by which variations in ERα expression and maternal care are transmitted from mother to daughter remain to be clarified, it is tempting to consider the role of epigenetics mechanisms (see[258]).

CONCLUSIONS

The results of the studies suggest that epigenetic mechanisms mediate the association between early childhood and gene expression, and thus explain, in part at least, individual differences in phenotypes that associate with vulnerability/resistance for specific forms of psychopathology. Studies in the rodent suggest direct effects of parental care on epigenetic regulation of individual differences in stress reactivity in the rat,[112,135,173] and findings with humans are consistent with this environmental epigenetic mechanism.[70,176] We focused on the regulation of hippocampal GR expression as a model and provide evidence for parental effects on hippocampal GR expression that associate with differences in the methylation of exon 1 promoters. There is now evidence for comparable effects at multiple regions of the genome.[135,172,173,259]

The value of the energetically-costly brain is to guide the function of the organism in accordance with its life history. The ability to mastermind such adaptation to circumstance relies upon the capacity of neurons and glia dynamically to adapt genomic structure and function.[171] The implicit hypothesis is that environmental signals alter chromatin modifications that then serve as the mechanism for the transcriptional "plasticity" that mediates sustained variation in neural function. Ironically, the dynamic nature and environmental sensitivity of DNA methylation in fully differentiated cells is somewhat at odds with the very stability that suggests DNA methylation as a mechanism for parental effects on gene expression. How do we square the dynamic nature of DNA methylation with the phenotypic "programming" associated with parental effects? We actually know rather little about the variation in methylation marks at specific loci over time within the same individual. Perhaps a similar caveat applies to parental effects, studies of which often take the form of characterizing a parental signal at one stage of development, and then examining epigenetic states and phenotype at a later phase of life. The process of cell specialization that defines neural development depends upon the silencing on non-neural genes. This process can be activated in vitro in stem cells and the resulting repression initially involves Histone 3 lysine 27 trimethylation, a polycomb-mediated, repressive histone modification.[260] Repression then comes to reflect increased DNA methylation as neural differentiation proceeds, which is then thought to stabilize gene silencing. However, multiple regions of the genome in neural tissues are enriched for bi-directional histone modifications (i.e. those associated with transcriptional activation, such as Histone 3 lysine 4 trimethylation, as well as repression, notably H3K27me3), a characteristic of pluripotent cells.[157] Such "bivalency" might define the potential for plasticity.[171] Repressive epigenetic contexts occur as a function of different epigenetic repertoires, which may vary in reversibility and confer variable environmental sensitivity.[261]

Our ability to address these issues will rely on studies of mechanism. The challenge is that of defining causal pathways between environmental event, epigenetic mark, genome function and functional outcome. As noted at the outset of this chapter, we need to know much more about the processes by which environmental events signal epigenetic remodeling. Advances in structural biology and biochemistry[151] have advanced our understanding of the meditators that lie between epigenetic signal and transcriptional activity. The challenge is to integrate such advances into neurobiology and thus move beyond the correlational analyses of epigenetic signal, transcriptional activity and function.

References

1. Bifulco A, Brown GW, Adler Z. Early sexual abuse and clinical depression in adult life. *Br J Psychiat*. 1991;159:115–122.
2. Brown GR, Anderson B. Psychiatric morbidity in adult inpatients with childhood histories of sexual and physical abuse. *Am J Psychiat*. 1991;148:55–61.

3. McCauley J, Kern DE, Kolodner K, et al. Clinical characteristics of women with a history of childhood abuse: unhealed wounds. *J Am Med Assoc*. 1997;277:1362–1368.
4. Felitti VJ, Anda RF, Nordenberg D, et al. Relationship of childhood abuse and household dysfunction to many of the leading causes of death in adults. *Am J Prevent Med*. 1998;14:245–258.
5. Fletcher JM. Childhood mistreatment and adolescent and young adult depression. *Soc Sci Med*. 2009;68:799–806.
6. Lenneke RAA, Chichetti D, Kim J, Rogosh FA. Mediating and moderating processes in the relation between maltreatment and psychopathology: mother-child relationship quality and emotion regulation. *J Abnorm Child Psychol*. 2009;37:831–843.
7. Shonkoff JP, Boyce WT, McEwen BS. Neuroscience, molecular biology, and the childhood roots of health disparities: building a new framework for health promotion and disease prevention. *J Am Med Assoc*. 2009;301:2252–2259.
8. Widom CS, DuMont K, Czaja SJ. A prospective investigation of major depressive disorder and comorbidity in abused and neglected children grown up. *Arch Gen Psychiatry*. 2007;64:49–56.
9. Ammerman RT, Cassisi JE, Hersen M, van Hasselt VB. Consequences of physical abuse and neglect in children. *Clin Psychol Rev*. 1986;6:291–310.
10. Trickett PK, McBride-Chang C. The developmental impact of different forms of child abuse and neglect. *Devel Rev*. 1995;15:311–337.
11. Repetti RL, Taylor SE, Seeman TE. Risky families: family social environments and the mental and physical health of offspring. *Psychol Bull*. 2002;128(2):330–366.
12. Lupien SJ, McEwen BS, Gunnar MR, Heim C. Effects of stress throughout the lifespan on the brain, behaviour and cognition. *Nature Rev Neurosci*. 2009;10:434–445.
13. Holmes SJ, Robins LN. The influence of childhood disciplinary experience on the development of alcoholism and depression. *J Child Psychol Psychiatry*. 1987;28:399–415.
14. Gottman JM. Psychology and the study of marital processes. *Ann Rev Psychol*. 1998;49:169–197.
15. Hill J, Pickles A, Burnside E, et al. Child sexual abuse, poor parental care and adult depression: evidence for difference mechanisms. *Br J Psychiatry*. 2001;179:104–109.
16. Canetti L, Bachar E, Galili-Weisstub E, De-Nour AK, Shalev AY. Parental bonding and mental health in adolescence. *Adolescence*. 1997;32:381–394.
17. Parker G. Parental representations of patients with anxiety neurosis. *Acta Psychiatrica Scand*. 1981; 63:33–36.
18. Parker G. The parental bonding index: psychometric properties. *Psychiatr Devel*. 1989;7:317–335.
19. Kendler KS, Gardner CO, Prescott CA. Towards a comprehensive developmental model for major depression in women. *Am J Psychiatry*. 2002;159:1133–1145.
20. Hill J, Davis R, Byatt M, Burnside E, Rollinson L, Fear S. Childhood sexual abuse and affective symptoms in women: a general population study. *Psychol Med*. 2000;30:1283–1291.
21. Russak LG, Schwartz GE. Feelings of parental care predict health status in midlife: a 35 year follow-up of the Harvard mastery of stress study. *J Behav Med*. 1997;20:1.
22. Rutter M. Protective factors in children's responses to stress and disadvantage. *Prim Prev Psychopathol*. 1979;3:49–74.
23. Smith J, Prior M. Temperament and stress resilience in school-age children: a within families study. *J Am Acad Child Adolesc Psychiatry*. 1995;34:168–179.
24. McLoyd VC. Socioeconomic disadvantage and child development. *Am Psychol*. 1998;53:185–204.
25. Conger RD, Donnellan MB. An interactionist perspective on the socioeconomic context of human development. *Annu Rev Psychol*. 2007;58:157–199.
26. Linver MR, Brooks-Gunn J, Kohen DE. Family processes as pathways from income to young children's development. *Dev Psychol*. 2002;38:719–734.
27. Hackman D, Farah MJ, Meaney MJ. Socioeconomic status and the brain: mechanistic insights from human and animal research. *Nature Rev Neurosci*. 2010;11:651–659.
28. Belsky J. Theory testing, effect-size evaluation, and differential susceptibility to rearing influence: the case of mothering and attachment. *Child Devel*. 1997;64:598–600.
29. Olds D, Henderson CR, Cole R, et al. Long-term effects of nurse home visitation on children's criminal and antisocial behavior: 15-year follow-up of a randomized controlled trial. *J Am Med Assoc*. 1998;280: 1238–1244.

30. Klein Velderman M, Bakermans-Kranenburg MJ, Juffer F, Van IJzendoorn MH. Effects of attachment-based interventions on maternal sensitivity and infant attachment: differential susceptibility of highly reactive infants. *J Fam Psychol.* 2006;20:266–274.
31. Van Zeijl J, Mesman J, Van IJzendoorn MH. Attachment based intervention for enhancing sensitive discipline in mothers of 1- to 3-year-old children at risk for externalizing behavior problems: a randomized controlled trial. *J Consult Clin Psychol.* 2006;74:994–1005.
32. Fleming AS. The neurobiology of mother–infant interactions: experience and central nervous system plasticity across development and generations. *Neursci Biobehav Rev.* 1999;23:673–685.
33. Belsky J. The determinants of parenting: a process model. *Child Devel.* 1984;55:83–96.
34. Elder GH, van Nguyen TV, Caspi A. Linking family hardship to children's lives. *Child Dev.* 1985;56:361–375.
35. Eisenberg L. Psychiatry and health in low-income populations. *Comp Psychiatry.* 1990;38:69–73.
36. McLoyd VC. The impact of economic hardship on black families and children: psychological distress, parenting, and socio-emotional development. *Child Dev.* 1990;61:311–346.
37. Dix T. The affective organization of parenting adaptive and maladaptive processes. *Psych Bull.* 1991;110:3–25.
38. Goldstein LH, Diener ML, Mangelsdorf SC. Maternal characteristics and social support across the transition to motherhood: associations with maternal behavior. *J Fam Psychol.* 1996;10:60–71.
39. Vaughn B, Egeland B, Sroufe LA, Waters E. Individual differences in infant–mother attachment at twelve and eighteen months: stability and change in families under stress. *Child Dev.* 1979;50:971–975.
40. Grolnick WS, Gurland ST, DeCource W, Jacob K. Antecedents and consequences of mothers' autonomy support: an experimental investigation. *Dev Psychol.* 2002;38:143–155.
41. Belsky J. Attachment and close relationships: an individual differences perspective. *Psychol Inquiry.* 1994:5.
42. Whipple EE, Webster-Stratton C. The role of parental stress in physically abusive families. *Child Abuse Neglect.* 1991;15:275–291.
43. Kessler RC, Davis CG, Kendler KS. Childhood adversity and adult psychiatric disorder in the US national comorbidity study. *Psychol Med.* 1997;27:1101–1119.
44. Sadowki H, Ugarte B, Kolvin L, Kaplan C, Barnes J. Early life family disadvantages and major depression in adulthood. *Br J Psychiatry.* 1999;174:112–120.
45. Ritsher JEB, Warner V, Johnson JG, Dohrenwend BP. Intergenerational longitudinal study of social class and depression: a test of social causation and social selection models. *Br J Psychiatry.* 2001;178:584–590.
46. Gilman SE, Kawachi I, Fitzmaurice GM, Buka SL. Family disruption in childhood and risk of adult depression. *Am J Psychiatry.* 2003;160:939–946.
47. Bosma H, Schrijvers C, Mackenbach JP. Socioeconomic inequalities in mortality and importance of perceived control: cohort study. *Br Med J.* 1999;319(7223):1469–1470.
48. Fleming AS. Factors influencing maternal responsiveness in humans: usefulness of an animal model. *Psychoneuroendocrinology.* 1988;13:189–212.
49. Field T. Maternal depression effects on infants and early interventions. *Prev Med.* 1998;27:200–203.
50. Rosenblum LA, Andrews MW. Influences of environmental demand on maternal behavior and infant development. *Acta Paediatr.* 1994(suppl 397):57–63.
51. Coplan JD, Andrews MW, Rosenblum LA, Nemeroff CB. Persistent elevations of cerebrospinal fluid concentrations of corticotropin-releasing factor in adult nonhuman primates exposed to early-life stressors: implications for the pathophysiology of mood and anxiety disorders. *Proc Natl Acad Sci USA.* 1996;93:1619–1623.
52. Coplan JD, Trost RC, Owens MJ, et al. Cerebrospinal fluid concentrations of somatostatin and biogenic amines in grown primates reared by mothers exposed to manipulated foraging conditions. *Arch Gen Psychiatry.* 1998;55:473–477.
53. Hinde RA. Some implications of evolutionary theory and comparative data for the study of human prosocial and aggressive behaviour. In: Olweus D, Block J, Radke-Yarrow M, eds. *Development of Anti-Social and Prosocial Behaviour.* Orlando: Academic Press; 1986. p. 13–32.
54. Meaney MJ. Maternal care, gene expression, and the transmission of individual differences in stress reactivity across generations. *Annu Rev Neurosci.* 2001;24:1161–1192.
55. Purves D. *A Trophic Theory of Neural Development.* Cambridge MA: Harvard University Press; 1988.
56. Gottlieb G. Experiential canalization of behavioral development: theory. *Dev Psychol.* 1991;27:4–13.
57. Mousseau TA, Fox CW. The adaptive significance of maternal effects. *Trends Evol Ecol.* 1998;13:403–407.
58. Rossiter MC. In: Mousseau TA, Fox CW, eds. *Maternal Effects as Adaptations.* London: Oxford University Press; 1999.

59. Qvarnstrom A, Price TD. Maternal effects, paternal effects and sexual selection. *Trends Ecol Evol.* 2001;16:95–100.
60. Cameron NM, Champagne FA, Parent C, Fish EW, Ozaki-Kuroda K, Meaney MJ. The programming of individual differences in defensive responses and reproductive strategies in the rat through variations in maternal care. *Neurosci Biobehav Rev.* 2005;29:843–865.
61. O'Steen S, Cullum AJ, Bennett AF. Rapid evolution of escape ability in Trinidadian guppies (Poecilia reticulata). *Evolution.* 2002;56:776–784.
62. Pollak SD, Kistler DJ. Early experience is associated with the development of categorical representations for facial expressions of emotion. *Proc Natl Acad Sci USA.* 2002;99:9072–9076.
63. Bagot RC, van Hasselt FN, Champagne DL, Meaney MJ, Krugers HJ, Joëls M. Maternal care determines rapid effects of stress mediators on synaptic plasticity in adult rat hippocampal dentate gyrus. *Neurobiol Learn Mem.* 2009;92:292–300.
64. Dallman MF, Strack AM, Akana SF, et al. Feast and famine: critical role of glucocorticoids with insulin in daily energy flow. *Front Neuroendocrinol.* 1993;14:303–347.
65. Shanks N, Larocque S, Meaney MJ. Neonatal endotoxin exposure alters the development of the hypothalamic-pituitary-adrenal axis: early illness and later responsivity to stress. *J Neurosci.* 1995;15:376–384.
66. Spencer SJ, Galic MA, Pittman QJ. Neonatal programming of innate immune function. *Am J Phsyiol.* 2011;300:E11–E18.
67. Shanks N, Windle RJ, Perks PA, et al. Early-life exposure to endotoxin alters hypothalamic–pituitary–adrenal function and predisposition to inflammation. *Proc Natl Acad Sci USA.* 2000;97:5645–5650.
68. Laban O, Dimitrijevic M, von Hoersten S, Markovic BM, Jankovic BD. Experimental allergic encephalomyelitis in adult DA rats subjected to neonatal handling or gentling. *Brain Res.* 1995;676:133–140.
69. O'Donnell D, Larocque S, Seckl JR, Meaney MJ. Postnatal handling alters glucocorticoid, but not mineralocorticoid messenger RNA expression in the hippocampus of adult rats. *Brain Res Mol Brain Res.* 1994;26: 242–248.
70. Heim C, Newport DJ, Heit S, et al. Pituitary-adrenal and autonomic responses to stress in women after sexual and physical abuse in childhood. *J Am Med Assoc.* 2000;284:592–597.
71. Pruessner JL, Champagne FA, Meaney MJ, Dagher A. Parental care and neuroendocrine and dopamine responses to stress in humans: a PET imaging study. *J Neurosci.* 2004;24:2825–2831.
72. Farrington DA, Gallagher B, Morley L, St Ledger RJ, West DJ. Are there any successful men from criminogenic backgrounds? *Psychiatry.* 1988;51:116–130.
73. Haapasap J, Tremblay RE. Physically aggressive boys from ages 6 to 12: family background. Parenting behavior, and prediction of delinquency. *J Consult Clin Psychol.* 1994;62:1044–1052.
74. Dingemanse NJ, Bloth C, Drent PJ, Tinbergen JM. Fitness consequences of avian personalities in a fluctuating environment. *Proc Roy Soc Lond B.* 2004;271:847–852.
75. Schwabl H. Yolk is a source of maternal testosterone for developing birds. *Proc Natl Acad Sci USA.* 1993;90:11446–11450.
76. Groothuis TGG, Muller W, von Engelhardt N, Carere C, Eising CM. Maternal hormones as a tool to adjust offspring phenotype in avian species. *Neurosci Biobehav Rev.* 2005;29:329–352.
77. Seckl JR, Meaney MJ. Early life events and later development of ischaemic heart disease. *Lancet.* 1994; 342:1236.
78. Nemeroff CB. The corticotropin-releasing factor (CRF) hypothesis of depression: new findings and new directions. *Molec Psychiatry.* 1996;1:336–342.
79. Sroufe LA. Psychopathology as an outcome of development. *Dev Psychopathol.* 1997;9:251–268.
80. Francis D, Diorio J, Liu D, Meaney MJ. Nongenomic transmission across generations of maternal behavior and stress responses in the rat. *Science.* 1999;286:1155–1158.
81. Francis DD, Caldji C, Champagne F, Plotsky PM, Meaney MJ. The role of corticotropin-releasing factor–norepinephrine systems in mediating the effects of early experience on the development of behavioral and endocrine responses to stress. *Biol Psychiatry.* 1999;46:1153–1166.
82. Fish EW, Shahrokh D, Bagot R, et al. Epigenetic programming of stress responses through variations in maternal care. *Ann N Y Acad Sci.* 2004;1036:167–180.
83. Klaassens ER, van Noorden MS, Giltay MS, van Pel J, van Veen T, Zitman FG. Effects of childhood trauma on HPA-axis reactivity in women free of lifetime psychopathology. *Prog Neuro-Psychopharmacol Biol Psychiatry.* 2009;33:889–894.

84. Cicchetti D, Rogosch FA, Gunnar MR, Toth SL. The differential impacts of early abuse on internalizing problems and diurnal cortisol activity in school-aged children. *Child Devel*. 2010;25:252–269.
85. DeBellis MD, Chrousos GP, Dorn LD, et al. Hypothalamic-pituitary-adrenal dysregulation in sexually abused girls. *J Clin Endocrinol Metab*. 1994;78:249–255.
86. Flinn MV, England BG. Childhood stress and family environment. *Curr Anthropol*. 1995;5:1569–1579.
87. Leucken LJ. Childhood attachment and loss experiences affect adult cardiovascular and cortisol function. *Psychosom Med*. 1998;60:765–770.
88. Luecken LJ, Lemery KS. Early caregiving and physiological stress responses. *Clin Psychol Rev*. 2004;24:171–191.
89. Taylor SE, Lerner JS, Sage RM, Lehman BJ, Seeman TE. Early environment, emotions, responses to stress, and health. Special issue on personality and health. *J Personal*. 2004;72:1365–1393.
90. Taylor SE, Eisenberger NI, Saxbe D, Lehman BJ, Lieberman MD. Neural responses to emotional stimuli are associated with childhood family stress. *Biol Psychiatry*. 2006;60:296–301.
91. Reid WJ, Crisafulli A. Marital discord and child behavior problems: a meta-analysis. *J Abnorm Child Psychol*. 1990;18:105–117.
92. Wichers M, Myin-Germeys I, Jacobs N, et al. Genetic risk of depression and stress-induced negative affect in daily life. *Br J Psychiatry*. 2007;191:218–223.
93. Wichers M, Geschwind N, Jacobs N, et al. Transition from stress sensitivity to a depressive state: longitudinal twin study. *Br J Psychiatry*. 2009;195:498–503.
94. Myers MM, Brunelli SA, Squire JM, Shindeldecker RD, Hoffer MA. Maternal behavior of SHR rats and its relationship to offspring blood pressures. *Dev Psychobiol*. 1989;22:29–53.
95. Champagne FA, Francis DD, Mar A, Meaney MJ. Variations in maternal care in the rat as a mediating influence for the effects of environment on development. *Physiol Behav*. 2003;79:359–371.
96. Champagne FA, Stevenson C, Gratton A, Meaney MJ. Individual differences in maternal behavior are mediated by dopamine release in the nucleus accumbens. *J Neurosci*. 2004;24:4113–4123.
97. Hellstrom IC, Dhir S, Diorio JC, Zhang TY, Meaney MJ. Maternal licking regulates hippocampal glucocorticoid receptor transcription through a thyroid hormone, serotonin, NGFI-A signalling cascade. *Philosoph Transact R Soc B Biol Sci*. 2012;367:2495–2510.
98. Ivy AS, Brunson KL, Sandman C, Baram TZ. Dysfunctional nurturing behavior in rat dams with limited access to nesting material: a clinically relevant model for early life stress. *Neuroscience*. 2008;152:1132–1142.
99. Champagne F, Weaver I, Diorio J, Dymov S, Szyf M, Meaney MJ. Maternal care associated with methylation of the estrogen receptor-alpha1b promoter and estrogen receptor-alpha expression in the medial preoptic area of female offspring. *Endocrinology*. 2006;147:2909–2915.
100. Smith JW, Seckl JR, Evans AT, Costall B, Smythe JW. Gestational stress induces post-partum depression-like behavior and alters maternal care in rats. *Psychoneuroendocrinology*. 2004;29:227–244.
101. Roth TL, Sweatt JD. Annual research review: epigenetic mechanisms and environmental shaping of the brain during senstive periods. *J Child Psychol Psychiatry*. 2011;52:398–408.
102. Francis DD, Champagne F, Meaney MJ. Variations in maternal behaviour are associated with differences in oxytocin receptor levels in the rat. *J Neuroendocrinol*. 2000;12:1145–1148.
103. Liu D, Diorio J, Tannenbaum B, et al. Maternal care, hippocampal glucocorticoid receptors, and hypothalamic-pituitary-adrenal responses to stress. *Science*. 1997;277:1659–1662.
104. Caldji C, Tannenbaum B, Sharma S, Francis D, Plotsky PM, Meaney MJ. Maternal care during infancy regulates the development of neural systems mediating the expression of behavioral fearfulness in adulthood in the rat. *Proc Natl Acad Sci USA*. 1998;95:5335–5340.
105. Menard J, Champagne D, Meaney MJ. Variations of maternal care differentially influence 'fear' reactivity and regional patterns of cFos immunoreactivity in response to the shock-probe burying test. *Neuroscience*. 2004;129:297–308.
106. Weaver ICG, Cervoni N, D'Alessio AC, et al. Epigenetic programming through maternal behavior. *Nat Neurosci*. 2004;7:847–854.
107. Weaver ICG, Champagne FA, Brown SE, et al. Reversal of maternal programming of stress responses in adult offspring through methyl supplementation: altering epigenetic marking later in life. *J Neurosci*. 2005;25:11045–11054.
108. Zhang TY, Bagot R, Parent C, et al. Maternal programming of defensive responses through sustained effects on gene expression. *Biol Psychiatry*. 2006;73:72–89.

109. Toki S, Morinobu S, Imanaka A, Yamamoto S, Yamawaki S, Honma K. Importance of early lighting conditions in maternal care by dam as well as anxiety and memory later in life of offspring. *Eur J Neurosci.* 2007;25:815–829.
110. Weaver IC, Meaney MJ, Szyf M. Maternal care effects on the hippocampal transcriptome and anxiety-mediated behaviors in the offspring that are reversible in adulthood. *Proc Natl Acad Sci USA.* 2006;103:3480–3485.
111. Kurata A, Morinobu S, Fuchikami M, Yamamoto S, Yamawaki S. Maternal postpartum learned helplessness (LH) affects maternal care by dams and responses to LH test in adolescent offspring. *Horm Behav.* 2009;56:112–120.
112. Weaver ICG, Diorio J, Seckl JR, Szyf M, Meaney MJ. Environmental regulation of hippocampal glucocorticoid receptor gene expression: characterization of intracellular mediators and potential genomic targets. *Ann NY Acad Sci.* 2004;1024:182–212.
113. Meaney MJ, Aitken DH. The effects of early postnatal handling on the development of hippocampal glucocorticoid receptors: temporal parameters. *Dev Brain Res.* 1985;22:301–304.
114. Barden N. Implication of the hypothalamic-pituitary-adrenal axis in the physiopathology of depression. *J Psychiatry Neurosci.* 2004;29:185–193.
115. Boyle MP, Brewer JA, Funatsu M, et al. Acquired deficit of forebrain glucocorticoid receptor produces depression-like changes in adrenal axis regulation and behavior. *Proc Natl Acad Sci USA.* 2005;102:473–478.
116. Ridder S, Chourbaji S, Hellweg R, et al. Mice with genetically altered glucocorticoid receptor expression show altered sensitivity for stress-induced depressive reactions. *J Neurosci.* 2005;25:6243–6250.
117. Reichardt HM, Tronche F, Bauer A, Schutz G. Molecular genetic analysis of glucocorticoid signaling using the Cre/loxP system. *Biol Chem.* 2000;381:961–964.
118. Caldji C, Diorio J, Meaney MJ. Variations in maternal care alter GABAA receptor subunit expression in brain regions associated with fear. *Neuropsychopharmacology.* 2003;28:150–159.
119. van Hasselt FN, Cornelisse S, Zhang TY, et al. Adult hippocampal glucocorticoid receptor expression and dentate synaptic plasticity correlate with maternal care received by individuals early in life. *Hippocampus.* 2012;22:255–266.
120. Meaney MJ, Diorio J, Francis D, et al. Postnatal handling increases the expression of cAMP-inducible transcription factors in the rat hippocampus: the effects of thyroid hormones and serotonin. *J Neurosci.* 2000;20:3926–3935.
121. Laplante P, Diorio J, Meaney MJ. Serotonin regulates hippocampal glucocorticoid receptor expression via a 5-HT7 receptor. *Brain Res Dev Brain Res.* 2002;139:199–203.
122. Mitchell JB, Rowe W, Boksa P, Meaney MJ. Serotonin regulates type II, corticosteroid receptor binding in cultured hippocampal cells. *J Neurosci.* 1990;10:1745–1752.
123. Meaney MJ, Diorio J, Francis D, et al. Environmental regulation of the development of glucocorticoid receptor systems in the rat forebrain: the role of serotonin. *Ann N Y Acad Sci.* 1994;746:260–275.
124. Weaver ICG, DiAlessio AC, Brown SE, et al. The transcription factor NGFI-A mediates epigenetic programming: altering epigenetic marking through immediate early genes. *J Neurosci.* 2007;27:1756–1768.
125. Meaney MJ, Aitken DH, Sapolsky RM. Thyroid hormones influence the development of hippocampal glucocorticoid receptors in the rat: a mechanism for the effects of postnatal handling on the development of the adrenocortical stress response. *Neuroendocrinology.* 1987;45:278–283.
126. Gross C, Zhuang X, Stark K, et al. Serotonin1A receptor acts during development to establish normal anxiety-like behaviour in the adult. *Nature.* 2002;416:396–400.
127. Gross C, Hen R. The developmental origins of anxiety. *Nat Rev Neurosci.* 2004;5:545–552.
128. Yan W, Wilson CC, Haring JH. Effects of neonatal serotonin depletion on the development of rat dentate gyrus granule cells. *Devel Brain Res.* 1997;98:177–184.
129. McCormick JA, Lyons V, Jacobson MD, et al. 5′-heterogeneity of glucocorticoid receptor messenger RNA is tissue specific: differential regulation of variant transcripts by early-life events. *Mol Endocrinol.* 2000;14:506–517.
130. Turner JD, Muller CP. Structure of the glucocorticoid receptor (NR3C1) gene 5′ untranslated region: identification, and tissue distribution of multiple new human exon 1. *J Mol Endocrinol.* 2005;35:283–292.
131. Crosby SD, Puetz JJ, Simburger KS, Fahrner TJ, Milbrandt J. The early response gene NGFI-C encodes a zinc finger transcriptional activator and is a member of the GCGGGGGCG (GSG) element-binding protein family. *Mol Cell Biol.* 1991;11:3835–3841.

132. Ogryzko VV, Schiltz RL, Russanova V, Howard BH, Nakatani Y. The transcriptional coactivators p300 and CBP are histone acetyltransferases. *Cell*. 1996;87:953–959.
133. Macleod D, Charlton J, Mullins J, Bird AP. Spl sites in the mouse aprt gene promoter are required to prevent methylation of the CpG island. *Genes Devel*. 1994;8:2282–2292.
134. Brandeis M, Frank D, Keshet I, et al. Sp1 elements protect a CpG island from de novo methylation. *Nature*. 1994;371:435–438.
135. Zhang T-Y, Hellstrom IC, Bagot RC, Wen X, Diorio J, Meaney MJ. Maternal care and DNA methylation of the a glutamic acid decarboxylase 1 promoter in rat hippocampus. *J Neurosci*. 2010;30:13130–13137.
136. Szyf M. Epigenetics DNA methylation, and chromatin modifying drugs. *Annu Rev Pharmacol Toxicol*. 2009;49:243–263.
137. Baylin SB, Herman JG. DNA hypermethylation in tumorigenesis: epigenetics joins genetics. *Trends Genet*. 2000;16:168–174.
138. Kriaucionis S, Heintz N. The nuclear DNA base 5-hydroxymethylcytosine is present in Purkinje neurons and the brain. *Science*. 2009;324:929–930.
139. Tahiliani M, Koh KP, Shen Y, et al. Conversion of 5-methylcytosine to 5-hydroxymethylcytosine in mammalian DNA by MLL partner TET1. *Science*. 2009;324:930–935.
140. Ito S, D'Alessio AC, Taranova OV, Hong K, Sowers LC, Zhang Y. Role of Tet proteins in 5mC to 5hmC conversion, ES-cell self-renewal and inner cell mass specification. *Nature*. 2010;466:1129–1133.
141. Guo U, Ma DK, Mo H, et al. Neuronal activity modifies the DNA methylation landscape in the adult brain. *Nat Neurosci*. 2011;14:1345–1351.
142. Williams K, Christensen J, Pedersen MT, et al. TET1 and hydroxymethylcytosine in transcription and DNA methylation fidelity. *Nature*. 2011;473:343–3438.
143. Zhang TY, Labonté B, Wen XL, Turecki G, Meaney MJ. Epigenetic mechanisms for the early environmental regulation of hippocampal glucocorticoid receptor gene expression in rodents and humans. *Neuropsychopharmacology*. doi:10.1038/npp.2012.149 [Epub ahead of print].
144. Song CX, Szulwach KE, Fu Y, et al. Selective chemical labeling reveals the genomewide distribution of 5-hydroxymethylcytosine. *Nat Biotechnol*. 2011;29:68–72.
145. Valinluck V, Tsai HH, Rogstad DK, Burdzy A, Bird A, Sowers LC. Oxidative damage to methyl-CpG sequences inhibits the binding of the methyl-CpG binding domain (MBD) of methyl-CpG binding protein 2 (MeCP2). *Nucleic Acids Res*. 2004;32:4100–4108.
146. Wu H, D'Alessio AC, Ito S, et al. Genome-wide analysis of 5-hydroxymethylcytosine distribution reveals its dual function in transcriptional regulation in mouse embryonic stem cells. *Genes Dev*. 2011;25:679–684.
147. Wu SC, Zhang Y. Active DNA demethylation: many roads lead to Rome. *Nat Rev Mol Cell Biol*. 2010;11:607–620.
148. Millar CB, Grunstein M. Genome-wide patterns of histone modifications in yeast. *Nat Rev Mol Cell Biol*. 2006;7:657–666.
149. Wade PA, Pruss D, Wolffe AP. Histone acetylation: chromatin in action. *Trends Biochem Sci*. 1997;22:128–132.
150. Turner BM. *Chromatin Structure and the Regulation of Gene Expression*. Oxford: Blackwell Science Ltd; 2001.
151. Taverna SD, Li H, Ruthenburg AJ, Allis CD, Patel DJ. How chromatin-binding modules interpret histone modifications: lessons from professional pocket pickers. *Nat Struct Mol Biol*. 2007;14:1025–1040.
152. Shahbazian MD, Grunstein M. Functions of site-specific histone acetylation and deacetylation. *Annu Rev Biochem*. 2007;76:75–100.
153. Grunstein M. Histone acetylation in chromatin structure and transcription. *Nature*. 1997;389:349–352.
154. Hake SB, Allis CD. Histone H3 variants and their potential role in indexing mammalian genomes: the "H3 barcode hypothesis". *Proc Natl Acad Sci USA*. 2006;103:6428–6435.
155. Jenuwein T, Allis CD. Translating the histone code. *Science*. 2001;293:1074–1080.
156. Bagot RC, Zhang TY, Wen XL, et al. Variations in postnatal maternal care and the epigenetic regulation of Grm1 expression and hippocampal function in the rat. *Proc Natl Acad Sci USA*. doi: 10.1073/pnas.1204599109 [Epub ahead of print].
157. Bernstein BE, Kamal M, Lindblad-Toh K, et al. Genomic maps and comparative analysis of histone modifications in human and mouse. *Cell*. 2005;120:169–181.
158. Berger SL. The complex language of chromatin regulation during transcription. *Nature*. 2007;447:407–412.
159. Kouzarides T. Chromatin modifications and their function. *Cell*. 2007;128:693–705.

160. Ruthenburg AJ, Allis CD, Wysocka J. Methylation of lysine 4 on histone H3: intricacy of writing and reading a single epigenetic mark. *Mol Cell*. 2007;25:15–30.
161. Pokholok DK, Harbison CT, Levine S, et al. Genome-wide map of nucleosome acetylation and methylation in yeast. *Cell*. 2005;122:517–527.
162. Ooi SK, Qiu C, Bernstein E, et al. DNMT3L connects unmethylated lysine 4 of histone H3 to de novo methylation of DNA. *Nature*. 2007;448:714–717.
163. Thompson JP, Skene PJ, Selfridge J, et al. CpG islands influence chromatin structure via the CpG-binding protein Cfp1. *Nature*. 2010;464:1082–1086.
164. Ciccone DN, Su H, Hevi S, et al. KDM1B is a histone H3K4 demethylase required to establish maternal genomic imprints. *Nature*. 2009;461:415–418.
165. Bird AP, Wolffe AP. Methylation-induced repression--belts, braces, and chromatin. *Cell*. 1999;99:451–454.
166. Bird A. Molecular biology. Methylation talk between histones and DNA. *Science*. 2001;294:2113–2115.
167. Klose RJ, Bird AP. Genomic DNA methylation: the mark and its mediators. *Trends Biochem Sci*. 2006;31:89–97.
168. Li E. Chromatin modification and epigenetic reprogramming in mammalian development. *Nat Rev Genet*. 2002;3:662–673.
169. Miranda TB, Jones PA. DNA methylation: the nuts and bolts of repression. *J Cell Physiol*. 2007;213: 384–390.
170. Nan X, Ng HH, Johnson CA, et al. Transcriptional repression by the methyl-CpG-binding protein MeCP2 involves a histone deacetylase complex. *Nature*. 1998;393:386–389.
171. Meaney MJ, Ferguson-Smith A. Epigenomic regulation of the neural transcriptome: the meaning of the marks. *Nat Neurosci*. 2010;13:1313–1318.
172. Roth TL, Lubin F, Funk A, Sweatt JD. Lasting epigenetic influence of early-life adversity on the BDNF gene. *Biol Psychiatry*. 2009;65:760–769.
173. Murgatroyd C, Patchev AV, Wu Y, et al. Dynamic DNA methylation programs persistent adverse effects of early-life stress. *Nat Neurosci*. 2009;12:1559–1566.
174. Caldji C, Francis DD, Sharma S, Plotsky PM, Meaney MJ. The effects of early rearing environment on the development of GABAA and central benzodiazepine receptor levels and novelty-induced fearfulness in the rat. *Neuropsychopharmacology*. 2000;22:219–229.
175. McGowan PO, Suderman M, Sasaki A, et al. Broad epigenetic signature of maternal care in the brain of adult rats. *PLoS One*. 2011;6:e14739.
176. McGowan PO, Sasaki A, D'Alessio AC, et al. Epigenetic regulation of the glucocorticoid receptor in human brain associates with childhood abuse. *Nat Neurosci*. 2009;12:342–348.
177. Labonte B, Yerko V, Moffatt K, Szyf M, Meaney MJ, Turecki G. Differential glucocorticoid receptor expression in the brain of suicide completers with a history of childhood abuse. *Arch Gen Psychiatry*. 2012;72:41–48.
178. Dumais A, Lesage AD, Alda M, et al. Risk factors for suicide completion in major depression: a case-control study of impulsive and aggressive behaviors in men. *Am J Psychiatry*. 2005;162:2116–2124.
179. McGirr A, Renaud J, Seguin M, Alda M, Turecki G. Course of major depressive disorder and suicide outcome: a psychological autopsy study. *J Clin Psychiatry*. 2008;69:966–970.
180. Zouk H, Tousignant M, Seguin M, Lesage A, Turecki G. Characterization of impulsivity in suicide completers: clinical, behavioral and psychosocial dimensions. *J Affect Dis*. 2006;92:195–204.
181. Radtke KM, Ruf M, Gunter HM, et al. Transgenerational impact of intimate partner violence on methylation in the promoter of the glucocorticoid receptor. *Transl Psychiat*. 2011;1:e21.
182. Tyrka AR, Prince LH, Marsit C, Walters OC, Carpenter LL. Childhood adversity and epigenetic modulation of the leukocyte glucocorticoid receptor: preliminary findings in healthy adults. *PloS One*. 2012;7:e30148.
183. Turner JD, Alt SR, Cao L, et al. Transcriptional control of the glucocorticoid receptor: CpG islands, epigenetics and more. *Biochem Pharmacol*. 2010;80:1860–1868.
184. Nobukuni Y, Smith CL, Hager GL, Detera-Wadleigh SD. Characterization of the human glucocorticoid receptor promoter. *Biochemistry*. 1995;34:8207–8214.
185. Adams AD, Choate DM, Thompson MA. NF1-L is the DNA-binding component of the protein complex at the peripherin negative regulatory element. *J Biol Chem*. 1995;270:6975–6983.
186. Brodin G, Ahgren A, ten Dijke P, Heldin CH, Heuchel R. Efficient TGF-beta induction of the Smad7 gene requires cooperation between AP-1, Sp1, and Smad proteins on the mouse Smad7 promoter. *J Biol Chem*. 2000;275:29023–29030.

187. Hurst HC, Jones NC. Identification of factors that interact with the E1A-inducible adenovirus E3 promoter. *Genes Dev*. 1987;1:1132–1146.
188. Inoue T, Tamura T, Furuichi T, Mikoshiba K. Isolation of complementary DNAs encoding a cerebellum-enriched nuclear factor I family that activates transcription from the mouse myelin basic protein promoter. *J Biol Chem*. 1990;265:19065–19070.
189. Kardassis D, Papakosta P, Pardali K, Moustakas A. C-Jun transactivates the promoter of the human p21(WAF1/Cip1) gene by acting as a superactivator of the ubiquitous transcription factor Sp1. *J Biol Chem*. 1999;274:29572–29581.
190. Laniel MA, Poirier GG, Guerin SL. Nuclear factor 1 interferes with Sp1 binding through a composite element on the rat poly(ADP-ribose) polymerase promoter to modulate its activity in vitro. *J Biol Chem*. 2001;276:20766–20773.
191. Rafty LA, Santiago FS, Khachigian LM. NF1/X represses PDGF A-chain transcription by interacting with Sp1 and antagonizing Sp1 occupancy of the promoter. *EMBO J*. 2002;21:334–343.
192. Roy RJ, Guerin SL. The 30-kDa rat liver transcription factor nuclear factor 1 binds the rat growth-hormone proximal silencer. *Eur J Biochem*. 1994;219:799–806.
193. Tapias A, Ciudad CJ, Noe V. Transcriptional regulation of the 5′-flanking region of the human transcription factor Sp3 gene by NF-1, c-Myb, B-Myb, AP-1 and E2F. *Biochim Biophys Acta*. 2008;1779:318–329.
194. Oberlander T, Weinberg J, Papsdorf M, Grunau R, Misri S, Devlin A. Prenatal exposure to maternal depression, neonatal methylation of human glucocorticoid receptor gene (NR3C1) and infant cortisol stress responses. *Epigenetics*. 2008;3:97–106.
195. Alt SR, Turner JD, Klok MD, et al. Differential expression of glucocorticoid receptor transcripts in major depressive disorder is not epigenetically programmed. *Psychoneuroendocrinology*. 2010;35:544–556.
196. de Kloet ER, Joels M, Holsboer F. Stress and the brain: from adaptation to disease. *Nat Rev Neurosci*. 2005;6:463–475.
197. Lee R, Geracioti Jr TD, Kasckow JW, Coccaro EF. Childhood trauma and personality disorder: positive correlation with adult CSF corticotropin-releasing factor concentrations. *Am J Psychiatry*. 2005;162:995–997.
198. Essex MJ, Klein MH, Cho E, Kalin NH. Maternal stress beginning in infancy may sensitize children to later stress exposure: effects on cortisol and behavior. *Biol Psychiatry*. 2002;52:776–784.
199. Teicher MH, Andersen SL, Polcari A, Anderson CM, Navalta CP. Developmental neurobiology of childhood stress and trauma. *Psychiatr Clin N Am*. 2002;25:397–426. [vii–viii].
200. Leucken LJ. Parental caring and loss during childhood and adult cortisol responses to stress. *Psychol Health*. 2000;15:841–851.
201. Fisher PA, Gunnar MR, Chamberlain P, Reid JB. Preventive intervention for maltreated preschool children: impact on children's behavior, neuroendocrine activity, and foster parent functioning. *J Am Acad Child Adolesc Psychiatry*. 2000;39:1356–1364.
202. Neigh GN, Nemeroff CB. Reduced glucocorticoid receptors: consequence or cause of depression? *Trends Endocrinol Metab*. 2006;17:124–125.
203. Webster MJ, Knable MB, O'Grady J, Orthmann J, Weickert CS. Regional specificity of brain glucocorticoid receptor mRNA alterations in subjects with schizophrenia and mood disorders. *Mol Psychiatry*. 2002;7:985–994. [924].
204. van Rossen EFC, Binder EB, Majer M, et al. Polymorphisms of the glucocorticoid receptor gene and major depression. *Biol Psychiatr*. 2006;59:681–688.
205. van West D, Van Den Eede F, JDel-Favero J, et al. Glucocorticoid receptor gene-based SNP analysis in patients with recurrent major depression. *Neuropsychopharmacol*. 2006;31:620–627.
206. Schatzberg AF, Rothschild AJ, Langlais PJ, Bird ED, Cole JO. A corticosteroid/dopamine hypothesis for psychotic depression and related states. *J Psychiatr Res*. 1985;19:57–64.
207. Holsboer F. The corticosteroid receptor hypothesis of depression. *Europsychopharmacology*. 2000;23:477–501.
208. DeBattista C. Augmentation and combination strategies for depression. *J Psychopharmacol*. 2006;20 (3 suppl):11–18.
209. Borghol N, Suderman M, McArdle W, et al. Associations with early-life socio-economic position in adult DNA methylation. *Int J Epidemiol*. 2012;41:62–74.
210. Mill J, Tang T, Kaminsky Z, et al. Epigenomic profiling reveals DNA-methylation changes associated with major psychosis. *Am J Hum Genet*. 2008;82:696–711.

211. Hernandez DG, Nalls MA, Gibbs JR, et al. Distinct DNA methylation changes highly correlated with chronological age in the human brain. *Hum Mol Genet*. 2011;20:1164–1172.
212. Tremolizzo L, Doueiri MS, Dong E, et al. Valproate corrects the schizophrenia-like epigenetic behavioral modifications induced by methionine in mice. *Biol Psychiatry*. 2005;57:500–509.
213. Pascale RM, Simile MM, Satta G, et al. Comparative effects of L-methionine, S-adenosyl-L-methionine and 5′-methylthioadenosine on the growth of preneoplastic lesions and DNA methylation in rat liver during the early stages of hepatocarcinogenesis. *Anticancer Res*. 1991;11:1617–1624.
214. Grayson DR, Jia X, Chen Y, et al. Reelin promoter hypermethylation in schizophrenia. *Proc Natl Acad Sci USA*. 2005;102:9341–9346.
215. Kundakovic M, Chen Y, Costa E, Grayson DR. DNA methyltransferase inhibitors coordinately induce expression of the human reelin and glutamic acid decarboxylase 67 genes. *Mol Pharmacol*. 2007;71:644–653.
216. Noh JS, Sharma RP, Veldic M, et al. DNA methyltransferase 1 regulates reelin mRNA expression in mouse primary cortical cultures. *Proc Natl Acad Sci USA*. 2005;102:1749–1754.
217. Veldic M, Caruncho HJ, Liu S, et al. DNA-methyltransferase 1 mRNA is selectively overexpressed in telencephalic GABAergic interneurons of schizophrenia brains. *Proc Natl Acad Sci USA*. 2004;101:348–353.
218. Martinowich K, Hattori D, Wu H, et al. DNA methylation-related chromatin remodeling in activity-dependent BDNF gene regulation. *Science*. 2003;302:890–893.
219. Vecsey CG, Hawk JD, Lattal KM, et al. Histone deacetylase inhibitors enhance memory and synaptic plasticity via CREB: CBP-dependent transcriptional activation. *J Neurosci*. 2007;27:6128–6140.
220. Lubin FD, Roth TL, Sweatt JD. Epigenetic regulation of bdnf gene transcription in the consolidation of fear memory. *J Neurosci*. 2008;28:10576–10586.
221. Miller CA, Sweatt JD. Covalent modification of DNA regulates memory formation. *Neuron*. 2007;53:857–869.
222. Sweatt JD. Experience-dependent epigenetic modifications in the central nervous system. *Biol Psychiatry*. 2009;65:191–197.
223. Day JJ, Sweatt JD. DNA methylation and memory formation. *Nat Neurosci*. 2010;13:1319–1323.
224. Brown SE, Weaver IC, Meaney MJ, Szyf M. Regional-specific global cytosine methylation and DNA methyltransferase expression in the adult rat hippocampus. *Neurosci Lett*. 2008;440:49–53.
225. Feng J, Zhou Y, Campbell SL, et al. Dnmt1 and Dnmt3a maintain DNA methylation and regulate synaptic function in adult forebrain neurons. *Nature Neurosci*. 2010;13:423–430.
226. Weber M, Hellmann I, Stadler MB, et al. Distribution, silencing potential and evolutionary impact of promoter DNA methylation in the human genome. *Nat Genetics*. 2007;39:457–466.
227. Grayson DR, Chen Y, Dong E, Kundakovic M, Guidotti A. From trans-methylation to cytosine methylation: evolution of the methylation hypothesis of schizophrenia. *Epigenetics*. 2009;4:144–149.
228. Ma DK, Jang MH, Guo JU, et al. Neuron activity-induced Gadd45b promotes epigenetic DNA demethylation and adult neurogenesis. *Science*. 2009;323:1074–1077.
229. Zhang D, Cheng L, Badner JA, et al. Genetic control of individual differences in gene-specific methylation in human brain. *Am J Hum Genet*. 2010;86:411–419.
230. Gatt J, Nemeroff C, Dobson-Stone C, et al. Interactions between BDNF Val66Met polymorphism and early life stress predict brain and arousal pathways to syndromal depression and anxiety. *Mol Psychiatry*. 2009;14:681–695.
231. Bakermans-Kranenburg MJ, van IJzendoorn MH, Pijlman FT, Mesman J, Juffer F. Experimental evidence for differential susceptibility: dopamine D4 receptor polymorphism (DRD4 VNTR) moderates intervention effects on toddlers' externalizing behavior in a randomized controlled trial. *Dev Psychol*. 2008;44:293–300.
232. Lovejoy MC, Graczyk PA, O'Hare E, Neuman G. Maternal depression and parenting behaviour: a meta-analytic review. *Clin Psychol Rev*. 2000;20:561–592.
233. Barrett J, Fleming AS. Annual research review: all mothers are not created equal: neural and psychobiological perspectives on mothering and the importance of individual differences. *J Child Psychol Psychiatry Allied Disc*. 2011;52:368–397.
234. Kendler KS, Kuhn J, Prescott CA. The interrelationship of neuroticism, sex, and stressful life events in the prediction of episodes of major depression. *Am J Psychiat*. 2004;161:631–636.
235. Field T. Prenatal depression effects on early development: a review. *Infant Behav Dev*. 2011;34:1–14.
236. Murray L, Fiori-Cowley A, Hooper R, Cooper P. The impact of postnatal depression and associated adversity on early mother-infant interactions and later infant outcome. *Child Dev*. 1996;67:2512–2526.

237. Goodman SH, Rouse MH, Connell AM, Broth MR, Hall CM, Heyward D. Maternal depression and child psychopathology: a meta-analytic review. *Clin Child Fam Psychol Rev*. 2011;14:1–27.
238. Fairbanks L. Individual differences in maternal style: causes and consequences for mothers and offspring. *Adv Study Behav*. 1996;25:579–611.
239. Maestripieri D. Emotions, stress, and maternal motivation in primates. *Am J Primatol*. 2011;73:516–529.
240. Maestripieri D, Mateo J. [ed]. In: Maestripieri D, Mateo JM, eds. *Maternal Effects in Mammals*. Chicago: The University of Chicago Press; 2009.
241. Miller L, Kramer R, Warner V, Wickramaratne P, Weissman M. Intergenerational transmission of parental bonding among women. *J Am Acad Child Adolesc Psychiatry*. 1997;36:1134–1139.
242. Champagne F, Diorio J, Sharma S, Meaney MJ. Naturally occurring variations in maternal behavior in the rat are associated with differences in estrogen-inducible central oxytocin receptors. *Proc Natl Acad Sci USA*. 2001;98:12736–12741.
243. Cameron N, Del Corpo A, Diorio J, McAllister K, Sharma S, Meaney MJ. Maternal programming of sexual behavior and hypothalamic-pituitary-gonadal function in the female rat. *PlosOne*. 2008;21:e2210.
244. Lovic V, Fleming AS. Artificially reared female rats show reduced prepulse inhibition and deficits in the attentional set shifting task – reversal of effects with maternal-like licking stimulation. *Behav Brain Res*. 2004;148:209–219.
245. Melo AI, Lovic V, Gonzalez A, Madden M, Sinopoli K, Fleming AS. Maternal and littermate deprivation disrupts maternal behavior and social-learning of food preference in adulthood: tactile stimulation, nest odor, and social rearing prevent these effects. *Dev Psychobiol*. 2006;48:209–219.
246. Fleming AS, Ruble DN, Flett GL, van Wagner V. Adjustments in first-time mothers: changes in mood and mood content during the first postpartum months. *Dev Psychol*. 2001;26:137–143.
247. Bridges RS. Biochemical basis of parental behavior in the rat. *Adv Study Behav*. 1996;25:215–242.
248. Numan M, Stolzenberg DS. Medial preoptic area interactions with dopamine neural systems in the control of the onset and maintenance of maternal behavior in rats. *Front Neuroendocrinol*. 2009;30:46–64.
249. Pedersen CA. Oxytocin control of maternal behavior. Regulation by sex steriods and offspring stimuli. *Ann NY Acad Sci*. 1995:126–145.
250. Shahrokh DK, Zhang TY, Diorio J, Gratton A, Meaney MJ. Oxytocin-dopamine interactions mediate variations in maternal behavior in the rat. *Endocrinology*. 2010;151:2276–2286.
251. Bale TL, Pedersen CA, Dorsa DM. CNS oxytocin receptor mRNA expression and regulation by gonadal steroids. *Adv Exp Med Biol*. 1995;395:269–280.
252. Fahrbach SE, Morrell JI, Pfaff DW. Possible role for endogenous oxytocin in estrogen-facilitated maternal behavior in rats. *Neuroendocrinology*. 1985;40:526–532.
253. Young LJ, Wang Z, Donaldson R, Rissman EF. Estrogen receptor alpha is essential for induction of oxytocin receptor by estrogen. *Neuroreport*. 1998;9:933–936.
254. Pedersen CA, Prange Jr AJ. Induction of maternal behavior in virgin rats after intracerebroventricular administration of oxytocin. *Proc Natl Acad Sci USA*. 1979;76:6661–6665.
255. Fahrbach SE, Pfaff DW. Effect of preoptic region implants of dilute estradiol on the maternal behavior of ovariectomized, nulliparous rats. *Horm Behav*. 1986;20:354–363.
256. Young LJ, Wang Z. The neurobiology of pair-bonding. *Nat Neurosci*. 2004;7:1048–1054.
257. Champagne FA, Weaver ICG, Diorio J, Sharma S, Meaney MJ. Natural variations in maternal care are associated with estrogen receptor alpha expression and estrogen sensitivity in the MPOA. *Endocrinology*. 2003;144:4720–4724.
258. Franklin TB, Russig H, Weiss IC, et al. Epigenetic transmission of the impact of early stress across generations. *Biol Psychiatry*. 2010;68:408–415.
259. Bagot RC, Tse YC, Wong AS, Meaney MJ, Wong TP. Maternal care influences hippocampal NMDA receptor function and dynamic regulation by corticosterone in adulthood. *BiolPsychiatry*. 2012;72:491–498.
260. Mohn F, Weber M, Rebhan M, et al. Lineage-specific polycomb targets and de novo DNA methylation define restriction and potential of neuronal progenitors. *Mol Cell*. 2008;30:755–766.
261. McEwen KR, Ferguson-Smith AC. Distinguishing epigenetic marks of developmental and imprinting regulation. *Epigenet Chromat*. 2010;3:2.

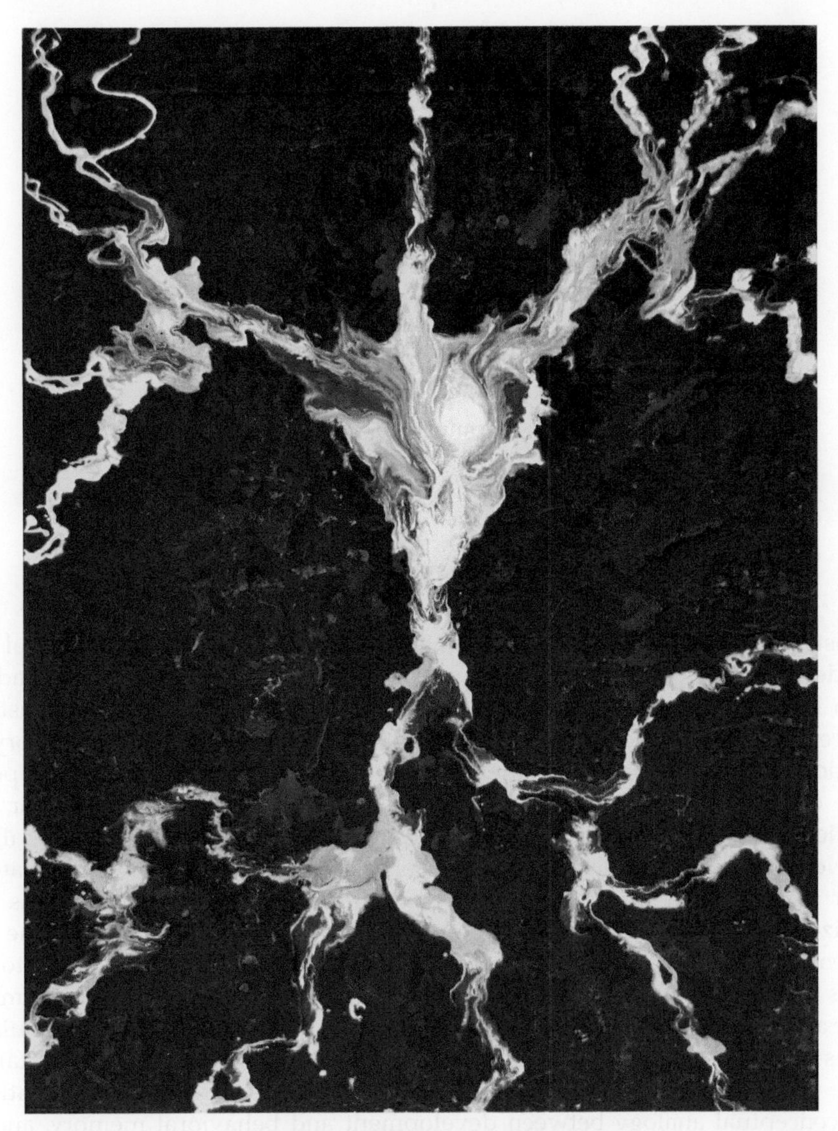

"Hippocampal Pyramidal Neuron"

J. David Sweatt, acrylic on canvas (36 x 48), 2012

CHAPTER

5

Epigenetic Mechanisms in Learning and Memory

Jeremy J. Day and J. David Sweatt

McKnight Brain Institute, Department of Neurobiology, University of Alabama at Birmingham, Birmingham, Alabama, USA

INTRODUCTION

Biologists have realized for decades that there are interesting conceptual parallels between development and behavioral memory formation.[1] Both development and memory formation involve the capacity of transient environmental signals to trigger lasting, even lifelong, cellular changes. Thus, a critical component of both behavioral memory and the perpetuation of cellular phenotype after developmental differentiation is their persistence over time despite the perpetual turnover of cellular proteins. At the molecular level, the perpetuation of a specific biochemical memory trace (be it cellular or behavioral) requires that individual molecules have the ability to self-perpetuate a given acquired state, despite molecular turnover and in the absence of the initial triggering stimulus. This has been hypothesized to occur via a specific type of biochemical reaction in which the active or mnemogenic form of a molecule is able to autoconvert a nascent form of itself into an active form. This type of reaction has long been hypothesized to underlie memory,[2–5] making the search for a molecule capable of autoconversion a critical step in elucidating cellular as well as systems-level memory storage. Interestingly, both DNA methylation and certain aspects of histone post-translational modifications exhibit the capacity for self-perpetuation. Thus, both the conceptual analogy between development and behavioral memory, and the fact that epigenetic molecular mechanisms can exhibit self-perpetuation over time, have been driving concepts motivating the specific investigation of the potential role of epigenetic molecular mechanisms in behavioral memory.[6–10]

Against this backdrop, in this chapter, we will describe findings generated over the past ten years implicating epigenetic molecular mechanisms in adult CNS function and experience-driven behavioral change. We will parse our discussion of recent discoveries

Epigenetic Regulation in the Nervous System
DOI: http://dx.doi.org/10.1016/B978-0-12-391494-1.00005-7

concerning the important role for epigenetic molecular mechanisms in behavioral learning and memory into four main sections:

1. In the first section, we will discuss the identities of the epigenetic molecular marks proposed to be involved in regulating learning and memory, and provide a discussion of the experimental evidence supporting their involvement in these processes.
2. In the second section, we will describe how the epigenetic marks may be read out in neurons at the molecular level in order to drive functional change in the CNS. In providing a discussion of the proximal functional targets of epigenetic marking in the CNS, this section will focus largely on regulation and modulation of gene transcription.
3. In the third section, we will briefly discuss epigenetically based human disorders impacting learning and memory. These considerations provide direct evidence implicating the relevance of epigenetic mechanisms to human cognition. This section will be fairly brief because the role of epigenetic mechanisms in drug addiction, psychiatric illness, neurodevelopmental disorders, aging-related cognitive dysfunction, and neurodegenerative diseases are covered in other specific chapters in this book.
4. In the fourth and final section, we will discuss a theoretical framework for a role for epigenetic molecular mechanisms in learning and memory; how the marks are maintained, are read out, and subsequently affect memory function at the cellular, circuit, and neural systems levels.

EPIGENETIC MARKS IN THE NERVOUS SYSTEM AND THEIR ROLES IN LEARNING AND MEMORY

A Review of Chromatin Structure

As has already been described (Chapter 1), within a cell nucleus, 147 base pairs of DNA are wrapped tightly around an octamer of histone proteins (two each of H2A, H2B, H3, and H4) to form the basic unit of chromatin called the nucleosome. Each histone protein is composed of a central globular domain and an N-terminal tail that contains multiple sites for potential modifications, including acetylation, phosphorylation, methylation, ubiquitination, and ADP-ribosylation (Figure 5.1). Each of these marks is bi-directionally catalyzed or removed by a specific set of enzymes.[11] Thus, histone acetyltransferases (HATs) catalyze the transfer of acetyl groups to histone proteins, whereas histone deacetylases (HDACs) cause the removal of acetyl groups. Likewise, histone methylation is initiated by histone methyltransferases (HMTs) such as G9a, whereas histone demethylases (HDMs) such as LSD1 remove methylation marks. Interestingly, a number of histone sites can undergo dimethylation or even trimethylation. Finally, phosphorylation of serine or threonine residues on histone tails can be accomplished by a broad range of nuclear kinases, such as MSK-1, and dephosphorylated by protein phosphatases such as protein phosphatase 1 (PP1).

Importantly, histone modifications are capable of being not only specifically targeted to individual genes within the genome, but also localized selectively to subdomains within the gene's exon–intron structure, meaning that they are in an ideal position subtly and selectively to influence gene readout. Site-specific modifications are known to alter directly

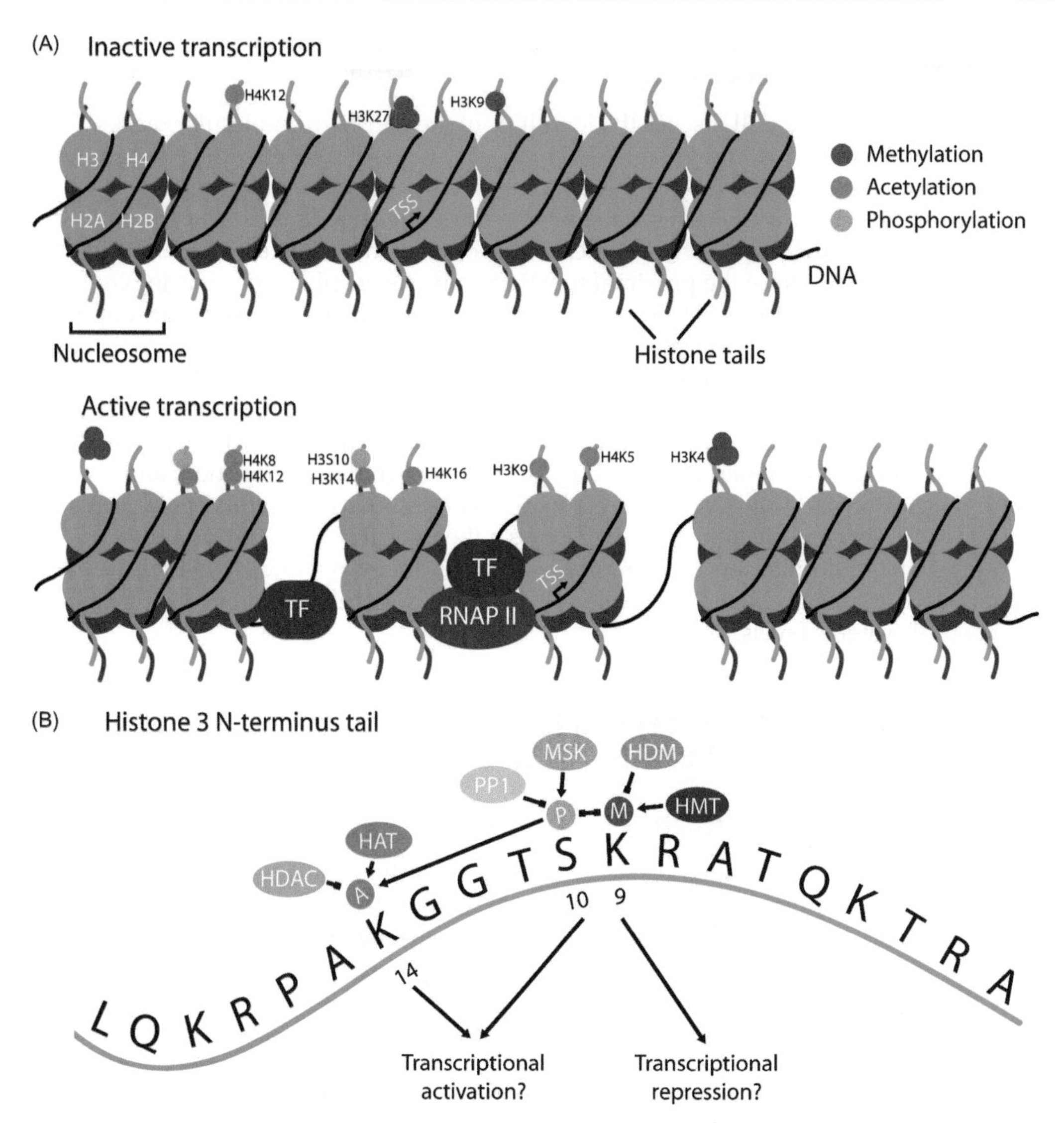

FIGURE 5.1 **Dynamic regulation of histone modifications directs transcriptional activity.** (A) Individual residues on histone tails undergo a number of unique modifications, including acetylation, phosphorylation, and mono-, di-, and trimethylation surround the transcription start site (TSS) for a given gene. These modifications in turn correlate with transcriptional repression (top), in which DNA is tightly condensed on the nucleosome and therefore inaccessible, or transcriptional activation (bottom), in which transcription factors (TF) or RNA polymerase II (RNAP II) can access the underlying DNA to promote gene expression. The specific marks listed correlate with transcriptional activation or repression, although this list is by no means exhaustive. (B) Expanded view of individual modifications on the tail of histone H3. (See text for details and acronyms.) The concept of a histone "code" suggests that individual marks interact with each other to form a combinatorial outcome. In this case, methylation at lysine 9 on H3 (a mark of transcriptional repression) and phosphorylation at serine 10 on H3 repress each other, whereas phosphorylation at serine 10 enhances acetylation on lysine 14 (a mark of transcriptional activation).

chromatin state and gene transcription through a number of mechanisms that were discussed in Chapter 1. Moreover, while some modifications such as histone acetylation or phosphorylation are generally associated with transcriptional activation, others are more closely correlated with transcriptional repression,[12,13] leading to the capacity of histone modifications bi-directionally to regulate gene readout.

Given that histone proteins can be modified at a number of sites, this raises the possibility that specific modifications could work together as a sort of "code", which would ultimately dictate whether a specific gene was transcribed. This idea, first formalized over a decade ago,[11,14,15] in essence suggests that certain combinations of modifications will lead to transcriptional activation whereas others would lead to transcriptional repression. This hypothesis has largely been supported experimentally[16] (see Figure 5.1) as discussed in more detail in Chapter 3.

Histone Post-translational Modifications in Memory Formation

Substantial and compelling evidence indicates that histone modifications in the CNS are essential components of memory formation and consolidation. Indeed, multiple types of behavioral experiences are capable of inducing histone acetylation in several brain regions.[7,10,17–23] For example, contextual fear conditioning, a hippocampus-dependent form of memory, increases H3S10/H3K14 phospho-acetylation in the CA1 region of the hippocampus.[20,24] Moreover, contextual fear conditioning induces increased acetylation at multiple sites on the tails of H3 and H4, including H3K9, H3K14, H4K5, H4K8, and H4K12 in the hippocampus.[23] This increased acetylation is rather transient, occurring within an hour of conditioning, but returning to baseline levels at 24 h post-conditioning. Additionally, none of these changes occur in control animals that are exposed to the same context but receive no fear conditioning, indicating that these modifications are specific to the formation of learned associations. Importantly, these histone modifications also appear to be necessary for the formation of memories, as interference with the molecular machinery that regulates histone acetylation is capable of altering associative learning as well as the cellular correlates of memory such as long-term potentiation (LTP).[7,18–21,25–27] Specifically, enhancing histone acetylation using HDAC inhibitors results in improved memory formation and superior LTP,[7] whereas mutations in the gene for CREB binding protein (CBP), a known HAT, results in deficits in memory formation and impaired LTP.[25]

The co-occurrence of acetylation at H3K9, H3K14, H4K5, H4K8, and H4K12 in the hippocampus following fear conditioning is associated with changes in the transcription of hundreds of genes in young mice.[23] In contrast, in elderly mice, which lack only one of these modifications in response to fear conditioning (acetylation at H4K12) and display poor fear learning, there are almost no conditioning-induced changes in gene expression. This suggests that a specific combination of histone modifications is necessary to initiate learning-related gene expression programs. Consistent with this hypothesis, treatment with an HDAC inhibitor selectively restored H4K12 acetylation, enabled the conditioning-induced changes in gene expression, and improved fear memory formation.[23]

Overall, these modifications are consistent with the concept of a "histone code" contributing to memory in which specific sets of changes are produced in response to specific types of behavioral experiences, and these modifications are necessary for memory formation and/or consolidation. However, in the context of learning and memory, the histone code hypothesis would also predict that it is the combination of histone modifications, rather than the sum of individual modifications, that produces unique changes in gene expression required for memory formation. This appears to be the case and, in the next several paragraphs, we will discuss evidence that other histone modifications besides acetylation also contribute to memory formation.

Histone Methylation

Unlike histone acetylation, histone methylation is not universally associated with either transcriptional repression or transcriptional activation.[28,29] Instead, certain modifications, such as dimethylation at H3K9, are associated with transcriptional repression whereas other modifications, such as dimethylation or trimethylation of H3K4, are associated with transcriptional activation.[13,29] Also unlike histone acetyl transferases, histone methyltransferases (HMTs), such as G9a and SUV39H1, catalyze histone methylation, but not in a global manner as appears to be the case for HATs (Figure 5.2 [29,30]). Thus, G9a methylates lysine 9 on H3 (H3K9), which is generally a mark of repressed transcription.[31,32] Likewise, MLL1 methylates H3K4, which is associated with transcriptional activation.[33] Similarly, enzymes that remove methyl groups from lysine residues, called histone demethylases (HDMs) also exhibit substrate specificity, with JHDM1 removing methyl groups from H3K36 and LSD1 demethylating H3K9.[34–36] Indeed, part of the promise for using HMTs and HDMs as a therapeutic treatment is that the writers and erasers of this mark are exceptionally specific.

Also, unlike other histone modifications, histone methylation is often not distinctly an either/or modification. In fact, lysine residues can be mono-, di, or even trimethylated, with each distinct methyl group addition producing unique results. These differential states are produced by tight regulation of histone methylation machinery, as HMTs and HDMs are known to catalyze the addition or removal of different numbers of methyl groups. For example, the histone demethylase LSD1 (also known as KDM1) requires a protonated lysine to function, and therefore cannot remove methyl groups from a trimethylated lysine.[37] In addition, it is clear that proteins which contain chromodomains (and therefore bind to methylated lysine) can possess different affinities for methylation levels at a specific histone target, thereby conferring unique methylation states with unique functional consequences.[29,35]

Overall, however, histone methylation and demethylation represent a second set of modifications beyond acetylation that is relevant to learning and memory. A role for histone methylation in memory is increasingly being appreciated. Within the hippocampus, trimethylation of H3K4 and dimethylation of H3K9 are both increased immediately after contextual fear conditioning.[20] This is interesting given that, as stated above, these marks are associated with opposite transcriptional regulation. Thus, this finding likely indicates that these modifications occur at gene-specific targets, whereby they can differentially control expression of memory-associated genes. Importantly, histone methylation also appears to be functionally relevant for memory formation, as mice that lack Mll (an H3K4 specific methyltransferase) exhibit impaired contextual fear conditioning.[20] Likewise, mice with neuronal deletion of the histone methyltransferase G9a/GLP (an H3K9 specific

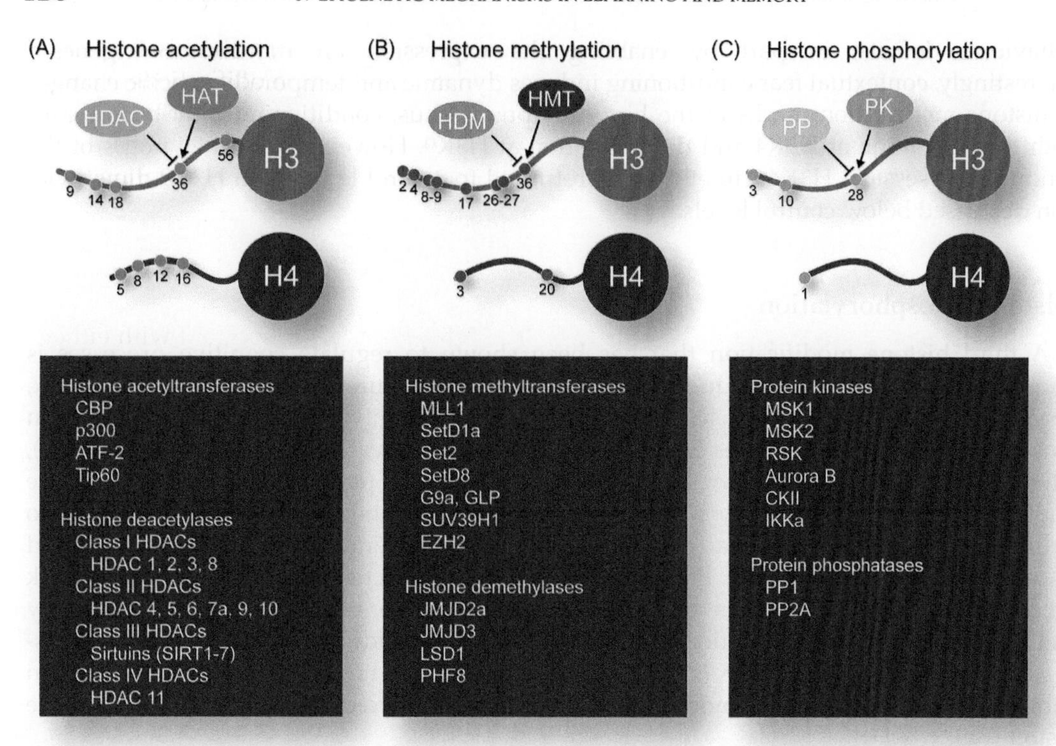

FIGURE 5.2 **Summary of well-understood histone modifications and histone-modifying enzymes.** (A) Histone acetylation at numerous lysine residues on histone tails is catalyzed by histone acetyltransferases (HATs) and removed by histone deacetlyases (HDACs). Histone acetylation is generally a transcriptionally permissive mark. Different HAT and HDAC enzymes are listed below. Importantly, specific HDACs isoforms are differentially expressed across brain structures and appear uniquely to regulate different aspects of cognition. (B) Histone methylation at lysine and arginine residues on histone tails is catalyzed by histone methyltransferases (HMTs) and removed by histone demethylases (HDMs). Histone methylation at different amino acid residues has been linked to both transcriptional activation and transcriptional repression. Methylation can occur in mono-, di-, or even trimethylated states. Many HDMs and HMTs are specific for modifications at individual amino acids on histone tails or even a specific number of methyl groups. (C) Histone phosphorylation at serine residues is catalyzed by protein kinases (PKs) such as mitogen- and stress-activated protein kinase 1 (MSK1), whereas phosphorylation marks are removed by protein phosphatases such as protein phosphatase 1 (PP1). Histone phosphorylation is generally linked to transcriptional activation.

methyltransferase) exhibit impaired motor behavior, decreased motivation to consume palatable sucrose, and severe learning and memory deficits.[38] Additionally, this methyltransferase also regulates cocaine reward and cocaine-induced neuronal plasticity in the striatum.[39]

Moreover, there are a number of relatively selective compounds capable of modifying specific methylation marks,[29,30,40] such as a small-molecule inhibitor of the G9a methyltransferase, which reverses H3K9 dimethylation.[41] In rodents, forebrain specific deletion of the GLP/G9a histone methyltransferase complex results in a number of learning-related behavioral deficits, in part by enabling the expression on non-neuronal genes.[38] Interestingly, contextual fear conditioning induces dynamic and temporally-specific changes in histone methylation marks in the hippocampus.[20] Thus, conditioning rapidly increases both trimethylation at H3K4 and dimethylation at H3K9. However, within 24 hours of the conditioning session, H3K4 trimethylation returned to control levels, and H3K9 dimethylation decreased below control levels.

Histone Phosphorylation

A third histone modification that has been shown to regulate cognitive processes is phosphorylation of serine residues on histone tails (see Figures 5.1 and 5.2C). This mark, normally associated with transcriptional activation, is catalyzed by a range of protein kinases.[42,43] This group includes the mitogen- and stress-activated protein kinase 1 (MSK1), which is activated downstream of the ERK/MAP kinase pathway. Conversely, histone phosphorylation is reversed by protein phosphatases PP1 and PP2a, which are known to be inhibited by other molecular cascades including dopamine and cyclic-AMP regulated phosphoprotein 32 (DARPP32). Perhaps the most well-characterized phosphorylation mark occurs at serine 10 on H3. This modification recruits GCN5, which contains HAT activity and therefore increases acetylation at neighboring lysine residues K9 and K14 and repressing histone methylation at H3K9.[44] In addition to recruiting HATs, H3S10 phosphorylation enhances transcription factor binding by modifying the interaction between DNA and the histone tail.[45,46]

Several studies have revealed a role for histone phosphorylation at H3S10 in regulating memory formation. Mutations in the gene encoding RSK2, which has been shown to phosphorylate H3, produces Coffin–Lowry syndrome, an X-linked disorder that is associated with psychomotor retardation and physical abnormalities.[47,48] In animal models, H3S10 phosphorylation and H3S10/K14 phosphoacetylation increase rapidly in the hippocampus following contextual fear conditioning, and these increases are blocked by ERK inhibition.[24] Likewise, mice lacking MSK1 exhibit impaired fear conditioning and spatial memory.[18] Interestingly, this deficit is not reversed by treatment with an HDAC inhibitor, revealing that the histone phosphorylation pathway occurs in parallel to (rather than downstream of) histone acetylation. Accordingly, inhibition of nuclear PP1, the major histone phosphatase, improves long-term object recognition memory and spatial memory without affecting short-term memory.[21] Together, these findings suggest that enhancing histone phosphorylation via inhibition of PP1 may be a distinct and even complementary treatment for learning and memory disorders. However, this remains to be examined experimentally.

Other Histone Modifications: Ubiquitination and Poly-ADP Ribosylation

Although the three modifications reviewed above (acetylation, methylation and phosphorylation) represent the bulk of ongoing research in the area of epigenetic control in cognitive processes, a number of additional histone modifications can occur in vivo and regulate many aspects of cellular signaling and function. Among these are histone ubiquitination, poly-ADP ribosylation, sumoylation, and O-GLCNAcylation. Below we will consider two specific modifications (ubiquitination and poly-ADP ribosylation) which, through recent discoveries, have been linked to cognition.

Histone ubiquitination occurs at all four histone proteins, but is most well characterized at the carboxy terminus of H2A and H2B. It is regulated by a three-step enzymatic process in which the enzyme E1 first activates ubiquitin, and the ubiquitin-carrier enzyme E2 directs ubiquitin to the substrate. Finally, ubiquitin is linked to a lysine residue by the E3 ubiquitin ligase family of enzymes.[49,50] Likewise, a wide family of deubiquitinases (DUBs) remove ubiquitin from lysine residues.[49,51] Histone ubiquitination is known to have a wide range of cellular functions, most notably controlling transcriptional initiation and elongation.[52] Ubiquitination/deubiquitination enzymes also interact with other epigenetic machinery, and thus correlate (or are even prerequisites for) other histone marks, specifically histone methylation.[52–55] However, perhaps the clearest link between histone ubiquitination and neurodegenerative impairments comes from Huntington's disease, which is characterized by a mutation in the gene encoding huntingtin protein and aberrant transcriptional regulation. Interestingly, huntingtin interacts with the ubiquitin ligase hPRC1L, suggesting a role for histone ubiquitination in Huntington's disease.[56] Moreover, ubiquitinated H2A is increased in a mouse model of Huntington's disease, whereas ubiquitinated H2B is decreased, resulting in modified histone methylation patterns and transcriptional dysregulation.[56] Accordingly, targeting ubiquitin ligases may be a potential therapeutic avenue for patients with Huntington's disease.

Poly-ADP ribosylation of histones occurs at all histone proteins as well,[57,58] and is catalyzed by poly (ADP-ribose) polymerases such as PARP1 and PARP2, with NAD^+ acting as a substrate. Likewise, poly-ADP ribosylation is reversed by poly(ADP-ribose)glycohydrolase.[58] Interestingly, although poly-ADP ribosylation and PARP1 activity are most well known for their role in DNA repair following damage, PARP1 activation has been shown to increase in cortical neurons following depolarization.[59] Likewise, PARP1 is required for long-term synaptic facilitation and long-term memory in *Aplysia*, suggesting a functional role in the nervous system.[60,61] Similarly, in mouse models, poly-ADP ribosylation is increased at the histone linker protein H1 following novel object exposure, and PARP1 is required for hippocampal LTP and novel object recognition memory.[62] These findings indicate that, in addition to the mechanisms reviewed above, histone poly-ADP ribosylation may be an interesting target for learning and memory.

The DNA Methylation Code

As has already been discussed in earlier chapters of this book, DNA methylation, or the addition of a methyl group to the 5′ position on a cytosine pyrimidine ring, can occur at multiple sites within a gene (Figure 5.3). While it has historically been held that methylation is generally limited to cytosine nucleotides followed by guanine nucleotides, or so-called CpG sites, recent studies are increasingly finding cytosine methylation at non-CpG sites as well. CpG sites are under-represented throughout the genome, yet are occasionally clustered in CpG "islands" (Figure 5.4). Interestingly, these CpG sites tend to exist in the promoter regions of active genes, suggesting the ability to control transcriptional output. A number of proteins with methyl binding domains bind to methylated cytosine sites on

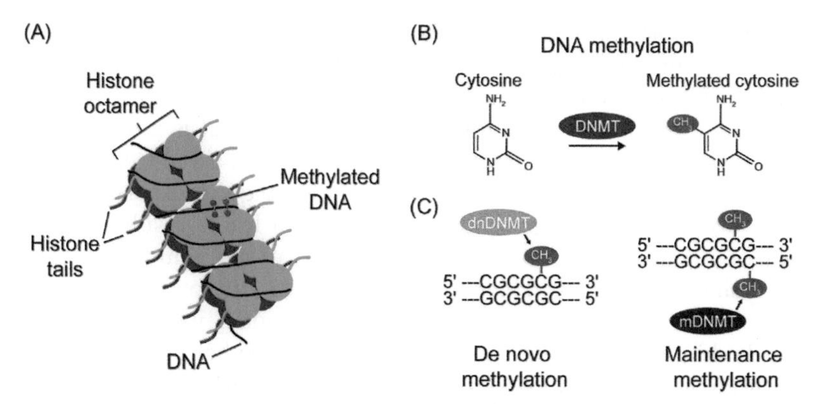

FIGURE 5.3 **DNA methylation.** (A) Inside a cell nucleus, DNA is wrapped tightly around an octamer of highly basic histone proteins to form chromatin. Epigenetic modifications can occur at histone tails, or directly at DNA via DNA methylation. (B) DNA methylation occurs at cytosine bases when a methyl group is added at the 5′ position on the pyrimidine ring by a DNA methyltransferase (DNMT). (C) Two types of DNMTs initiate DNA methylation. De novo DNMTs methylate previously non-methylated cytosines, whereas maintenance DNMTs methylate hemi-methylated DNA at the complementary strand.

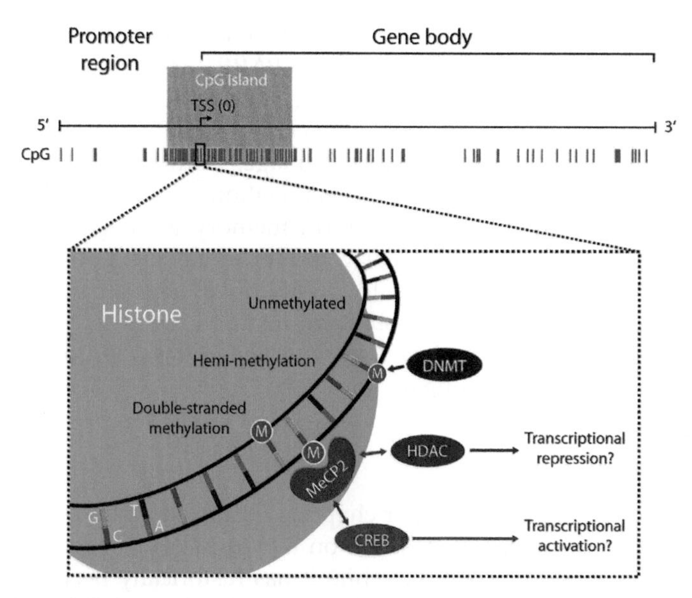

FIGURE 5.4 **DNA methylation status affects gene transcription.** A number of plasticity-related genes in the brain possess large CpG islands within the gene promoter region. Each CpG dinucleotide in the DNA sequence can undergo methylation by DNA methyltransferases (DNMTs), resulting in hemi-methylation and/or double-stranded DNA methylation. Proteins with methyl binding domains bind to methylated DNA and associate with other co-factors, such as HDACs or transcription factors like CREB, to alter gene expression. It is presently unclear if the specific combination of CpG methylation marks constitutes a "code" for unique outcomes, or if the overall or average density of methylation is a larger determinant of transcriptional efficacy.

DNA, including MeCP2. DNA methylation is catalyzed by two groups of enzymes, known as DNA methyltransferases (DNMTs, Figure 5.5). The first group, the de novo DNMTs, methylate cytosines that are not methylated on either DNA strand. The second group, maintenance DNMTs, recognize hemi-methylated DNA and attach a methyl group to the complementary cytosine base. Through this process, DNA methylation can be self-perpetuating in the face of ongoing passive demethylation, allowing the potential for a chemical modification that exists throughout the lifetime of a single cell.[6] The self-perpetuating nature of DNA methylation will be discussed in greater detail in later sections of this chapter – for now, we will focus on evidence linking DNA methylation to learned behaviors and memory formation and storage.

Like histone modifications, DNA methylation has also been considered to constitute an epigenetic code,[63] although this idea is more recent and has been less fully explored. Clearly, methylation at promoter regions is capable of altering transcription due to the affinity of certain proteins for methylated cytosine (the methyl binding domain proteins, or MBDs). The prototypical example of this is MeCP2, which is mutated in the neurodevelopmental disorder Rett syndrome and dramatically affects synaptic plasticity in the hippocampus and memory formation.[64–66] Mechanistically, MeCP2 is capable of recruiting both repressive and activating transcription factors or chromatin remodeling complexes such as HDACs.[67] Importantly, MBDs like MeCP2 have different affinities for fully methylated and

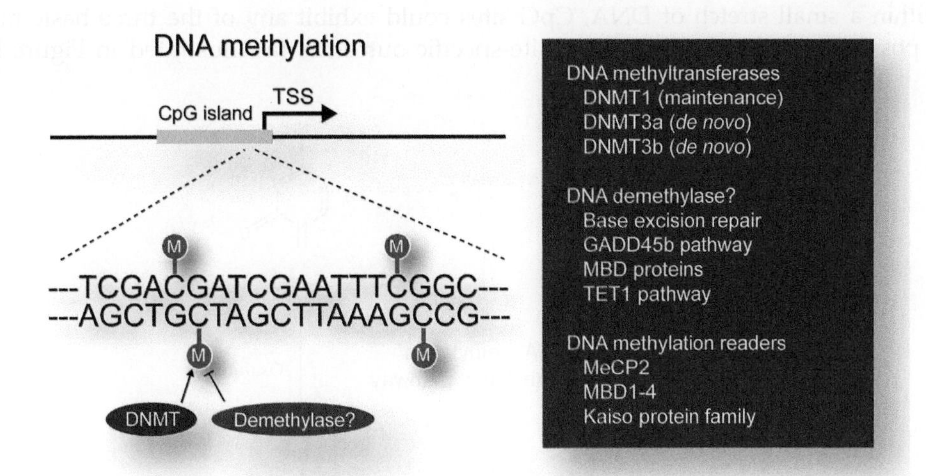

FIGURE 5.5 **DNA methylation and demethylation.** A majority of mammalian gene promoters contain dense clusters of cytosine-guanine dinucleotides, called CpG islands, at which methylation can occur to influence dramatically gene transcription. In this example, the CpG island (green bar) overlaps the transcription start site. At CpG dinucleotides, methylation is catalyzed by DNA methyltransferases (DNMTs). De novo DNMTs direct the methylation of unmethylated CpGs, whereas maintenance DNMTs recognize hemi-methylated DNA and methylate the complementary strand. The existence of a direct demethylase is controversial, but a number of different mechanisms has been proposed to regulate removal of the methyl moiety, including excision and replacement of the entire base pair. DNA methylation marks are read by a family of proteins with methyl binding domains (MBD proteins), which includes MeCP2. Each of these targets may represent candidates for therapeutic treatments of disorders characterized by aberrant DNA methylation. (See text for additional details.)

hemi-methylated DNA, meaning that the difference between these two states may actually be a critical component of this code.[68]

In addition to simple methylation, cytosine residues can also undergo hydroxymethylation (OHMeC) in the adult brain.[69,70] Interestingly, this modification severely decreases the binding affinity of methyl-binding domain proteins such as MeCP2 for the cytosine base.[68] Moreover, although the carbon bond between a methyl group and the cytosine base is likely too strong to undergo direct enzymatic removal, this is not the case for OHMeC. Thus, it is possible that OHMeC acts as an intermediate stage that marks individual MeC modifications for demethylation. Therefore, hydroxymethylation could conceptually reverse the effects of DNA methylation, or at least make the original modification ineffective at recruiting binding proteins that alter gene transcription. However, despite the obvious appeal of this mechanism in regulating gene expression, little is known about how it influences the formation and maintenance of behavioral memories. It also is not completely clear if hydroxymethylation represents a completely separate mark or an intermediate stage that marks methyl groups for removal from cytosines, as will be discussed in more detail shortly.

What might a DNA methylation code look like? Given that methylation/demethylation machinery can produce at least three different outcomes for each CpG in question (no methylation, hemi-methylation, and full methylation of both DNA strands), that these individual methyl groups can then be differentially hydroxylated, and that the promoter and intragenic regions of a plasticity gene may contain hundreds of CpG sites, the potential combinatorial complexity of a DNA methylation code is astounding. Indeed, it is conceivable that even within a small stretch of DNA, CpG sites could exhibit any of the three basic methylation possibilities, thereby leading to site-specific outcomes (as illustrated in Figure 5.6).

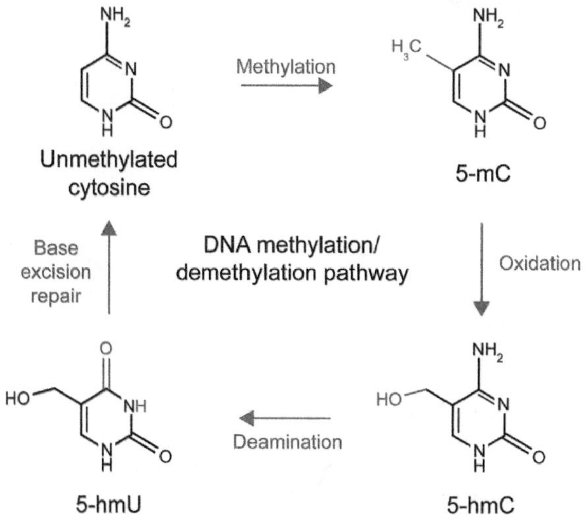

FIGURE 5.6 **Potential mechanism for demethylation of methylated DNA.** After methylation, cytosine bases can be oxidized by TET proteins, creating 5-hydroxymethylcytosine (5hmC). Subsequently, 5hmC can undergo deamination by cytidine deaminases, creating 5-hydroxymethyluracil (5hmU). Base or nucleotide excision repair processes are then able to replace 5hmU with unmethylated cytosine. It is unclear how this potential mechanism would affect methylation status on the complementary DNA strand.

Therefore, much as understanding how DNA codes for genes requires knowing the underlying nucleotide sequence, understanding how DNA methylation contributes to transcriptional efficacy will require examination of DNA methylation changes at the single nucleotide level. However, it is also important to note that the context in which the DNA methylation modification occurs – i.e. where it is positioned relative to a transcription factor binding site or transcription state site, as well as the underlying nucleotide sequence of DNA – may dramatically influence its potential effect on gene transcription.[71,72] However, to date, existing studies have typically only examined CpG methylation at relatively small stretches of DNA near gene transcription start sites.

Dynamic DNA Methylation Changes in Memory Formation

Much like histone modifications, there is now compelling evidence indicating that changes in DNA methylation represent a critical molecular component for both the formation and maintenance of long-term memories.[73–77] Initial studies investigating the potential role of DNA methylation in regulating plasticity and memory formation tested the hypothesis that DNMT activity might regulate the induction of hippocampal long-term potentiation (LTP), and this was indeed found to be the case.[8,9] Subsequent studies examined the role of DNA methylation in hippocampus-dependent forms of learning, such as contextual fear conditioning. These studies revealed that DNMT expression is significantly enhanced within the hippocampus shortly after contextual fear conditioning, and blocking DNMT activity in the hippocampus correspondingly disrupts the formation of context-shock associations.[77] Interestingly, contextual fear conditioning produces both increases and decreases in methylation at memory-related genes within the hippocampus, indicating that both methylation and demethylation of DNA may be important learning mechanisms.[6,76,77] Consistent with the idea that these changes are necessary for memory formation, inhibition of DNMTs within the hippocampus, which produces a hypomethylated state in naïve animals, results in impaired expression of contextual fear memories.[74,77] Likewise, forebrain- and neuron-specific deletion of DNMT1 and DNMT3a impairs performance on the Morris Water Maze and fear learning.[73] In contrast, selective inhibition of DNA methylation in the prefrontal cortex impairs the recall of existing memories, but not the formation of new memories, indicating circuit-specific roles for DNA methylation in memory formation and maintenance.[76] Finally, DNMT inhibitors impair the induction of LTP at hippocampal synapses, providing an important cellular correlate to the learning deficits induced by blocking DNA methylation.[8,9]

Moreover, DNA methylation at the *reelin* gene (which is associated with memory formation) is decreased following fear conditioning, whereas DNA methylation at the *PP1* gene (which is viewed as a memory repressor) is enhanced following fear conditioning (Figure 5.7). Importantly, both of these changes returned to baseline levels 24 hours after conditioning, suggesting dynamic temporal regulation of both DNA methylation and demethylation in the hippocampus.[78]

Additionally, the gene encoding brain-derived neurotrophic factor (BDNF), which has recently been linked to the persistence of fear memories,[79–82] also undergoes unique changes in DNA methylation as a result of fear conditioning.[74] Specifically, exon IV of the *BDNF* gene undergoes significant suppression in DNA methylation in its promoter region

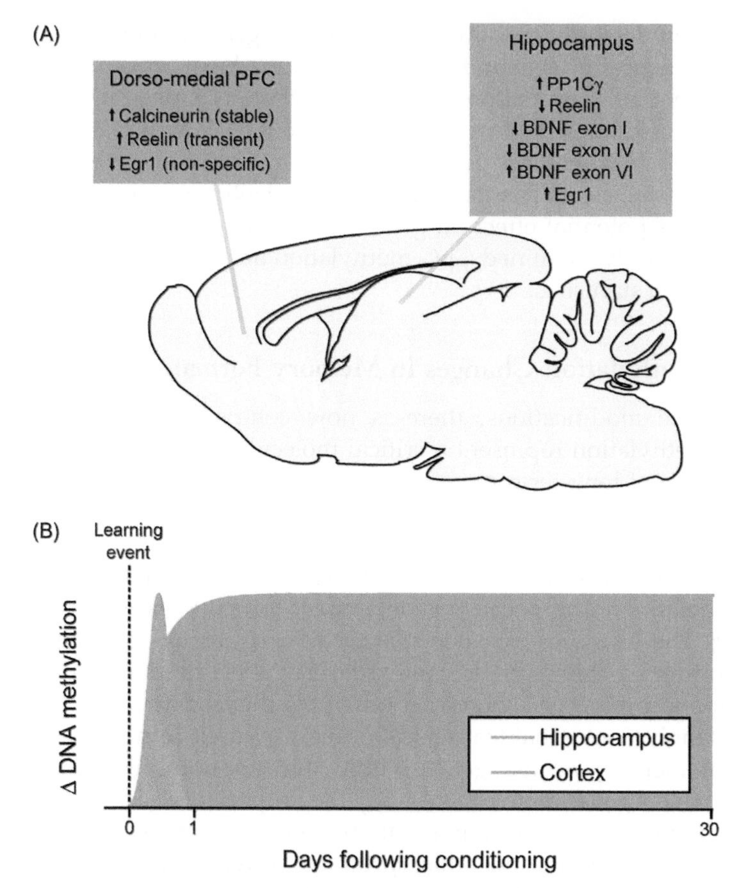

FIGURE 5.7 **Unique targets and timelines for DNA methylation/demethylation in the hippocampus and cortex.** (A) Sagittal section of rat brain representing methylation changes occurring at CpG islands in the promoter regions of selected genes in response to contextual fear conditioning. Both DNA methylation and demethylation are observed at different genes and even within the same gene. Brain section graphic taken from.[77] (B) Temporal dynamics of DNA methylation changes in response to contextual fear conditioning. Although all of the changes observed in the hippocampus return to basal levels within 24 hours of learning, at least one change (increased *calcineurin* methylation) is maintained within the dorsomedial PFC. Importantly, blocking DNA methylation in the hippocampus immediately after conditioning disrupts memory consolidation. In contrast, inhibiting DNA methylation in the dmPFC immediately after learning has no effect on retrieval of remote memories 30 days later, but blocking DNA methylation several days immediately prior to remote retrieval significantly impairs performance, indicating a necessity for ongoing DNA methylation in the maintenance of fear memories.[76]

following contextual fear conditioning, and this corresponds to a substantial increase in exon IV mRNA that returns to baseline levels within 24 hours.[74] This change in DNA methylation was reversed by intrahippocampal infusions of the DNMT inhibitor zebularine as well as the NMDA receptor blocker MK-801, which both impaired memory formation.[74] In contrast, methylation at *BDNF* exons I and VI is substantially decreased following context exposure alone, resulting in significantly increased mRNA for these exons. These findings in studies of the *BDNF* gene suggest that changes in DNA methylation, even at the same gene within the same brain structure, are regulated in a highly complex fashion by different stimuli during learning, and that only some of these changes correspond to and are important for memory formation.

Recent studies of cued fear conditioning have also begun to implicate DNA methylation and histone acetylation in supporting amygdala-dependent learning and memory. Specifically, Glenn Schafe's group has shown that cued fear conditioning is associated with increased expression of DNMT3A and acetylated histone H3 in the lateral amygdala (LA). As in the hippocampus, administration of DNMT inhibitors into the amygdala impaired the formation of a cue-shock association and this effect was rescued by pretreatment with an intra-amygdala injection of an HDAC inhibitor after training.[83] In a separate set of experiments, the same group showed that DNMT inhibitors administered into the LA impaired memory reconsolidation and that this deficit was reversed by pretreatment with an HDAC inhibitor immediately after recall.[84] Importantly, these authors showed that DNMT inhibition impaired memory rather than simply enhancing extinction, indicating that DNA methylation is required for the persistence of amygdala-dependent memory for cued fear conditioning. These studies did not examine changes in DNA methylation at specific genes and the duration of epigenetic changes in the amygdala remains to be determined. Specifically, the upregulation of histone acetylation and DNMT expression was evident only at the latest time point tested (90 min), thus it is not clear whether these changes are transient in the amygdala as they are in the hippocampus. This point is of particular relevance in the amygdala because this region has been implicated in the initial memory formation and in the maintenance of remote memories for cued fear. It remains to be determined whether epigenetic marks established after fear conditioning persist for the duration of memory for cued fear.

Taken together, these experiments strongly support the hypothesis of a role for active changes in DNA methylation in memory. But they also raise important questions. For example, although most plasticity-permissive genes undergo *decreased* methylation (and *increased* expression) during memory formation, compounds that theoretically decrease DNA methylation (DNMT inhibitors) were found to impair memory formation. While this appears to create a paradox, it is important to note that, in the case of BDNF, DNMT inhibitors actually prevent (instead of enhance) the learning-associated hypomethylation and subsequent increase in mRNA expression of exon IV.[74] This occurs despite the fact that, in experimentally naïve animals, DNMT inhibitors decrease *BDNF* methylation at all gene exons examined. Thus, DNMT inhibitors may produce different results when learning-related signaling mechanisms are activated compared to when they are not activated. In fact, there is evidence that DNMTs are involved in both active methylation and demethylation of DNA, depending on the cellular context.[85] One possibility is that DNMTs play a homeostatic role in DNA methylation by adding methyl groups when methylation is too low and removing them when methylation is too high. Another explanation for this discrepancy is that, although DNMTs presumably demethylate pro-memory genes such as *BDNF* and *reelin*, they also demethylate memory repressor genes such as *PP1*, producing an increase in the expression of these genes.[78] Therefore, DNMT inhibitors could potentially activate genes which serve to counteract or prevent memory formation and maintenance. Thus, the effects produced by DNMT inhibitors may result more from activation of memory suppressor genes, which could in turn downregulate the expression of memory activator genes.[74]

The Mysterious Demethylating Mechanism

Interestingly, studies of nervous system function and of physiologic regulation of non-neuronal cells indicate that demethylation of cytosine bases in DNA occurs and is critically important to cellular function.[74,78,86] The mechanism through which this occurs is presently unclear and controversial.[87–89] Given that the methyl group and cytosine base at a methylated CpG are linked via a covalent carbon–carbon bond, it is unlikely that an enzyme directly removes the methyl group due to energetic constraints. However, a number of alternative methods have been proposed that may account for rapid DNA demethylation.[86,90–92] One particularly attractive model proposes that a methylated cytosine may undergo deamination, which produces a thymine residue. Subsequently, base excision repair machinery would recognize the T-G mismatch in DNA, and substitute an unmethylated cytosine for thymine.[90] This process appears to occur within the brain in response to neuronal activity, and is assisted by growth arrest and DNA damage-inducible protein 45 (GADD45).[91,93,94] In a variation on this idea, Song and colleagues[90] have proposed a specific model for DNA demethylation in neurons based on recent exciting results from their laboratory[91] (see Figures 5.5 and 5.6 and Chapter 12). The model involves the conversion of methylated cytosine to thymine through a hydroxymethyl-cytosine intermediate, prior to "net" deamination through mismatched base excision. However, multiple mechanisms have been implicated in demethylation of DNA, including direct demethylation by proteins with methyl-binding domains (e.g. MBD2), oxidative demethylation, complete excision of the methylated cytosine by DNA glycosylases, and deamination by RNA editing enzymes (for a detailed review of the mechanisms, see[86]). Thus, although the exact mechanisms underlying demethylation of DNA remain controversial, it is clear that decreases in methylation can and do play an important role in brain function.

It also remains unclear whether this model could account for demethylation of both DNA strands, nevertheless, this mechanism would enable selective demethylation at specific sites in DNA, allowing: (1) transience of methylation; (2) active demethylation; and (3) a route for entry for the nucleoside analog inhibitors of DNMTs into the DNA in non-dividing cells. Specifically, after becoming phosphorylated by cytidine kinases, prodrugs like 5-aza-2′-deoxycytidine or zebularine may operate by substituting for cytosine during base excision repair (BER). This altered base is resistant to methylation and also traps DNMTs,[95] resulting in the demethylation of the newly repaired strand as well as a decrease in DNMT activity. This provides a satisfying explanation for the unanticipated results described above regarding active DNA demethylation – the model of Song and colleagues provides a mechanism for reversal of DNA demethylation, a mechanism for active demethylation in non-dividing cells, and a molecular basis for nucleoside DNMT inhibitors to act in the mature CNS.

A Role for DNA Methylation in Memory Maintenance

Interestingly, the transience of DNA methylation via these DNA demethylating and remethylating processes negates the broad initial hypothesis motivating the first studies to investigate the role of DNA methylation in memory formation and storage. The initial idea was that a DNA methylation reaction would underlie memory maintenance. However, the initial studies actually demonstrated *plasticity* of DNA methylation in the mature CNS, including reversibility of DNA methylation. These findings were exciting because they implied novel mechanisms like experience-dependent DNA demethylation and a role for chemical modification of DNA in memory formation. However, they refuted the potential role of these mechanisms as a long-term molecular storage device, thus revealing that DNA de/methylation is a much more dynamic process than previously thought (at least in the hippocampus).

However, these early studies all focused on the hippocampus, hippocampal synaptic plasticity, and hippocampal neuron function.[8,9,75,78] Although the hippocampus is critical for memory consolidation, it is not essential for most forms of long-term memory storage. Thus, the observations of plasticity of DNA methylation in the hippocampus are consistent with the behavioral and systems role of this neuronal circuit and brain subregion. For these reasons, new studies have turned their attention to the cortex, which is a site of long-term memory storage.[76,96–98]

In this vein, recent observations have shown that contextual fear conditioning can induce robust, long-lasting changes in DNA methylation in the anterior cingulate cortex (ACC, see Figure 5.7).[76] In fact, such changes were found to last at least 30 days following conditioning, the longest time point that was investigated. Moreover, remote (very long-lasting) memory for contextual fear conditioning can be reversed by infusion of DNMT inhibitors into the ACC, demonstrating that ongoing perpetuation of DNA methylation occurs in the cortex and is necessary as a memory stabilizing mechanism. Thus these findings demonstrate that, indeed, a learning experience can drive the generation of persisting changes in DNA methylation in the adult CNS. However, the reversibility of the memory upon DNMT inhibitor infusion suggests that at least some of these marks that are necessary for memory storage are undergoing turnover and self-regeneration. How this is likely to happen will be a central topic of the final section of this chapter.

Other Types of Epigenetic Marks in Learning and Memory

Epigenetic mechanisms are traditionally defined as mitotically and/or meiotically heritable changes in gene expression that do not involve changes in DNA sequence.[99] This definition has undergone revision in recent years, in large part to include marks that occur at chromosomes in non-dividing cells such as neurons. Thus, a more inclusive definition considers epigenetic mechanisms to be: "the structural adaptation of chromosomal regions so as to register, signal or perpetuate altered activity states".[100] Although this definition includes the mechanisms discussed above, it excludes a number of additional molecular mechanisms that, while not considered "epigenetic" in the strict sense, are capable of exerting powerful control over gene expression or protein function, and therefore are of potential interest when considering the broad role of epigenetically related mechanisms in memory formation. Additionally, these mechanisms have considerable interplay with traditionally defined epigenetic mechanisms,[101] such as DNA methylation, and it is therefore important to address these mechanisms in the context of epigenetics.

MicroRNAs

The first of these mechanisms derives from the activity of small RNA molecules, which include microRNAs, small interfering RNA (siRNA), and small nuclear RNA (snRNA).

Small RNAs have multiple functions within a cell, including activation, repression, or interference with gene expression,[102] and have been implicated in a number of cognitive disorders. For example, microRNA binds to 3′ untranslated regions of messenger RNA and thereby either cleaves and degrades the messenger RNA or controls its expression through translational mechanisms.[103,104] MicroRNAs thereby control expression of the majority of genes within the genome, and represent a critical component of normal physiology and function in the developing and adult nervous system.[103,105]

An example of the importance of microRNA regulation of gene expression comes from Fragile X syndrome, a commonly inherited disorder characterized by mental retardation and autism-like behavioral abnormalities. Fragile X is caused by a mutation in the *FMR1* gene, which causes gene hypermethylation and resulting absence of its product, Fragile X mental retardation protein (FMRP).[106] Interestingly, FMR1 is an RNA binding protein that interacts with several microRNAs to regulate gene expression that is critical for neural development.[107] In fact, FMR1's association with a specific microRNA, miR125b, has been shown to underlie NMDA receptor expression in mouse models, providing a specific mechanism for behavioral impairments in Fragile X syndrome.[108]

Further implicating a role for microRNAs in psychiatric diseases, recent evidence has revealed that Alzheimer's disease is associated with an altered expression of several microRNAs, some of which interact with the amyloid precursor protein's cleavage machinery which has long been associated with this disease.[109] Similarly, in an animal model of drug addiction, prolonged experience with cocaine self-administration dramatically increased expression of miR-212 in the striatum.[110] However, in this case, this overexpression was associated with a dampening of the motivational properties of the drug by modulating CREB pathways linked to the stimulatory effects of cocaine. These data, together with the substrate specificity of microRNAs, suggest that microRNA targets may be excellent therapeutic candidates for cognitive disorders.

Moreover, several recent studies in laboratory animal models have directly implicated a role for miRNAs in long-term synaptic plasticity and memory formation. In pioneering studies, Priya Rajasethupathy in the Kandel laboratory investigated the role of miRNAs in transcription-dependent long-term facilitation in *Aplysia* sensory neurons.[111] These studies investigated the role of small regulatory RNAs in learning-related synaptic plasticity by first performing massive parallel sequencing to profile the small RNAs of *Aplysia californica*. They found several miRNAs unique to *Aplysia* that were brain enriched and were rapidly downregulated by transient exposure to the facilitatory neurotransmitter serotonin. They further discovered that miR-124 was exclusively present in sensory neurons and that it constrains serotonin-induced synaptic facilitation through regulation of the transcriptional factor CREB, controlling long-term synaptic plasticity in this system.

Additional recent seminal studies from the laboratories of Tim Bredy and Li-Huei Tsai have directly implicated miRNAs in mammalian learning and memory as well. For example, the Tsai laboratory has found that that SIRT1 modulates synaptic plasticity and memory formation via a microRNA-mediated mechanism.[112] They found that unchecked miR-134 expression following SIRT1 deficiency results in the downregulated expression of CREB and brain-derived neurotrophic factor (BDNF), thereby impairing synaptic plasticity. Thus, they demonstrated the existence of a novel microRNA-based mechanism by which SIRT1 regulates plasticity and memory. The Bredy laboratory has focused largely on fear extinction memory and the role of miRNAs in those memory processes.[113] They found that gene silencing through increased expression of the brain-specific microRNA miR-128b disrupted stability of several plasticity-related target genes and that miR-128b is necessary for fear-extinction learning. Thus, emerging evidence indicates that miR-128b activity facilitates the formation of a new fear-extinction memory in the mammalian CNS.

Prion-Like Proteins

A second set of mechanisms, prion-like proteins, involves proteins capable of two conformational states. Interestingly, prion proteins are able not only to undergo a conformational change, but are also then capable of converting normal isoforms into the conformationally altered version. Prion-like proteins have recently been shown to play functional roles in memory maintenance in the *Aplysia* memory model system. In this system, cytoplasmic polyadenylation element binding protein (CPEB), which exhibits prion characteristics, is essential for synaptic facilitation in *Aplysia*. Moreover, in Aplysia synaptic facilitation there is evidence that CPEB undergoes a prion-like self-perpetuation of a polymerized state. Thus far, the existence of prion-like conformers of CPEB in other model systems has not been extensively investigated. However, the normal "baseline" function of CPEB, which is a regulator of mRNA translation, is necessary for memory function in both mice and drosophila,[114–119] opening up the possibility that prion-like function of CPEB may be a necessary and common component of synaptic plasticity. Finally, another prion protein, PrPC, is critical for inhibitory avoidance memory in rats,[120] and mice that overexpress PrPC fail to show age-related declines in social recognition memory.[121] Together, these tantalizing results suggest the possibility that prion proteins may have a normal functional role in the establishment of long-term memory in the mammalian CNS.

HOW DOES THE EPIGENETIC CODE MANIFEST AS A FUNCTIONAL CHANGE?

Testing the Hypothesis of an Epigenetic Code in Memory

Several initial predictions of the hypothesis that an epigenetic code may operate in memory formation have already been tested experimentally. As discussed above, changes in histone modifications and DNA methylation in the CNS have been observed to occur in association with memory formation, blocking DNA and histone methylation and histone acetylation has been shown to attenuate memory, and augmenting histone acetylation has been found to enhance memory formation. These findings are strongly supportive of the idea that an epigenetic code might operate in learning and memory. However, the vast majority of the experiments undertaken thus far have not attempted to address the epigenetic code concept directly, that is the idea that specific patterns of histone and DNA chemical modifications are written and read in a combinatorial fashion to subserve specific aspects of memory. Moreover, presently there is essentially no understanding of how epigenetic marks and mechanisms get translated into functional synaptic, anatomical, and circuit changes in memory-associated circuits in the CNS. No doubt, testing these defining aspects of the epigenetic code concept will be a large undertaking and require multiple independent lines of experimentation. In this section, we will briefly comment on a few of the challenges that stand in the way of testing the epigenetic code hypothesis, with the idea in mind that defining some of the challenges may help in conceptualizing advances designed to overcome the challenges.

If an epigenetic code operates to control aspects of memory formation and storage, then testing this idea will require generating data that a combinatorial set of epigenetic marks is generated in neurons within a memory-encoding circuit, in response to a memory-evoking experience. Further more refined experimental testing would then require that disrupting the specific combinatorial pattern, without altering the overall sum of the marks across the epigenome, would lead to a diminution of memory capacity. Moreover, the combinatorial code would be presumed to occur at the level(s) of a single gene or allele, perhaps at a single CpG island, at an individual chromatin particle, or even at a single histone amino-terminal tail. Finally, all contemporary models of memory storage posit sparse encoding of memories within a memory circuit, meaning that changes at the level of individual neurons would be a relevant parameter.

Taken in sum, these considerations indeed would represent an immense set of technical hurdles to be overcome in order to test the epigenetic code hypothesis. However, a reasonable number of memory-associated genes have been identified and shown to be epigenetically modified in response to experience, including *bdnf, reelin, zif268, PP1, arc,* and *calcineurin*. This list provides a set of candidates for both assessing combinatorial epigenetic changes at the single-gene or single-exon/intron level, and candidates for highly selective gene engineering approaches. Chromatin immunoprecipitation (ChIP) procedures also allow assessment of the presence of methyl-DNA binding proteins and specific histone modifications at the level of these and other specific gene loci.

Moreover, new methods are also becoming available for assessing whether changes in DNA methylation have occurred at the single-cell and single-allele level. Bisulfite sequencing is considered the "gold standard" for assaying DNA methylation, as it provides single-nucleotide information about a cytosine's methylation state. Bacterial subcloning of single pieces of DNA, which originate from single alleles within a single cell, allows isolation of DNA from single CNS cells. Thus, the use of direct bisulfite sequencing combined with DNA subcloning allows for the quantitative interrogation of single-allele molecular changes in single cells from brain tissue.[76] Subsequent bisulfite sequencing of the clones allows sequencing precision at the single-nucleotide level and quantification of methylation from individual copies of a gene. This type of approach may be especially powerful for neuroscience applications such as interrogating the sparse encoding of environmentally-induced neuronal changes, such as memory.

Overall these several recent and emerging techniques allow entrée to beginning to test for the presence of a set of experience-driven epigenetic marks potentially constituting an epigenetic code for memory formation.

Finally, the principal limitation of the studies undertaken so far to test the concept of an epigenetic code for memory is that these studies have not determined in a comprehensive fashion how DNA methylation and chromatin modification at the cellular level gets translated into altered circuit and behavioral function. These studies have identified a small set of candidate genes whose function might be altered in association with memory formation, but no comprehensive model exists for how these (and other) genes are epigenetically regulated, and how these marks are read out to alter neural and circuit function.

How Does the Epigenetic Code Manifest Itself? Gene Readout

The principal mechanism by which epigenetic modifications might result in functional consequences within a cell or a circuit is via modulation of overall gene expression, and accumulating evidence already supports the hypothesis that this is a functional readout of epigenetic marking in the CNS in memory formation. As highlighted above, histone modifications or changes in DNA methylation may result in both increased and decreased transcription of a specific gene. This is most clear in terms of the differentiation and development of individual cells, which requires specific patterns of gene expression to initiate and perpetuate cellular phenotypes. To produce the diverse array of cell classes despite working with identical underlying genetic material, cells must be capable expressing or repressing a given set of genes to generate a neuron, a hepatocyte, or a hematocyte. The major mechanisms underlying this differentiation appear to be epigenetic in nature,[122] and ensure that a given cell lineage can be maintained through multiple rounds of cell division or prolonged life in the case of non-dividing cells.

Contemporary models consider the histone code to be more of a "language" that controls gene expression instead of a strict code where certain combinations of modifications always generate an identical response. For example, a recent paper employed a Chip-Seq technique to analyze histone modifications across the human genome, and found that a specific combination of 17 modifications tended to co-occur at the level of the individual nucleosome and was associated with increased gene expression.[13] Importantly, this group of modifications was observed at thousands of gene promoters, indicating that it is a relatively general mechanism by which histone modifications may alter gene transcription.[13] However, this study also makes it clear that, although small groups of certain modifications tend to occur together, these modifications are only correlated with (rather than explicitly predictive of) increased gene expression. Moreover, exact combinations of modifications across a nucleosome are seldom repeated at different genes, indicating complex and gene-specific regulation of histone modifications. Thus, the histone code hypothesis has since been modified to consider both the context of a specific modification as well as the final outcome.[15,123]

Theoretically, the incorporation of multiple histone modifications into a code could occur in a number of ways. For instance, a specific modification may recruit other histone modifying enzymes that either repress or facilitate nearby marks (see Figure 5.1B). This appears to be the case with phosphorylation at Ser10 on H3, which both represses methylation at lysine 9 and encourages acetylation at lysine 14.[44–46] Interestingly, this type of interaction may occur between different histone tails as well as on the same tail.[124] Another possibility is that, although certain marks may act as transcriptional repressors under some cases, they may facilitate transcription in the presence of another mark on the same histone tail. This would explain why a number of histone modifications have been associated with both transcriptional activation and transcriptional repression, and sets of marks that are both independently correlated with transcriptional activation do not necessarily always occur together.[12] Yet another means by which specific histone modifications could combine to produce a unique epigenetic signature is via the inherent kinetics underlying each reaction. Acetylation and phosphorylation are likely reversed very rapidly after undergoing alteration, whereas histone methylation may persist for longer periods of time. This would allow two mechanisms to control synergistically gene expression despite having no direct interactions.

As discussed above, it is well understood that certain histone modifications interact with each other by preventing access to or recruiting histone-modifying enzymes. However, it is less clear how DNA methylation affects histone modifications and vice-versa. One possibility is that specific DNA methylation patterns are established and maintained by specific combinations of chromatin modifications. For example, HDACs are known to interact with DNMTs, whereas transcription factors that recruit HAT enzymes can trigger demethylation of DNA.[125,126] Likewise, HDAC inhibitors are capable of inducing DNA demethylation.[95,127]

Conversely, it is also possible that DNA methylation regulates important aspects of chromatin state, indicating a bi-directional relationship between histone and DNA modifications. Consistent with this hypothesis, MeCP2, which binds preferentially to fully methylated DNA, can associate with both HDAC machinery as well as histone methyltransferases to alter specific histone modifications.[128–130] However, regardless of the direction of this relationship, it is now clear that these modifications can and do interact with each other within the brain. For example, DNMT inhibition blocks increased H3 acetylation associated with memory formation.[75] Moreover, although DNMT inhibitors also block memory formation as well as synaptic plasticity in the hippocampus, each of these deficits is reversed by pretreatment with an HDAC inhibitor.[75] Taken together, these results reveal a complex relationship between histone modifications and DNA methylation, and suggest that simple considerations of a "histone code" or "DNA methylation code" will each be inadequate in terms of predicting transcriptional output. The interactions of both mechanisms need to be fully understood in order to formulate a more comprehensive epigenetic code hypothesis for transcriptional regulation in memory.

Similarly, it is increasingly clear that epigenetic mechanisms in neurons are also involved in regulating overall transcription rates of specific genes in the establishment, consolidation, and maintenance of behavioral memories. Thus, contextual fear conditioning induces a rapid but reversible methylation of the memory suppressor gene *PP1* within the hippocampus, and demethylation of *reelin*, a gene involved in cellular plasticity and memory.[78] Importantly, each of these DNA methylation changes is functionally relevant, leading to decreased expression of PP1 and increased expression of reelin.[78] Moreover, consistent with the finding that blocking DNA methylation in the anterior cingulate cortex prevents remote memory maintenance, another study reported long-lasting changes in methylation of the memory suppressor gene *calcineurin* within this brain area following contextual fear conditioning.[76] Specifically, changes in methylation at the *calcineurin* gene were not observed immediately after conditioning, but once generated persisted at least 30 days following conditioning, which was the longest time point tested. Clearly, this change is stable enough to maintain a memory over time despite ongoing cellular activity and molecular turnover, indicating that it is an excellent candidate for a molecular storage device. Likewise, histone modifications have repeatedly been associated with changes in gene transcription and expression in multiple organisms, systems, and brain subregions.[20,21,87,131–133] Thus, these results reveal that even within non-dividing neurons in the adult CNS, epigenetic mechanisms regulate patterns of gene expression in a functionally relevant manner.

Alternative Splicing

A related means for epigenetic control of gene expression involves the unique regulation of specific protein isoforms, or differently spliced versions of the same protein. This can occur in multiple ways, such as increased expression of one exon over another competing exon or silencing of an entire exon. By regulating the expression of splice variants with different cellular functions or different affinities for effector proteins, this mechanism is able to expand the potential uses of the same gene locus in a multiplicative fashion.[134]

The mechanisms that regulate alternative splicing are currently unclear. However, histone modifications appear to modulate this process by recruiting different splicing regulators that determine splicing outcome.[135] DNA methylation is also likely involved in the differential expression of *BDNF* exons following fear learning.[74] Contextual fear conditioning produces a rapid increase in mRNA for BDNF exon IV and a concomitant decrease in methylation at this gene site in area CA1 of the hippocampus. Interestingly, context exposure alone (but not conditioning) produced increases in BDNF exon I and VI mRNA, which also corresponded to decreased CpG methylation at these sites. Moreover, intra-hippocampal infusions of the DNMT inhibitors zebularine or RG108 impaired fear memory expression, despite the fact that they *increase* expression of all BDNF exons in naïve animals. Importantly, the same study reported that in animals that underwent contextual fear conditioning, zebularine blocked the learning-related decreases in *BDNF* exon IV methylation. Together, these results reveal that DNA methylation regulates BDNF expression in a complex, experience-dependent manner, and that the effects of DNMT inhibitors likely depend on the overall behavioral and cellular context. Indeed, this experience-dependent regulation of specific BDNF isoforms by DNA methylation represents the clearest evidence supporting the use of a CpG methylation "code" in the formation and consolidation of behavioral memories.

These observations also reveal a general mechanism by which DNA methylation may contribute to neuronal function. By regulating the expression of specific protein isoforms, DNA methylation could influence the creation of differently spliced versions of the same protein. This so-called alternative splicing can account for the differential regulation of multiple proteins with potentially distinct functions within a cell, essentially expanding the amount of information encoded in the proteome without altering the number of genes within the genome.[134] Alternative splicing can take a number of different forms, such as selection of one exon over another competing exon and exclusion of an entire exon within a gene. These variations can alter the efficiency of the protein in modulation of specific targets, thereby changing protein function. Indeed, other studies have also reported that specific BDNF transcripts are used differently in other brain regions as well.[136–138] Moreover, DNA methylation status at different exons at other genes clearly contributes to splice expression profiles.[139] Future studies will be required to determine how this rather amazing temporal and genetic selectivity is achieved.

Imprinting and Allelic Tagging

Adult fully differentiated cells in placental mammals can manifest differential handling of paternal and maternal copies of somatic genes, a phenomenon referred to as *imprinting*. Thus specific genes expressed in non-germline cells including neurons, genes which are not on the X or Y chromosome, can be "imprinted" with DNA methylation marks identifying them as originating from the mother's versus the father's genome. These imprinting marks cross the generations through the germline, and mark a particular copy (allele) of a gene as having originated with the mother versus the father. In traditional cases of genetic imprinting, one copy of the gene is fully silenced, leaving one parent's copy of the gene the exclusive source of cellular mRNA product of that gene.

One prominent example of an imprinted gene involved in controlling cognition is the *ube3a* gene, which encodes a ubiquitin E3 ligase. Imprinted alleles of the *ube3a* gene are preferentially expressed in a brain subregion-specific fashion: for example, the maternal copy is selectively expressed in neurons in the cerebellum and forebrain, including the hippocampus.[140] When the maternal copy of the *ube3a* gene is mutated, this results in Angelman syndrome, a profound disability manifesting autism, severe learning and memory deficits, and a virtually complete absence of speech learning. Studies of Angelman syndrome were the first to implicate an epigenetic mechanism (imprinting) in controlling learning, memory, and synaptic plasticity.[140] Thus, mice with a *maternal* deficiency in UBE3A function have deficits in hippocampus-dependent learning and memory and a loss of hippocampal long-term potentiation at Schaffer/collateral synapses.

For many years imprinting of genes in the adult CNS was assumed to be restricted to a few genes, 30–50 or so being a common assumption. However, gene imprinting has recently been found to occur at much higher levels than this: a recent pair of exciting papers from Catherine Dulac's laboratory has greatly expanded our view of the importance of gene imprinting in CNS function in the adult nervous system.[141,142] This work from Dulac and colleagues demonstrated that over 1300 gene loci in the adult CNS manifest differential read-out of the paternal versus maternal allele. Many of these differentially regulated genes also exhibited brain subregion-selective expression as well. These findings identify parental expression bias as a major mode of epigenetic regulation in the adult CNS, and one important implication of these studies is that epigenetic control of the expression of parent-specific alleles is a driving factor for regulating gene transcription broadly in the brain.

The control of the specific expression of one parental allele versus another through imprinting of genes in the mature CNS may greatly increase the complexity and subtlety of transcriptional control that may be operating in cognition. The traditional view of imprinting has been based on inheritance patterns and involved essentially all-or-none silencing of one allele. The work of Dulac and colleagues may necessitate a re-definition of imprinting to incorporate the concept of widespread partial attenuation of one allele with differential readout and handling of paternal versus maternal alleles.

We speculate that one function of this genetic parent-of-origin effect may be "allelic marking" of specific copies of a gene within a cell. Thus, we propose that, in the case of some genes within the CNS, evolution may have selected for a mechanism whereby one allele of a gene (e.g. the paternal copy) can be tagged and identified so that it can be regulated differentially from another copy of the same gene in the same cell. This powerful mechanism would allow a single cell to in effect have two copies of an identical gene (i.e. protein-encoding DNA sequence), so that it could selectively regulate one copy (allele) in one way and the other copy (the other allele) another way. For example, one copy might be used selectively during development, epigenetically marked as appropriate for regulating cellular phenotype during differentiation, and then silenced in the adult CNS. The second copy could then be used in the adult as a fresh template of the gene that would be available for epigenetic and transcriptional regulation uniquely needed for ongoing function in the adult. A second example might be that the two alleles of the same gene within a single cell might be differentially regulated through selective regulation of alternative splicing of products produced from each copy of the gene – this would allow a cell, using epigenetic mechanisms, to produce one specific pattern of exon expression from one allele and a different pattern of exon readout from the other allele.

This concept is motivated by the recent emergence of understanding the overall importance of epigenetic transcription-regulating mechanisms in the function of the adult CNS. Epigenetic mechanisms must of necessity operate directly upon the nucleotides encoding a gene, through either chemical covalent modification or regulation of its three-dimensional structure. Moreover, a single allele of a gene lasts the entire lifetime of a cell. Thus, in any cell a given gene is available with at most two copies for the cell to use. Treating both copies identically would greatly limit the options that the cell has available to it in terms of regulating the readout of that gene product. After all, a gene is not like an mRNA or a protein that can have thousands of individual copies of itself present. A single allele of a gene is *the only chemical copy of that allele* that is present in the cell, and epigenetic molecular mechanisms operate directly on that single copy. Treating the two alleles of a gene differentially then would provide two different gene templates that could be differentially regulated by epigenetic mechanisms. Epigenetic marking of the paternal versus maternal alleles would be a prerequisite for this sort of differential epigenetic handling.

One speculative example, which we use strictly for illustrative purposes, would be the following. Within a single neuron, the tagged paternal allele of the BDNF gene might be used exclusively during development, and epigenetically regulated as appropriate for its role during early life. The maternal BDNF allele might then be reserved for use in the adult, wherein memory-associated epigenetic mechanisms might operate upon a fresh template of the gene as necessary for triggering short- or long-term activity-dependent changes in BDNF transcription. Imprinting-dependent epigenetic marking of the paternal versus maternal alleles would be a prerequisite for this sort of differential epigenetic handling. While these ideas are completely speculative at present, the underlying thinking is driven by recent discoveries concerning the widespread utilization of epigenetic transcription-regulating mechanisms in experience-dependent processes in the CNS, and the conservation of utilization of homologous epigenetic mechanisms in both development and ongoing function of the adult CNS.

While speculative, this model is consistent with Dulac's finding that the "imprinted" genes were not completely silenced in most cases, but were instead differentially regulated in a relative but not all-or-none fashion as would be the case with typical imprinting. These ideas are also consistent with the observation that parent-of-origin effects appear to be particularly prominent in the CNS, where active epigenetic transcriptional regulation appears to be particularly important for mental function. It is an interesting possibility that the existence of these mechanisms would greatly increase the capacity of ongoing epigenetic molecular mechanisms in the CNS to regulate cognitive function.

EPIGENETICALLY BASED DISORDERS OF COGNITION AND NOVEL THERAPEUTIC TARGETS

Recent advances in our understanding of the role of epigenetic molecular mechanisms in CNS development and ongoing adult cognitive function have allowed the conceptualization of epigenetically based CNS disorders (Table 5.1). In terms of the cognitive domains of CNS

TABLE 5.1 Selected List of Psychiatric Disorders and Syndromes with Epigenetic Origins or Treatments

Disorder/disease	Epigenetic Dysregulation	Potential Treatments
Rett syndrome	Mutation in *MeCP2* gene	HDAC inhibitors
Age-associated cognitive decline	Impaired H4K12 acetylation in response to learning event	HDAC inhibitors
Schizophrenia	Hypermethylation of *reelin* gene, decreased H3K4me3 and increased H3K27me3 at GAD 67 promoter	DNMT inhibitors, HDAC inhibitors
Rubinstein–Taybi syndrome	Mutation of *CBP* gene	HDAC inhibitors
Drug addiction	Multiple changes in DNA methylation and histone modifications at striatal plasticity genes, increased MeCP2	HDAC inhibitors specifically during extinction training
Coffin–Lowry syndrome	Mutation in *RSK2* gene	PP1 inhibitor
Alzheimer's disease	Aberrant histone acetylation and phosphorylation, DNA hypomethylation at several genes	HDAC inhibitors, S-adenosylmethionine, methyl-donor rich diets
Depression	Increased H3K27me2 at specific promoters of *BDNF* gene, downregulation of HDAC5 in hippocampus, DNA methylation differences in catecholamine-signaling genes	HDAC inhibitors, specific HMTs or HDM inhibitors
Angelman syndrome	Abnormal DNA methylation-related imprinting of maternal alleles	–
Prader–Willi syndrome	Abnormal DNA methylation-related imprinting of paternal alleles	–
Fragile X syndrome	Hypermethylation of DNA at *FMR* genes	–

function, several diseases and disorders of cognitive function have received a considerable amount of attention, and we will focus on these in this section. One major category of cognitive disorders are intellectual disabilities and, in several instances, epigenetically driven mechanisms have been implicated as playing a role in learning and memory disruption in these clinical syndromes. We have already touched on several relevant examples such as Angelman syndrome and Rubinstein–Taybi syndrome, Fragile X mental retardation (FMR) and, finally, what could be described as the prototypical epigenetic disorder, Rett syndrome. In addition, a considerable amount of recent work has implicated disruption of epigenetic molecular mechanisms as a player in cognitive disorders at the other end of the developmental spectrum, that is aging-related cognitive disorders such as Alzheimer's disease and various other neurodegenerative conditions. Other disease categories wherein an explosion of recent progress has implicated derangement of epigenetic mechanisms are drug addiction and various psychiatric disorders, which are covered in other chapters of this book. Given the protracted and often devastating nature of these disorders, drugs that target any underlying epigenetic mechanisms could provide potentially groundbreaking therapeutic avenues.

In this section, we will first discuss recent exciting findings that implicate disruptions of histone-modifying processes in memory dysfunction specifically, and use this as an example wherein manipulating histone acetylation might prove therapeutically beneficial. At the end of this section, we will touch on the emerging possibility of histone methylation and other histone marking enzymes as potential therapeutic targets. These topics are discussed in much greater detail in Chapter 8. However, we provide a brief overview here because of the specific relevance of these results to learning and memory processes. Indeed, the pharmacologic effects upon memory of various agents that affect the epigenome provide some of the most convincing evidence supporting a role for epigenetics in the cognitive function of the adult CNS, the central topic of this chapter.

Histone Acetylation and HDAC Inhibitors

A major potential target for therapeutic intervention involving the epigenetic machinery of the CNS involves drugs that inhibit the removal of acetyl groups on histone tails, specifically HDAC inhibitors.[95,143] This class of drugs, which includes trichostatin A (TSA), suberoylanilide hydroxamic acid (SAHA), and sodium butyrate, inhibit several isoforms of HDAC enzymes, and result in histone hyperacetylation in a global manner. A number of these drugs have already been approved for clinical use in patients or are currently in clinical trials in the cancer arena.[95]

As discussed above, histone acetylation is robustly associated with "activated" gene transcription, and the formation of new memories produces increases in histone acetylation in the hippocampus.[23] In this context, pioneering studies in this area demonstrated that treatment with HDAC inhibitors improves memory formation in hippocampus-dependent tasks, and enhances the formation of LTP in the hippocampus.[7] Moreover, HDAC inhibitors have been shown selectively to reverse deficits in histone acetylation in aged animals, effectively restoring the ability to learn new associations.[23] Finally, even after the induction of severe neuronal atrophy, HDAC inhibitors restore memory formation and even enable access to previously formed long-term memories.[19] This result is especially exciting given that a number of patients who present with dementia or Alzheimer's disease are already affected by an inability to retrieve previously formed memories.[144]

Importantly, the memory-enhancing effects of HDAC inhibitors may be mediated by specific HDAC isoforms. Selective overexpression of HDAC2 in neurons produces a decrease in spine density and impairs synaptic plasticity and memory formation, whereas overexpression of HDAC1 had little effect.[145] Likewise, deficiency in HDAC2 or chronic treatment with HDAC inhibitors resulted in increased spine density and improved memory function.[145] In contrast, another study indicated that systemic inhibition of HDACs (and specifically class 1 HDACs) dramatically improved contextual memory function in a mouse model of Alzheimer's disease.[146] Finally, Marcelo Wood and his colleagues have presented compelling evidence for the involvement of the HDAC3 isoform in regulation of synaptic plasticity and memory.[147,148] Thus, future research will be required to parse the effects of isoform-specific HDAC inhibitors on memory function in normal, aged, and diseased mouse models. Nevertheless, the use of HDAC inhibitors in treating learning and memory disorders or neurodegenerative diseases possesses clear therapeutic potential.

Histone Methylation

Emerging results also indicate that treatments that increase histone methylation at specific sites may be useful candidates for a range of learning and memory disorders. Recent studies have indicated that aberrant histone demethylase signaling appears to be relevant for human learning disorders, as mutations in the JARID1c gene (a lysine demethylase) have been associated with autism and X-linked mental retardation.[149–151] Thus, although the therapeutic potential of histone methylation modifying enzymes is relatively unexplored at the present time, these results indicate that selective antagonists of H3K4 demethylating enzymes such as LSD1[34] may be interesting candidates for learning and memory disorders. Currently, our understanding of the role of histone methylation in cognition and cognitive disorders remains in its infancy. Thus, future research will be required to elucidate the complex functional consequences of altered histone methylation and determine which sites and which histone methylation enzymes are good candidates for therapeutic intervention.

DNA Methylation and Cognitive Dysfunction: Insights from Rett Syndrome

Rett syndrome is a disorder that affects around 1 in 10000 to 15000 females. Typically, females with Rett syndrome appear developmentally normal until between 6 and 18 months of age, at which time development stagnates and subsequently regresses. Classic Rett syndrome is characterized by profound cognitive impairment, communication dysfunction, stereotypic movements, and pervasive growth failure.[152] In a breakthrough discovery, mutations in the gene encoding MeCP2 were found to be responsible for at least 95% of classic Rett syndrome cases.[64] This seminal finding provided a link between DNA methylation, specifically involving the methyl-DNA binding protein MeCP2, and intellectual dysfunction.

The identification of *mecp2* as the mutated gene in Rett syndrome led to the creation of several transgenic mouse models of Rett syndrome. Initial attempts to create MeCP2-null mice resulted in embryonic lethality.[153] To circumvent this problem, two groups independently used the Cre/LoxP recombination system to delete portions of the MeCP2 gene. The Jaenisch lab used a targeted construct that deleted exon 3, which encodes for most of the MBD, while the Bird lab deleted exons 3 and 4, which encode for all but the first eight amino acids of the protein.[154–156] MeCP2-null mice from the Jaenisch and Bird labs have impairments in hippocampal physiology and behavior, as well as a number of more general physical deficits including early lethality. Symptomatic male mice have altered hippocampal NMDA receptor expression and impairments in LTP and LTD.[157] Male mutant mice also display deficits in cued fear conditioning, while mutant mice of both sexes display deficits in object recognition and altered anxiety.[158]

The MeCP2-null mice generated by the Bird and Jaenisch labs display an early onset of symptoms and short lifespan that differentiates them from classic Rett syndrome and limits the analysis of symptoms. Two groups have developed models that attempted to address these limitations. The Zoghbi group created a mouse model with a truncated mutation of MeCP2, called MeCP2308 mutants, as the premature stop codon was inserted after codon 308, an area where mutations are located in many humans with Rett syndrome. These mice exhibit a milder phenotype, presumably because the truncated protein retains partial function. MeCP2308/Y mice display slightly impaired motor function, reduced activity, and a stereotypic forelimb-clasping movement. These mice also display hyperacetylation of H3.[159] MeCP2308/Y mice also display abnormal social interactions in the tube test and in some aspects of the resident-intruder test,[160] as well as deficits in spatial memory, contextual fear conditioning, and long-term social memory. They also have impaired LTP, increased basal synaptic transmission, and deficits in the induction of LTD.[66] The Tam group generated another line of MeCP2-null mice by replacing the sequence for the region that binds methylated DNA with a floxed Pgk-neo cassette and producing mutant mice by germline transmission; the resulting mutant mice are called Mecp2tm1Tam mice. Behavioral testing of these mice revealed deficits in cerebellar learning on the rotarod as well as impairments in both cued and contextual fear conditioning and contextual association.[161]

In a collaborative effort, the Zoghbi and Sweatt laboratories showed that MeCP2-deficient animals have deficits in spatial learning, contextual fear conditioning, and LTP deficits.[66] Moreover, they also showed that overexpression of MeCP2 resulted in enhanced fear conditioning, and enhanced LTP.[162] Since Rett syndrome is caused by mutations in MeCP2, enhancing MeCP2 levels could therefore be a therapeutic option. Overall, these findings strongly support the idea that MeCP2 might be involved in regulation of LTP and hippocampus-dependent memory formation. The principal caveat to these studies is that the genetic alterations were present throughout the animals' development. However, Adrian Bird's laboratory recently observed that inducible expression of MeCP2 in adult animals could provide extensive rescue of neurological phenotypes in MeCP2-deficient animals. This finding strongly suggests a dynamic role for MeCP2 in the adult animal. Altogether, these knockout and knock-in mouse studies involving manipulation of MeCP2 activity in the CNS suggest that MeCP2 plays an important role in normal learning and memory and in regulating long-term synaptic plasticity.

Rett syndrome has classically been viewed as a neurodevelopmental disorder, the underlying genetic basis of which is mutation/deletion of the MeCP2 gene and resultant disruption of normal MeCP2 function during prenatal and early postnatal development. This model is consistent with the fact that the mutated gene product is present throughout development. However, the mutant gene product is also present in the fully developed adult CNS. Thus, it is unclear if Rett syndrome is caused exclusively by disruption of MeCP2 function during development, or whether loss of MeCP2 in the mature CNS might also contribute to neurobehavioral and cognitive dysfunction in Rett patients. Indeed, as described above, recent data from Adrian Bird's group has suggested that loss of normal MeCP2 function in the adult nervous system contributes to neurobehavioral dysfunction in Rett syndrome. If MeCP2 functions to control cognition in the mature CNS, cognitive dysfunction in Rett syndrome might in significant part be due to disruption of MeCP2's actions in the fully developed CNS. A new understanding of the role of MeCP2 in the adult CNS might allow the development of new therapeutic approaches to Rett treatment based on restoration or augmentation of MeCP2 function after CNS development is largely finished. An intriguing possibility is that MeCP2 is necessary for normal cognitive function in the fully developed brain; that is, that MeCP2 function is necessary in an ongoing fashion for normal learning and memory and synaptic plasticity in the mature CNS. The central hypothesis concerning a role for MeCP2 in adult CNS function is that the methyl-CpG-binding protein MeCP2 regulates the brain's ability to store memories by actively regulating transcriptional activity during learning and memory formation.

The findings in Rett syndrome patients and in the genetically engineered mouse models are also consistent with other data described above implicating DNA methylation in general as playing an important role in adult memory formation. Thus, one prediction of the hypothesis of a role for active DNA methylation and demethylation in controlling memory formation in the adult CNS is that animals deficient in methyl-DNA binding proteins should have deficits in memory and long-term synaptic plasticity. Indeed, the findings of learning and memory deficits in both Rett syndrome patients and genetically engineered mice is consistent with this hypothesis.

Finally, these observations are consistent with the overall theme we are developing in this chapter, that is a co-opting of developmental molecular mechanisms to subserve long-lasting functional changes in the adult CNS.

SUMMARY AND CONCLUSIONS FOR PARTS 1–3 OF THE CHAPTER

In writing this section of the chapter, we have endeavored to provide an overview of an emerging area at the cross-section of developmental biology and cognitive neuroscience. We have attempted to provide a novel synthesis of ideas across the modalities of DNA methylation and histone post-translational modifications and apply them to the learning and memory field. There are interesting and compelling new questions that arise from recent work in both these areas concerning unanticipated biochemical control mechanisms that may be operating to regulate epigenetic processes in non-dividing neurons in the adult CNS. Both DNA methylation and histone regulation are emerging as critical sites contributing to memory consolidation and memory storage. These molecular epigenetic mechanisms are also beginning to receive attention as general experience-driven molecular mechanisms.[163] Indeed, in a broader sense, behavioral epigenetics is emerging as a subfield in its own right (Table 5.2).

TABLE 5.2 Why *Behavioral* Epigenetics?

Active demethylation and remethylation in non-dividing cells
Non-heritable but yet persisting methylation marks
Only a small subset of methylation marks may be modifiable in neurons, with preservation of methylation patterns encoding cell phenotype
Unique kinetic properties of methylation/demethylation in different brain subregions, according to their cognitive function
Perpetuation of epigenetically-based changes following from neural connectivity
Potentially neuron-specific epigenetic molecular mechanisms such as hydroxymethylcytosine in DNA and selective GADD45 regulation
Allelic tagging in the mature CNS

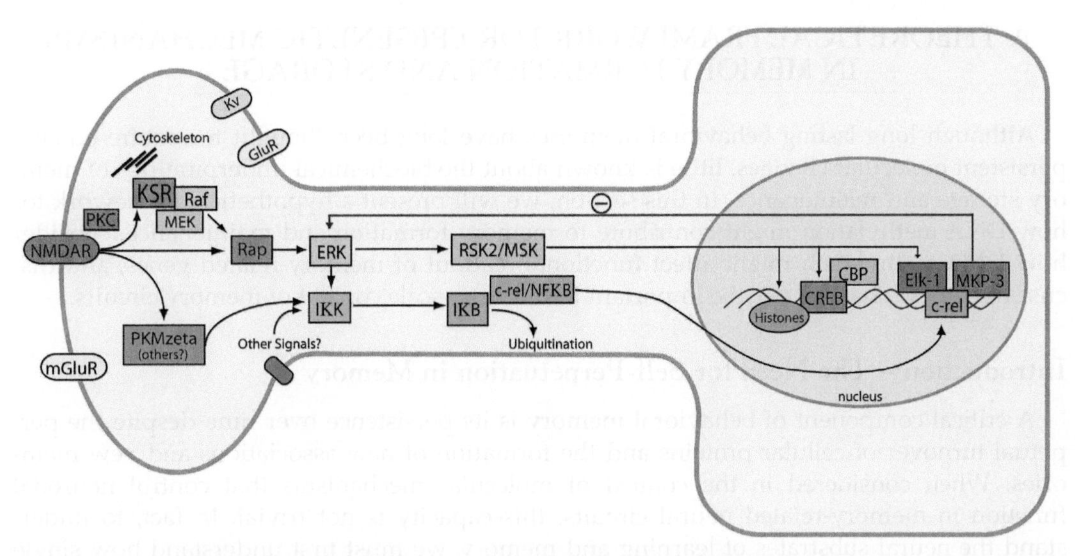

FIGURE 5.8 **Model for roles and regulation of ERK MAP kinase and NFkappaB signaling in epigenetic regulation in mature neurons.** (See text and[22] for discussion.) The ERK/MAP kinase cascade can be activated by a number of receptors and protein kinases within the hippocampus. As such, it can integrate a wide variety of signals and result in a final common output. The ERK cascade is initiated by the activation of Raf kinase via the small GTP-binding protein, ras, or the ras-related protein, rap-1. Activated Raf then phosphorylates MEK, a dual specific kinase. MEK phosphorylates ERK 1 and 2 on a tyrosine and threonine residue. Once activated, ERK exerts many downstream effects, including the regulation of cellular excitability and the activation of transcription factors leading to altered gene expression. Each MAP kinase cascade (ERK, JNK, and p38 MAPK) is composed of three distinct kinases activated in sequence and, despite the fact that many separate MAP kinase families exist, there is limited cross-talk between these highly homologous cascades. While many of the steps of the ERK cascade have been elucidated, the mechanisms by which the components of the MAP kinase cascade come into physical contact have not been investigated. In this context, it is interesting to note that there are multiple upstream regulators of ERK in the hippocampus: NE, DA, nicotinic ACh, muscarinic ACh, histamine, estrogen, serotonin, BDNF, NMDA receptors, metabotropic glutamate receptors, AMPA receptors, voltage-gated calcium channels, reactive oxygen species, various PKC isoforms, PKA, NO, NF1, and multiple ras isoforms and homologs.

Investigators in this area have begun to synthesize a broader "epigenetic code" hypothesis that incorporates both mechanisms of histone modification and mechanisms of DNA methylation/demethylation, and hypothesize how these changes might be integrating signals at the epigenomic level and subsequently triggering lasting changes in memory formation. In this brief review, we have attempted to highlight a few of what we feel are some of the most interesting questions arising from this line of thought.

Finally, one overarching theme of this review is that there is a conservation of "developmental" molecular mechanisms, e.g. growth factor regulation, MAPK signaling and epigenetic mechanisms, to subserve long-term plasticity and memory formation in the adult CNS[1,164,165](Figure 5.8). The growing realization is that beyond being simply *analogous* to each other by virtue of the common ability of transient signals driving persisting changes, development and adult memory are molecular *homologs*, i.e. share identical molecular and biochemical mechanisms.

A THEORETICAL FRAMEWORK FOR EPIGENETIC MECHANISMS IN MEMORY FORMATION AND STORAGE

Although long-lasting behavioral memories have long been thought to require equally persistent molecular changes, little is known about the biochemical underpinnings of memory storage and maintenance. In this section, we will present a hypothetical framework for how DNA methylation might contribute to memory formation and maintenance, consider how DNA methylation might affect functional readout of memory-related genes, and discuss how these changes may be important in the large-scale context of memory circuits.

Introduction – The Need for Self-Perpetuation in Memory

A critical component of behavioral memory is its persistence over time despite the perpetual turnover of cellular proteins and the formation of new associations and new memories. When considered in the context of molecular mechanisms that control neuronal function in memory-related neural circuits, this capacity is not trivial. In fact, to understand the neural substrates of learning and memory, we must first understand how single molecules or groups of molecules in the brain can store information across time despite the relatively rapid turnover of every protein within a cell.[166,167] At the molecular level, the perpetuation of a specific activation state that contributes to cellular memory requires that individual molecules have the ability to self-perpetuate a given type of activity despite turnover and in the absence of the stimulus which initiated that activation state. This has been hypothesized to occur via a specific type of biochemical reaction in which the active or mnemogenic form of a molecule is able to autoconvert a nascent form of itself into an active form. This type of reaction has long been hypothesized to underlie memory,[2–5] making the search for a molecule capable of autoconversion a critical step in elucidating cellular as well as systems-level memory storage.

Searching for Self-Perpetuating Molecular Activity: The Case of CaMKII

The historical search for a molecular storage device for long-term memory is perhaps best encapsulated by the story of Ca^{2+}/calmodulin (CaM)-dependent protein kinase II (CaMKII). As its name implies, CaMKII is activated by calcium ions as well as calmodulin. When activated, CaMKII alters its physical structure, such that the catalytic domain (which contains the substrate binding site) can become active (Figure 5.9). Increased synaptic calcium causes CaMKII to translocate to post-synaptic sites, where it can bind to the NMDA receptor, placing it in an ideal location to regulate synaptic strength.[168,169] However, in the case of very strong synaptic stimulation, prolonged activation by calcium also allows a site on the inhibitory domain of CaMKII to undergo phosphorylation at threonine-286 by a nearby subunit.[170] This phosphorylation prevents the inhibitory subunit from binding to the catalytic domain, effectively increasing the duration of CaMKII activity. Moreover, because this site can continue to be phosphorylated by neighboring subunits within the holoenzyme in the absence of Ca^{2+}/calmodulin, CaMKII also essentially possesses a unique form of "autonomous" activity in which the catalytic domain can remain in an active state in the absence of the initial stimulus.[171] In essence, this auto-activation would conceptually serve as a cellular

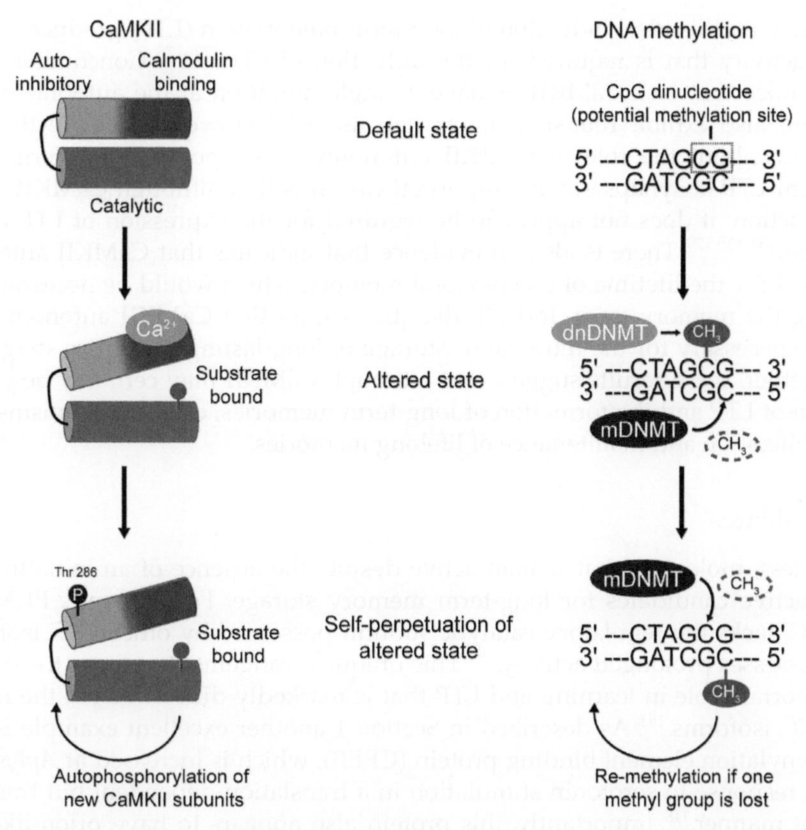

FIGURE 5.9 **Self-perpetuation mechanisms in learning and memory.** Left: CaMKII possesses a catalytic region that is usually bound by the autoinhibitory subunit (first panel), resulting in an inability to act at possible substrates such as the NMDA receptor. When activated by calcium or calmodulin, this autoinhibition gate is opened, allowing the catalytic subunit to bind substrates (second panel). Subsequently, in the absence of the original calcium stimulus, autophosphorylation of threonine 286 on the autoinhibitory domain prevents binding to the catalytic domain, preserving catalytic activity (third panel). Right: A hypothetical model for how DNA methylation can be self-perpetuating through the combined activities of de novo and maintenance DNMT activities. In this model DNA possesses CpG dinucleotides capable of undergoing methylation (first panel). De novo DNMTs methylate previously unmethylated CpG sites. Then, maintenance DNMTs recognize hemi-methylated DNA and methylate the complementary strand (second panel). Together, these reactions produce a change in transcriptional processes. Even with spontaneous or enzymatic removal of methyl group at one cytosine residue, maintenance DNMTs have the capacity to remethylate DNA to preserve transcriptional change. It is important to note that this model of methyl-cytosine self-perpetuation in non-dividing neurons is still somewhat speculative at this point. However, there exist data for "cycling" of cytosine methylation, i.e. sequential demethylation/remethylation of cytosine (see main text), consistent with the idea that demethylation/remethylation occurs in non-dividing cells of several sorts, including neurons. Also, it is well established that this self-perpetuating hemi-methylating activity of DNMT1 occurs at the DNA replication fork during cell division.

memory trace by which the previous calcium stimulation is "remembered". This type of autophosphorylation event has long been proposed to underlie the molecular storage of memory in part because it can outlast the relatively rapid protein turnover within a cell, and thereby perpetuate a biochemical signal long after the initiating stimulus has ceased.[171,175]

In keeping with this idea, induction of long-term potentiation (LTP) produces an increase in CaMKII activity that is required for the induction of LTP.[172–175] Concomitant with this LTP deficit, mice that lack CaMKII or have a single mutation at the autophosphorylation site (Thr 286) also exhibit robust memory deficits.[176–178] However, despite these results, doubts remain about the role of CaMKII autonomy in supporting long-term memories and persistent LTP at synapses. One important caveat is that although CaMKII is required for LTP induction, it does not appear to be required for the expression of LTP once it has been induced.[154,155,179] There is also no evidence that indicates that CaMKII autonomy can be maintained for the lifetime of a behavioral memory, which would be necessary if it was perpetuating the memory trace. Indeed, the observation that CaMKII autonomy does not appear to be necessary for the retrieval or storage of long-lasting memories suggests otherwise.[179] Together, these results suggest that although CaMKII may certainly be involved in the initiation of LTP and the formation of long-term memories, other mechanisms may support the stabilization and maintenance of lifelong memories.

Other Candidates

Nevertheless, molecules that remain active despite the absence of an initiating stimulus remain attractive candidates for long-term memory storage. For example, PKMζ, an isoform of PKC, lacks the regulatory catalytic subunit possessed by other PKC isoforms, and therefore possesses prolonged activity.[180] This unique arrangement appears to confer PKMζ with an important role in learning and LTP that is markedly different from the role played by other PKC isoforms.[181] As described in Section 1 another excellent example is cytoplasmic polyadenylation element binding protein (CPEB), which is increased at *Aplysia* sensory synapses in response to serotonin stimulation in a translation-dependent but transcription-independent manner.[118] Importantly, this protein also appears to have prion-like capabilities, with two different conformational forms.[118,120] In this case, the active conformation is self-perpetuating, as it is able to convert inactive CPEB into active CPEB. More recent work has revealed that the active conformation of CPEB is necessary for the long-term maintenance of synaptic facilitation at these synapses, providing an interesting corollary to long-term behavioral responses.[117] Analogous proteins have also been discovered in the mouse hippocampus,[182] suggesting the potential for a conserved role in synaptic plasticity across species. However, future studies will be required to examine more thoroughly a role for CPEB in mammalian memory storage.

A Role for Epigenetics?

The epigenome sits at the interface of the cellular environment and the genome, enabling epigenetic changes to exert robust control over transcriptional processes. In somatic cells, epigenetic influences are responsible for cellular differentiation and the perpetuation of the cellular phenotype over time and across cell division.[183,184] Unlike a number of cytoplasmic protein modifications, epigenetic mechanisms possess a number of features that are consistent with a molecular storage device for long-term memory. First, these modifications are believed to be relatively stable in comparison to other alterations. Secondly, this class of modifications is capable of altering gene expression directly, and is therefore able to modulate gene programs known to be involved in learning and memory. As changes in gene transcription are known to be necessary for memory stabilization, these characteristics raise the intriguing possibility that epigenetic mechanisms within non-dividing cells in the CNS are co-opted to support long-lasting behavioral changes in response to particular types of experience. Although these epigenetic changes include a number of unique mechanisms (including histone modifications, microRNA activity, and proteins with prion-like activity), for the purposes of this section of the chapter, we will focus our discussion on the potential role of DNA methylation in learning and memory.

DNA Methylation

Perhaps the most intriguing epigenetic modification is the methylation of individual cytosine residues on DNA itself. Indeed, one of the earliest theoretical papers concerning the molecular mechanisms that might underlie long-term memory[185] speculated that DNA methylation might contribute to memory encoding. Although their specific model, "DNA Ticketing", is untenable in light of our current understanding of the molecular biology of memory, this early paper accurately envisions the potential of a role for chemical modification of DNA in memory formation and storage. As an aside, the paper is also fascinating to read because the writing style and freedom to speculate that are manifest in the manuscript are no longer seen in "high-profile" journals in the modern era.

A quick reiteration of the biochemistry of DNA methylation is called for at this point to provide a refresher for how this type of covalent modification might participate in a self-perpetuating chemical reaction underlying memory. You will recall that DNA methylation is catalyzed by two different forms of DNA methyltransferases (DNMTs), both of which are expressed in mature neurons.[73,186] De novo DNMTs are able to methylate previously unmethylated cytosines, whereas *maintenance* DNMTs recognize hemi-methylated DNA and methylate the unmethylated strand (see Figure 5.9). For both of these reactions, S-adenosyl methionine serves as the methyl donor. The result of these two mutually reinforcing enzymatic activities is an incredibly powerful biochemical reaction which has the ability to restore itself if lost. For example, even if one methyl group on a cytosine is somehow removed from one strand, maintenance DNMTs will recognize the methylated complementary strand and re-establish a methylated cytosine on both DNA strands. This peculiar arrangement highlights an important feature of DNA methylation, which is that despite ongoing events that may remove a methylation mark on one strand, a substrate exists on the other strand to perpetuate effectively that mark through time (see Figure 5.9). In fact, this reaction has been proposed to serve as one of the mechanisms that underlie lifelong inactivation of one X chromosome in females.[187,188] Thus, this reaction has the same basic form as self-perpetuating CaMKII autophosphorylation, but has been observed to persist across time and despite ongoing cellular stimulation.[76] Therefore, we view this mechanism as a key candidate for the molecular storage of long-term memory.

DNA Methylation as a Molecular Component of the Engram

Memory is a multicomponent process consisting of distinct phases that require distinct changes within the CNS. Unlike short-term memory (STM), the formation and maintenance of long-term memory (LTM) requires the synthesis of new protein within the brain,[189] suggesting that it is mediated by a mechanism or group of mechanisms that operate within the nucleus of a cell. Given that DNA methylation plays a major role in transcriptional regulation and is capable of perpetuation of potentially life-long changes in gene expression, we and others became interested in determining whether changes in DNA methylation are required for the formation and storage of behavioral memories.

Stable DNA Methylation Changes in Memory Maintenance

Although the data discussed above in Section 1 of this chapter demonstrate that DNA methylation is important for the establishment of hippocampus-dependent memories, all of these changes reverse to baseline within 24 hours of conditioning. Clearly, this rapid removal of altered methylation marks is not consistent with the memory-related biochemical reaction discussed above. However, although the hippocampus has been implicated in the formation and consolidation of memories, its precise role appears to be temporally limited, and it is not believed to be the ultimate storage site for multiple forms of long-term memory.[190–192] Rather, the maintenance of long-term remote memories is thought to occur within the cortex. Indeed, a number of studies have now revealed that neural activity in some cortical areas increases in a manner consistent with memory storage, and remote memories can be disrupted by manipulation of these cortical areas.[193–196]

With this in mind, one might predict that if patterns of DNA methylation in the brain are involved in the regulation of behavioral memories, they should reflect or incorporate this system's consolidation. For example, given that the formation (but not long-term storage) of certain forms of fear-related memories are dependent upon the hippocampus, we may expect that DNA methylation in this region is uniquely conducive to memory formation as compared to methylation patterns in other regions. The findings discussed above indicate that this is the case: DNA methylation is rapidly altered in the hippocampus following the formation of a fear memory, but returns to naïve control levels within 24 hours. Likewise, if cortical areas such as the prefrontal cortex (PFC) underlie the maintenance and/or storage of memories, it is possible that cortical DNA methylation profiles differ from those in the hippocampus.

To determine if this is the case, a recent report examined methylation at the several memory-related genes in the dorsomedial PFC (dmPFC) at different time points following contextual fear conditioning, the same paradigm used in the studies discussed above.[76] This study found that *calcineurin* underwent robust methylation in its CpG-rich promoter region within 1 day following learning. Moreover, this effect persisted for up to 30 days following conditioning (this was the longest time point tested), and corresponded to changes in calcineurin mRNA and protein levels. Interestingly, other changes in DNA methylation were also observed, but were either more transient (e.g. increased *reelin* methylation) or not association-specific (e.g. decreased *Egr1* methylation, which also occurred in the context-only group). Thus, fear learning was found to induce long-lasting but gene-specific changes in DNA methylation.

Finally, this study also revealed that DNA methylation in the dmPFC is required for the expression of the contextual fear memory, as infusion of three different DNMT inhibitors into this area prior to a probe test 30 days after contextual fear conditioning all significantly weakened the retrieval of fear memory. Conversely, the same infusions had no effect on 30-day memory retrieval when delivered within 2 days after the initial learning experience. In combination with the reports reviewed above, these findings suggest that DNA methylation changes are used to support both the formation and maintenance of fear memories in a time-dependent, regionally specific manner. Moreover, the difference between DNA methylation changes in the cortex and hippocampus are entirely consistent with the function of these structures in the establishment and long-term consolidation of memory. Taken together, these observations are highly consistent with the hypothesis of self-perpetuating methylation in the dmPFC, and suggest an ongoing need for methylation maintenance, and the existence of a true self-perpetuating reaction (see Figure 5.9) in this brain region for the maintenance of memory.

DNA Methylation in Other Memory Systems

Importantly, the memory deficits induced by DNMT inhibitors do not appear to be specific to the class of memories mentioned above. Indeed, recent studies have elegantly demonstrated that conditioned place preference memory is also regulated by DNA methylation in a spatially and temporally distinct fashion. Thus, microinfusions of the DNMT inhibitor 5-aza-2-deoxycytidine into the hippocampus essentially abolished the formation of a conditioned place preference (CPP) for cocaine reward, but had no effect on the retrieval of a previously established cocaine CPP.[197] Interestingly, this effect was reversed in the prelimbic cortex, where infusions of 5-aza-2-deoxycytidine failed to abolish the formation of cocaine CPP, but eliminated the retrieval of an established CPP.[197] This observation lends credence to the idea that memory formation and storage are supported by unique changes in DNA methylation operating in the hippocampus and cortex, respectively (see Figure 5.7). However, the effect of changes in DNA methylation on the establishment of memory does not appear to be limited to these regions. Thus, another recent report found that DNMT inhibition in the nucleus accumbens (NAc) actually boosted the formation of a cocaine CPP, whereas virally-induced overexpression of DNMT3a in the NAc impaired CPP.[198] Moreover, DNMT methylation also regulates spine density in the NAc,[198] a region which has previously been implicated in reward learning and the formation of addiction-related memories.[199,200] Thus, these diverse results indicate that DNA methylation is important if not critical for the formation and storage of many different behavioral memories, but likely plays very unique roles within different neural structures and in relation to different types of memory.

DNA Methylation in the Context of Other Epigenetic Modifications

We would also note that these data have built on a large literature implicating post-translational modification of histone proteins in learning and memory.[7,10,20,23,27,201–203] Thus, we do not feel that DNA methylation represents the only nuclear change capable of modulating memory formation and maintenance. Rather, we believe that changes in DNA methylation may interact with other epigenetic changes to regulate the expression of genes that are critical to memory stabilization. Consistent with this idea, a number of studies have revealed that DNA methylation changes act in concert with histone acetylation and methylation,[20,204–207] and that this interaction is important for the consolidation of memory.[75]

Specificity and Integration of DNA Methylation

Taken together, these findings pose several important questions. First, how are specific gene loci targeted over others for changes in methylation during the formation and/or storage of a memory? Indeed, although a number of studies have reported changes in DNMT levels that are presumably cell-wide phenomena, the observation that DNA methylation can both increase and decrease at different genes following the same behavioral time point is initially perplexing. In fact, very little is known about how changes in DNA methylation can target specific genes or even specific exons within a gene. One possibility is that specificity in the sequence of DNA itself confers this trait via interactions with transcription factors that target different sets of DNMTs or DNA binding proteins to MeC sites. Additionally, gene selectivity in DNA methylation may be related to the fact that transcription-factor dependent transcription can potentially establish other epigenetic marks (such as histone methylation or acetylation) that in turn interact with subsequent DNA methylation profiles. For example, Cfp1, and zinc-finger binding domain protein, is associated with lysine 4 trimethylation on histone 3 as well as paucity of methylation on the corresponding DNA.[206] However, the implication of this gene (and within-gene) selectivity indicates that methylated DNA sites can rapidly and selectively control transcriptional function in a way that other mechanisms may not be able to.

A second question concerns the functional role of methyl binding proteins in transcriptional regulation. Although methylation of DNA was long thought to inhibit transcription by association with transcriptional repressors like MeCP2, it is now clear that this simple description is not adequate. In fact, MeCP2 binding can be associated with both transcriptional activation and transcriptional repression, depending on its association with different transcription factors.[67] Thus, recent conceptualizations have argued that single modifications and binding proteins should be viewed within a complex epigenetic landscape that operates to integrate multiple cellular signals that combine to control gene transcription.[208] Nevertheless, future studies will be required to determine under what circumstances methylated DNA binding proteins result in transcriptional repression or activation, and whether these circumstances differ from gene to gene within the central nervous system.

DNA Methylation in Relation to Synaptic and Circuit Plasticity

Memories are formed and stored in unique neural circuits that involve several brain structures, many different types of neurons, and a multitude of distinct plasticity mechanisms. In fact, it is plasticity at synapses (such as long-term potentiation and long-term depression) that is believed to underlie memory formation and maintenance. LTP and memory formation share many important features, including associativity, input specificity, and a requirement for a sufficiently strong stimulus. Moreover, like the formation of a long-term memory, long-lasting LTP cannot be maintained in the presence of protein synthesis inhibitors,[209] indicating that protein synthesis is required for persistent LTP.

Therefore, if DNA methylation represents a molecular mechanism of memory storage, it is important to demonstrate a link between synaptic plasticity and changes in methylation. This important issue has now been addressed in several different ways. As mentioned above, in an early study, DNMT inhibitors were used to block DNA methylation in hippocampal slices during the delivery of a theta-burst stimulation of Schaeffer Collateral synapses projecting from area CA3 to CA1 of the hippocampus.[8,9] In vehicle treated slices, this stimulation protocol resulted in long-lasting LTP within the CA1 region of the hippocampus. However, in slices treated with the DNMT inhibitors zebularine or 5-aza-2-deoxycytidine, LTP was abolished within 1–2 hours of induction. Moreover, these drugs produced a robust demethylation at the *reelin* and *BDNF* genes, which are known to play a role in synaptic plasticity and were subsequently shown to undergo similar methylation changes during fear conditioning.[74,78] Importantly, DNMT inhibition produced no effect on baseline synaptic transmission in slices treated with DNMT inhibitors, indicating that the LTP deficit induced was specific and experience-dependent.[8,9] Moreover, a follow-up to this report demonstrated that the disruption of long-lasting LTP induced by DNMT inhibitors could be reversed by pretreatment with the histone deacetylase (HDAC) inhibitor trichostatin-A, suggesting a potential cross-talk between histone acetylation and DNA methylation.[75]

Subsequent genetic approaches have provided an important confirmation of these findings. Given that prodrugs like zebularine or 5-aza-2-deoxycytidine are only thought to be effective after being chemically incorporated into DNA (where they can trap DNMTs), it is also important to demonstrate a more direct link to DNMT activity and synaptic plasticity. In keeping with this, a recent paper examined hippocampal LTP and LTD in mice with a double knockout (KO) of both DNMT isoforms (DNMT1 and DNMT3a) in post-mitotic neurons in the forebrain.[73] This report revealed that double KO mice (unlike single KO mice) exhibit a significant deficit in hippocampal LTP that mimics the deficit observed after DNMT inhibition in the aforementioned studies. Moreover, these mice also exhibit an enhancement in hippocampal LTD, but little or no change in basal synaptic transmission. These data indicate that DNMT1 and DNMT3a may have overlapping roles in adult neurons, and that at least one form is required to maintain normal hippocampal LTP.[73]

How does the persisting change in methylation get translated into a functional memory-subserving change in the cortex? This question is especially important since a subtext here is that the readout of DNA methylation is presumed to be cell-wide, whereas current models of memory maintenance emphasize synapse-specific changes in function. In terms of how the epigenetic marks are transformed into functional consequences in the cell, there are three broad possibilities (Figure 5.10). First, DNA methylation changes may drive a change in the response state of the neuron that is permissive for other mechanisms to establish and maintain more permanent changes. Second, methylation events may actively participate in the altered gene readout that contributes to ongoing memory, e.g. by enhancing synaptic strength. Third, the most unusual concept is that epigenetic mechanisms might actually render the cell totally aplastic, stabilizing a given distribution of synaptic weights as a necessary condition for memory stability. It is possible that changes in DNA methylation serve to stabilize the inputs that a given cell receives after a plastic change has occurred, in effect removing the cell from further synaptic competition by rendering it aplastic. This explanation seems plausible given that memory stabilization should require that a cell maintain its response to a given set of inputs across weeks or years despite the occurrence of new and different stimulation patterns in the intervening period. Layered on all three possibilities is the conundrum of how cell-wide changes (driven by epigenetic marks) can participate in the face of the apparent necessity of a role for synapse specificity in memory circuits. The last mechanism addresses this in a simple fashion, which is an appealing aspect of this novel idea.

However, it is worth noting that the first two ideas are not mutually exclusive, even within the same cell. In terms of the entire memory storage circuit, all three mechanisms could possibly play a role at different sites or at different times. Since epigenetic changes occur downstream of synaptic activity, they have the ability to integrate multiple cellular signals and modulate the long-term responsiveness of a neuron by controlling gene expression. In terms of memory storage, epigenetic changes may therefore enable cells to cement effectively a specific response to a given set of inputs by controlling the degree of plasticity

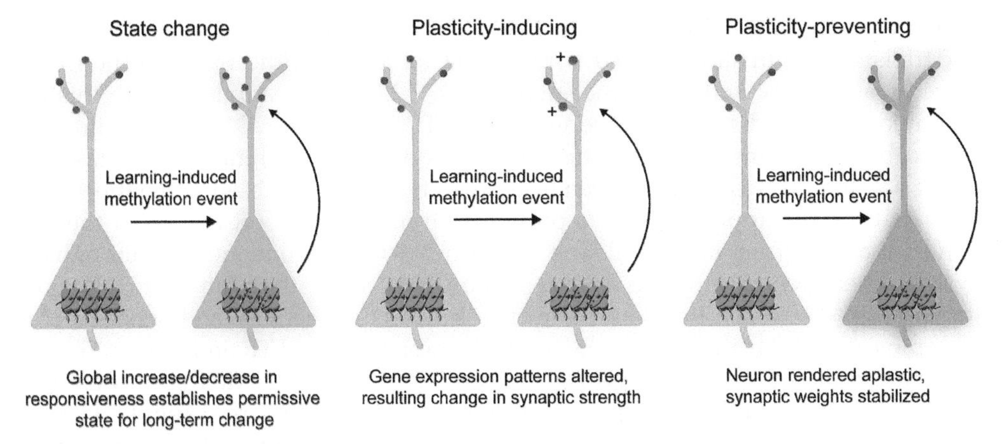

FIGURE 5.10 **Putative actions of cell-wide DNA methylation changes on neuronal function.** Changes in DNA methylation could induce a state change (left panel) which alters responsivity to existing inputs and acts permissively to enable other long-term changes which are ultimately responsible for memory. Altered patterns of DNA methylation could also directly or indirectly alter gene expression and contribute to changes in synaptic strength that are thought to underlie the formation and maintenance of memories (center panel). Alternatively, changes in methylation status within a cell may act to render it aplastic, in effect stabilizing the current synaptic weights and responsivity (right panel). Critically, these changes may occur in different brain regions or at different time points as part of the overall process of learning, memory consolidation, and memory maintenance. It is important to note that the changes in DNA methylation driving altered neuronal function are likely to occur at a small subset of the total methylation sites in the cell, in order that the overall neuronal phenotype be preserved. It also is worth considering that because the methyl-DNA binding proteins do not effectively recognize hemi-methylated DNA, hemi-demethylation of DNA is likely just as effective as double-stranded demethylation in triggering functional changes in the neuron.

that occurs at all synapses. In this way, memory storage may be conceptually thought of as both a synaptic process that controls the nature of signals that a cell receives and an epigenetic process that controls subsequent expression of memory-related genes.

Cortex-Specific Epigenetic Regulation?

Given that the cortex is involved in the long-term storage of remote memories, it is especially important to understand how DNA methylation changes may contribute to the circuit-level phenomena occurring in regions like the PFC. The PFC receives inputs from a number of sensory and limbic systems and projects to areas implicated in motivation and action selection, making it the ideal storage site for specific memories.[210,211] Neuronal processing in the PFC can be extremely selective, with individual cells responding selectively to unique events, sensory and spatial properties of different stimuli, specific outcomes, and expectations.[212–217] Thus, one possibility is that DNA methylation provides a regulatory mechanism prior to a learning event in order to generate cell-specific responses. For example, DNA methylation could modulate cell excitability both positively and negatively, thereby determining which neurons are eligible for participation in a given memory trace. This mechanism would ensure that only a small fraction of neurons are capable of undergoing plastic changes at any given time. Such a function would be reminiscent of the role played by CREB during the formation of fear memories in the lateral amygdala.[218–221] Alternatively, the inherent connectivity of cortical circuits could engender selectivity in responsiveness, while DNA methylation serves to stamp in a specific distribution of synaptic weights after learning in order to preserve certain memories. Finally, given that behavioral representations of a memory likely depend on multiple interacting circuits, another possibility is that DNA methylation may silence an entire node within a circuit to produce a given behavioral outcome. Critically, each of these hypotheses could potentially explain the remote memory deficits induced by DNMT inhibitors.

Regardless of the potential circuit-level roles played by DNA methylation, it is possible (or even likely) that unique epigenetic mechanisms exist in the cortex to support the long-term stabilization and maintenance of specific memories. Indeed, it is already evident that epigenetic changes in the hippocampus tend to be rapid in both onset and offset,[74,75,77] whereas changes within the cortex may be particularly stable over time.[76,96] Further, there are differences in both the genes selected for methylation changes and the direction of these changes following the same behavioral experience (see Figure 5.7). However, the mechanisms that may support such differential function are unclear. One possible explanation for the temporal differences in DNA methylation changes could lie in region-selective expression of certain DNA modifying enzymes or related transcriptional machinery. For example, DNMT expression is known to vary within different subregions of the hippocampus, and is correlated with the degree of methylation in those regions.[222] Nevertheless, at this time, very little is known about regional differences in the expression of DNMTs or the demethylation apparatus throughout the brain.

Relationship to Systems Neuroscience

The idea that epigenetic modifications regulate the formation, maintenance, and expression of memories does not diminish the importance of circuit-level phenomena in learning and memory. In fact, to understand how DNA methylation could contribute to memory, it is first necessary to understand how neural circuits encode, consolidate, and store memory-related information. For example, contextual fear conditioning produces transient changes in DNA methylation in the hippocampus, but prolonged changes in DNA methylation in the cortex. Our speculation is that there are actually two different mechanisms in play; one that participates in consolidation (hippocampus) and one that participates in storage (cortex). Together, these mechanisms could allow for plasticity in hippocampal circuits to enable rapid consolidation, and stability in cortical circuits to promote the long-term maintenance of memory. As the hippocampus is needed to form new, subsequent memories, its epigenetic mechanisms may have to be plastic in order to allow the system to reset after it has served its function. We speculate that how a brain region uses epigenetic modifications to regulate memory will differ based on the functional roles of that structure.

SUMMARY AND CONCLUSIONS – COGNITIVE EPIGENETICS

Although the epigenetic changes reviewed above demonstrate dynamic epigenetic regulation of transcription, it is worth noting that none of these changes could be included in the traditional definition of epigenetics, which requires that a change be heritable across cell division.[100] As adult neurons are post-mitotic, the epigenetic changes established during learning and maintained across days and months are impressive, but not heritable. In response to this discrepancy, recent conceptualizations have sought to redefine the field of epigenetics so as to include all adaptations to chromatin which mark and perpetuate altered activity states.[100] In the same spirit, we argue that the findings discussed above highlight the emergence of "cognitive epigenetics" as a subfield of traditional epigenetics. We emphasize this subfield not to exclude or escape the rigor associated with traditional epigenetics, but rather to recognize the unique challenges and phenomena associated with epigenetic changes in neuronal systems (see Table 5.2). For example, the dynamic changes in DNA methylation observed in neurons indicate that unique mechanisms may regulate this epigenetic mark. Further, given that neurons do not divide, the epigenetic changes that occur in neurons likely target many different genes that do not undergo epigenetic alteration in somatic cells, and vice-versa. Indeed, the set of epigenetic alterations that regulate cellular differentiation after division may be completely partitioned from the epigenetic changes that dynamically regulate synaptic and cellular plasticity, learning, and memory maintenance. Thus, epigenetic changes in somatic cells may differ in both degree and scope from changes in non-dividing neuronal cells.

The study of epigenetics in neural systems is also associated with a number of unique challenges. One particular difficulty arises from the complexity of neuronal subtypes, even within the same brain region. It would not be surprising if the various types of interneurons and projection neurons exhibit drastically different epigenetic profiles, even at the same moment in time. Thus, an important challenge will be to unfold the many layers of epigenetic regulation that occur within separate cell types, and link each of these layers to overall neuronal and systems function in a meaningful way. A second and related challenge will be to understand how different neural circuits use epigenetic mechanisms in different ways for behavioral adaptation. This will be important not only for understanding the molecular mechanisms underlying the function of a single brain region, but also to understand how multiple regions integrate and store information at different speeds. Finally, it will be necessary to understand fully how the genes that undergo epigenetic modification following learning in turn influence synaptic and circuit plasticity, and understand how and why this may differ in discrete brain regions.

As highlighted above, epigenetic mechanisms in non-dividing CNS cells are subject to sometimes unique regulation that is consistent with a molecular component of memory storage. However, a number of studies now indicate that epigenetic mechanisms (and specifically DNA methylation) are involved in a wide range of behavioral outcomes other than memory formation, including responses to drugs of abuse,[198,223,224] psychiatric disorders and syndromes,[64,225–227] and the long-lasting effects of early life experiences.[96–98] Thus, similar to memory formation, many types of experience result in long-term changes in the epigenetic environment that contributes to subsequent behavioral output. Therefore, understanding the mechanisms discussed here will help to elucidate not only basic molecular principles of learning and memory, but will also provide key insights into the molecular nature of numerous behaviors and behavioral disorders.

We speculate that the new understanding of the role of neuroepigenetic molecular mechanisms in memory formation can answer the long-standing question in neuroscience of why neurons cannot divide. The fact that neurons have co-opted epigenetic mechanisms to subserve long-term functional changes may preclude their use of these same mechanisms to perpetuate cellular phenotype with cell division. In a sense, the neuron can't have its cake and eat it too – it can either use epigenetic molecular mechanisms to perpetuate cell fate across cell division, or use a subset of them to perpetuate acquired functional changes across time, but not both. Obviously, this remains our speculation, and future investigations will be required to address fully this hypothesis. Interestingly, accumulating evidence indicates that DNA methylation is also involved in the development, survival, and function of newborn neurons in the subventricular and subgranular zones of adult animals,[91,228,229] revealing yet another potential locus for neuroepigenetic mechanisms to influence the function of the mature CNS. Nevertheless, it remains unclear whether the epigenetic modifications that underlie conversion of neural stem cells into mature adult neurons overlap with the mechanisms responsible for long-term maintenance of functional change.

It is clear that we have not yet begun to determine in a comprehensive fashion *how* DNA methylation at the cellular level gets translated into altered circuit and behavioral function. Thus far, most studies have been restricted to using a candidate target gene approach to identify specific sites of methylation changes. However, these data only allow the assessment of a small subset of changes in DNA methylation. It is not yet possible to try mechanistically to tie these specific changes at single gene exons to complex multicellular, multicomponent processes like LTP, hippocampal circuit stabilization, and behavioral memory at this point, because of the limitation that the molecular approaches are sampling such a small subset of genes. Thus, a future challenge for neuroepigenetics researchers will be to expand the level of analysis by incorporating sophisticated epigenome-wide screens into the technical repertoire,[73] potentially revealing a myriad of functional effector genes subjected to epigenetic control and perhaps identify novel mnemogenic molecules.

Acknowledgments

The authors wish to thank Tom Carew, Huda Zoghbi, Art Beaudet, and Eric Kandel for many helpful discussions. We apologize to the many authors whose primary work was not directly cited, owing to limitations of space. Research in the authors' laboratory is supported by funds from the NINDS, NIMH, NIA, NIDA, the Rett Syndrome Foundation, the Ellison Medical Foundation, and the Evelyn F. McKnight Brain Research Foundation.

References

1. Marcus EA, Emptage NJ, Marois R, Carew TJ. A comparison of the mechanistic relationships between development and learning in Aplysia. *Prog Brain Res*. 1994;100:179–188.
2. Bailey CH, Kandel ER, Si K. The persistence of long-term memory: a molecular approach to self-sustaining changes in learning-induced synaptic growth. *Neuron*. 2004;44:49–57.
3. Crick F. Memory and molecular turnover. *Nature*. 1984;312:101.
4. Lisman JE. A mechanism for memory storage insensitive to molecular turnover: a bistable autophosphorylating kinase. *Proc Natl Acad Sci USA*. 1985;82:3055–3057.
5. Roberson ED, Sweatt JD. Memory-forming chemical reactions. *Rev Neurosci*. 2001;12:41–50.
6. Day JJ, Sweatt JD. DNA methylation and memory formation. *Nat Neurosci*. 2010;13:1319–1323.
7. Levenson JM, O'Riordan KJ, Brown KD, Trinh MA, Molfese DL, Sweatt JD. Regulation of histone acetylation during memory formation in the hippocampus. *J Biol Chem*. 2004;279:40545–40559.
8. Levenson JM, Sweatt JD. Epigenetic mechanisms: a common theme in vertebrate and invertebrate memory formation. *Cell Mol Life Sci*. 2006;63:1009–1016.

9. Levenson JM, Roth TL, Lubin FD, et al. Evidence that DNA (cytosine-5) methyltransferase regulates synaptic plasticity in the hippocampus. *J Biol Chem*. 2006;281:15763–15773.
10. Swank MW, Sweatt JD. Increased histone acetyltransferase and lysine acetyltransferase activity and biphasic activation of the ERK/RSK cascade in insular cortex during novel taste learning. *J Neurosci*. 2001;21:3383–3391.
11. Strahl BD, Allis CD. The language of covalent histone modifications. *Nature*. 2000;403:41–45.
12. Barski A, Cuddapah S, Cui K, et al. High-resolution profiling of histone methylations in the human genome. *Cell*. 2007;129:823–837.
13. Wang Z, Zang C, Rosenfeld JA, et al. Combinatorial patterns of histone acetylations and methylations in the human genome. *Nat Genet*. 2008;40:897–903.
14. Jenuwein T, Allis CD. Translating the histone code. *Science*. 2001;293:1074–1080.
15. Turner BM. Histone acetylation and an epigenetic code. *Bioessays*. 2000;22:836–845.
16. Campos EI, Reinberg D. Histones: annotating chromatin. *Annu Rev Genet*. 2009;43:559–599.
17. Bredy TW, Wu H, Crego C, Zellhoefer J, Sun YE, Barad M. Histone modifications around individual BDNF gene promoters in prefrontal cortex are associated with extinction of conditioned fear. *Learn Mem*. 2007;14:268–276.
18. Chwang WB, Arthur JS, Schumacher A, Sweatt JD. The nuclear kinase mitogen- and stress-activated protein kinase 1 regulates hippocampal chromatin remodeling in memory formation. *J Neurosci*. 2007;27:12732–12742.
19. Fischer A, Sananbenesi F, Wang X, Dobbin M, Tsai LH. Recovery of learning and memory is associated with chromatin remodelling. *Nature*. 2007;447:178–182.
20. Gupta S, Kim SY, Artis S, et al. Histone methylation regulates memory formation. *J Neurosci*. 2010;30:3589–3599.
21. Koshibu K, Graff J, Beullens M, et al. Protein phosphatase 1 regulates the histone code for long-term memory. *J Neurosci*. 2009;29:13079–13089.
22. Lubin FD, Sweatt JD. The IkappaB kinase regulates chromatin structure during reconsolidation of conditioned fear memories. *Neuron*. 2007;55:942–957.
23. Peleg S, Sananbenesi F, Zovoilis A, et al. Altered histone acetylation is associated with age-dependent memory impairment in mice. *Science*. 2010;328:753–756.
24. Chwang WB, O'Riordan KJ, Levenson JM, Sweatt JD. ERK/MAPK regulates hippocampal histone phosphorylation following contextual fear conditioning. *Learn Mem*. 2006;13:322–328.
25. Alarcon JM, Malleret G, Touzani K, et al. Chromatin acetylation, memory, and LTP are impaired in CBP+/− mice: a model for the cognitive deficit in Rubinstein-Taybi syndrome and its amelioration. *Neuron*. 2004;42:947–959.
26. Korzus E, Rosenfeld MG, Mayford M. CBP histone acetyltransferase activity is a critical component of memory consolidation. *Neuron*. 2004;42:961–972.
27. Vecsey CG, Hawk JD, Lattal KM, et al. Histone deacetylase inhibitors enhance memory and synaptic plasticity via CREB:CBP-dependent transcriptional activation. *J Neurosci*. 2007;27:6128–6140.
28. Ng HH, Robert F, Young RA, Struhl K. Targeted recruitment of Set1 histone methylase by elongating Pol II provides a localized mark and memory of recent transcriptional activity. *Mol Cell*. 2003;11:709–719.
29. Scharf AN, Imhof A. Every methyl counts – Epigenetic calculus. *FEBS Lett*. 2010;585(13):2001–2007.
30. Greiner D, Bonaldi T, Eskeland R, Roemer E, Imhof A. Identification of a specific inhibitor of the histone methyltransferase SU(VAR)3–9. *Nat Chem Biol*. 2005;1:143–145.
31. Tachibana M, Matsumura Y, Fukuda M, Kimura H, Shinkai Y. G9a/GLP complexes independently mediate H3K9 and DNA methylation to silence transcription. *Embo J*. 2008;27:2681–2690.
32. Tachibana M, Sugimoto K, Fukushima T, Shinkai Y. Set domain-containing protein, G9a, is a novel lysine-preferring mammalian histone methyltransferase with hyperactivity and specific selectivity to lysines 9 and 27 of histone H3. *J Biol Chem*. 2001;276:25309–25317.
33. Akbarian S, Huang HS. Epigenetic regulation in human brain – focus on histone lysine methylation. *Biol Psychiatry*. 2009;65:198–203.
34. Shi Y, Lan F, Matson C, et al. Histone demethylation mediated by the nuclear amine oxidase homolog LSD1. *Cell*. 2004;119:941–953.
35. Shi Y, Whetstine JR. Dynamic regulation of histone lysine methylation by demethylases. *Mol Cell*. 2007;25:1–14.
36. Tsukada Y, Fang J, Erdjument-Bromage H, et al. Histone demethylation by a family of JmjC domain-containing proteins. *Nature*. 2006;439:811–816.
37. Stavropoulos P, Blobel G, Hoelz A. Crystal structure and mechanism of human lysine-specific demethylase-1. *Nat Struct Mol Biol*. 2006;13:626–632.

38. Schaefer A, Sampath SC, Intrator A, et al. Control of cognition and adaptive behavior by the GLP/G9a epigenetic suppressor complex. *Neuron*. 2009;64:678–691.
39. Maze I, Covington III HE, Dietz DM, et al. Essential role of the histone methyltransferase G9a in cocaine-induced plasticity. *Science*. 2010;327:213–216.
40. Allis CD, Berger SL, Cote J, et al. New nomenclature for chromatin-modifying enzymes. *Cell*. 2007;131:633–636.
41. Kubicek S, O'Sullivan RJ, August EM, et al. Reversal of H3K9me2 by a small-molecule inhibitor for the G9a histone methyltransferase. *Mol Cell*. 2007;25:473–481.
42. Berger SL. The complex language of chromatin regulation during transcription. *Nature*. 2007;447:407–412.
43. Deng H, Bao X, Cai W, et al. Ectopic histone H3S10 phosphorylation causes chromatin structure remodeling in Drosophila. *Development*. 2008;135:699–705.
44. Fischle W, Tseng BS, Dormann HL, et al. Regulation of HP1-chromatin binding by histone H3 methylation and phosphorylation. *Nature*. 2005;438:1116–1122.
45. Cheung P, Allis CD, Sassone-Corsi P. Signaling to chromatin through histone modifications. *Cell*. 2000;103:263–271.
46. Cheung P, Tanner KG, Cheung WL, Sassone-Corsi P, Denu JM, Allis CD. Synergistic coupling of histone H3 phosphorylation and acetylation in response to epidermal growth factor stimulation. *Mol Cell*. 2000;5:905–915.
47. Delaunoy J, Abidi F, Zeniou M, et al. Mutations in the X-linked RSK2 gene (RPS6KA3) in patients with Coffin-Lowry syndrome. *Hum Mutat*. 2001;17:103–116.
48. Merienne K, Pannetier S, Harel-Bellan A, Sassone-Corsi P. Mitogen-regulated RSK2-CBP interaction controls their kinase and acetylase activities. *Mol Cell Biol*. 2001;21:7089–7096.
49. Higashi M, Inoue S, Ito T. Core histone H2A ubiquitylation and transcriptional regulation. *Exp Cell Res*. 2010;316:2707–2712.
50. Wang H, Wang L, Erdjument-Bromage H, et al. Role of histone H2A ubiquitination in Polycomb silencing. *Nature*. 2004;431:873–878.
51. Atanassov BS, Koutelou E, Dent SY. The role of deubiquitinating enzymes in chromatin regulation. *FEBS Lett*. 2011;585:2016–23.
52. Zhu P, Zhou W, Wang J, et al. A histone H2A deubiquitinase complex coordinating histone acetylation and H1 dissociation in transcriptional regulation. *Mol Cell*. 2007;27:609–621.
53. Jason LJ, Moore SC, Lewis JD, Lindsey G, Ausio J. Histone ubiquitination: a tagging tail unfolds? *Bioessays*. 2002;24:166–174.
54. Lee JS, Shukla A, Schneider J, et al. Histone crosstalk between H2B monoubiquitination and H3 methylation mediated by COMPASS. *Cell*. 2007;131:1084–1096.
55. Sadri-Vakili G, Bouzou B, Benn CL, et al. Histones associated with downregulated genes are hypo-acetylated in Huntington's disease models. *Hum Mol Genet*. 2007;16:1293–1306.
56. Kim MO, Chawla P, Overland RP, Xia E, Sadri-Vakili G, Cha JH. Altered histone monoubiquitylation mediated by mutant huntingtin induces transcriptional dysregulation. *J Neurosci*. 2008;28:3947–3957.
57. Messner S, Altmeyer M, Zhao H, et al. PARP1 ADP-ribosylates lysine residues of the core histone tails. *Nucleic Acids Res*. 2010;38:6350–6362.
58. Quenet D, El Ramy R, Schreiber V, Dantzer F. The role of poly(ADP-ribosyl)ation in epigenetic events. *Int J Biochem Cell Biol*. 2009;41:60–65.
59. Homburg S, Visochek L, Moran N, et al. A fast signal-induced activation of Poly(ADP-ribose) polymerase: a novel downstream target of phospholipase c. *J Cell Biol*. 2000;150:293–307.
60. Cohen-Armon M, Visochek L, Katzoff A, et al. Long-term memory requires polyADP-ribosylation. *Science*. 2004;304:1820–1822.
61. Hernandez AI, Wolk J, Hu JY, et al. Poly-(ADP-ribose) polymerase-1 is necessary for long-term facilitation in Aplysia. *J Neurosci*. 2009;29:9553–9562.
62. Fontan-Lozano A, Suarez-Pereira I, Horrillo A, del-Pozo-Martin Y, Hmadcha A, Carrion AM. Histone H1 poly[ADP]-ribosylation regulates the chromatin alterations required for learning consolidation. *J Neurosci*. 2010;30:13305–13313.
63. Turner BM. Defining an epigenetic code. *Nat Cell Biol*. 2007;9:2–6.
64. Amir RE, Van den Veyver IB, Wan M, Tran CQ, Francke U, Zoghbi HY. Rett syndrome is caused by mutations in X-linked MECP2, encoding methyl-CpG-binding protein 2. *Nat Genet*. 1999;23:185–188.
65. Chao HT, Zoghbi HY, Rosenmund C. MeCP2 controls excitatory synaptic strength by regulating glutamatergic synapse number. *Neuron*. 2007;56:58–65.

66. Moretti P, Levenson JM, Battaglia F, et al. Learning and memory and synaptic plasticity are impaired in a mouse model of Rett syndrome. *J Neurosci*. 2006;26:319–327.
67. Chahrour M, Jung SY, Shaw C, et al. MeCP2 a key contributor to neurological disease, activates and represses transcription. *Science*. 2008;320:1224–1229.
68. Valinluck V, Tsai HH, Rogstad DK, Burdzy A, Bird A, Sowers LC. Oxidative damage to methyl-CpG sequences inhibits the binding of the methyl-CpG binding domain (MBD) of methyl-CpG binding protein 2 (MeCP2). *Nucleic Acids Res*. 2004;32:4100–4108.
69. Kriaucionis S, Heintz N. The nuclear DNA base 5-hydroxymethylcytosine is present in Purkinje neurons and the brain. *Science*. 2009;324:929–930.
70. Tahiliani M, Koh KP, Shen Y, et al. Conversion of 5-methylcytosine to 5-hydroxymethylcytosine in mammalian DNA by MLL partner TET1. *Science*. 2009;324:930–935.
71. Klose RJ, Sarraf SA, Schmiedeberg L, McDermott SM, Stancheva I, Bird AP. DNA binding selectivity of MeCP2 due to a requirement for A/T sequences adjacent to methyl-CpG. *Mol Cell*. 2005;19:667–678.
72. Weber M, Hellmann I, Stadler MB, et al. Distribution, silencing potential and evolutionary impact of promoter DNA methylation in the human genome. *Nat Genet*. 2007;39:457–466.
73. Feng J, Zhou Y, Campbell SL, et al. Dnmt1 and Dnmt3a maintain DNA methylation and regulate synaptic function in adult forebrain neurons. *Nat Neurosci*. 2010;13:423–430.
74. Lubin FD, Roth TL, Sweatt JD. Epigenetic regulation of BDNF gene transcription in the consolidation of fear memory. *J Neurosci*. 2008;28:10576–10586.
75. Miller CA, Campbell SL, Sweatt JD. DNA methylation and histone acetylation work in concert to regulate memory formation and synaptic plasticity. *Neurobiol Learn Mem*. 2008;89:599–603.
76. Miller CA, Gavin CF, White JA, et al. Cortical DNA methylation maintains remote memory. *Nat Neurosci*. 2010;13:664–666.
77. Paxinos G, Watson C. *The Rat Brain in Stereotaxic Coordinates*, 5th ed. New York: Elsevier; 2005.
78. Miller CA, Sweatt JD. Covalent modification of DNA regulates memory formation. *Neuron*. 2007;53:857–869.
79. Alonso M, Bekinschtein P, Cammarota M, Vianna MR, Izquierdo I, Medina JH. Endogenous BDNF is required for long-term memory formation in the rat parietal cortex. *Learn Mem*. 2005;12:504–510.
80. Bekinschtein P, Cammarota M, Izquierdo I, Medina JH. BDNF and memory formation and storage. *Neuroscientist*. 2008;14:147–156.
81. Bekinschtein P, Cammarota M, Katche C, et al. BDNF is essential to promote persistence of long-term memory storage. *Proc Natl Acad Sci USA*. 2008;105:2711–2716.
82. Bekinschtein P, Cammarota M, Igaz LM, Bevilaqua LR, Izquierdo I, Medina JH. Persistence of long-term memory storage requires a late protein synthesis- and BDNF- dependent phase in the hippocampus. *Neuron*. 2007;53:261–277.
83. Monsey MS, Ota KT, Akingbade IF, Hong ES, Schafe GE. Epigenetic alterations are critical for fear memory consolidation and synaptic plasticity in the lateral amygdala. *PLoS One*. 2011;6:e19958.
84. Maddox SA, Schafe GE. Epigenetic alterations in the lateral amygdala are required for reconsolidation of a Pavlovian fear memory. *Learn Mem*. 2011;18:579–593.
85. Metivier R, Gallais R, Tiffoche C, et al. Cyclical DNA methylation of a transcriptionally active promoter. *Nature*. 2008;452:45–50.
86. Wu SC, Zhang Y. Active DNA demethylation: many roads lead to Rome. *Nat Rev Mol Cell Biol*. 2010;11:607–620.
87. Dulac C. Brain function and chromatin plasticity. *Nature*. 2010;465:728–735.
88. Gehring M, Reik W, Henikoff S. DNA demethylation by DNA repair. *Trends Genet*. 2009;25:82–90.
89. Ooi SK, Bestor TH. The colorful history of active DNA demethylation. *Cell*. 2008;133:1145–1148.
90. Ma DK, Guo JU, Ming GL, Song H. DNA excision repair proteins and Gadd45 as molecular players for active DNA demethylation. *Cell Cycle*. 2009;8:1526–1531.
91. Ma DK, Jang MH, Guo JU, et al. Neuronal activity-induced Gadd45b promotes epigenetic DNA demethylation and adult neurogenesis. *Science*. 2009;323:1074–1077.
92. Niehrs C. Active DNA demethylation and DNA repair. *Differentiation*. 2009;77:1–11.
93. Barreto G, Schäfer A, Marhold J, et al. Gadd45a promotes epigenetic gene activation by repair-mediated DNA demethylation. *Nature*. 2007;445:671–675.
94. Jin SG, Guo C, Pfeifer GP. GADD45A does not promote DNA demethylation. *PLoS Genet*. 2008;4:e1000013.
95. Szyf M. Epigenetics DNA methylation, and chromatin modifying drugs. *Annu Rev Pharmacol Toxicol*. 2009;49:243–263.

96. Roth TL, Lubin FD, Funk AJ, Sweatt JD. Lasting epigenetic influence of early-life adversity on the BDNF gene. *Biol Psychiatry*. 2009;65:760–769.
97. Weaver IC, Cervoni N, Champagne FA, et al. Epigenetic programming by maternal behavior. *Nat Neurosci*. 2004;7:847–854.
98. Weaver IC, Champagne FA, Brown SE, et al. Reversal of maternal programming of stress responses in adult offspring through methyl supplementation: altering epigenetic marking later in life. *J Neurosci*. 2005;25:11045–11054.
99. Russo VEA, Martienssen RA, Riggs AD, eds. *Epigenetic Mechanisms of Gene Regulation*. Woodbury: Cold Spring Harbor Laboratory Press; 1996.
100. Bird A. Perceptions of epigenetics. *Nature*. 2007;447:396–398.
101. Rouhi A, Mager DL, Humphries RK, Kuchenbauer F. MiRNAs epigenetics, and cancer. *Mamm Genome*. 2008;19:517–525.
102. He L, Hannon GJ. MicroRNAs: small RNAs with a big role in gene regulation. *Nat Rev Genet*. 2004;5:522–531.
103. Bartel DP. MicroRNAs: genomics, biogenesis, mechanism, and function. *Cell*. 2004;116:281–297.
104. Vasudevan S, Tong Y, Steitz JA. Switching from repression to activation: microRNAs can up-regulate translation. *Science*. 2007;318:1931–1934.
105. Miranda KC, Huynh T, Tay Y, et al. A pattern-based method for the identification of MicroRNA binding sites and their corresponding heteroduplexes. *Cell*. 2006;126:1203–1217.
106. Penagarikano O, Mulle JG, Warren ST. The pathophysiology of fragile x syndrome. *Annu Rev Genomics Hum Genet*. 2007;8:109–129.
107. Jin P, Zarnescu DC, Ceman S, et al. Biochemical and genetic interaction between the fragile X mental retardation protein and the microRNA pathway. *Nat Neurosci*. 2004;7:113–117.
108. Edbauer D, Neilson JR, Foster KA, et al. Regulation of synaptic structure and function by FMRP-associated microRNAs miR-125b and miR-132. *Neuron*. 2010;65:373–384.
109. Provost P. MicroRNAs as a molecular basis for mental retardation, Alzheimer's and prion diseases. *Brain Res*. 2010;1338:58–66.
110. Hollander JA, Im HI, Amelio AL, et al. Striatal microRNA controls cocaine intake through CREB signalling. *Nature*. 2010;466:197–202.
111. Rajasethupathy P, Fiumara F, Sheridan R, et al. Characterization of small RNAs in Aplysia reveals a role for miR-124 in constraining synaptic plasticity through CREB. *Neuron*. 2009;63(6):803–817.
112. Gao J, Wang WY, Mao YW, et al. A novel pathway regulates memory and plasticity via SIRT1 and miR-134. *Nature*. 2010;466(7310):1105–1109.
113. Lin Q, Wei W, Coelho CM, et al. The brain-specific microRNA miR-128b regulates the formation of fear-extinction memory. *Nat Neurosci*. 2011;14(9):1115–1117.
114. Berger-Sweeney J, Zearfoss NR, Richter JD. Reduced extinction of hippocampal-dependent memories in CPEB knockout mice. *Learn Mem*. 2006;13:4–7.
115. Keleman K, Kruttner S, Alenius M, Dickson BJ. Function of the Drosophila CPEB protein Orb2 in long-term courtship memory. *Nat Neurosci*. 2007;10:1587–1593.
116. Miniaci MC, Kim JH, Puthanveettil SV, et al. Sustained CPEB-dependent local protein synthesis is required to stabilize synaptic growth for persistence of long-term facilitation in Aplysia. *Neuron*. 2008;59:1024–1036.
117. Si K, Choi YB, White-Grindley E, Majumdar A, Kandel ER. Aplysia CPEB can form prion-like multimers in sensory neurons that contribute to long-term facilitation. *Cell*. 2010;140:421–435.
118. Si K, Giustetto M, Etkin A, et al. A neuronal isoform of CPEB regulates local protein synthesis and stabilizes synapse-specific long-term facilitation in aplysia. *Cell*. 2003;115:893–904.
119. Si K, Lindquist S, Kandel ER. A neuronal isoform of the aplysia CPEB has prion-like properties. *Cell*. 2003;115:879–891.
120. Coitinho AS, Freitas AR, Lopes MH, et al. The interaction between prion protein and laminin modulates memory consolidation. *Eur J Neurosci*. 2006;24:3255–3264.
121. Rial D, Duarte FS, Xikota JC, et al. Cellular prion protein modulates age-related behavioral and neurochemical alterations in mice. *Neuroscience*. 2009;164:896–907.
122. Ng RK, Gurdon JB. Epigenetic inheritance of cell differentiation status. *Cell Cycle*. 2008;7:1173–1177.
123. Lee JS, Smith E, Shilatifard A. The language of histone crosstalk. *Cell*. 2010;142:682–685.

124. Zippo A, Serafini R, Rocchigiani M, Pennacchini S, Krepelova A, Oliviero S. Histone crosstalk between H3S10ph and H4K16ac generates a histone code that mediates transcription elongation. *Cell.* 2009;138:1122–1136.
125. D'Alessio AC, Szyf M. Epigenetic tete-a-tete: the bilateral relationship between chromatin modifications and DNA methylation. *Biochem Cell Biol.* 2006;84:463–476.
126. D'Alessio AC, Weaver IC, Szyf M. Acetylation-induced transcription is required for active DNA demethylation in methylation-silenced genes. *Mol Cell Biol.* 2007;27:7462–7474.
127. Cervoni N, Szyf M. Demethylase activity is directed by histone acetylation. *J Biol Chem.* 2001;276:40778–40787.
128. Bird A. DNA methylation patterns and epigenetic memory. *Genes Dev.* 2002;16:6–21.
129. Fuks F, Hurd PJ, Deplus R, Kouzarides T. The DNA methyltransferases associate with HP1 and the SUV39H1 histone methyltransferase. *Nucleic Acids Res.* 2003;31:2305–2312.
130. Fuks F, Hurd PJ, Wolf D, Nan X, Bird AP, Kouzarides T. The methyl-CpG-binding protein MeCP2 links DNA methylation to histone methylation. *J Biol Chem.* 2003;278:4035–4040.
131. Brami-Cherrier K, Valjent E, Herve D, et al. Parsing molecular and behavioral effects of cocaine in mitogen- and stress-activated protein kinase-1-deficient mice. *J Neurosci.* 2005;25:11444–11454.
132. Guan Z, Giustetto M, Lomvardas S, et al. Integration of long-term-memory-related synaptic plasticity involves bidirectional regulation of gene expression and chromatin structure. *Cell.* 2002;111:483–493.
133. Renthal W, Nestler EJ. Epigenetic mechanisms in drug addiction. *Trends Mol Med.* 2008;14:341–350.
134. Nilsen TW, Graveley BR. Expansion of the eukaryotic proteome by alternative splicing. *Nature.* 2010;463:457–463.
135. Luco RF, Pan Q, Tominaga K, Blencowe BJ, Pereira-Smith OM, Misteli T. Regulation of alternative splicing by histone modifications. *Science.* 2010;327:996–1000.
136. Ou LC, Gean PW. Transcriptional regulation of brain-derived neurotrophic factor in the amygdala during consolidation of fear memory. *Mol Pharmacol.* 2007;72:350–358.
137. Rattiner LM, Davis M, Ressler KJ. Differential regulation of brain-derived neurotrophic factor transcripts during the consolidation of fear learning. *Learn Mem.* 2004;11:727–731.
138. Rattiner LM, Davis M, French CT, Ressler KJ. Brain-derived neurotrophic factor and tyrosine kinase receptor B involvement in amygdala-dependent fear conditioning. *J Neurosci.* 2004;24:4796–4806.
139. Maunakea AK, Nagarajan RP, Bilenky M, et al. Conserved role of intragenic DNA methylation in regulating alternative promoters. *Nature.* 2010;466:253–257.
140. Jiang YH, Armstrong D, Albrecht U, et al. Mutation of the Angelman ubiquitin ligase in mice causes increased cytoplasmic p53 and deficits of contextual learning and long-term potentiation. *Neuron.* 1998;21:799–811.
141. Gregg C, Zhang J, Butler JE, Haig D, Dulac C. Sex-specific parent-of-origin allelic expression in the mouse brain. *Science.* 2010;329:682–685.
142. Gregg C, Zhang J, Weissbourd B, et al. High-resolution analysis of parent-of-origin allelic expression in the mouse brain. *Science.* 2010;329:643–648.
143. Kazantsev AG, Thompson LM. Therapeutic application of histone deacetylase inhibitors for central nervous system disorders. *Nat Rev Drug Discov.* 2008;7:854–868.
144. American Psychological Association. *Diagnostic and Statistical Manual of Mental Disorders (Revised 4th Edition).* Washington, D.C.: American Psychological Association; 2000.
145. Guan JS, Haggarty SJ, Giacometti E, et al. HDAC2 negatively regulates memory formation and synaptic plasticity. *Nature.* 2009;459:55–60.
146. Kilgore M, Miller CA, Fass DM, et al. Inhibitors of class 1 histone deacetylases reverse contextual memory deficits in a mouse model of Alzheimer's disease. *Neuropsychopharmacology.* 2010;35:870–880.
147. McQuown SC, Barrett RM, Matheos DP, et al. HDAC3 is a critical negative regulator of long-term memory formation. *J Neurosci.* 2011;31(2):764–774.
148. McQuown SC, Wood MA. HDAC3 and the molecular brake pad hypothesis. *Neurobiol Learn Mem.* 2011;96(1):27–34.
149. Adegbola A, Gao H, Sommer S, Browning M. A novel mutation in JARID1C/SMCX in a patient with autism spectrum disorder (ASD). *Am J Med Genet A.* 2008;146A:505–511.
150. Iwase S, Lan F, Bayliss P, et al. The X-linked mental retardation gene SMCX/JARID1C defines a family of histone H3 lysine 4 demethylases. *Cell.* 2007;128:1077–1088.
151. Tzschach A, Lenzner S, Moser B, et al. Novel JARID1C/SMCX mutations in patients with X-linked mental retardation. *Hum Mutat.* 2006;27:389.

152. Wan M, Lee SS, Zhang X, et al. Rett syndrome and beyond: recurrent spontaneous and familial MECP2 mutations at CpG hotspots. *Am J Hum Genet*. 1999;65:1520–1529.
153. Tate P, Skarnes W, Bird A. The methyl-CpG binding protein MeCP2 is essential for embryonic development in the mouse. *Nat Genet*. 1996;12:205–208.
154. Chen HX, Otmakhov N, Strack S, Colbran RJ, Lisman JE. Is persistent activity of calcium/calmodulin-dependent kinase required for the maintenance of LTP? *J Neurophysiol*. 2001;85:1368–1376.
155. Chen RZ, Akbarian S, Tudor M, Jaenisch R. Deficiency of methyl-CpG binding protein-2 in CNS neurons results in a Rett-like phenotype in mice. *Nat Genet*. 2001;27:327–331.
156. Guy J, Hendrich B, Holmes M, Martin JE, Bird A. A mouse Mecp2-null mutation causes neurological symptoms that mimic Rett syndrome. *Nat Genet*. 2001;27:322–326.
157. Asaka Y, Jugloff DG, Zhang L, Eubanks JH, Fitzsimonds RM. Hippocampal synaptic plasticity is impaired in the Mecp2-null mouse model of Rett syndrome. *Neurobiol Dis*. 2006;21:217–227.
158. Stearns NA, Schaevitz LR, Bowling H, Nag N, Berger UV, Berger-Sweeney J. Behavioral and anatomical abnormalities in Mecp2 mutant mice: a model for Rett syndrome. *Neuroscience*. 2007;146:907–921.
159. Shahbazian M, Young J, Yuva-Paylor L, et al. Mice with truncated MeCP2 recapitulate many Rett syndrome features and display hyperacetylation of histone H3. *Neuron*. 2002;35:243–254.
160. Moretti P, Bouwknecht JA, Teague R, Paylor R, Zoghbi HY. Abnormalities of social interactions and home-cage behavior in a mouse model of Rett syndrome. *Hum Mol Genet*. 2005;14:205–220.
161. Pelka GJ, Watson CM, Radziewic T, et al. Mecp2 deficiency is associated with learning and cognitive deficits and altered gene activity in the hippocampal region of mice. *Brain*. 2006;129:887–898.
162. Collins A, Levenson JM, Vilaythong AP, et al. Mild overexpression of MeCP2 causes a progressive neurological disorder in mice. *Hum Mol Genet*. 2004;13:2679–2689.
163. Barrès R, Yan J, Egan B, et al. Acute exercise remodels promoter methylation in human skeletal muscle. *Cell Metab*. 2012;15:405–411.
164. Ehninger D, Li W, Fox K, Stryker MP, Silva AJ. Reversing neurodevelopmental disorders in adults. *Neuron*. 2008;60:950–960.
165. Weeber EJ, Sweatt JD. Molecular neurobiology of human cognition. *Neuron*. 2002;33:845–848.
166. Mammen AL, Huganir RL, O'Brien RJ. Redistribution and stabilization of cell surface glutamate receptors during synapse formation. *J Neurosci*. 1997;17:7351–7358.
167. Price JC, Guan S, Burlingame A, Prusiner SB, Ghaemmaghami S. Analysis of proteome dynamics in the mouse brain. *Proc Natl Acad Sci USA*. 2010;107:14508–14513.
168. Bayer KU, De Koninck P, Leonard AS, Hell JW, Schulman H. Interaction with the NMDA receptor locks CaMKII in an active conformation. *Nature*. 2001;411:801–805.
169. Leonard AS, Lim IA, Hemsworth DE, Horne MC, Hell JW. Calcium/calmodulin-dependent protein kinase II is associated with the N-methyl-D-aspartate receptor. *Proc Natl Acad Sci USA*. 1999;96:3239–3244.
170. Rich RC, Schulman H. Substrate-directed function of calmodulin in autophosphorylation of Ca2+/calmodulin-dependent protein kinase II. *J Biol Chem*. 1998;273:28424–28429.
171. Lisman J, Schulman H, Cline H. The molecular basis of CaMKII function in synaptic and behavioural memory. *Nat Rev Neurosci*. 2002;3:175–190.
172. Barria A, Muller D, Derkach V, Griffith LC, Soderling TR. Regulatory phosphorylation of AMPA-type glutamate receptors by CaM-KII during long-term potentiation. *Science*. 1997;276:2042–2045.
173. Fukunaga K, Stoppini L, Miyamoto E, Muller D. Long-term potentiation is associated with an increased activity of Ca2+/calmodulin-dependent protein kinase II. *J Biol Chem*. 1993;268:7863–7867.
174. Malinow R, Schulman H, Tsien RW. Inhibition of postsynaptic PKC or CaMKII blocks induction but not expression of LTP. *Science*. 1989;245:862–866.
175. Silva AJ, Stevens CF, Tonegawa S, Wang Y. Deficient hippocampal long-term potentiation in alpha-calcium-calmodulin kinase II mutant mice. *Science*. 1992;257:201–206.
176. Frankland PW, O'Brien C, Ohno M, Kirkwood A, Silva AJ. Alpha-CaMKII-dependent plasticity in the cortex is required for permanent memory. *Nature*. 2001;411:309–313.
177. Giese KP, Fedorov NB, Filipkowski RK, Silva AJ. Autophosphorylation at Thr286 of the alpha calcium-calmodulin kinase II in LTP and learning. *Science*. 1998;279:870–873.
178. Silva AJ, Paylor R, Wehner JM, Tonegawa S. Impaired spatial learning in alpha-calcium-calmodulin kinase II mutant mice. *Science*. 1992;257:206–211.

179. Buard I, Coultrap SJ, Freund RK, et al. CaMKII "autonomy" is required for initiating but not for maintaining neuronal long-term information storage. *J Neurosci*. 2010;30:8214–8220.
180. Sacktor TC. PKMzeta LTP maintenance, and the dynamic molecular biology of memory storage. *Prog Brain Res*. 2008;169:27–40.
181. Sacktor TC, Osten P, Valsamis H, Jiang X, Naik MU, Sublette E. Persistent activation of the zeta isoform of protein kinase C in the maintenance of long-term potentiation. *Proc Natl Acad Sci USA*. 1993;90:8342–8346.
182. Theis M, Si K, Kandel ER. Two previously undescribed members of the mouse CPEB family of genes and their inducible expression in the principal cell layers of the hippocampus. *Proc Natl Acad Sci USA*. 2003;100:9602–9607.
183. Feinberg AP. Phenotypic plasticity and the epigenetics of human disease. *Nature*. 2007;447:433–440.
184. Reik W. Stability and flexibility of epigenetic gene regulation in mammalian development. *Nature*. 2007;447:425–432.
185. Griffith JS, Mahler HR. DNA ticketing theory of memory. *Nature*. 1969;223:580–582.
186. Bird A. DNA methylation de novo. *Science*. 1999;286:2287–2288.
187. Chow JC, Brown CJ. Forming facultative heterochromatin: silencing of an X chromosome in mammalian females. *Cell Mol Life Sci*. 2003;60:2586–2603.
188. Chow JC, Yen Z, Ziesche SM, Brown CJ. Silencing of the mammalian X chromosome. *Annu Rev Genomics Hum Genet*. 2005;6:69–92.
189. Davis HP, Squire LR. Protein synthesis and memory: a review. *Psychol Bull*. 1984;96:518–559.
190. Bontempi B, Laurent-Demir C, Destrade C, Jaffard R. Time-dependent reorganization of brain circuitry underlying long-term memory storage. *Nature*. 1999;400:671–675.
191. Frankland PW, Bontempi B. The organization of recent and remote memories. *Nat Rev Neurosci*. 2005;6:119–130.
192. Squire LR, Knowlton B, Musen G. The structure and organization of memory. *Annu Rev Psychol*. 1993;44:453–495.
193. Cui Z, Wang H, Tan Y, Zaia KA, Zhang S, Tsien JZ. Inducible and reversible NR1 knockout reveals crucial role of the NMDA receptor in preserving remote memories in the brain. *Neuron*. 2004;41:781–793.
194. Frankland PW, Bontempi B, Talton LE, Kaczmarek L, Silva AJ. The involvement of the anterior cingulate cortex in remote contextual fear memory. *Science*. 2004;304:881–883.
195. Maviel T, Durkin TP, Menzaghi F, Bontempi B. Sites of neocortical reorganization critical for remote spatial memory. *Science*. 2004;305:96–99.
196. Shema R, Sacktor TC, Dudai Y. Rapid erasure of long-term memory associations in the cortex by an inhibitor of PKM zeta. *Science*. 2007;317:951–953.
197. Han J, Li Y, Wang D, Wei C, Yang X, Sui N. Effect of 5-aza-2-deoxycytidine microinjecting into hippocampus and prelimbic cortex on acquisition and retrieval of cocaine-induced place preference in C57BL/6 mice. *Eur J Pharmacol*. 2010;642:93–98.
198. LaPlant Q, Vialou V, Covington HE, et al. Dnmt3a regulates emotional behavior and spine plasticity in the nucleus accumbens. *Nat Neurosci*. 2010;13:1137–1143.
199. Berke JD, Hyman SE. Addiction, dopamine, and the molecular mechanisms of memory. *Neuron*. 2000;25:515–532.
200. Day JJ, Carelli RM. The nucleus accumbens and Pavlovian reward learning. *Neuroscientist*. 2007;13:148–159.
201. Graff J, Mansuy IM. Epigenetic codes in cognition and behaviour. *Behav Brain Res*. 2008;192:70–87.
202. Wood MA, Attner MA, Oliveira AM, Brindle PK, Abel T. A transcription factor-binding domain of the coactivator CBP is essential for long-term memory and the expression of specific target genes. *Learn Mem*. 2006;13:609–617.
203. Wood MA, Hawk JD, Abel T. Combinatorial chromatin modifications and memory storage: a code for memory? *Learn Mem*. 2006;13:241–244.
204. Angrisano T, Lembo F, Pero R, et al. TACC3 mediates the association of MBD2 with histone acetyltransferases and relieves transcriptional repression of methylated promoters. *Nucleic Acids Res*. 2006;34:364–372.
205. Jones PL, Veenstra GJ, Wade PA, et al. Methylated DNA and MeCP2 recruit histone deacetylase to repress transcription. *Nat Genet*. 1998;19:187–191.
206. Thomson JP, Skene PJ, Selfridge J, et al. CpG islands influence chromatin structure via the CpG-binding protein Cfp1. *Nature*. 2010;464:1082–1086.

207. Wade PA, Gegonne A, Jones PL, Ballestar E, Aubry F, Wolffe AP. Mi-2 complex couples DNA methylation to chromatin remodelling and histone deacetylation. *Nat Genet*. 1999;23:62–66.
208. Meaney MJ, Ferguson-Smith AC. Epigenetic regulation of the neural transcriptome: the meaning of the marks. *Nat Neurosci*. 2010;13:1313–1318.
209. Frey U, Krug M, Reymann KG, Matthies H. Anisomycin an inhibitor of protein synthesis, blocks late phases of LTP phenomena in the hippocampal CA1 region in vitro. *Brain Res*. 1988;452:57–65.
210. Groenewegen HJ, Berendse HW, Wolters JG, Lohman AH. The anatomical relationship of the prefrontal cortex with the striatopallidal system, the thalamus and the amygdala: evidence for a parallel organization. *Prog Brain Res*. 1990;85:95–116. [discussion 116–118].
211. Groenewegen HJ, Uylings HB. The prefrontal cortex and the integration of sensory, limbic and autonomic information. *Prog Brain Res*. 2000;126:3–28.
212. Funahashi S, Bruce CJ, Goldman-Rakic PS. Mnemonic coding of visual space in the monkey's dorsolateral prefrontal cortex. *J Neurophysiol*. 1989;61:331–349.
213. Goldman-Rakic PS. Cellular and circuit basis of working memory in prefrontal cortex of nonhuman primates. *Prog Brain Res*. 1990:325–335. [discussion 335–326].
214. Goldman-Rakic PS, Bates JF, Chafee MV. The prefrontal cortex and internally generated motor acts. *Curr Opin Neurobiol*. 1992;2:830–835.
215. Pasternak T, Greenlee MW. Working memory in primate sensory systems. *Nat Rev Neurosci*. 2005;6:97–107.
216. Rolls ET, Baylis LL. Gustatory, olfactory, and visual convergence within the primate orbitofrontal cortex. *J Neurosci*. 1994;14:5437–5452.
217. Watanabe M. Reward expectancy in primate prefrontal neurons. *Nature*. 1996;382:629–632.
218. Han JH, Kushner SA, Yiu AP, et al. Neuronal competition and selection during memory formation. *Science*. 2007;316:457–460.
219. Han JH, Kushner SA, Yiu AP, et al. Selective erasure of a fear memory. *Science*. 2009;323:1492–1496.
220. Silva AJ, Zhou Y, Rogerson T, Shobe J, Balaji J. Molecular and cellular approaches to memory allocation in neural circuits. *Science*. 2009;326:391–395.
221. Zhou Y, Won J, Karlsson MG, et al. CREB regulates excitability and the allocation of memory to subsets of neurons in the amygdala. *Nat Neurosci*. 2009;12:1438–1443.
222. Brown SE, Weaver IC, Meaney MJ, Szyf M. Regional-specific global cytosine methylation and DNA methyltransferase expression in the adult rat hippocampus. *Neurosci Lett*. 2008;440:49–53.
223. Anier K, Malinovskaja K, Aonurm-Helm A, Zharkovsky A, Kalda A. DNA methylation regulates cocaine-induced behavioral sensitization in mice. *Neuropsychopharmacology*. 2010;35(12):2450–2461.
224. Im HI, Hollander JA, Bali P, Kenny PJ. MeCP2 controls BDNF expression and cocaine intake through homeostatic interactions with microRNA-212. *Nat Neurosci*. 2010;13:1120–1127.
225. Grayson DR, Jia X, Chen Y, et al. Reelin promoter hypermethylation in schizophrenia. *Proc Natl Acad Sci USA*. 2005;102:9341–9346.
226. Petronis A. Epigenetics as a unifying principle in the aetiology of complex traits and diseases. *Nature*. 2010;465:721–727.
227. Sutcliffe JS, Nelson DL, Zhang F, et al. DNA methylation represses FMR-1 transcription in fragile X syndrome. *Hum Mol Genet*. 1992;1:397–400.
228. Wu H, Coskun V, Tao J, et al. Dnmt3a-dependent nonpromoter DNA methylation facilitates transcription of neurogenic genes. *Science*. 2010;329:444–448.
229. Zhao X, Ueba T, Christie BR, et al. Mice lacking methyl-CpG binding protein 1 have deficits in adult neurogenesis and hippocampal function. *Proc Natl Acad Sci USA*. 2003;100:6777–6782.

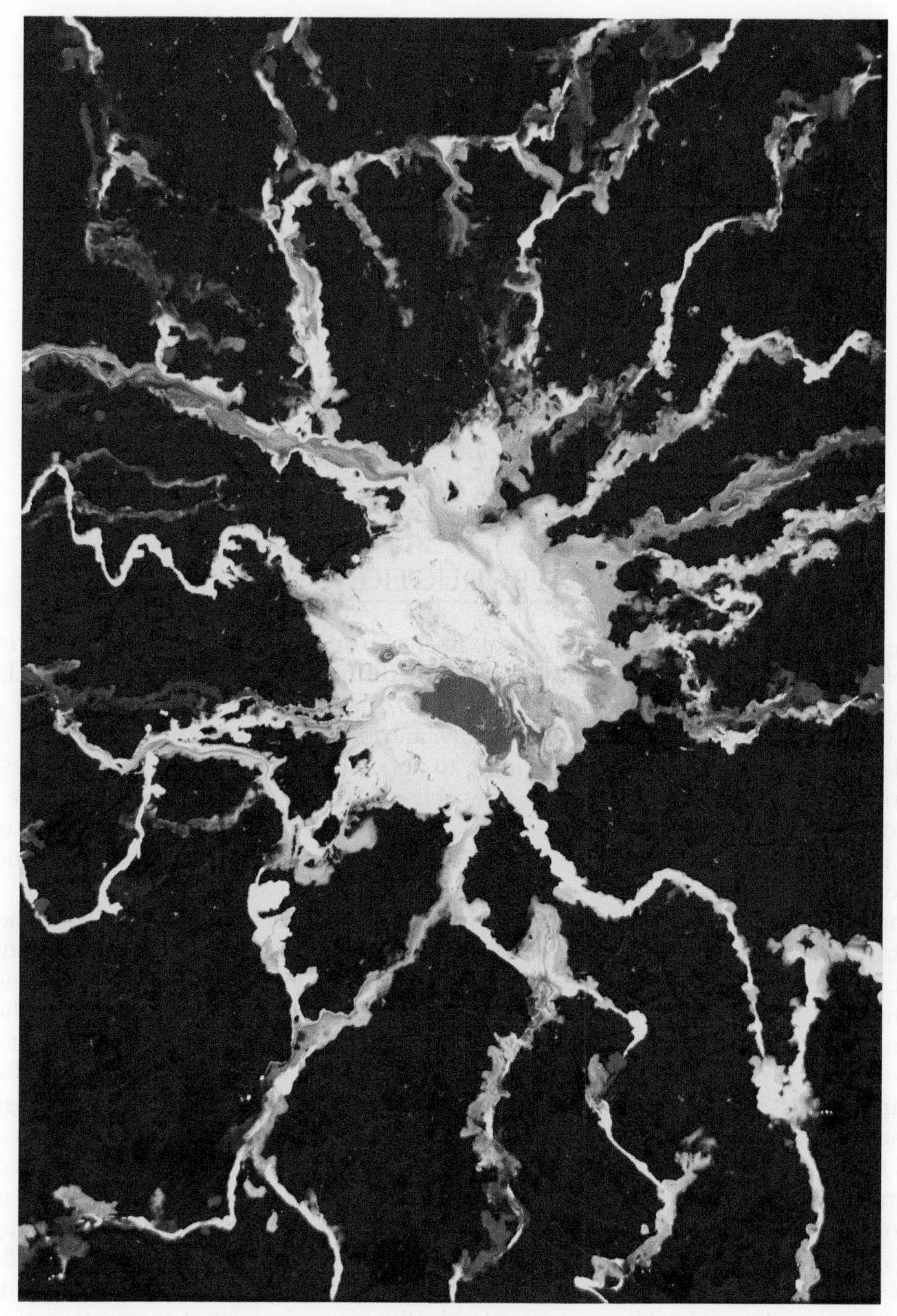

"Medium Spiny Neuron"
J. David Sweatt, acrylic on canvas (36 x 48), 2012

CHAPTER

6

Drug Addiction and Reward

Alfred J. Robison, Jian Feng and Eric J. Nestler

Fishberg Department of Neuroscience, Friedman Brain Institute, Mount Sinai School of Medicine, New York, USA

INTRODUCTION

Drug abuse exacts an enormous medical, financial, and emotional toll on society in the form of overdose and health complications, family disintegration, loss of employment, and crime. The National Institute of Drug Abuse (NIDA) estimates that the total cost of drug abuse in the USA is over $600 billion annually, and though abuse of illicit drugs is widespread and steady, it is particularly alarming to note a sharp increase in abuse of prescription drugs ([1], www.nida.nih.gov/), with new abuse among teenagers increasing more than fivefold between 1992 and 2003.[2] These data substantiate the need for increased study of the effects of drugs of abuse on the brain and the mechanisms of addiction in the expectation of uncovering novel targets for treating and preventing addictive disorders.

Although most individuals are exposed to abused drugs, only a subset experiences the loss of control over drug use and compulsion for drug seeking and taking that defines the addicted state. Entrance into this state is strongly influenced by both an individual's genetic constitution and the psychological and social context in which drug exposure occurs.[3–5] Although the genetic risk for addiction is roughly 50%, the specific genes involved remain almost completely unknown.[4] The addictive phenotype can persist for the length of an individual's life, with drug craving and relapse occurring even after decades of abstinence. This persistence suggests that drugs induce long-lasting changes in the brain that underlie addiction behaviors.

It is becoming apparent that many of the same processes of epigenetic regulation involved in the normal differentiation of cells and tissues during development are also engaged in the adult organism to mediate cellular adaptation to environmental stimuli, including drugs of abuse. At the same time, certain findings derived from adult brain may not be in line with discoveries in simpler systems. Considering the post-mitotic nature and intricate cell–cell interactions of neurons, it would not be surprising to identify new principles of epigenetic regulation different from the ones established in dividing cells. Meanwhile, epigenetic studies of stem cells and cancer biology have relied largely on in vitro systems, with very different insights

Epigenetic Regulation in the Nervous System
DOI: http://dx.doi.org/10.1016/B978-0-12-391494-1.00006-9

likely to be obtained from intact brain. Studies of epigenetic mechanisms of drug addiction offer particular promise in advancing our appreciation of chromatin regulation in the brain, given the robustness of altered gene expression seen in addiction models and the longevity of the associated behavioral abnormalities and the relative ease of measuring them in animals.

The epigenetic processes engaged by drugs in the regulation of transcriptional potential are varied and highly complex, and include changes in chromatin packing, covalent modification of DNA bases, and induction of non-coding RNAs. Increasing evidence suggests that such adaptations are one of the main processes by which drugs induce the stable changes in the brain that mediate the addicted phenotype. This chapter summarizes the findings that support this hypothesis, and highlights areas where future research will extend this fundamental knowledge of addiction and exploit it for new therapeutics.

DRUG ADDICTION AND GENE TRANSCRIPTION

A seemingly equivalent syndrome of addiction can occur with exposure to a wide variety of chemical substances or even rewarding activities, from cocaine to gambling to sex. One common mechanism across these various forms of addiction is thought to be activation of the brain's reward circuitry, which centers on dopaminergic neurons in the ventral tegmental area (VTA) of the midbrain and their projections to the limbic system, in particular, the nucleus accumbens (NAc; also known as ventral striatum), dorsal striatum, amygdala, hippocampus, and regions of prefrontal cortex (Figure 6.1), among several other regions.[6–8]

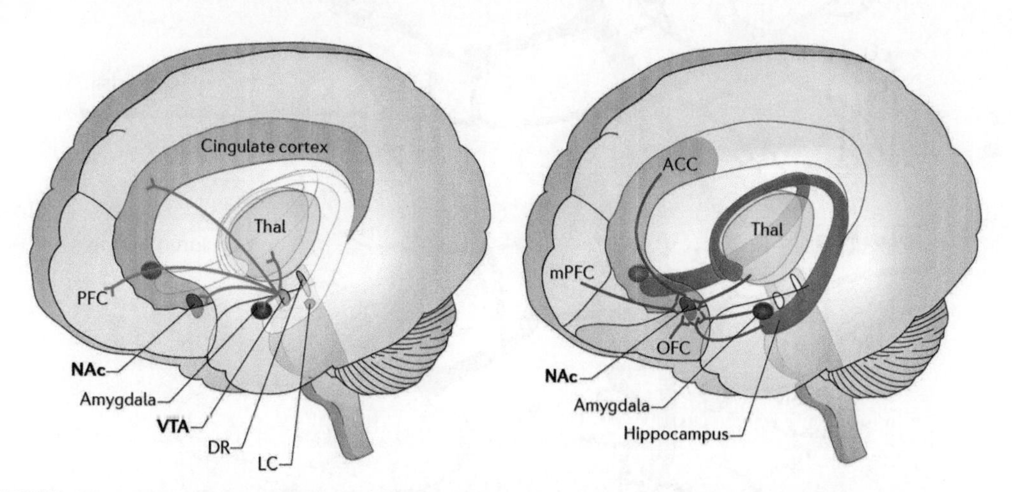

FIGURE 6.1 **Brain reward circuitry.** The brain on the left depicts dopaminergic afferents (green arrows) which originate in the ventral tegmental area (blue) and release dopamine in the nucleus accumbens (red) and many other limbic targets. Also shown are other monoaminergic nuclei – the noradrenergic locus coeruleus (blue) and serotonergic dorsal raphe (green) – which modulate drug reward and other actions. The brain on the right highlights glutamatergic regions that are important for reward: medial prefrontal cortex (yellow), orbitofrontal cortex (green), anterior cingulate cortex (orange), thalamus (blue), hippocampus (purple), and amygdala (dark red), all of which send excitatory projections to the nucleus accumbens (red). Drugs of abuse alter this reward circuitry in complex ways, which lead to addiction.

This reward circuitry is activated by stimuli or pursuits that promote evolutionary fitness of the organism, like nutrient-rich foods, sex, and social stimulation. As drugs of abuse activate this circuitry far more strongly and persistently than natural rewards, and without association with productive behavioral outcomes, chronic exposure to drugs modulates brain reward regions in part through a homeostatic desensitization that renders the individual unable to attain sufficient feelings of reward in the absence of drug. An alternate, but not mutually exclusive, hypothesis focuses on sensitization, whereby drugs alter the reward circuitry to cause increased assignment of incentive salience to drug cues, effectively making drug-associated environmental stimuli more difficult to ignore and leading to intense drug craving and relapse.[9] Pathological drug-induced changes in the reward circuitry further impair behavioral control.

Virtually all rewarding drugs or activities increase dopaminergic transmission from the VTA to the NAc and other target limbic regions, though they each employ partly distinct mechanisms and, in some cases, involve other neurotransmitter systems as well.[6–8] The actions of drugs on the NAc are further complicated by the cellular heterogeneity of this brain region, as the NAc is composed of multiple neuronal cell types (Figure 6.2), with each cell type

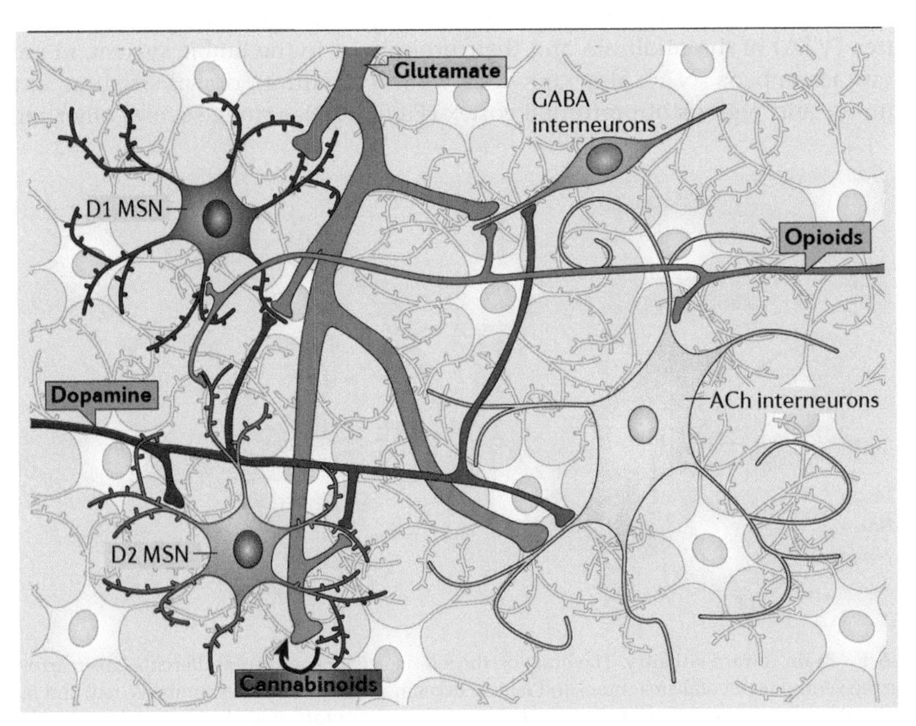

FIGURE 6.2 **Cellular organization of the nucleus accumbens.** The NAc is composed of multiple neuronal cell types, including GABAergic medium spiny neurons that express D1 or D2 dopamine receptors (D1- or D2-MSNs) and acetylcholinergic and GABAergic interneurons. These cells receive glutamatergic inputs from multiple brain regions, and dopaminergic inputs from the ventral tegmental area. Endogenous opioids are released locally and from input neurons, and cannabinoids act primarily at glutamate synapses to repress activity.

seeming to exhibit different transcriptional responses to drugs of abuse and to mediate distinct aspects of drug reward and addiction. Glutamatergic afferents from the hippocampus, prefrontal cortex, amygdala, and other regions excite all subtypes of NAc neurons,[10] with such excitation differentially regulating drug reward and motivation, as shown by recent optogenetic experiments.[11,12] These excitatory inputs are modulated by dopamine afferents from the VTA and psychostimulant drugs such as cocaine and amphetamine act by directly prolonging the effects of these dopamine signals. Excitatory inputs to the NAc are also modulated by endogenous opioid peptides that are both expressed locally and released by input neurons. Opiate drugs thus act directly on NAc neurons that express opioid receptors, and they also promote dopamine release in the NAc indirectly by inhibiting VTA γ-aminobutyric acid (GABA)-ergic interneurons. Cannabinoids also have a role in regulating NAc neurons – they act primarily by locally repressing the function of glutamatergic synapses.

Much work is needed to understand further the cellular specificity of drug action in the NAc. Ninety-five percent of NAc neurons are GABAergic medium spiny neurons (MSNs), which can be further differentiated into those that express the D1 dopamine receptor (D1-type MSNs) along with dynorphin and substance P, and those that express the D2 dopamine receptor (D2-type MSNs) along with encephalin,[13] and neuronal activity of these two cell types causes opposing effects on the rewarding properties of cocaine.[11] In addition, the intracellular signaling cascades activated by drug exposure (Figure 6.3) differ by cell type. For instance, acute cocaine causes extracellular signal-regulated kinase (ERK)-dependent phosphorylation of mitogen-and stress-activated kinase 1 (MSK1) and of histone 3 specifically in D1-type MSNs,[14] although the functional consequences of this histone modification are not yet known. By contrast, the effects of cannabinoids seem to predominate at glutamatergic synapses on D2-type MSNs.[15] About 1–2% of NAc neurons are spiny large cholinergic interneurons, which have been shown to play an important part in cocaine reward,[12] and a similar number are GABAergic interneurons, the function of which is less well understood.

Although drugs differ in their acute mechanisms of action, the common syndrome of addiction suggests that chronic exposure to these distinct acute mechanisms induces some shared molecular adaptations in brain reward regions that mediate the lasting nature of the addictive phenotype. It has long been hypothesized that changes in the transcriptional potential of genes contribute importantly to many of the neuroadaptations that result from chronic exposure to drugs of abuse.[16] We know that many mRNAs display altered expression in brain reward regions after chronic drug exposure, which suggests that transcription of individual genes is differentially regulated under these conditions. Over the past several years, studies at the chromatin level have confirmed the involvement of such transcriptional mechanisms in vivo. Moreover, beyond stable changes in steady-state mRNA levels, this work has demonstrated that the "inducibility" of a gene – its ability to be induced or repressed in response to the next drug exposure or some other environmental stimulus – is also altered by chronic drug exposure and that such gene "priming" or "desensitization" is mediated by stable drug-induced changes in the chromatin state around individual genes (Figure 6.4).

This transcriptional and epigenetic model of chronic drug action provides a plausible mechanism for how environmental influences during development can increase (or decrease) the risk for addiction later in life. For example, there is mounting evidence that

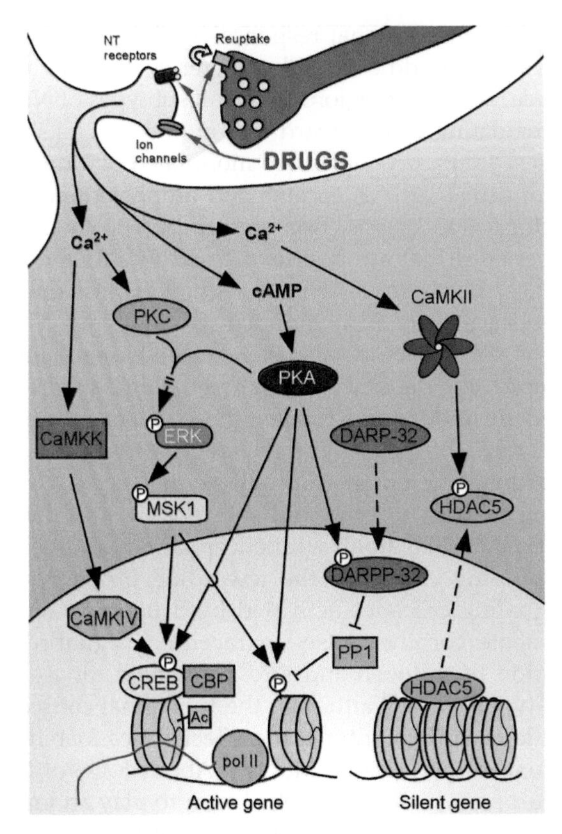

FIGURE 6.3 **Intracellular signaling induced by drugs.** Drugs or abuse act primarily through synapses to increase intracellular levels of second messengers, including Ca^{2+} and cAMP. Ca^{2+} activates of Ca^{2+}/calmodulin-dependent protein kinases (red/orange), resulting in phosphorylation/activation of CREB and phosphorylation/nuclear export of HDAC5. Phospho-CREB recruits CBP to increase histone acetylation and transcription. cAMP activates PKA which phosphorylates CREB and H3, increasing transcription, and phosphorylates DARPP-32, causing it to enter the nucleus and inhibit PP1 dephosphorylation of H3. PKA also acts indirectly to promote the nuclear translocation of HDAC5 (not shown). PKA, PKC, and CaMKs can activate the MAPK cascade (green), which also leads to phosphorylation of CREB and H3.

stress during adolescence increases the risk of addiction, and that exposure to drugs in utero increases the risk in adolescence and adulthood.[17,18] Long-lasting changes in gene transcription or in transcriptional potential that result from early-life stress or drug exposure – mediated at the chromatin level in the absence of genetic differences in primary DNA sequence – might render an adult brain more vulnerable to the addictive process. As alterations in transcriptional potential can last for many years, this model also explains how relapse can occur despite decades of abstinence. While largely conjectural, recent work supports the role of epigenetic modifications in mediating life-long changes in addiction vulnerability.[19]

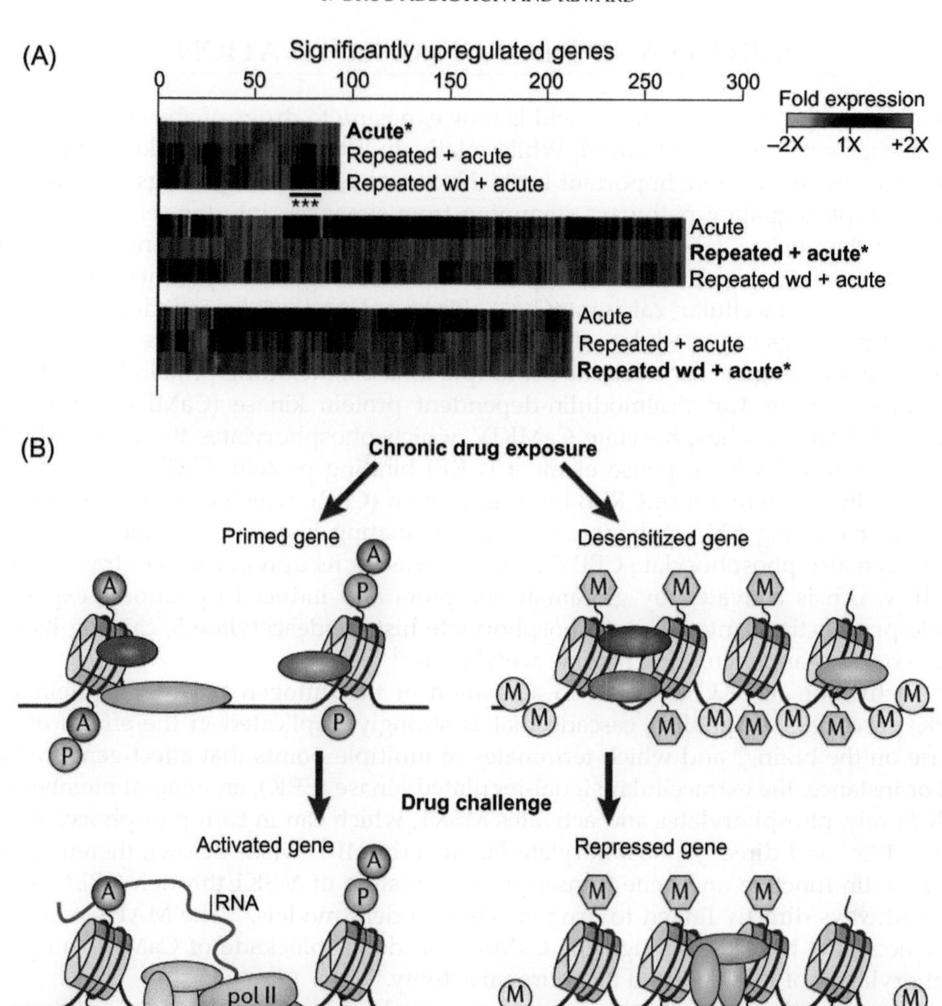

FIGURE 6.4 **Gene priming and desensitization.** In addition to regulating the steady-state expression levels of certain genes, cocaine induces latent effects at many other genes, which alter their inducibility in response to a subsequent stimulus. (A) Analysis of mRNA expression after acute or chronic cocaine. Heat maps marked with an asterisk (*) show all genes that are upregulated in the NAc one hour after a cocaine challenge in naïve animals (acute), in animals treated repeatedly with cocaine (repeated + acute), or in animals after 1 week of withdrawal from repeated cocaine (repeated wd + acute). Associated heat maps show how the same genes were affected under the other two conditions. Desensitized transcriptional responses after repeated cocaine are indicated (***). (B) Early evidence suggests that epigenetic mechanisms are important in mediating such gene priming and desensitization and that many such changes are latent, meaning that they are not reflected by stable changes in steady-state mRNA levels. Rather, such changes alter chromatin structure such that later drug challenge induces a given gene to a greater (primed) or lesser (desensitized) extent based on the epigenetic modifications engendered by previous chronic drug exposure. A major goal of current research is to identify the chromatin signatures that underlie gene priming and desensitization.

DRUGS AND NEURONAL ACTIVATION

One of the major questions in the field is how exposure to drugs of abuse regulates chromatin mechanisms within cell nuclei. While relatively little is still as yet known about this crucial issue, there are some important leads. Neuronal activation by drugs of abuse sets in motion multiple signaling pathways stemming from synapses, into the cell body, and ultimately into the nucleus (see Figure 6.3). Glutamate release activates several types of ionotropic and metabotropic glutamate receptors which, through multiple mechanisms, leads to increases in intracellular calcium (Ca^{2+}). Increased and prolonged dopamine signaling caused by drugs can modulate this glutamatergic activation, and directly regulate the same second messengers. Ca^{2+} activates multiple kinases, including protein kinase C (PKC) and members of the Ca^{2+}/calmodulin-dependent protein kinase (CaMK) family. CaMK kinase (CaMKK) can phosphorylate CaMKIV, which phosphorylates the cyclic adenosine monophosphate (cAMP) response element (CRE) binding protein (CREB) at serine 133.[20] This allows the recruitment of CREB binding protein (CBP), which can affect transcription by directly recruiting RNA polymerase, or by regulating histone acetylation (see below). CaMKIV can also phosphorylate CBP directly, increasing its upregulation of transcription.[21] CaMKII, which is activated by glutamate receptors and induced by chronic exposure to multiple psychostimulants,[22,23] can phosphorylate histone deacetylase 5, causing its export from the nucleus and increasing histone acetylation.[24]

Ca^{2+} activation of PKC can lead to activation of the mitogen-activated protein kinase (MAPK) cascade,[25] a signaling cascade that is strongly implicated in the effects of drugs of abuse on the brain,[26] and which terminates in multiple points that affect gene transcription. For instance, the extracellular signal-regulated kinase (ERK), an integral member of the MAPK family, phosphorylates and activates MSK1, which can in turn phosphorylate CREB at serine 133[27] and directly phosphorylate H3 at serine 10[28,29] (see below), thereby regulating chromatin function and gene transcription. Signaling of MSK1 through CREB is one of many pathways directly linked to drug effects in rodent models.[30] The MAPK cascade can also be activated by Ca^{2+} through the CaMK cascade, as blockade of CaMKI can prevent phosphorylation of ERK induced by neuronal activity.[31]

D1 receptor activation caused by drug exposure leads to increased intracellular cAMP and subsequent activation of protein kinase A (PKA). PKA has many downstream targets, and can directly phosphorylate H3 serine 10[32] in a manner induced by exposure to psychostimulants and regulated by cell type.[14,33] Interestingly, PKA also phosphorylates DARPP-32 (dopamine and cAMP-regulated phosphoprotein, 32 kDa), a molecule strongly implicated in addiction,[34] which can then downregulate protein phosphatase 1 (PP1) activity, preventing PP1 from dephosphorylating H3 at serine 10. PKA can also directly activate the MAPK cascade,[35] leading to the CREB/CBP and histone changes discussed above. Finally, a recent study demonstrated that PKA activation in NAc by cocaine leads to the nuclear import of HDAC5 and subsequent regulation of histone acetylation.[36]

Contributions by multiple labs indicate that many neurotrophic factors regulated by drugs of abuse can cause epigenetic changes in the brain. Brain-derived neurotrophic factor (BDNF), nerve growth factor (NGF), and other neurotrophins act at various receptors including Trk and p75, to initiate multiple signaling cascades in the NAc in response drugs of abuse.[37–39] BDNF, acting through the TrkB receptor, can activate phospholipase C to engage the MAPK cascade,[40] which can regulate chromatin as described above.

The regulation of transcription through neuronal activation is brain region- and cell type-specific, and the same signaling cascades can have differential effects on different gene promoters, resulting in complex molecular underpinnings that are only now beginning to be unraveled. In the remainder of this chapter, we discuss epigenetic mechanisms regulated by drugs of abuse, some of which result from signaling mechanisms discussed above, but most of which are the result of as yet undiscovered signals induced by drug exposure.

EPIGENETICS OF ADDICTION

Over the past decade, research into the regulation of transcriptional potential through modification of DNA and chromatin structure has grown exponentially. As it became clear that epigenetic change underlies adaptations in the adult organism, investigations of epigenetic mechanisms have proven fruitful in numerous fields, including drug addiction.[41,42] Here, we introduce three major mechanisms of epigenetic regulation – histone tail modification, DNA methylation, and microRNAs (miRNAs) – and summarize the major findings that have linked each of these mechanisms to addiction.

Histone Tail Modification

As described in previous chapters, many residues in the tails of histones are covalently modified in many ways, resulting in a complex "code" that is thought to control the accessibility of the genome to the transcriptional machinery.[43,44] Histone acetylation, which negates the positive charge of lysine residues in the histone tail, is associated with transcriptional activation. This process is controlled by histone acetyltransferases (HATs) and histone deacetylases (HDACs), each of which comprises multiple enzyme classes whose expression and activity are exquisitely regulated.[45] Histone methylation has been associated with both transcriptional activation and repression depending on the particular residue and the extent of methylation:[46,47] both lysine and arginine residues can be methylated by several families of histone methyltransferases (HMTs), and this reaction can be reversed by equally diverse histone demethylases. Histone tail modifications also include phosphorylation, ubiquitination, sumoylation, ADP ribosylation, among many others,[45] whose roles in drug addiction are only now being investigated. For instance, we find that global levels of poly(ADP-ribose), or PAR, and the activity of poly(ADP-ribose) polymerase-1 (PARP-1) are elevated in the NAc of mice exposed to chronic cocaine.[48] Moreover, pharmacological blockade of PARP-1 in the NAc reduces locomotor sensitization to cocaine, whereas overexpression of PARP-1 in this region increases locomotor sensitization and cocaine place conditioning,[48] suggesting that ADP ribosylation in the reward circuitry is important for drug responses.

Multiple drugs of abuse induce changes in histone acetylation in brain, and evidence has begun to accumulate that these modifications underlie some of the functional abnormalities found in addiction models.[42] By use of chromatin-immunoprecipitation (ChIP) technology, it was found that acute and chronic cocaine induced different histone acetylation changes in striatum, whose enrichment at promoter sites is gene specific.[49] Interestingly, acute cocaine caused robust induction of H4 acetylation, but not H3 acetylation, at the *cFos* promoter. However, no acetylation change was seen at the same site after chronic cocaine, which is consistent with cocaine's ability to induce *cFos* acutely but desensitize it chronically.[50] In contrast, at promoters of genes that are only induced by chronic cocaine, such as *BDNF* and *Cdk5*, H3 acetylation was observed after chronic but not acute cocaine. These findings demonstrate that different histone acetylation changes likely encode distinct functional changes in a gene-specific manner. Furthermore, drug-induced histone acetylation changes can be either labile or stable in accordance with the corresponding transcriptional change. For example, H4 acetylation at the *cFos* promoter appeared at 30 min after acute cocaine and lasted less than 3 hours, whereas H3 acetylation at the *BDNF* promoter after chronic cocaine persisted and even increased further over a week of drug withdrawal,[49] consistent with the delayed and stable induction of BDNF by cocaine.[49]

Gene promoters that show increased H3 vs H4 acetylation have been mapped genome-wide on promoter arrays by use of ChIP-chip.[51] Many more genes display increased histone acetylation than those that show reductions. Unexpectedly, very little overlap was seen between genes with altered levels of H3 and H4 acetylation.[51] These findings emphasize a key question as to what governs gene-specific acetylation changes in the face of global modifications. Another key question, as noted earlier, concerns the precise intracellular signaling cascades through which cocaine induces changes in histone acetylation – there is some information that such changes may be specific to D1-type MSNs and involve regulation of growth factor-associated kinases.[14,52] Meanwhile, alcohol withdrawal has been demonstrated to increase HDAC activity and reduce histone acetylation in the mouse amygdala,[53] and the commonly abused inhalant benzyl alcohol regulates potassium channels that are tied to alcohol tolerance via H4 acetylation in *Drosophila*.[54] In addition, exposure to Δ^9-THC, the active ingredient in marijuana, increases HDAC3 in trophoblast cells.[55] However, this alteration was absent in a genome-wide screen of brain tissue from Δ^9-THC-treated mice,[56] demonstrating that experiments on cell culture can yield effects that are very different from those found in a complex heterogeneous tissue like the brain. These data highlight the need for further research to define the effects of drugs of abuse on histone acetylation in brain in a region- and cell type-specific manner and to identify the specific HAT and HDAC subtypes and intracellular signaling pathways that mediate this regulation in vivo.

Since we hypothesize that the changes in histone acetylation induced by chronic exposure to drugs underlie the concurrent behavioral maladaptations, it is important to determine whether experimental alterations in histone acetylation affect addiction-related behaviors. Indeed, a great deal of recent research demonstrates exactly this. For instance, short-term administration of non-specific HDAC inhibitors, either systemic or intra-NAc, potentiates place conditioning and locomotor responses to psychostimulants and to opiates.[41,49,57] More prolonged HDAC inhibition has been reported to induce changes in the opposite direction,[58,59] perhaps through adaptations that oppose initial enzyme inhibition. Studies of specific HDAC isoforms have yielded interesting information: overexpression of HDAC4 or HDAC5 decreases behavioral responses to cocaine,[49,57] whereas genetic deletion of HDAC5 hypersensitizes mice to the chronic (but not acute) effects of the drug.[57] Likewise, mutant mice with reduced expression of CBP, a major HAT in brain, exhibit decreased sensitivity to chronic cocaine.[60] Much additional work is needed to define the influence of specific HAT and HDAC subtypes on addiction-related phenomena, perhaps through a genome-wide study to explore their global effects as well as their unique targets. In sum, these studies suggest that histone acetylation can potently regulate drug responses and provide a possible path (e.g. HDAC inhibitors) for future clinical intervention.

The potential complexity involved is indicated by recent findings on sirtuins, which are considered Class III HDACs but in reality influence many non-histone proteins. For example, SIRT1 deacetylates the brain-specific helix-loop-helix transcription factor NHLH2 on lysine 49 to increase its activation of the monoamine oxidase A (MAO-A) promoter, which in turn activates transcription of MAO-A to reduce serotonin levels and mediates levels of anxiety.[61] Genome-wide studies of chromatin alterations in the NAc after chronic cocaine revealed upregulation of two sirtuins, SIRT1 and SIRT2.[51] Pharmacological inhibition of sirtuins in the NAc decreases cocaine place preference and self-administration, whereas systemic activation of sirtuins increases rewarding responses to cocaine. SIRT1 and SIRT2 induction is associated with increased H3 acetylation,[51] suggesting that sirtuin genes are themselves epigenetically regulated. Work is now needed to identify the proteins that are affected in NAc by cocaine-induced regulation of these sirtuins. For example, sirtuins deacetylate several transcription factors such as forkhead box (FoxO) proteins, bind and deacetylate the transcription co-factor p300 and downregulate transcription,[62] and serve scaffolding functions by contributing to transcriptional repressive complexes,[63] processes which now warrant study in cocaine models.

Histone methylation is directly regulated by drugs of abuse as well: global levels of histone 3 lysine 9 dimethylation (H3K9me2) are reduced in the NAc after chronic cocaine[64] and a genome-wide screen revealed alterations in H3K9me2 binding on the promoters of numerous genes in this brain region;[51] both increases and decreases were observed, indicating again that epigenetic modifications at individual genes often defy global changes. The global decrease in H3K9me2 in the NAc is likely mediated by cocaine-induced downregulation of two HMTs, G9a and G9a-like protein (GLP), which catalyze H3K9me2.[64] These adaptations mediate enhanced responsiveness to cocaine, as selective knockout or pharmacological inhibition of G9a in the NAc promotes cocaine-induced behaviors, whereas G9a overexpression has the opposite effect. G9a likewise mediates the ability of cocaine to increase the spine density of NAc MSNs.[64] Interestingly, there is a functional feedback loop between G9a and ΔFosB, a key transcription factor in addiction:[65] ΔFosB seems to be responsible for cocaine-induced suppression of G9a, and G9a binds to and represses the *fosb* promoter, such that G9a downregulation may promote the accumulation of ΔFosB observed after chronic cocaine.[64] In addition, G9a and ΔFosB share many of the same target genes.

Chronic cocaine also downregulates H3K9me3, a mark of heterochromatin, specifically in the NAc and this change is associated with a decrease in the total amount of heterochromatin in NAc MSN nuclei and an increase in the volume of these nuclei.[66] ChIP of H3K9me3 followed by deep sequencing (ChIP-seq) identified the pattern of H3K9me3 in NAc and demonstrated predominant binding at non-genic regions, including many repetitive sequences. Cocaine decreased H3K9me3 enrichment at specific genomic repeats (e.g. long interspersed nuclear element [LINE]-1 repeats), and such changes were associated with increased expression of LINE-1 retrotransposon-associated repetitive elements in NAc. The increase may reflect global patterns of genomic destabilization in this brain region after repeated cocaine administration.[66]

Studies are now needed to examine the actions of other drugs of abuse on these histone endpoints, as well as the effect of drugs on many other types of histone modifications known to regulate eukaryotic gene expression in other systems, in addiction models. Examples include recent, preliminary observations of chronic cocaine regulation of histone arginine methylation and poly-ADP ribosylation, of several families of chromatin remodeling proteins, and of histone variant subunits in the NAc, all of which illustrate the complexity of epigenetic changes associated with drug exposure.[67–70] A major goal of this research is, by integrating it with global measures of RNA expression (e.g. by RNA-seq), to crack the histone modification code and understand how drugs of abuse regulate the genome within a given brain reward region.

An important part of this research is to relate drug-induced modifications of histones, occurring at specific drug-regulated genes, with the recruitment of numerous additional proteins that ultimately constitute the transcriptional activation or repression complexes that mediate such regulation. For example, early studies have demonstrated that cocaine induction of CDK5 in the NAc involves a cascade of events which includes binding of ΔFosB to the *Cdk5* gene promoter, followed by the recruitment of CBP, increased H3 acetylation, and the recruitment of specific chromatin remodeling factors, such as BRG1[49] (Figure 6.5). Such activation also involves reduced repressive histone methylation at this promoter, which is mediated via cocaine suppression of G9a. In contrast, a very different cascade mediates chronic amphetamine repression of the *c-fos* gene. Here, ΔFosB binds to the *c-fos* promoter and recruits HDAC1 and SIRT1, and presumably numerous other proteins.[50] Also, chronic amphetamine induces increased repressive histone methylation at the *c-Fos* promoter, perhaps mediated via increased G9a binding.[64] It is interesting that such increased G9a binding occurs despite the global decrease in G9a expression, once again highlighting gene-specific changes that occur on top of global modifications. Understanding the molecular basis of such gene-specific modifications – e.g. why ΔFosB triggers a cascade of transcriptional activation when it binds to one promoter, but a cascade of transcriptional repression when it binds to another – is a crucial goal of current research. To date, these efforts have been pursued on a protein-by-protein basis, which is experimentally painstaking. A major need in the field is to develop tools to analyze the complete protein complexes recruited to individual genes in concert with drug exposure.

DNA Methylation

DNA methylation involves the addition of a methyl group to the C5 position of cytosine (5-mC), a reaction catalyzed by DNA methyltransferases (DNMTs). It is an important epigenetic modification in higher eukaryotes and is required for several key physiological processes, such as cell differentiation/reprogramming, genetic imprinting, X inactivation, and silencing of repetitive elements.[71,72] Cytosines at CpG dinucleotides are highly methylated (60–80% in mammals). In contrast, non-CpG cytosines are generally unmethylated in differentiated cells. Cytosine methylation changes at non-CpG sites appear to be dynamic in early development and in pluripotent stem cells,[73] and so are less studied in the adult brain. DNA methylation is covered in detail in other chapters of this book. Hence, we highlight a few key questions here.

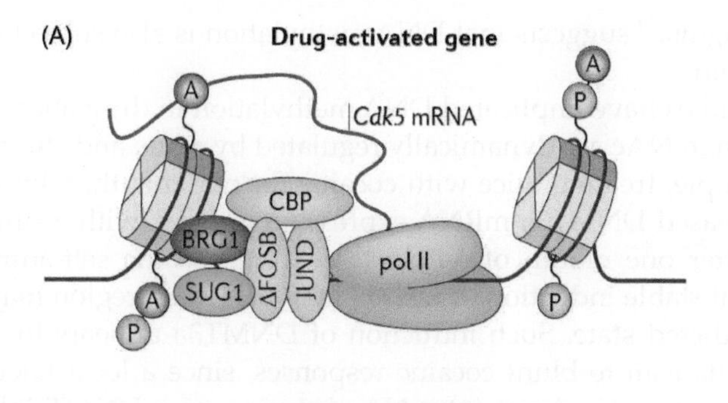

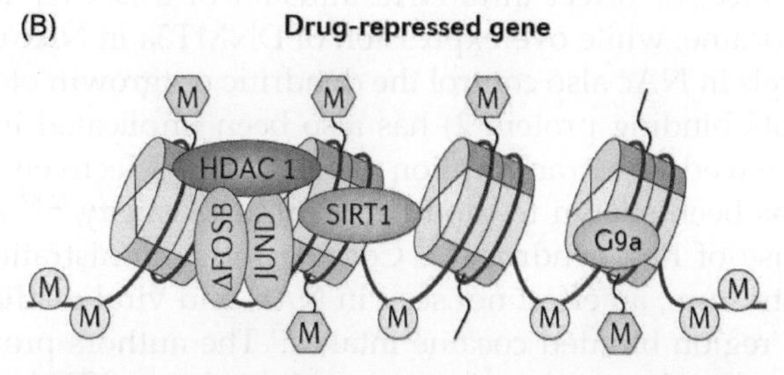

FIGURE 6.5 Epigenetic basis of drug regulation of gene expression. The figure is based on the mechanisms by which chronic cocaine, through ΔFosB, activates the *cdk5* gene (A) and represses the *c-fos* gene (B). (A) ΔFosB binds to the *cdk5* gene and recruits several co-activators, including CBP (CREB binding protein) – a type of histone acetyltransferase (HAT) – leading to increased histone acetylation, transcription factor BRG1 (also known as brahma-related gene 1) – a type of chromatin remodeling factor – and SUG1 (proteasome 26S ATPase subunit 5), another type of chromatin regulatory protein. ΔFosB also represses G9a expression, leading to reduced repressive histone methylation at the *cdk5* gene. The net result is gene activation and increased CDK5 expression. (B) In contrast, ΔFosB binds to the *c-fos* gene and recruits several co-repressors, including HDAC1 (histone deacetylase 1) and SIRT 1 (sirtuin 1). The gene also shows increased G9a binding and repressive histone methylation (despite global decreases in these markers). The net result is *c-fos* gene repression. As transcriptional regulatory complexes contain dozens or hundreds of proteins, much further work is needed to define further the activational and repressive complexes that cocaine recruits to particular genes to mediate their transcriptional regulation and to explore the range of distinct activational and repressive complexes involved in cocaine action.

The role of DNA methylation on gene transcription is generally repressive. The methylated DNA can either directly prevent the association of DNA-binding factors with their target sequence, or it can bind to methyl-CpG-binding proteins (MBP) which recruit transcriptional co-repressors (such as HDACs) to modify surrounding chromatin into a silenced state.[74] Though most focus to date has been on promoter CpG methylation, a considerable number of methylated CpGs are found in gene bodies and their effects on transcription remain controversial.[75] Previous technology could not distinguish 5-mC from several other forms of methylated DNA, such as 5-hydroxymethylcytosine (5-hmC), which is enriched in gene bodies and positively correlates with transactivation.[76,77] Three DNMTs have been studied in the nervous system, which include the maintenance enzyme DNMT1 and the de novo methyltransferases DNMT3a and DNMT3b.[78] Although DNA methylation is considered to be more stable than histone tail modifications, recent research, largely from learning and memory paradigms,[79] suggests that DNA methylation is also subject to highly dynamic regulation in the brain.

Several recent studies have implicated DNA methylation in drug abuse models. Levels of DNMT3a expression in NAc are dynamically regulated by acute and chronic cocaine administration.[80] For example, treating mice with cocaine for one month, followed by one month of withdrawal, increased DNMT3a mRNA expression in NAc, with a similar lasting induction seen in rats after one month of withdrawal from cocaine self-administration. These findings suggest that stable induction of DNMT3a in this brain region may be important for maintaining the addicted state. Such induction of DNMT3a appears to serve as a homeostatic feedback mechanism to blunt cocaine responses, since a local knockout of DNMT3a in the NAc of adult mice, or direct intra-NAc infusion of a DNMT inhibitor, attenuated behavioral effects of cocaine, while overexpression of DNMT3a in NAc exerted the opposite effects.[80] DNMT3a levels in NAc also control the dendritic outgrowth of NAc MSNs.[80,81]

MeCP2 (methyl CpG binding protein 2) has also been implicated in stimulant action.[81] MeCP2 is generally viewed as a transcription repressor by selectively binding to methylated DNA. MeCP2 has been shown to modulate neural plasticity,[82,83] and loss-of-function mutations are the cause of Rett syndrome.[84] Cocaine self-administration increases MeCP2 expression in dorsal striatum, an effect not seen in NAc, and viral-mediated knock-down of MeCP2 in the former region blunted cocaine intake.[85] The authors provided evidence that MeCP2 produces this effect by suppressing expression of an miRNA, miR132/212, which normally suppresses expression of BDNF. According to this scheme, induction of MeCP2 thus leads to induction of BDNF and enhanced cocaine intake. Another study showed that mice with a hypomorphic mutation in MeCP2 display reduced behavioral responses to amphetamine, although these authors related this phenotype to MeCP2 action in the NAc.[81]

The next step in understanding these phenomena is to identify the specific genes whose expression levels are controlled in NAc or dorsal striatum by stimulant-induced alterations in DNMT3a or MeCP2 expression. An important question in the field is whether DNMT3a and MeCP2 might influence gene expression without corresponding changes in DNA methylation. A recent study on stem cells suggests that DNMT3a binding at non-promoter sites is functionally important,[86] which emphasizes the need for genome-wide investigations of DNA methylation in addiction models. The story for MeCP2 appears even more complex. Even though MeCP2 binding predominates at methylated CpG sites, it can also bind to unmethylated DNA. Moreover, the main regulatory effect of MeCP2 may be to exert more global effects on the genome instead of repressing expression of specific genes; for instance, to reduce "transcriptional noise".[87,88] This might explain why MeCP2 can activate transcription under certain circumstances.

To date, only a few studies have reported changes in the methylation status of a small number of candidate genes in addiction models.[89,90] The major need in the field is for genome-wide characterizations, but this remains challenging for several reasons. Methodologically, the only reliable means for obtaining such methylation maps is through sequencing, and this is very costly. Also, the highly dynamic temporal changes reported for DNMT3a regulation in NAc by cocaine[80] demonstrate the need for carrying out such analyses across broad time points. Through evolution, methylated cytosines appear susceptible to deamination and conversion to thymines. This may explain why methylated cytosines are generally underrepresented in the genome and often grouped into dense patches called

CpG islands, which are generally unmethylated and less vulnerable to spontaneous deamination.[91] However, most studies of DNA methylation in brain have focused on CpG islands, where the changes are usually subtle and the overall DNA methylation pattern remains hypomethylated across the whole CpG island. It is thus hard to interpret how such DNA methylation changes can trigger robust transcription changes. Rather, some sporadic CpG sites[92] or less densely packed CpG regions[93] seem to be more dynamic and have a greater impact on transcription. It is hoped that genome-wide DNA methylation profiling paired with transcriptome analyses will provide answers to these important questions.

Thus, much more work is needed to explore the involvement of DNA methylation in drug addiction. In addition to the genome-wide mapping studies noted above, it will also be important to explore a potential role for DNA demethylation in drug action. As well, the field must consider potential drug regulation of many forms of DNA methylation, far beyond 5-mc or 5-hmc stated earlier. For example, it was recognized recently that 5-hmC can be further oxidized into 5-formylcytosine (5-fC) and 5-carboxylcytosine (5-caC), which can then be converted into unmethylated cytosine.[94,95] Interestingly, 5-hmC, 5-fC, and 5-caC are all highly enriched in neurons, which promises essential roles of these newly defined nucleotide bases in brain function.

Gene Priming and Desensitization

Ongoing studies of chromatin regulation in addiction models support the view that epigenetic modifications at individual genes, in addition to underlying stable changes in the steady-state levels of mRNA expression of certain genes, alter the inducibility of many additional genes in response to some subsequent stimulus in the absence of changes in baseline expression levels.[44] Although such studies are still in relatively early stages of development, these types of latent epigenetic changes can be viewed as "molecular scars" that dramatically alter an individual's adaptability and contribute importantly to the addicted state.

Such priming and desensitization of genes is evident in a recently published microarray study.[64] Numerous desensitized genes were identified: ≈10% of genes whose transcription is induced acutely in the NAc by cocaine are no longer induced by a cocaine challenge after prior chronic exposure to the drug (see Figure 6.4A).[37] Conversely, numerous genes are primed: genes that are not affected by acute cocaine become inducible after a chronic course of cocaine, with approximately threefold more genes being induced in cocaine-experienced animals. The mechanisms underlying such gene desensitization and priming remain incompletely understood; our hypothesis is that epigenetic mechanisms are crucial (see Figure 6.4B). A subset of primed genes show reduced binding of G9a and H3K9me2 at their promoters in the NAc, suggesting the involvement of this epigenetic mark.[64] Desensitization of the *c-fos* gene in the NAc, discussed above and depicted in Figure 6.5, involves stable increases in the binding of ΔFosB, G9a, and related co-repressors, which – although not affecting steady-state levels of *c-Fos* mRNA – dramatically repress its inducibility to subsequent drug exposure.[50] More recently, we have demonstrated priming of the *fosB* gene in NAc after chronic cocaine exposure, which persists for at least one month after withdrawal.[96] This leads to the more rapid re-accumulation of ΔFosB upon re-exposure to cocaine and to enhanced behavioral responses to the drug. Stable priming of *fosB* seems to involve the pausing of a specific phosphorylated form of RNA polymerase II at the *fosB* promoter, which is then relieved in response to a cocaine challenge, enabling more rapid and robust transcription of the gene.

A major need for the field is now to investigate many additional chromatin mechanisms that are recruited by drug exposure to mediate gene priming and desensitization and to understand the detailed mechanisms that target those particular genes. The goal of such studies would be to identify "chromatin signatures" that underlie such long-lasting regulation. The prominence of gene priming and desensitization indicates that studies of steady-state mRNA levels per se would miss important aspects of drug regulation that are not captured at the particular time point examined. For example, the aforementioned microarray study[64] measured mRNA levels one hour after a cocaine challenge, and preliminary evidence suggests that a partly distinct set of genes show evidence of priming and desensitization at 4 hours. These observations highlight the unique utility of genome-wide assays of chromatin regulation, as such assays would reveal priming and desensitization more globally.[51]

microRNAs

Increasing attention has focused on a variety of non-coding RNAs that are important in biological regulation.[97] These include miRNAs, which are the best studied small non-coding RNAs in brain. They are generally around 22bp long, are found in all mammalian cells, and are post-translational regulators that bind to complementary sequences on target mRNAs to repress translation and thus silence gene expression.[98] Like histone modifications and DNA methylation, expression of miRNAs can alter the transcriptional potential of a gene in the absence of any change to the DNA sequence, and thus can be considered an epigenetic phenomenon. Several recent studies have implicated miRNAs in addiction behaviors, and miRNAs altered by drugs of abuse have been shown to regulate the expression of many proteins strongly linked to addiction.[99,100]

Cocaine self-administration in rats reportedly increases expression of miR-212 in striatum, and experimentally increasing miR-212 levels in this region decreases cocaine reward.[100] The actions of miR-212 depend on upregulation of CREB, which is known to decrease the rewarding effects of cocaine,[16,42] and more recent work demonstrates that cocaine induction of MeCP2 may repress transcription of miR-132/miR-212, which reduces miRNA repression of BDNF and leads to increased BDNF expression and cocaine intake, as noted above.[85] It has been proposed that this CREB–MeCP2–miR-212–BDNF mechanism is at least partially responsible for cocaine tolerance and escalating intake.[82] This finding provides a potential bridge between miRNA regulation and DNA methylation in addiction models. miR-124 and miR-181a are also regulated in brain by chronic cocaine, where they are decreased and increased, respectively.[101] miR-124 overexpression in the NAc reduces cocaine place conditioning, whereas overexpression of miR-181a has the opposite effect,[102] suggesting that drug regulation of these miRNAs may also act as mechanisms of tolerance and escalating intake. Like miR-212, miR-124 and miR-181a may operate through the CREB–BDNF pathway, since miR-124 overexpression downregulates both of these genes. However, these miRNAs have also been shown to affect the expression of the dopamine transporter, so their mechanisms of action are likely to be complex.[103] Finally, arginine exporter protein ARGO2 – which is important in miRNA-mediated gene silencing – along with several specific miRNAs have recently been implicated in cocaine regulation of gene expression selectively in the D2 subclass of striatal MSNs.[104]

Other drugs of abuse have been linked to miRNAs as well. Opioid receptor activation downregulates miR-190 in cultured rat hippocampal neurons in a beta-arrestin2-dependent manner,[105] and the *let-7* family of miRNA precursors is upregulated by chronic morphine exposure in mice.[106] Interestingly, the μ opioid receptor is itself a direct target for let-7, and the resulting repression of the receptor has been suggested as a novel mechanism for opiate tolerance.[106] In zebrafish and in cultured immature rat neurons, morphine decreases miR-133b expression, and this might influence dopamine neuron differentiation.[103] Additionally, both acute and chronic alcohol exposure upregulates miR-9 in cultured striatal neurons, and this may contribute to alcohol tolerance through regulation of large-conductance Ca^{2+} activated K^+ (BK) channels.[107] miR-9 seems preferentially to downregulate BK channel isoforms that are sensitive to alcohol potentiation, perhaps shifting BK channel expression toward more tolerant subytpes.[108] miR-9 also targets the D2 dopamine receptor,[108] and so probably influences alcohol reward.

In the future, next-generation sequencing of miRNAs in several brain regions after exposure to drugs of abuse will be essential to uncover regulation of specific miRNAs and eventually the genes they regulate. Compared with older microarray technology, small RNA sequencing also enables identification and characterization of novel miRNAs, as well as other less studied small RNAs, in addiction. Indeed, this process has already begun, as such screens are revealing numerous miRNAs regulated in the NAc after chronic cocaine.[82,104,109] For example, cocaine regulation of the miR-8 family suggests novel mechanisms for drug-induced alterations in the neuronal cytoskeletal and synaptic structure.[109] Exploring this mechanism in drug-induced regulation of NAc dendritic morphology is an important line of future investigation.

FUTURE DIRECTIONS

This chapter has summarized the increasing array of findings that support a role for regulation of the transcriptional potential of myriad genes in the brain's maladaptations to drugs of abuse. The mechanisms of transcriptional and epigenetic regulation are themselves varied and highly complex, and future studies are needed to catalog the vast number of regulatory events that occur as well as to understand the precise underlying mechanisms involved. Key questions include: What controls the recruitment or expulsion of individual transcriptional regulatory proteins to a particular target gene? Our hypothesis is that the underlying epigenetic state of that gene is a crucial determining factor, but then what controls the formation and maintenance of distinct epigenetic states at particular genes? Also, what are the intracellular signaling cascades that transduce the initial drug action at the neurotransmitter-receptor level to the neuronal nucleus to regulate the epigenetic state of specific subsets of genes?

The existing literature on transcriptional and epigenetic mechanisms of addiction is limited in several key ways. Most studies to date have employed conditioned place preference and locomotor sensitization paradigms. While these behavioral assays provide useful insight into an animal's sensitivity to the actions of drugs of abuse on the brain's reward circuitry, they do not provide direct measures of drug reinforcement or addiction per se. Rather, the field needs to make greater use of drug self-administration and relapse assays, which are considered the best available animal models of addiction.[110–112] Likewise, most studies have utilized experimenter-administered drugs of abuse, even though we know that drugs exert some distinct actions when self-administered or given within a particular environmental context. Work is also needed to move beyond the relatively short time frames of most current experiments to examine transcriptional and epigenetic endpoints after much longer periods of drug exposure and longer periods of withdrawal from drug exposure, as well as to extend what has largely been studies of cocaine action in NAc to studies of several other drugs and several other reward-related regions. Future studies of gene regulation will better inform drug discovery efforts as they increasingly incorporate experimental paradigms that better model human addiction.

Another limitation of the existing literature is the reliance of many studies on overexpression systems, viral or transgenic, which often induce levels of expression far greater than those seen under normal conditions or even after drug treatment. Such overexpression of transcription factors, chromatin regulatory proteins, or their dominant negative mutants can lead to artifactual changes in gene expression and subsequent alterations in cell morphology, physiology, or behavior. It is reassuring that many of the phenomena described here that utilize overexpression systems have been validated with other methods: e.g. those genes regulated by overexpression of ΔFosB in the NAc of inducible bitransgenic mice[113] overlap extensively with genes that show enrichment of endogenous ΔFosB binding after cocaine.[51] Similar caveats exist for the use of constitutive knockout animals, where loss of a gene in early development and in all tissues makes it difficult to interpret any changes observed in drug regulation involving a single brain region of an adult animal. Ultimately, a truly accurate understanding of the transcriptional and epigenetic regulation of the addiction process will require the generation of novel tools that control protein expression with greater spatial, temporal, *and* accumulation precision.

Methodological advances in epigenetics are needed as well. Current levels of experimental proof of epigenetic mechanisms of drug action have to date involved the overexpression or deletion of a given epigenetic protein (HAT, HDAC, HMT, DNMT, etc.) within a brain region of interest. However, such manipulations affect the epigenetic states of perhaps thousands of genes without targeting those genes that are specifically altered by drug exposure. Being able to manipulate experimentally the epigenetic state of an individual gene within a discrete brain region of an adult animal would represent a major advance for the field. Tools such as artificially designed zinc-finger proteins[114] or sequence-specific transcription activator-like effectors (TALEs),[115] designed to bind specific DNA sequences in vivo, would offer exciting possibilities for future studies. Indeed, we have had recent preliminary success with this approach.[116] Similarly, all genome-wide studies of drug-induced epigenetic changes in brain have thus far utilized total extracts of brain regions, even though we know that drugs produce very different effects on distinct neuronal and non-neuronal cell types within a given region. Genome-wide epigenetic analyses in a cell type-specific manner are a critical need in addiction research. Though ChIP protocols for <1000 cells (vs millions of cells for a regular ChIP) have become available,[117] they may only be suitable for high abundance proteins. A recently developed FACS-like neuronal nuclear sorting assay permits isolation of neuronal and non-neuronal populations, but the required amount of tissue is currently prohibitive for a study on small structures such as mouse NAc.[118] In addition, all sequencing-based DNA methylation profiling methods have their own limitations.[119] Whole genome bisulfite sequencing, in addition to its high cost, cannot distinguish 5-hmC from 5-mC, and there is no generally available method to characterize 5-fC, 5-caC.

Advances in bioinformatics are also needed. Next generation sequencing techniques provide unparalleled access to chromatin changes at non-coding regions, which account for >90% of the genome, however, reference databases and other tools are urgently needed to analyze these regions. Genome-wide studies of transcription factor binding and chromatin modifications generate enormous datasets, which require the development of more optimal tools to mine effectively the resulting data. For example, it will be crucial moving forward to overlay such epigenetic analyses with genome-wide changes in RNA expression and to compare data obtained in animal models with those from human post-mortem brain tissue. In a similar vein, the studies reviewed here on drug regulation of gene expression must be integrated over several other levels of analysis. How do individual differences in genome sequences relate to individual differences in epigenetic regulation? Do drug-induced epigenetic modifications occur in peripheral tissues such as blood and do any such changes reflect addiction-relevant phenomena? Recent studies, for example, have found altered levels of methylation of the monoamine oxidase-A (MAOA) and MAOB gene promoters in blood of smokers.[120,121] Additionally, altered methylation of the MAOA gene is associated with nicotine and alcohol dependence in women but not men,[114,122] emphasizing the need for studies of sex differences in epigenetic regulation in addiction models, which heretofore have focused almost exclusively on male animals.

As information on transcriptional and epigenetic mechanisms of addiction accumulates, it is essential to integrate it with equally important information regarding post-transcriptional (translational and post-translational) regulation to obtain a complete understanding of how chronic exposure to a drug of abuse changes the brain to cause addiction. The ultimate goal of this research is to understand basic principles of neuronal and behavioral adaptation and, ultimately, to identify new targets for the treatment of addictive disorders and new methods for their prevention.

Acknowledgments

Portions of this chapter are based on a recent review by Robison and Nestler (2011)[42] with permission.

References

1. Manchikanti L, Fellows B, Ailinani H, Pampati V. Therapeutic use, abuse, and nonmedical use of opioids: a ten-year perspective. *Pain Physician*. 2010;13(5):401–435.
2. Manchikanti L. National drug control policy and prescription drug abuse: facts and fallacies. *Pain Physician*. 2007;10(3):399–424.
3. Kendler KS, Myers J, Prescott CA. Specificity of genetic and environmental risk factors for symptoms of cannabis, cocaine, alcohol, caffeine, and nicotine dependence. *Arch Gen Psychiatry*. 2007;64(11):1313–1320.
4. Volkow N, Rutter J, Pollock JD, Shurtleff D, Baler R. One SNP linked to two diseases-addiction and cancer: a double whammy? Nicotine addiction and lung cancer susceptibility. *Mol Psychiatry*. 2008;13(11):990–992.
5. Goldman D, Oroszi G, Ducci F. The genetics of addictions: uncovering the genes. *Nat Rev Genet*. 2005;6(7):521–532.
6. Koob GF, Le Moal M. *Neurobiology of Addiction*. Maryland Heights, MO: Academic Press; 2005.

7. Kalivas PW, Volkow ND. The neural basis of addiction: a pathology of motivation and choice. *Am J Psychiatry*. 2005;162(8):1403–1413.
8. Hyman SE, Malenka RC, Nestler EJ. Neural mechanisms of addiction: the role of reward-related learning and memory. *Annu Rev Neurosci*. 2006;29:565–598.
9. Robinson TE, Berridge KC. The neural basis of drug craving: an incentive-sensitization theory of addiction. *Brain Res Brain Res Rev*. 1993;18(3):247–291.
10. Kalivas PW. The glutamate homeostasis hypothesis of addiction. *Nat Rev Neurosci*. 2009;10(8):561–572.
11. Lobo MK, Covington III HE, Chaudhury D, et al. Cell type-specific loss of BDNF signaling mimics optogenetic control of cocaine reward. *Science*. 2010;330(6002):385–390.
12. Witten IB, Lin SC, Brodsky M, et al. Cholinergic interneurons control local circuit activity and cocaine conditioning. *Science*. 2010;330(6011):1677–1681.
13. Self DW, In: Neve KA, editor. *The Dopamine Receptors*, vol. 2010. New York: Humana Press; 2010. p. 479–524.
14. Bertran-Gonzalez J, Bosch C, Maroteaux M, et al. Opposing patterns of signaling activation in dopamine D1 and D2 receptor-expressing striatal neurons in response to cocaine and haloperidol. *J Neurosci*. 2008;28(22):5671–5685.
15. Singla S, Kreitzer AC, Malenka RC. Mechanisms for synapse specificity during striatal long-term depression. *J Neurosci*. 2007;27(19):5260–5264.
16. Nestler EJ. Molecular basis of long-term plasticity underlying addiction. *Nat Rev Neurosci*. 2001;2(2):119–128.
17. Andersen SL, Teicher MH. Desperately driven and no brakes: developmental stress exposure and subsequent risk for substance abuse. *Neurosci Biobehav Rev*. 2009;33(4):516–524.
18. Malanga CJ, Kosofsky BE. Does drug abuse beget drug abuse? Behavioral analysis of addiction liability in animal models of prenatal drug exposure. *Brain Res Dev Brain Res*. 2003;147(1-2):47–57.
19. Tomasiewicz HC, Jacobs MM, Wilkinson MB, Wilson SP, Nestler EJ, Hurd YL. Proenkephalin mediates the enduring effects of adolescent cannabis exposure underlying adult addiction vulnerability. *Biol Psychiatr*. 2012 [In Press].
20. Shaywitz AJ, Greenberg ME. CREB: a stimulus-induced transcription factor activated by a diverse array of extracellular signals. *Annu Rev Biochem*. 1999;68:821–861.
21. Impey S, Fong AL, Wang Y, et al. Phosphorylation of CBP mediates transcriptional activation by neural activity and CaM kinase IV. *Neuron*. 2002;34(2):235–244.
22. Loweth JA, Baker LK, Guptaa T, Guillory AM, Vezina P. Inhibition of CaMKII in the nucleus accumbens shell decreases enhanced amphetamine intake in sensitized rats. *Neurosci Lett*. 2008;444(2):157–160.
23. Robison AJ, Vialou V, Collins M, et al. Chronic cocaine engages a feedback loop involving ΔFosB and CaMKII in the nucleus accumbens. *Society for Neuroscience Annual Meeting*. 2011;909.23.
24. Linseman DA, Bartley CM, Le SS, et al. Inactivation of the myocyte enhancer factor-2 repressor histone deacetylase-5 by endogenous Ca(2+) //calmodulin-dependent kinase II promotes depolarization-mediated cerebellar granule neuron survival. *J Biol Chem*. 2003;278(42):41472–41481.
25. Goldin M, Segal M. Protein kinase C and ERK involvement in dendritic spine plasticity in cultured rodent hippocampal neurons. *Eur J Neurosci*. 2003;17(12):2529–2539.
26. Zhai H, Li Y, Wang X, Lu L. Drug-induced alterations in the extracellular signal-regulated kinase (ERK) signalling pathway: implications for reinforcement and reinstatement. *Cell Mol Neurobiol*. 2008;28(2):157–172.
27. Deak M, Clifton AD, Lucocq LM, Alessi DR. Mitogen- and stress-activated protein kinase-1 (MSK1) is directly activated by MAPK and SAPK2/p38, and may mediate activation of CREB. *EMBO J*. 1998;17(15):4426–4441.
28. Thomson S, Clayton AL, Hazzalin CA, Rose S, Barratt MJ, Mahadevan LC. The nucleosomal response associated with immediate-early gene induction is mediated via alternative MAP kinase cascades: MSK1 as a potential histone H3/HMG-14 kinase. *EMBO J*. 1999;18(17):4779–4793.
29. Soloaga A, Thomson S, Wiggin GR, et al. MSK2 and MSK1 mediate the mitogen- and stress-induced phosphorylation of histone H3 and HMG-14. *EMBO J*. 2003;22(11):2788–2797.
30. Brami-Cherrier K, Valjent E, Herve D, et al. Parsing molecular and behavioral effects of cocaine in mitogen- and stress-activated protein kinase-1-deficient mice. *J Neurosci*. 2005;25(49):11444–11454.
31. Schmitt JM, Guire ES, Saneyoshi T, Soderling TR. Calmodulin-dependent kinase kinase/calmodulin kinase I activity gates extracellular-regulated kinase-dependent long-term potentiation. *J Neurosci*. 2005;25(5):1281–1290.
32. DeManno DA, Cottom JE, Kline MP, Peters CA, Maizels ET, Hunzicker-Dunn M. Follicle-stimulating hormone promotes histone H3 phosphorylation on serine-10. *Mol Endocrinol*. 1999;13(1):91–105.

33. Bertran-Gonzalez J, Hakansson K, Borgkvist A, et al. Histone H3 phosphorylation is under the opposite tonic control of dopamine D2 and adenosine A2A receptors in striatopallidal neurons. *Neuropsychopharmacology*. 2009;34(7):1710–1720.
34. Nairn AC, Svenningsson P, Nishi A, Fisone G, Girault JA, Greengard P. The role of DARPP-32 in the actions of drugs of abuse. *Neuropharmacology*. 2004;47(suppl 1):14–23.
35. Morozov A, Muzzio IA, Bourtchouladze R, et al. Rap1 couples cAMP signaling to a distinct pool of p42/44MAPK regulating excitability, synaptic plasticity, learning, and memory. *Neuron*. 2003;39(2): 309–325.
36. Taniguchi M, Carreira MB, Smith LN, Zirlin BC, Neve RL, Cowan CW. Histone deacetylase 5 limits cocaine reward through cAMP-induced nuclear import. *Neuron*. 2012;73(1):108–120.
37. Russo SJ, Mazei-Robison MS, Ables JL, Nestler EJ. Neurotrophic factors and structural plasticity in addiction. *Neuropharmacology*. 2009;56(suppl 1):73–82.
38. Bolanos CA, Nestler EJ. Neurotrophic mechanisms in drug addiction. *Neuromolecular Med*. 2004;5(1):69–83.
39. Pierce RC, Bari AA. The role of neurotrophic factors in psychostimulant-induced behavioral and neuronal plasticity. *Rev Neurosci*. 2001;12(2):95–110.
40. Numakawa T, Suzuki S, Kumamaru E, Adachi N, Richards M, Kunugi H. BDNF function and intracellular signaling in neurons. *Histol Histopathol*. 2010;25(2):237–258.
41. McQuown SC, Wood MA. Epigenetic regulation in substance use disorders. *Curr Psychiatry Rep*. 2010;12(2):145–153.
42. Robison AJ, Nestler EJ. Transcriptional and epigenetic mechanisms of addiction. *Nat Rev Neurosci*. 2011;12(11):623–637.
43. Jenuwein T, Allis CD. Translating the histone code. *Science*. 2001;293(5532):1074–1080.
44. LaPlant Q, Nestler EJ. CRACKing the histone code: cocaine's effects on chromatin structure and function. *Horm Behav*. 2011;59(3):321–330.
45. Borrelli E, Nestler EJ, Allis CD, Sassone-Corsi P. Decoding the epigenetic language of neuronal plasticity. *Neuron*. 2008;60(6):961–974.
46. Maze I, Nestler EJ. The epigenetic landscape of addiction. *Ann N Y Acad Sci*. 2011;1216:99–113.
47. Su IH, Tarakhovsky A. Lysine methylation and 'signaling memory. *Curr Opin Immunol*. 2006;18(2):152–157.
48. Scobie KS, Damez-Werno D, Sun HS, et al. Role of Poly-ADP Ribosylation in Addiction. *Society for Neuroscience Annual Meeting*. New Orleans, LA 2012.
49. Kumar A, Choi KH, Renthal W, et al. Chromatin remodeling is a key mechanism underlying cocaine-induced plasticity in striatum. *Neuron*. 2005;48(2):303–314.
50. Renthal W, Carle TL, Maze I, et al. Delta FosB mediates epigenetic desensitization of the c-fos gene after chronic amphetamine exposure. *J Neurosci*. 2008;28(29):7344–7349.
51. Renthal W, Kumar A, Xiao G, et al. Genome-wide analysis of chromatin regulation by cocaine reveals a role for sirtuins. *Neuron*. 2009;62(3):335–348.
52. Schroeder FA, Penta KL, Matevossian A, et al. Drug-induced activation of dopamine D(1) receptor signaling and inhibition of class I/II histone deacetylase induce chromatin remodeling in reward circuitry and modulate cocaine-related behaviors. *Neuropsychopharmacology*. 2008;33(12):2981–2992.
53. Pandey SC, Ugale R, Zhang H, Tang L, Prakash A. Brain chromatin remodeling: a novel mechanism of alcoholism. *J Neurosci*. 2008;28(14):3729–3737.
54. Wang Y, Krishnan HR, Ghezzi A, Yin JC, Atkinson NS. Drug-induced epigenetic changes produce drug tolerance. *PLoS Biol*. 2007;5(10):e265.
55. Khare M, Taylor AH, Konje JC, Bell SC. Delta9-tetrahydrocannabinol inhibits cytotrophoblast cell proliferation and modulates gene transcription. *Mol Hum Reprod*. 2006;12(5):321–333.
56. Parmentier-Batteur S, Jin K, Xie L, Mao XO, Greenberg DA. DNA microarray analysis of cannabinoid signaling in mouse brain in vivo. *Mol Pharmacol*. 2002;62(4):828–835.
57. Renthal W, Maze I, Krishnan V, et al. Histone deacetylase 5 epigenetically controls behavioral adaptations to chronic emotional stimuli. *Neuron*. 2007;56(3):517–529.
58. Romieu P, Host L, Gobaille S, Sandner G, Aunis D, Zwiller J. Histone deacetylase inhibitors decrease cocaine but not sucrose self-administration in rats. *J Neurosci*. 2008;28(38):9342–9348.
59. Kim WY, Kim S, Kim JH. Chronic microinjection of valproic acid into the nucleus accumbens attenuates amphetamine-induced locomotor activity. *Neurosci Lett*. 2008;432(1):54–57.

60. Levine AA, Guan Z, Barco A, Xu S, Kandel ER, Schwartz JH. CREB-binding protein controls response to cocaine by acetylating histones at the fosB promoter in the mouse striatum. *Proc Natl Acad Sci USA.* 2005;102(52):19186–19191.
61. Libert S, Pointer K, Bell EL, et al. SIRT1 activates MAO-A in the brain to mediate anxiety and exploratory drive. *Cell.* 2011;147(7):1459–1472.
62. Bouras T, Fu M, Sauve AA, et al. SIRT1 deacetylation and repression of p300 involves lysine residues 1020/1024 within the cell cycle regulatory domain 1. *J Biol Chem.* 2005;280(11):10264–10276.
63. Finkel T, Deng CX, Mostoslavsky R. Recent progress in the biology and physiology of sirtuins. *Nature.* 2009;460(7255):587–591.
64. Maze I, Covington III HE, Dietz DM, et al. Essential role of the histone methyltransferase G9a in cocaine-induced plasticity. *Science.* 2010;327(5962):213–216.
65. Nestler EJ. Review. Transcriptional mechanisms of addiction: role of DeltaFosB. *Philos Trans R Soc Lond B Biol Sci.* 2008;363(1507):3245–3255.
66. Maze I, Feng J, Wilkinson MB, Sun H, Shen L, Nestler EJ. Cocaine dynamically regulates heterochromatin and repetitive element unsilencing in nucleus accumbens. *Proc Natl Acad Sci USA.* 2011;108(7):3035–3040.
67. Sun H, Damez-Werno D, Kennedy PJ, et al. Cocaine and stress regulates ATPase-containing chromatin remodelers. *Society for Neuroscience Annual Meeting.* 2011;909.14.
68. Damez-Werno D, Scobie KN, Sun H, Dietz DM, Kennedy PJ, Nestler EJ. Histone arginine methylation in the nucleus accumbens in response to chronic cocaine and social stress. *Society for Neuroscience Annual Meeting.* 2011;909.16.
69. Kennedy PJ, Sun H, Damez-Werno D, et al. Differential histone H2A variant expression in the nucleus accumbens following repeated exposure to cocaine or morphine. *Society for Neuroscience Annual Meeting.* 2011;909.15.
70. Scobie K, Damez-Werno D, Sun H, Kennedy PJ, Nestler EJ. Role of poly(ADP-ribosyl)ation in drug-seeking behavior and resiliency to stress. *Society for Neuroscience Annual Meeting.* 2011;909.18.
71. Bird AP, Wolffe AP. Methylation-induced repression – belts, braces, and chromatin. *Cell.* 1999;99(5):451–454.
72. Jaenisch R, Bird A. Epigenetic regulation of gene expression: how the genome integrates intrinsic and environmental signals. *Nat Genet.* 2003;33(suppl):245–254.
73. Lister R, Pelizzola M, Dowen RH, et al. Human DNA methylomes at base resolution show widespread epigenomic differences. *Nature.* 2009;462(7271):315–322.
74. Goll MG, Bestor TH. Eukaryotic cytosine methyltransferases. *Annu Rev Biochem.* 2005;74:481–514.
75. Lorincz MC, Dickerson DR, Schmitt M, Groudine M. Intragenic DNA methylation alters chromatin structure and elongation efficiency in mammalian cells. *Nat Struct Mol Biol.* 2004;11(11):1068–1075.
76. Wu H, D'Alessio AC, Ito S, et al. Genome-wide analysis of 5-hydroxymethylcytosine distribution reveals its dual function in transcriptional regulation in mouse embryonic stem cells. *Genes Dev.* 2011;25(7):679–684.
77. Szulwach KE, Li X, Li Y, et al. 5-hmC-mediated epigenetic dynamics during postnatal neurodevelopment and aging. *Nat Neurosci.* 2011;14(12):1607–1616.
78. Feng J, Fan G. The role of DNA methylation in the central nervous system and neuropsychiatric disorders. *Int Rev Neurobiol.* 2009;89:67–84.
79. Miller CA, Sweatt JD. Covalent modification of DNA regulates memory formation. *Neuron.* 2007;53(6): 857–869.
80. LaPlant Q, Vialou V, Covington III HE, et al. Dnmt3a regulates emotional behavior and spine plasticity in the nucleus accumbens. *Nat Neurosci.* 2010;13(9):1137–1143.
81. Deng JV, Rodriguiz RM, Hutchinson AN, Kim IH, Wetsel WC, West AE. MeCP2 in the nucleus accumbens contributes to neural and behavioral responses to psychostimulants. *Nat Neurosci.* 2010;13(9):1128–1136.
82. Chen WG, Chang Q, Lin Y, et al. Derepression of BDNF transcription involves calcium-dependent phosphorylation of MeCP2. *Science.* 2003;302(5646):885–889.
83. Martinowich K, Hattori D, Wu H, et al. DNA methylation-related chromatin remodeling in activity-dependent BDNF gene regulation. *Science.* 2003;302(5646):890–893.
84. Chahrour M, Zoghbi HY. The story of Rett syndrome: from clinic to neurobiology. *Neuron.* 2007;56(3):422–437.
85. Im HI, Hollander JA, Bali P, Kenny PJ. MeCP2 controls BDNF expression and cocaine intake through homeostatic interactions with microRNA-212. *Nat Neurosci.* 2010;13(9):1120–1127.
86. Wu H, Coskun V, Tao J, et al. Dnmt3a-dependent nonpromoter DNA methylation facilitates transcription of neurogenic genes. *Science.* 2010;329(5990):444–448.

87. Skene PJ, Illingworth RS, Webb S, et al. Neuronal MeCP2 is expressed at near histone-octamer levels and globally alters the chromatin state. *Mol Cell*. 2010;37(4):457–468.
88. Feng J, Nestler EJ. MeCP2 and drug addiction. *Nat Neurosci*. 2010;13(9):1039–1041.
89. Anier K, Malinovskaja K, Aonurm-Helm A, Zharkovsky A, Kalda A. DNA methylation regulates cocaine-induced behavioral sensitization in mice. *Neuropsychopharmacology*. 2010;35(12):2450–2461.
90. Nielsen DA, Yuferov V, Hamon S, et al. Increased OPRM1 DNA methylation in lymphocytes of methadone-maintained former heroin addicts. *Neuropsychopharmacology*. 2009;34(4):867–873.
91. Weber M, Hellmann I, Stadler MB, et al. Distribution, silencing potential and evolutionary impact of promoter DNA methylation in the human genome. *Nat Genet*. 2007;39(4):457–466.
92. Weaver IC, Cervoni N, Champagne FA, et al. Epigenetic programming by maternal behavior. *Nat Neurosci*. 2004;7(8):847–854.
93. Hansen KD, Timp W, Bravo HC, et al. Increased methylation variation in epigenetic domains across cancer types. *Nat Genet*. 2011;43(8):768–775.
94. He YF, Li BZ, Li Z, et al. Tet-mediated formation of 5-carboxylcytosine and its excision by TDG in mammalian DNA. *Science*. 2011;333(6047):1303–1307.
95. Ito S, Shen L, Dai Q, et al. Tet proteins can convert 5-methylcytosine to 5-formylcytosine and 5-carboxylcytosine. *Science*. 2011;333(6047):1300–1303.
96. Damez-Werno D, LaPlant Q, Dietz DM, et al. Drug experience epigenetically primes fosB gene inducibility in rat nucleus accumbens and caudate putamen. *J Neurosci*. 2012;32(30):10267–10272.
97. Taft RJ, Pang KC, Mercer TR, Dinger M, Mattick JS. Non-coding RNAs: regulators of disease. *J Pathol* 2010;220(2):126–139.
98. Bartel DP. MicroRNAs: target recognition and regulatory functions. *Cell*. 2009;136(2):215–233.
99. Li MD, van der Vaart AD. MicroRNAs in addiction: adaptation's middlemen? *Mol Psychiatry*. 2011;16:1159–1168.
100. Hollander JA, Im HI, Amelio AL, et al. Striatal microRNA controls cocaine intake through CREB signalling. *Nature*. 2010;466(7303):197–202.
101. Chandrasekar V, Dreyer JL. microRNAs miR-124, let-7d and miR-181a regulate cocaine-induced plasticity. *Mol Cell Neurosci*. 2009;42(4):350–362.
102. Chandrasekar V, Dreyer JL. Regulation of MiR-124, Let-7d, and MiR-181a in the accumbens affects the expression, extinction, and reinstatement of cocaine-induced conditioned place preference. *Neuropsychopharmacology*. 2011;36(6):1149–1164.
103. Sanchez-Simon FM, Zhang XX, Loh HH, Law PY, Rodriguez RE. Morphine regulates dopaminergic neuron differentiation via miR-133b. *Mol Pharmacol*. 2010;78(5):935–942.
104. Schaefer A, Im HI, Veno MT, et al. Argonaute 2 in dopamine 2 receptor-expressing neurons regulates cocaine addiction. *J Exp Med*. 2010;207(9):1843–1851.
105. Zheng H, Zeng Y, Zhang X, Chu J, Loh HH, Law PY. mu-Opioid receptor agonists differentially regulate the expression of miR-190 and NeuroD. *Mol Pharmacol*. 2010;77(1):102–109.
106. He Y, Yang C, Kirkmire CM, Wang ZJ. Regulation of opioid tolerance by let-7 family microRNA targeting the mu opioid receptor. *J Neurosci*. 2010;30(30):10251–10258.
107. Pietrzykowski AZ, Friesen RM, Martin GE, et al. Posttranscriptional regulation of BK channel splice variant stability by miR-9 underlies neuroadaptation to alcohol. *Neuron*. 2008;59(2):274–287.
108. Pietrzykowski AZ. The role of microRNAs in drug addiction: a big lesson from tiny molecules. *Int Rev Neurobiol*. 2010;91:1–24.
109. Eipper-Mains JE, Kiraly DD, Palakodeti D, Mains RE, Eipper BA, Graveley BR. microRNA-Seq reveals cocaine-regulated expression of striatal microRNAs. *RNA*. 2011;17:1529–1543.
110. Pelloux Y, Everitt BJ, Dickinson A. Compulsive drug seeking by rats under punishment: effects of drug taking history. *Psychopharmacology (Berl)*. 2007;194(1):127–137.
111. Pickens CL, Airavaara M, Theberge F, Fanous S, Hope BT, Shaham Y. Neurobiology of the incubation of drug craving. *Trends Neurosci*. 2011;34(8):411–420.
112. O'Connor EC, Chapman K, Butler P, Mead AN. The predictive validity of the rat self-administration model for abuse liability. *Neurosci Biobehav Rev*. 2011;35(3):912–938.
113. McClung CA, Nestler EJ. Regulation of gene expression and cocaine reward by CREB and DeltaFosB. *Nat Neurosci*. 2003;6(11):1208–1215.

114. Laganiere J, Kells AP, Lai JT, et al. An engineered zinc finger protein activator of the endogenous glial cell line-derived neurotrophic factor gene provides functional neuroprotection in a rat model of Parkinson's disease. *J Neurosci*. 2010;30(49):16469–16474.
115. Zhang F, Cong L, Lodato S, Kosuri S, Church GM, Arlotta P. Efficient construction of sequence-specific TAL effectors for modulating mammalian transcription. *Nat Biotechnol*. 2011;29(2):149–153.
116. Heller E, Sun HS, Cates H, et al. Bidirectional regulation of the fosB gene using synthetic zinc-finger transcription factors for the study of addiction and depression. *Society for Neuroscience Annual Meeting*. New Orleans, LA2012.
117. Dahl JA, Collas P. A rapid micro chromatin immunoprecipitation assay (microChIP). *Nat Protoc*. 2008;3(6):1032–1045.
118. Cheung I, Shulha HP, Jiang Y, et al. Developmental regulation and individual differences of neuronal H3K4me3 epigenomes in the prefrontal cortex. *Proc Natl Acad Sci USA*. 2010;107(19):8824–8829.
119. Harris RA, Wang T, Coarfa C, et al. Comparison of sequencing-based methods to profile DNA methylation and identification of monoallelic epigenetic modifications. *Nat Biotechnol*. 2010;28(10):1097–1105.
120. Philibert RA, Beach SR, Gunter TD, Brody GH, Madan A, Gerrard M. The effect of smoking on MAOA promoter methylation in DNA prepared from lymphoblasts and whole blood. *Am J Med Genet B Neuropsychiatr Genet*. 2010;153B(2):619–628.
121. Launay JM, Del Pino M, Chironi G, et al. Smoking induces long-lasting effects through a monoamine-oxidase epigenetic regulation. *PLoS One*. 2009;4(11):e7959.
122. Philibert RA, Gunter TD, Beach SR, Brody GH, Madan A. MAOA methylation is associated with nicotine and alcohol dependence in women. *Am J Med Genet B Neuropsychiatr Genet*. 2008;147B(5):565–570.

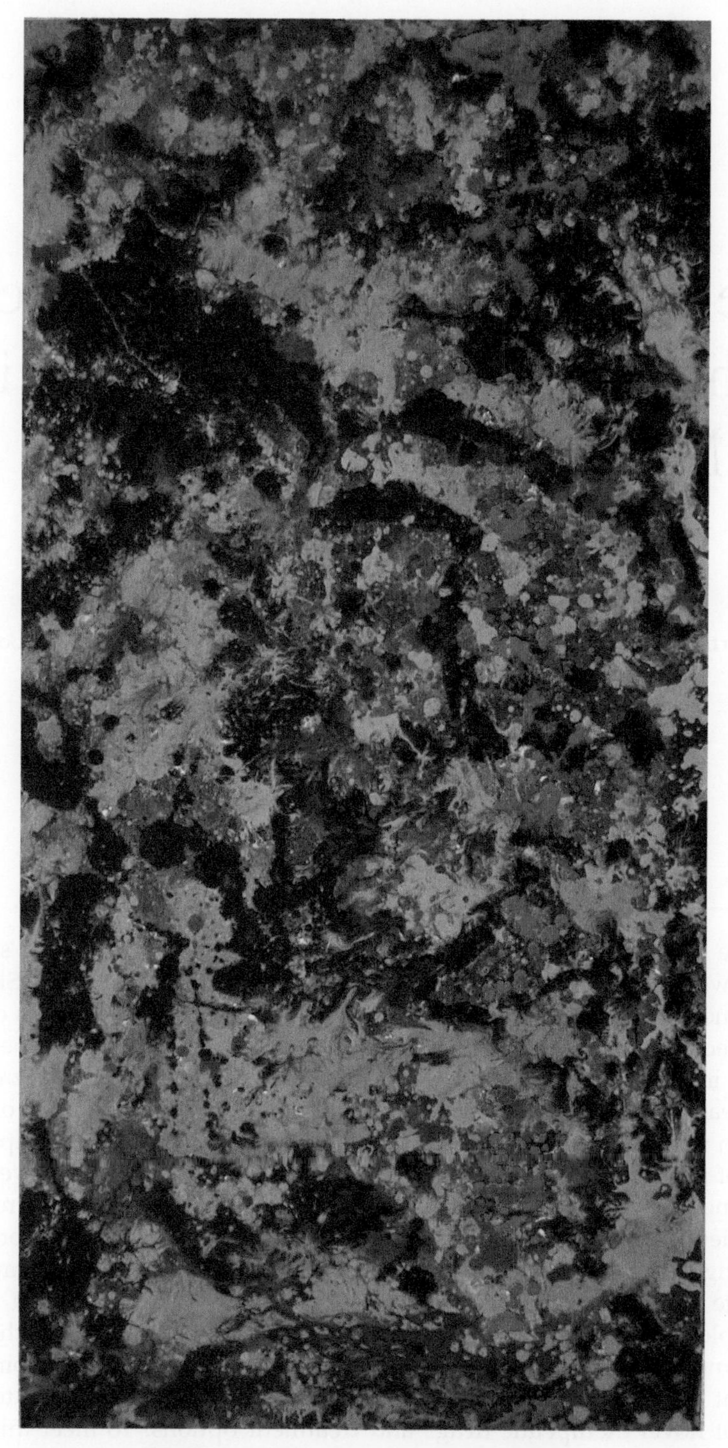

"Epigenetic Mechanisms in Psychiatric Disorders"

J. David Sweatt, acrylic on wood panel (24 x 48), 2012

CHAPTER

7

The Mind and its Nucleosomes – Chromatin (dys)Regulation in Major Psychiatric Disease

Rahul Bharadwaj, Cyril J. Peter and Schahram Akbarian

Department of Psychiatry, Friedman Brain Institute, Mount Sinai School of Medicine, New York, New York, USA

INTRODUCTORY REMARKS

Many of the more common psychiatric syndromes, including autism, schizophrenia, depression, as well as anxiety and post-traumatic stress disorder-related (PTSD) conditions, are becoming increasingly well defined by structural or functional measures obtained from the living subject. To mention just two prominent examples from this type of literature, the alterations in higher frequency (20–200 Hz) neural network synchronization affecting widespread areas of cerebral cortex play a central role in the neurobiology of psychosis.[1,2] Furthermore, in vivo neuroimaging studies revealed a decrease in hippocampal volumes in a wide range of psychiatric conditions including anxiety and PTSD.[3] However, the absence of a unifying neuropathology or a straightforward genetic risk architecture in the large majority of patients diagnosed with the aforementioned conditions poses a formidable challenge to investigators interested in the underlying molecular and cellular mechanisms of disease. Despite these challenges, progress in this type of research – which still is mostly confined to exploration of the diseased tissue itself (human brain from psychiatric subjects obtained post-mortem) or preclinical animal models which often mimic some but not all aspects relevant to the human condition[4] – is critical because it is expected to drive future development of novel therapeutic drugs and treatment options. To meet this challenge is

DOI: http://dx.doi.org/10.1016/B978-0-12-391494-1.00007-0

particularly important given that conventional psychopharmacology, including drugs targeting monoamine signaling, e.g. dopaminergic, serotonergic and noradrenergic pathways, elicit an insufficient therapeutic response in a significant portion, or roughly one half of cases on the psychosis,[5] or mood and anxiety spectrum.[6]

BOX 7.1

THE BUILDING BLOCKS OF THE EPIGENOME

The basic unit of chromatin is the nucleosome, which is 146bp of genomic DNA wrapped around an octamer composed of the core histones, H2A, H2B, H3 and H4. Chromatin fibers are defined as arrays of these nucleosomes (see Figure 7.1), which are connected by linker DNA and linker histones.

DNA methylation. Two types of DNA modifications, methylation (m) and hydroxymethylation (hm) of cytosine carbon 5 (C5) mostly in CpG dinucleotides, occur primarily within CpG enriched islands (CGI, often defined by a GC percentage >50% across a minimum of 200bp).[7] The mC5 and hmC5 markings show a strikingly different distribution, with hmC5 mostly confined to the 5′ end of genes, at levels that overall correlate with gene expression activity.[8,9] In contrast, only a minute portion (<3%) of mC5 locates to CGIs at the 5′ end of genes where it is thought to function as a repressive mark, while the remaining 97% are found in intra- and intergenic sequences and within DNA repeats.[10]

Post-translational histone modifications (PTMs). According to recent studies, the number of amino acid residue-specific PTMs in a typical vertebrate cell could be far higher than 100.[11] Examples of histone PTM include mono- (me1), di- (me2) and tri- (me3) methylation, acetylation and crotonylation, polyADP-ribosylation and small protein (ubiquitin, SUMO) modification of specific lysine residues, as well as arginine (R) methylation and "citrullination", serine (S) phosphorylation, tyrosine (T) hydroxylation, and several others.[11–13] Various combinations of these site- and residue-specific PTMs define different types of chromatin. There are epigenetic signatures specific for proximal promoters and gene bodies at sites of actual or potential transcription, for enhancer and other regulatory sequences and for condensed and silenced chromatin.[14]

Histone variants. In addition to the canonical core histones H2A/H2B/H3/H4, metazoan genomes encode a number of histone variants that provide another layer of epigenetic regulation. Some of the well-known variants include H3.3, H2A.Z and H2A.X that, in contrast to the canonical histones, are subject to replication-independent expression and assembly,[15] and are thought to regulate nucleosome mobility and compaction during gene transcription or silencing.[16]

Linker histones, including histone H1, contribute to the three-dimensional architecture of chromatin and the "zigzag" arrangement of nucleosomes by regulating linker DNA folding and levels of linker DNA strongly correlate with nucleosome repeat length (NRL).[17] Importantly, H1 levels in neurons are much lower compared to in most other cell types,[17] which explains that the average NRL in rat neurons is 40bp shorter compared to glia (162 vs 201bp).[18]

Epigenome readers, writers and erasers. Most or virtually all epigenetic markings that have been studied to date in brain are subject to bi-directional and potentially highly dynamic regulation in the context of neuronal activity and various other paradigms.[19,20] The underlying molecular machineries are often complex; for example, three DNA methyltransferases *DNMT1, DNMT3a DNMT3b* establish and maintain DNA methylation marks, which is counterbalanced by active demethylation pathways involving mC5 hydroxylation and oxidation via ten-eleven translocation (TET) dioxygenases, or activation-induced deaminase (AID)/APOBEC-mediated deamination of mC5 or hmC5, followed by base excision repair-mediated replacement with (unmethylated) cytosine.[21,22] Other systems show a surprising degree of diversity, or perhaps redundancy, at the genetic level. For example, the collective set of histone methyltransferases (KMTs) and demethylases (KDMs) together could easily account for >100 genes in a mammalian genome.[23,24] Proteins that bind to a specific epigenetic mark are defined by their characteristic "reader module"; well-studied examples include the MBD domain for mC5-DNA, the bromodomain for lysine acetylation, and "chromo", "Tudor", "MBT", 2WD40repeat", "PHD finger" domains targeting methylated lysines or arginines in a residue-specific manner.[13] Conversely, specific methyl-lysine marks could become the target of 50–100 reader proteins. For the "open chromatin" mark histone H3-trimethyl-lysine 4 (H3K4me3), these include many components of the RNA polymerase II-associated transcriptional initation complex, while other marks such as H3K9me3 are primarily targeted by transcriptional repressors and regulators of chromatin condensation.[25] To date, there are at least 15 monogenetic forms of neurodevelopmental or adult onset neuropsychiatric disease that are caused by deleterious mutations in genes that encode either "writers", "erasers" or "readers" of DNA methylation and histone modifications, or histone variants and proteins with a critical role for proper formation of higher order chromatin structures.[26,27]

EPIGENETICS IN PSYCHIATRY – WHY BOTHER?

Epigenetic (dys)regulation of gene expression and, more generally, genome organization and function is one avenue of research that has gained significant traction in the field over the course of the last few years, and there is a rapidly, if not exponentially growing knowledge base both in the clinical and preclinical psychiatric literature on this topic.[28] Epi- (Greek for "over", "above") genetics – a term traditionally confined to phenotypes resulting from heritable changes in gene expression and function that do not involve changes in DNA sequence – is in the field of neuroscience and translational medicine broadly defined in the context of regulatory mechanisms for chromatin structure and functions, including transcription, genome stability and retrotransposon suppression during the course of normal and diseased development and neuroplasticity.[29] The renewed interest in these topics in the field of biological psychiatry and human brain research in general can be traced back to six partially independent developments (Table 7.1) that are now beginning to coalesce.

Technological Advances

First, the basic sciences made rapid advances in defining the epigenome, which is majorly comprised of two types of DNA cytosine modifications, (hydroxy-)methylation,

TABLE 7.1 Epigenetic Approaches in Psychiatry: A House on Six Pillars

1. Epigenetic markings, including the genome-wide distribution of DNA and histone methylation markings, remain stable for many hours after death and therefore are amenable to analyses in post-mortem brain tissue
2. Epigenome mapping informs about chromatin function and thereby could provide valuable clues about the disease relevance of DNA mutations and structural variants in regulatory non-coding sequences
3. Genome-wide dysregulation of brain chromatin structure and function may not be limited to a dozen of mostly very rare monogenetic disorders but may affect a subset of so-called "sporadic" cases diagnosed with major psychiatric disease (including addiction, autism, depression, schizophrenia)
4. Gene expression defects in diseased brain tissue could be associated with alterations in level and distribution of epigenetic regulators of transcription
5. Epigenetic markings may contribute to transgenerational heritability of psychiatric disease
6. A rapidly growing cache of chromatin modifying drugs may hold the key to novel therapeutic treatment options for patients not responsive to conventional psychopharmacological regimens

and more than 100 site- and residue-specific histone modifications and multiple histone variants that in toto define genome organization at sites of active and condensed chromatin[30,31] (Figure 7.1). Many of the techniques and assays used to map the molecular components of the epigenome are applicable not only to cell cultures but also brain itself, including human post-mortem specimens where the intranuclear environment is thought to be less sensitive to the effects of autolysis time and subject to lesser degradation in comparison to many structures located in the cell membrane or cytoplasm.[32]

Analysis of Chromatin Structure and Function Could Help to Define Genetic Susceptibility

Second, it is now recognized that the genetic risk architecture of psychosis and mood and anxiety disorders is highly heterogeneous, with each individual likely to be defined by his or her "unique" set of disease (risk)-contributing common single nucleotide polymorphisms and other variants, which only in a small subset of cases (<10%) is complemented by a more straightforward "genetic explanation" of the illness, such as a rare copy number variant at a risk locus associated with high penetrance/likelihood for disease when mutated.[33] Furthermore, whole exome sequencing designed to capture disease-causing mutations in coding sequences (which comprise less than 1.5% of the entire genome) is thought to uncover disease-associated mutations probably in not more than one third or one half of cases even by the most optimistic estimates.[34,35] Therefore, mapping epigenetic landscapes, particularly in regulatory and non-coding sequences, will become an important strategy to identify genomic loci with aberrant chromatin structures, which in turn could then serve as a "red flag" to indicate the potential significance of mutations or other structural variants of the corresponding DNA sequence. In principle, this rationale follows in the footsteps of much earlier studies on monogenetic mental retardation syndromes such as Fragile X, where the expansion of a promoter-associated CGG codon upstream of the *FMR* transcription start site from (normally) 5–40 repeats from 50 to over 200 was found to be associated with excessive DNA methylation at the promoter, effectively silencing the surrounding chromatin.[36] In case of the *CGG* triplet expansion at the *FMR1* (fragile X) gene promoter,[36] or the *GAA* triplet repeat expansion in the first intron of the *FRATAXIN* gene associated with Friedreich

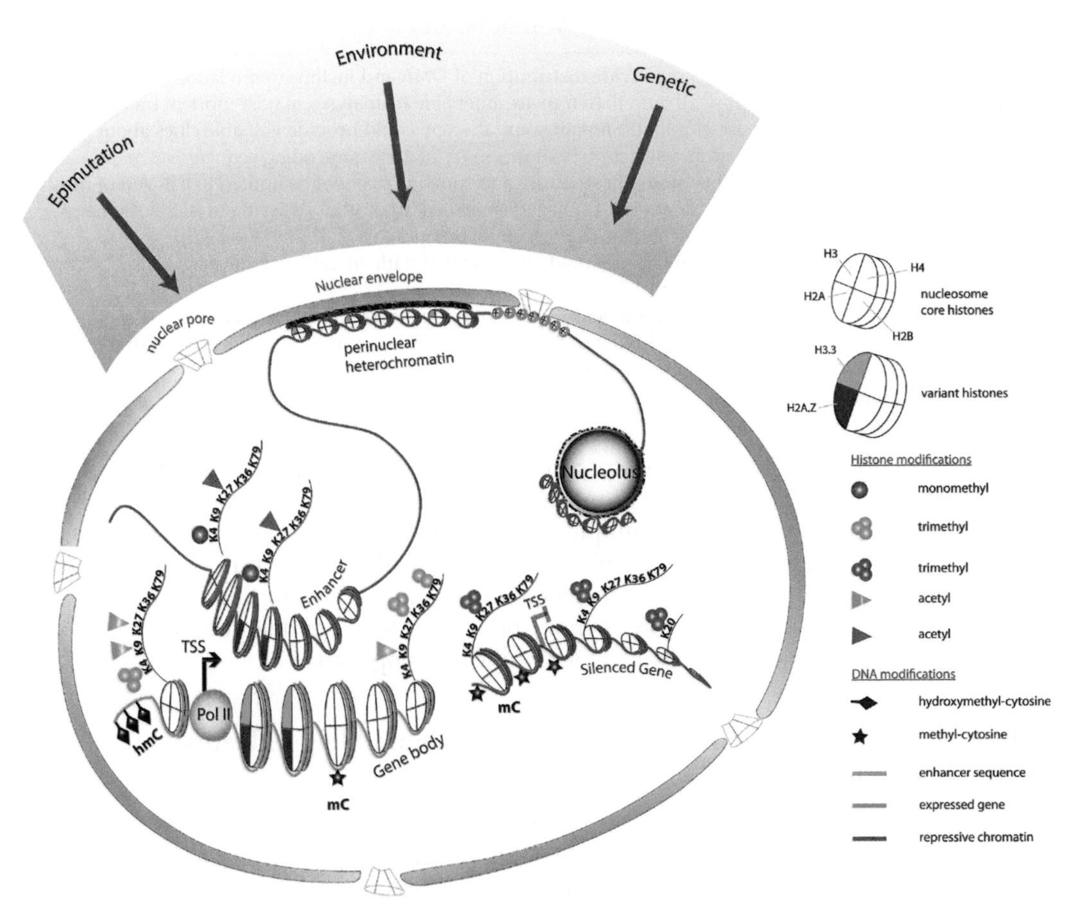

FIGURE 7.1 **Introduction to the epigenome.** The epigenome, broadly defined, involves DNA cytosine methylation and hydroxymethylation, and a large number (>100) site and residue-specific histone modifications and histone variants which together provide the major building blocks for the epigenetic landscapes that mold and organize DNA inside a cell nucleus into myriads of transcriptional units and clusters of condensed chromatin that distinguish between various cell types and developmental stages sharing the same genome,[30,31] and define chromatin structure and function including enhancer and promoter sequences, gene bodies, repressive chromatin etc. (See also Box 7.1 for additional information and definitions.) The drawing shows simplified illustration of a (green) gene poised for transcription by polymerase II (Pol II) initiation complex at transcription start site (TSS), interacting with a (blue) distal enhancer sequence which moves into close proximity to active gene promoter (red). Heterochromatic portions of the genome include facultatively silenced genes and constitutive heterochromatin such as pericentromeric or telomeric sequences at the extreme ends of the short and long arms of chromosomes. Many heterochromatic sequences are tethered to the nuclear envelope and pore complex, and also enriched at the periphery of the nucleolus which is a key structure for ribosomal biogenesis. A small but representative subset of histone variants and histone H3 site-specific lysine (K) residues at N-terminal tail (K4, K9, K27, K36, K79) and H4K20 residue are shown as indicated, together with their mono- and trimethyl, or acetyl modifications that show differential enrichment at active promoters, transcribed gene bodies, and repressive chromatin, as indicated. DNA cytosines that are hydroxymethylated at the C5 position are mostly found at active promoters, while methylated cytosines are positioned within the body of actively transcribed genes and around repressed promoters and in constitutive heterochromatin. The epigenome, including the patterning of DNA methylation and histone modifications and the three-dimensional organization of chromatin, is shaped by genetic variation and environmental influences. Another suspected source of between subject epigenomic variability are epimutations, which are defined by heritable, transgenerational alterations in chromatin structure and function that do not involve changes in DNA sequence.

ataxia,[37] dysregulation of DNA and histone methylation markings provided the critical intersect linking the genetic alterations to abnormal gene silencing and the pathophysiology of neurological disease.[37,38] Another set of insightful examples for the knowledge gained by combining genome sequencing with epigenetic studies is provided by recent studies in various cancer types which linked copy number DNA structural variants to abnormal DNA methylation signatures at chromosomal breakpoints with a critical role in the etiology of disease.[39] It will be interesting to follow the next wave of developments in the field of neurogenetics and to find out whether these general principles from epigenetic dysregulaton of monogenetic neurological disease or from cancer epigenetics will indeed emerge as useful *Leitmotiv* and guiding concept in the work up of cases diagnosed with more common psychiatric conditions including the ones that are the focus of this review.

Neuropsychiatric Disease Due to Defects in Chromatin Regulation May be More Widespread than Previously Thought

Third, contrasting the traditional view that chromatin defects in brain are static lesions of early development that occur in the context of rare genetic syndromes, it is now clear that deleterious mutations in key regulatory genes of the epigenetic machinery cover a much wider continuum, including psychosis with onset in child- or adulthood, and even adult-onset neurodegenerative disease. Examples include mutations in the gene encoding Methyl-CpG-binding protein 2 (*MECP2*, chr. Xq28) which, in addition to its classical manifestation – Rett syndrome with onset a few months after birth – may present as autism or childhood-onset schizophrenia,[40] or the histone H3-lysine 9 methyltransferase *KDM1D* (9q34.3) which may present not only as a mental retardation syndrome but also as adult/adolescent onset-schizophrenia[41] or non-specific psychiatric phenotypes with neurodegeneration in the post-adolescence period.[42] Likewise, a subset of DNA methyltransferase 1 (*DNMT1*, 19p13.2) mutations – including those affecting the targeting-sequence (TS) domain important for nuclear localization and DNMT activity at pericentric repeats – were recently linked to hereditary sensory and autonomy neuropathy type 1 (HSAN1),[43] a rare neurodegenerative condition characterized by various neuropathies and early onset dementia in the third or fourth decade. Figure 7.2 summarizes known monogenetic neuropsychiatric disorders associated with widespread dysregulation of brain chromatin structures. Such types of monogenetic CNS (chromatin) disorders associated with delayed or later-onset cognitive impairment and psychosis most certainly provide strong rationale to perform "epigenome scans" of brain or other tissues from subjects affected by the "sporadic" form of schizophrenia or depression. Any case with evidence for chromatin dysregulation across widespread portions of the genome may have been affected by a mutation in either one of the aforementioned molecular cogwheels driving the epigenetic machinery, or by some other hitherto unknown chromatin regulator important for normal brain function.

Transcriptional Dysregulation in Diseased Brain Tissue is Associated with Changes in Surrounding Chromatin Structures

Fourth, a large number of studies conducted in human (post-mortem) brain point to a distinct set of gene transcripts that frequently, albeit never consistently, show altered

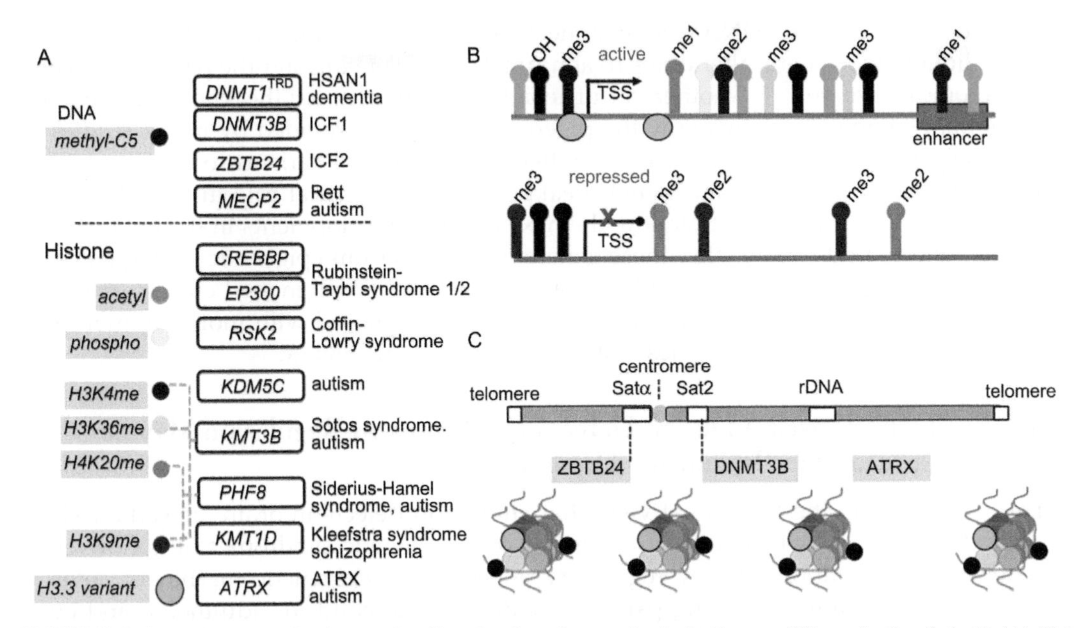

FIGURE 7.2 **"Monogenetic chromatin disorders" and neurological disease: Where is the defect?** (A) List of neurodegenerative (hereditary sensory-autonomic neuropathy type 1 (HSAN1) with early onset dementia) and various neurodevelopmental syndromes, including some cases on the autism and psychosis spectrum. Each condition is associated with a deleterious mutation in either (top) DNA methyl-writer and reader proteins, or (bottom) various histone acetyl- and methyl-transferases, demethylases, and ARX protein regulating variant histone H3.3. (B) Simplified scheme for distribution of disease-associated epigenetic markings and variant histone H3.3 in euchromatin, including active and repressed transcription start sites (TSS) and downstream gene bodies. Notice complex and differential distribution even for related modifications, such as methyl- vs hydroxy(OH)-methylcytosine and mono- vs di-/tri-methylH4K20me (see also Box 7.1). (C) Nearly all monogenetic disorders (from A) also affect heterochromatic repeat sequences. Three examples are shown: mutations in *DNMT3B* (ICF1) or *ZBTB24* (ICF2) result in hypo(DNA mC5) methylation of different types of pericentric satellite repeats.[44,45] The ATRX regulates variant histone H3.3 incorporation not only into nucleosomes surrounding the TSS of active genes (see B) but also at various repeat sequences, as indicated.

expression in post-mortem brain tissue collected from diseased individuals in comparison to control cohorts. For example, cerebral cortex in many subjects on the mood and psychosis spectrum appears to be affected by transcriptional or post-transcriptional dysregulation of key genes implicated in the development and maintenance of γ-aminobutyric acid (GABA)-ergic circuitry, or myelination and other oligodendrocyte-specific function; more generalized transcriptome changes compromising metabolic activities or pre- and post-synaptic neurotransmission have also been reported.[46–60] While the role of upstream regulators, including neuronal activity, dopamine receptor activation and many other factors becomes increasingly well defined, especially for RNA molecules such as GAD67 encoding a GABA synthesis enzyme which are among the most frequently reported alterations in subjects on the autism, mood or psychosis spectrum,[46,61–66] little is known about the molecular mechanisms that lead directly to these alterations in mRNA levels, which often occur in the absence of neurodegeneration.[67] Specifically, it is not known whether the alterations in the level of a specific mRNA are brought about by a change in transcription or mRNA turnover.

Therefore, to determine whether transcription itself is altered, additional molecular assays and profiling of epigenetic regulators of gene expression, such as promoter-associated DNA and repressive histone methylation in combination with quantification of the target mRNA and protein, would be extremely useful for this type of post-mortem (brain) study. However, transcription is an extremely complex and dynamic process involving multiple steps (transcriptional initiation, post-initiation, elongation, termination etc.) with multimeric transcription factor complexes, RNA polymerase subunits, and other molecules[68] that for mostly technical reasons are difficult to study in "real time" in the (animal) brain, let alone in human post-mortem brain. In contrast, DNA and histone modifications involved in epigenetic control of gene expression are amenable for study in both animal models and in human post-mortem brain.[69] Indeed, chromatin-based approaches greatly increased our knowledge about dysregulated gene expression in pre-clinical models for depression, addiction and psychosis,[70–75] as well as in brain tissue of patients with these or other psychiatric conditions.[76–81]

Vertical Transmission of Epigenetic Markings Could Play a Role in the Heritability of Disease

Fifth, there is the challenge of the "missing heritability" of complex disorders including schizophrenia and depression for which to date causative DNA sequence alterations remain unidentified for the majority of cases. While some scholars have pointed out that "missing heritability" could still turn out to be an illusion pending further advancements in uncovering genetic risk architectures,[82] epigenetics has the potential to deliver valid and testable alternatives to genetic heritability models. With other words, could heritable (epi)genome modifications that would not necessarily be associated with changes in DNA sequence play a role in the etiology of major psychiatric disease?[83] Presently, conclusive evidence in support of this hypothesis is still lacking but plenty of circumstantial findings open up the possibility that epimutations (a type of heritable information on gene expression and function that is not reflected by the DNA sequence) play a potentially important but hitherto not fully appreciated role in neuropsychiatric disease. For example, during spermiogenesis, the canonical histones (including their epigenetic modifications) are largely exchanged for protamines,[84] small basic proteins that form tightly packed DNA structures important for normal sperm function.[85] Because of this, the epigenetic contributions of sperm chromatin to embryo development have been considered extremely limited. However, it is now clear that nucleosomes are retained in 4% of the haploid genome in sperm, and these retained nucleosomes are enriched significantly at loci of developmental importance, including imprinted gene clusters, microRNA clusters, homeobox (HOX) gene clusters, and the promoters of stand-alone developmental transcription and signaling factors.[86] Furthermore, DNA methylation patterns and histone modifications are also preserved in the portion of the genome that maintains nucleosomal organization in sperm cells.[86,87] Therefore, in principle, chromatin templates, including their epigenetic markings, could be passed on to the next generation for perhaps as much as 4% of the human genome.[86] If this is indeed correct then the portion of the genome subject to transgenerational epigenetic inheritance would indeed be larger than the entire protein coding sequence ("exome") which amounts to "only" 1–1.5% of the genome. While epimutations in the context of neuropsychiatric disease remain to be identified, such types of heritability model are beginning to emerge for other medical conditions, including cancer. For example, there have been a number of studies reporting epimutations in the mismatch-repair genes MLH1 and MSH2 in humans, which are transmitted in specific cases of hereditary non-polyposis colorectal cancer. However, some of these findings have been controversial, highlighting the difficulties in establishing this type of inheritance in humans.[88–91]

Future Treatment of Neuropsychiatric Disorders May Involve Novel Chromatin Modifying Drugs

Sixth, and last but not least, the field eagerly anticipates the development of innovative therapies with chromatin modifying drugs. Indeed, preclinical work ascribes for histone deacetylases (HDAC) a wide range of therapeutic benefits in various brain disorders, including neurodegenerative conditions such as Parkinson's, Alzheimer's, triplet repeat (including Huntington and spinocerebellar ataxias) and motor neuron disease,[92–95] as well as depression and other psychiatric illnesses.[96–98] Notably, valproic acid, which is known as one of the most frequently prescribed mood-stabilizer and anticonvulsant drugs, broadly inhibits HDAC activity when administered at higher doses.[99] The clinical potential of other types of histone modifying enzymes, including those regulating histone lysine methylation, is also a promising but yet largely unexplored field given that there are approximately 100 histone methyl-transferases (KMTs) and demethylases (KDMs) encoded in the human genome.[100] Like for the HDAC, a subset of histone methyltransferase (KMT) inhibitors are in clinical trials for cancer treatment, and are likely to be explored in the context of neurological disease in the not too distant future.[100] An interesting candidate is the small molecule and diazepin-quinazolin-amine derivative BIX-01294, an inhibitor for the histone H3K9-specific methyltransferases *G9a/Glp*.[101] This drug de-represses neuronal gene expression,[101] and when administered directly into the ventral striatum (a key structure in the brain's addiction circuitry) strongly enhances the development of reward behaviors in stimulant drug (cocaine)-exposed mice.[102] Furthermore, DNA methylation inhibitors such as the cytidine analogs 5-azacytidine (5-Aza-CR), zebularine that sequester DNMT enzymes[103] reportedly disrupt synaptic plasticity and hippocampal learning and memory when directly infused into the brain. Interestingly, several of these drugs elicit strong effects on motivational behaviors in animal models for reward and addiction.[104–107] Another promising type of chromatin modifying drug involves topoisomerases (topos), that function as DNA cleaving enzymes important for replication and recombination, transcription and chromatin remodeling.[108] Recently, it was discovered that topoisomerase I or II enzyme inhibitors unlock expression of the epigenetically silenced paternal allele of the ubiquitin protein ligase E3A (UBE3A), by reducing expression of the imprinted UBE3A antisense RNA (*Ube3a-ATS*).[109] This anti-sense RNA is normally repressed on the maternal chromosome in conjunction with allele-specific DNA methylation of an imprinting center (defined as the DNA/chromatin structures that carry epigenetic information about parental orgin)[110] which, like hundreds of other loci, is defined by parent-of-origin-specific gene expression and was until now considered epigenetically stable throughout life.[111] Strikingly, however, a single intrathecal infusion of the FDA-approved topo inhibitor topotecan suffices to relieve silencing of the paternal *Ube3a* (sense) transcript in lumbar spinal neurons for an extended period of at least 3 months.[109] Given these promising leads, we anticipate that it will be only a matter of time until epigenetics-based drug therapies will significantly enrich the neuropsychopharmacological toolbox of present day clinicians.

EPIGENETIC ALTERATIONS IN SCHIZOPHRENIA AND AUTISM

A Brief Primer on Autism and Schizophrenia

Schizophrenia is a major psychiatric disorder affecting 1% of the general population, often with onset in young-adult years but, at least in some patients, subtle, cognitive and neurological deficits precede the full blown syndrome for many years.[112] The disease is viewed as highly heterogeneous in terms of genetics, and is primarily defined by partially independent symptom complexes such as (1) psychosis with delusions and other loss of reality testing, hallucinations and disorganized thought, (2) cognitive dysfunction including deficits in attention, memory and poor executive functioning and (3) depressed mood and negative symptoms including inability to experience pleasure (anhedonia), social withdrawal and poverty of thought and speech output.[113] Currently prescribed antipsychotics, which are mainly aimed at dopaminergic and/or serotonergic receptor systems, exert therapeutic effects on psychosis in approximately 75% of patients but often do not adequately address the cognitive impairment which is a potentially disabling feature of schizophrenia.[113] Currently, there are no established pharmacological treatments for this symptom complex. Cognitive dysfunction is an important predictor for long-term outcome and, therefore, this area is considered a high priority in schizophrenia research and gave rise to cross-institutional initiatives such as MATRICS (*Measurement and Treatment Research to Improve Cognition in Schizophrenia*).[113]

Autism is primarily a disorder of social communication and interaction with onset in early childhood often prior to age 3, and occurs with or without deficits in speech and language, repetitive behaviors and a host of other neurological symptoms. Treatment is largely symptomatic, and the social core deficits are poorly addressed by current psychoactive drugs.[114] Schizophrenia and autism show similarities in their genetic risk architecture, including a significant overlap in certain copy number and other rare structural DNA variants that carry high disease penetrance.[115–117] The etiology of both schizophrenia and autism is complex but defects in pre- and early postnatal neurodevelopment, including adverse influences such as maternal infection and malnutrition are thought to play an important role in a substantial portion of cases.[118–121] Furthermore, according to some studies, there is a striking contribution of "environmental" factors, as evidenced by a highly significant shared risk even in dizygotic twins which is only slightly less than the shared disease risk in monozygotic twin pairs.[122]

Epigenetic Studies in Post-Mortem Brain Tissue of Subjects with Schizophrenia

Given the prominent role of neurodevelopmental theories in the etiology of schizophrenia,[123,124] it seems not too surprising that the early "wave" of epigenetic studies in schizophrenia post-mortem brain was focused on candidate genes with a critical role in neurodevelopment such as the glycoprotein, reelin, important for migration and positioning of young neurons,[76,77] the transcription factor SOX10 which orchestrates transcription in myelin-producing oligodendrocytes,[80] and the 67kDa glutamic acid decarboxylase (GAD1/GAD67) GABA synthesis enzyme that is essential for proper formation of inhibitory neuronal circuitry,[78,79,125,126] and so on. These studies, taken together, indeed suggest that some subjects on the psychosis spectrum are affected by subtle DNA methylation and histone acetylation and methylation changes at the promoters of these and various other genes. Moreover, the DNA methylation and histone modification markings at *REELIN*, *GAD1*, *BDNF* (brain derived neurotrophic factor) and other genes with a key regulatory role during development are highly regulated in human cerebral cortex during the course of development and aging, including positive correlations between age and DNA methylation (REELIN, GAD1)[125,127] and a decline in GAD1-associated histone acetylation in older brains.[126] Consistent with such wide windows of epigenetic vulnerability, robust differences in the trajectories of age-associated DNA methylation and histone modifications emerge when disease versus control groups are compared, even with fairly limited cohort sizes of a few dozen brains or less.[79,126,127] Indeed, some of the observed epigenetic changes in schizophrenia post-mortem brain may have little to do with the etiology of disease but simply could reflect differences in lifestyles and diet or medication effect, because it becomes increasingly clear that exposure to alcohol,[128] nicotine,[129,130] psychostimulant[131–133] and antipsychotic drugs,[134–137] and many other external and internal factors affect brain DNA methylation and histone modifications.

Complicating matters further, many of the observed epigenetic alterations in chromatin surrounding the aforementioned candidate gene studies in cerebral cortex of subjects with schizophrenia are comparatively subtle, with only a small subset of CpG dinucleotides (or nucleosomes in case of histone modifications) at specific gene promoters showing significant changes on the group level. These are unlikely to be experimental artifacts related to the idiosyncrasies of post-mortem brain with its hosts of confounds related to the autolysis process or cause of death etc, because promoter DNA methylation changes in hippocampus of rats exposed to different rearing conditions are equally selective for specific sets of CpGs of, for example, the aforementioned *Gad1* gene.[138] Similarly, more extensive "genome-wide" surveys for differential DNA methylation at thousands of CpG islands revealed approximately 100 loci with differential methylation between subjects on the psychosis spectrum and controls, but the magnitude of change was comparatively subtle, such as 17% vs 25% for the *WD Repeat Domain 18* (*WDR18*) as one of the most significantly changed genes.[136] These changes in chronic neuropsychiatric disease then appear much more subtle when compared to the several fold (>100%) increases (or decreases) in DNA methylation at tumor suppressor and other genes epigenetically implicated in gliomas,[139–142] neuroectodermal tumors[139,143,144] and many other cancers in the CNS or peripheral tissues. Given the subtle nature of the observed epigenetic changes on the group level ("patients versus controls"), the uncertainties whether such type of alteration is "state or trait" and related to the disease process as opposed to some unrelated and perhaps even irrelevant environmental factor, the reader may start to wonder at this point whether there is indeed any scientific merit to study epigenetics in diseased (brain) tissue!

To answer such type of skepticism, we would argue that chromatin studies in human brain, when combined with genetic approaches, are likely to provide very meaningful insights into the mechanisms of disease of individual cases and may offer a way forward from current approaches in the field that are merely based on uncovering significant differences on the group level. Notably, DNA methylation analyses in blood chromatin collected across three generations from the same pedigrees provided evidence that, at some loci, more than 92% of the differences in methylcytosine load between alleles is explained by haplotype, suggesting a dominant role of genetic variation in the establishment of epigenetic markings, as opposed to environmental influences.[145] Furthermore, in the human cerebral and cerebellar cortex, methylation of several hundred CpG enriched sequences is significantly affected by genetic variations, including single nucleotide polymorphisms (SNPs) separated from the CpG site by more than one megabase[146,147] and an even larger number of SNPs "drive" between-subject gene expression differences in the prefrontal cortex.[148] Extrapolating from these general findings then, it is very likely that genotype is a critical variable in the context of epigenetic (dys)regulation and gene expression changes in schizophrenia brain. Therefore, we predict that the intersection between epigenetic dysregulation and genetic variation will turn out to be very important for a large number of psychiatric susceptibility genes. An early example is provided by the proximal promoter of the already mentioned *GAD1* gene (encoding the 67 kDa glutamic acid decarboxylase GABA synthesis enzyme). There is evidence that the decline in the GAD67 RNA and protein – which has been reported in ≈20 post-mortem studies on cerebral cortex of subjects with schizophrenia and related disease[46,149–153] – is accompanied by a deficit in *GAD1* promoter-associated open chromatin-associated histone methylation, including H3K4me3. However, these features appear not to be consistently present in all the clinical samples, but instead may occur mainly in conjunction with a specific haplotype, comprised by multiple SNPs, spanning ±3 kb from the *GAD1* transcription start site.[79] Interestingly, subjects with schizophrenia biallelic for the risk haplotype not only show decreased H3K4me3, but also an increased level of H3K27me3,[79] a mark regulated by repressive Polycomb group chromatin remodeling complexes associated with inhibition of transcription.[154] Furthermore, some of the same polymorphisms around the *GAD1* promoter were previously associated with schizophrenia and accelerated loss of frontal lobe gray matter[155,156] and emerged as genetic determinants for cognitive performance in an epistatic interaction with common variants of the catechyl-O-methyltransferase (COMT) gene, which regulates monoamine metabolism and signaling.[157] Therefore, at least three layers of epigenetic regulation emerge for the GAD1 promoter (Figure 7.3): (1) a developmentally regulated increase during the extended course of prefrontal maturation for the first two decades of postnatal life; (2) common polymorphisms around the GAD1 promoter which, in susceptible individuals diagnosed with schizophrenia, are associated with a shift in open and repressive chromatin-associated histone modifications; and (3) a subset of antipsychotic and mood-stabilizer drugs, including clozapine and valproate, upregulate transcription-associated histone methylation and acetylation at the *GAD1* promoter.[79] Thus, as illustrated by the *GAD1* example, which probably is representative for the large pool of "common variant, small contribution to disease risk" schizophrenia candidate genes, a complex pathophysiology exists with multiple, independent determinants converging on the same epigenetic phenotypes in diseased brain tissue.

Epigenetic Signatures of Autism

Epigenetic dysregulation of DNA methylation and histone modifications could play a role in autism, perhaps even a much more prominent role as it does in the cases of schizophrenia discussed above. This is because numerous neurodevelopmental syndromes with onset

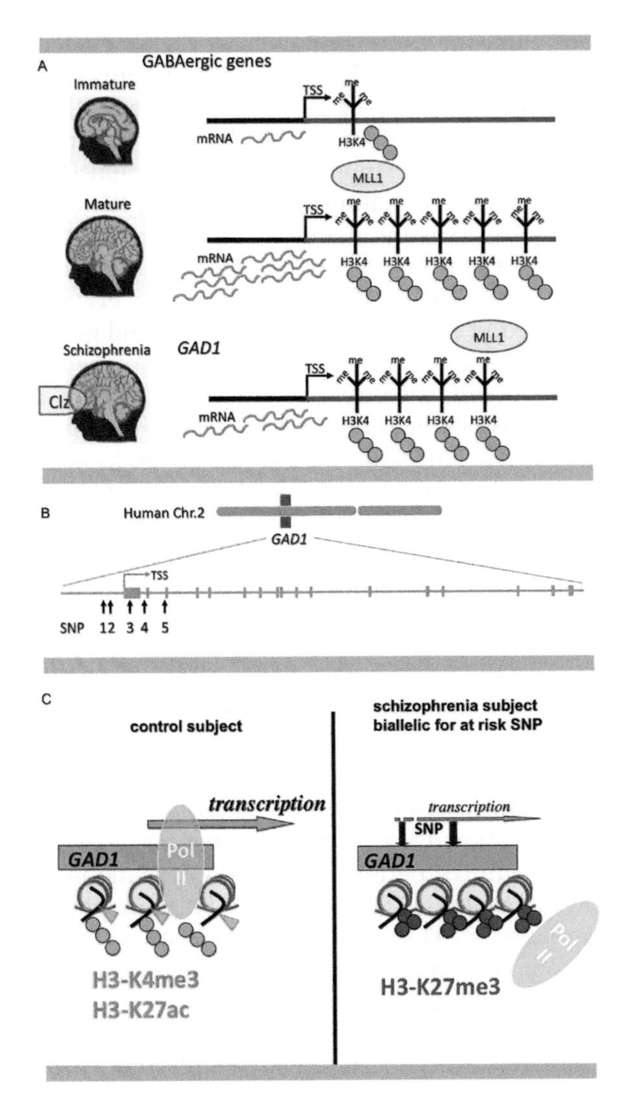

FIGURE 7.3 **Three layers of epigenetic regulation for the GAD1 promoter in human prefrontal cortex.** (A) In immature prefrontal cortex (neonates, child), *GAD1* gene expression activity, and epigenetic decoration with transcriptional histone marks, including H3K4me3, is comparatively low while, in normal adults, *GAD1* expression and promoter-associated H3K4me3 are much higher.[79] Some subjects with schizophrenia show prefrontal deficits in GAD1 RNA and GAD1-H3K4me3, and some of these effects could be mediated by mixed-lineage leukemia 1 (MLL1) methyl-transferase, or one of the other 12 or so H3K4-selective methyltransferases encoded in the genome. After treatment with the atypical antipsychotic clozapine (but not the conventional antipsychotic and dopamine D2 receptor antagonist, haloperidol), cortical levels of GAD1-H3K4me3 are increased. (B) A haplotype, comprised of at least 5 SNPs within a few kb from GAD1's transcription start site confers (i) genetic risk for childhood onset schizophrenia and accelerated loss of gray matter, (ii) is associated with decreased GAD1 gene expression in cerebral cortex of subjects on the psychosis spectrum, and (iii) may interact in epistatic interactions with other schizophrenia candidate genes. (C) The at-risk haplotype, through yet unknown mechanisms, is in diseased individuals associated with decreased *GAD1* gene expression, and a shift from open (H3K4me3) to repressive (H3K27me3) chromatin-associated histone methylation. In preclinical model systems, decreased H3K4me3 and lower GAD1 expression occurs in conjunction with a lower presence of phospho-activated RNA II polymerase at the proximal portions of the GAD1 gene.[79]

in early childhood are now linked to mutations of genes with pivotal importance for orderly regulation of chromatin structure and function. Examples including MECP2 and other methyl-CpG-binding proteins,[158–160] the histone deacetylase *HDAC4*,[161] the histone H3 lysine 9 (H3K9)-specific-methyltransferase *EHMT1/KMT1D/GLP*,[162] and the H3K4 demethylase *JARID1C/KDM5C/SMCX*[163] (see Figure 7.2). Epigenetic fine-tuning of H3K4, including the trimethylated form (H3K4me3) which, on a genome-wide level is primarily located at transcription start sites (TSS) and CpG enriched sequences, appears to be particularly important for neuronal health because of neuronal differentiation, and dependent upon the mixed-lineage leukemia (MLL) methyltransferases and transcriptional activation of RE-1 silencing transcription factor (REST) sensitive genes.[164,165] Hippocampal learning and memory requires *Mll1*-mediated H3K4 trimethylation (H3K4me3) at growth and plasticity-regulating genes.[166,167] Therefore, to explore epigenetic signatures in autism, a recent study undertook a genome-wide survey of H3K4me3 in neuronal and non-neuronal chromatin from prefrontal cortex of 16 subjects on the autism spectrum and a similar number of age-matched controls, using chromatin immunoprecipitation in conjunction with next generation sequencing methods.[81] On the group level, not a single locus was significantly altered in the autism cohort. However, altogether 700 loci were identified for which at least one case, or variable subsets of autism cases, showed a robust change in histone methylation, in comparison to each of the controls. Remarkably, among these 700 loci there was a two- to threefold overrepresentation of autism susceptibility genes, including the synaptic vesicle gene *RIMS3*, the retinoic acid signaling regulated gene *RAI1*,[168] the histone demethylase *JMJD1C*,[169] the astrotactin *ASTN2*,[170] the adhesion molecules *NRCAM*[171–173] and *SEMA5A*,[174,175] the ubiquitin ligase *PARKIN2 (PARK2)*[170,176] and other genes for which rare structural DNA variations were previously associated with high disease penetrance. Intriguingly, for most of these high-risk genes, the H3K4me3 changes occurred selectively in prefrontal neurons, but not their surrounding non-neuronal cells, and in the absence of a copy number or other significant DNA structural variant. These findings imply that a subset of autism susceptibility genes carrying strong penetrance for disease could be epigenetically dysregulated in a cell (type)-specific manner, independent of the occurrence of any rare structural DNA variations at these sites that were previously implicated in neurological disease. More broadly, these findings strongly suggest that there is significant overlap between the genetic and epigenetic risk architectures of autism.[81] Whether genetic polymorphisms, mutations in *cis* or *trans* (DNA sequence at the site of histone methylation change or at a different locus) or some other genetic mechanisms "drives" the observed epigenetic alterations, or whether these chromatin changes occurred independently from alterations in DNA sequence, remains subject to future investigations.

EPIGENETIC ALTERATIONS IN MOOD AND ANXIETY DISORDERS, INCLUDING PTSD

A Brief Primer on Depression and PTSD

Depression, with its core syndrome of excessive sadness, in conjunction with affective and cognitive imbalances and neurovegetative dysfunction (including sleep, appetite and general level of activity), is considered one of the most serious disorders in today's society, with a lifetime prevalence as high as 16.2% in the US adult population.[177] The introduction of the first pharmacological antidepressant medications in the 1950s, and subsequent development of drugs with lower side-effect profiles has greatly improved the therapeutic outlook for depressed patients.[178] However, conventional antidepressant drugs essentially all modulate monoamine neurotransmission and potentially take 6–8 weeks to exert their effects.[179] Still, inadequate response to at least one antidepressant trial of adequate doses and duration is encountered in up to 50–60% of patients.[180,181] Given this background, it is necessary to develop conceptually novel antidepressants that act rapidly and effectively in a much higher proportion of patients.[179] The genetic risk architecture of depression appears to be even less tractable than for some of the other aforementioned conditions such as autism and schizophrenia.[115,182,183] However, affective disorders as a group show, in terms of genetics, significant overlap with schizophrenia. For example, rare structural variants, including the balanced translocation at the Disrupted-in-Schizophrenia 1 (DISC-1) locus (1q42) or the 22q11 deletion are in different individuals associated with either mood disorder or schizophrenia.[184,185] Further, the genomic coordinates of copy number variants associated with manic-depressive (bipolar) illness show significant overlap with the spectrum of microdeletions and -duplications encountered in schizophrenia.[33]

PTSD describes a psychiatric syndrome following the experience of traumatic event(s), including intrusive recollections, avoidance behaviors, numbing, hyperarousal, and other psychological and somatic symptoms that in themselves eventually become a significant source of distress.[186] PTSD carries a significant human cost, including depression and suicide, substance abuse and many other serious sequels. A significant amount of PTSD research is focused on resilience, which is defined as a person's ability to adapt and cope with various forms of acute or chronic trauma and stress without developing psychiatric disease including PTSD and PTSD sequels. A better understanding of the psychosocial, neurochemical, genetic and perhaps also epigenetic determinants of resilience will have significant implications for the prevention of stress- and adversity-related illnesses.[187]

Epigenetic Dysregulation in Depression and PTSD

Similar to the aforementioned examples in autism and schizophrenia, epigenetic alterations are suspected to play an important role for the pathophysiology of disease in subjects diagnosed with a mood and anxiety spectrum disorder, including PTSD. Obviously, from a heuristic perspective, it is very attractive to design working hypotheses that attribute a key role for epigenetic markings inside the nucleus of neurons and other brain cells. Such molecular adaptations could subserve a memory function that, in response to an intense "environmental" influence (e.g. trauma), conveys a lasting change in a subject's neuronal plasticity, and ultimately is associated with changes in his or her emotional and physical health and resilience. Indeed, early evidence points to the promising potential of such types of working model. For example, a recent study on peripheral cells collected pre- and postdeployment of US military service members identified global DNA methylation levels in repetitive DNA sequences, including LINE-1 and Alu repeat elements, as biomarkers that were significantly associated with resilience or, conversely, vulnerability to PTSD.[188] Similarly, studies in civilian/urban populations discovered that changes in blood DNA methylation signatures in PTSD subjects selectively affected cytokine and steroid signaling, immune defense and inflammation related genes, consistent with various lines of evidence implicating some degree of peripheral immune dysregulation in this disorder.[189,190]

Extrapolating from the aforementioned studies in blood, one would predict that brain chromatin too is involved in the neurobiology of mood and anxiety and PTSD. Indeed, post-mortem studies in case and control cohorts reported DNA and histone methylation changes in hippocampus, ventral striatum and other anatomical structures with a critical role for emotion, affect and memory. For example, methylation of a glucocorticoid receptor gene promoter, *NR3C1*, and of ribosomal DNA repeats is increased, and RNA expression decreased, in hippocampus of adult suicide victims who also suffered childhood abuse.[191,192] Furthermore, the prefrontal cortex of suicide completers exhibits a shift from open to repressive chromatin-associated histone methylation for the *TRKB* neurotrophin high affinity receptor, and for various genes regulating polyamine metabolism.[193,194] Downregulated histone deacetylase 2 (HDAC2) expression in ventral striatum of subjects diagnosed with depression is thought to lead to an overall increase in histone H3 acetylation in this mesolimbic structure.[96] At least some of the reported epigenetic changes in post-mortem brain of subjects on the mood and anxiety spectrum are not likely to be epiphenomena due to medication etc. To mention just two examples, the hippocampal glucocorticoid receptor gene, *Nr3c1*, shows excessive methylation not only in patients,[191] but also in adults rats brought up with suboptimal maternal care ("low licking" vs "high licking" mothers).[195] Furthermore, chronic social defeat stress in mice and rats elicits histone acetylation changes in ventral striatum and hippocampus very similar to those encountered in depressed human subjects.[96,196]

Chromatin Modifying Drugs Show Promise in Preclinical Models of Depression and PTSD

Notwithstanding these extremely interesting chromatin-associated phenotypes in brain tissue of subjects diagnosed with mood disorder and in animal models for depression and PTSD, perhaps the most notable finding, as it pertains to epigenetics, are the antidepressant-like phenotypes in animals treated with histone deacetylase inhibitor (HDACi), as reported in multiple, independent studies. Systemic or localized infusions of butyrate and short chain fatty derivatives broadly inhibiting class I and II HDAC into prefrontal cortex, hippocampus or striatum reportedly improve behavioral despair and other psychomotor parameters associated with the depressed state,[96,197,198] enhance the therapeutic response of psychotropic agents including serotonin reuptake inhibitors[96,98] and sex steroids such as estrogen/estradiol.[199] The mechanism of action may include increased transcription of genes with a key regulatory role in neuronal growth and differentiation, including brain-derived neurotrophic factor (*Bdnf*) as one of the more prominent examples.[98,198] Negative findings were also reported but differences in doses and schedules could have been a factor.[200] On the one hand, these results may not appear too surprising given that HDACi seemingly exert a very broad therapeutic potential, including improved memory and cognition in a wide range of neurological conditions, from depression to acute brain injury and stroke, and neurodegenerative illness including Parkinson's, Alzheimer's, Triplet Repeat (including Huntington and spinocerebellar ataxias) and motor neuron disease.[92–95,201–203] These very broad therapeutic profiles across neurodegenerative and neuropsychiatric illness, however, do not speak against the potential of the HDACs to provide a novel antidepressant drug target. Or, to use an analogy, electroconvulsive therapy (ECT) elicits beneficial effects in mania (abnormal elevation of mood and activity), catatonia and some cases of Parkinsonism and various other neurological conditions. However, despite this broad therapeutic profile of ECT, this type of treatment is primarily reserved for cases with severe or life-threatening depression.

While HDACi and other chromatin modifying drugs are showing promise in preclinical studies, it is difficult to pinpoint the critical mechanism of action. For example, beneficial effects of HDACi in neurodegenerative and cognitive disease could reflect their broad effects on histone modifications, but changes in the acetylation of non-histone proteins such as the presynaptic molecule and regulator of vesicle release, Bruchpilot,[204] were also implicated in HDACi-mediated therapeutic effects. Both the class I/II HDACs (which are defined by their zinc-containing catalytic domain) and the (nicotinamide NAD+ dependent) class III HDACs, or sirtuins, regulate lysine acetylation of non-histone proteins.[205,206] Notably, SIRT1, via deacetylation of the helix-loop-helix transcription factor NHLH2, upregulates monoamine oxidase A (MAO-A), a critical regulator of serotonin catabolism and, furthermore, inhibition or downregulated expression of SIRT1 improves depression and anxiety-related behaviors in the mouse.[207]

Like for the above mentioned HDACs, the clinical potential of drugs interfering with histone methylation appears promising based on early studies but, as already mentioned above, this is a largely unexplored field in psychiatric epigenetics, with the exception of the small molecule and diazepin-quinazolin-amine derivative BIX-01294 which inhibits histone H3K9-specific methyltransferase *G9a/Glp*[101] and strongly enhances the development of reward behaviors in stimulant drug-exposed mice.[102] The drug's mechanism of action could, at least in part, involve inhibition of G9a/Glp-mediated repressive chromatin remodeling at the promoters of *Bdnf, Cdk5, Arc* and other genes acting as key regulators for spine density and synaptic connectivity in the brain and in the context of motivational and affective behaviors.[102]

SYNOPSIS AND OUTLOOK

Over the course of only a few years, we have witnessed a proliferation of epigenetic studies in brain of subjects diagnosed with neuropsychiatric disease and the preclinical model systems, respectively. There is evidence for large scale remodeling of DNA and histone methylation landscapes during the course of maturation and aging of the human cerebral cortex.[146,208] Still, hundreds of promoters are subject to epigenetic changes that seemingly continue into old age, and these data, taken together, leave little doubt that chromatin structures undergo remodeling throughout the lifespan of the human brain,[125,146,209] including neurons and other terminally differentiated cells.[208] Based on this and other post-mortem brain work, epigenetic risk architectures are beginning to emerge for a number of common psychiatric conditions and disorders, including autism,[81] schizophrenia,[210] depression and bipolar disorder[126,211] and alcoholism.[212] Enthusiasm in academia and industry to develop novel epigenetic therapies recently received a boost by the US Food and Drug Administration approval for HDACi drugs, including suberoylanilide hydroxamic acid (SAHA, trade name Vorinostat) and romidepsin (trade name Istodax). While these medications were tested and approved for the treatment of cancer, a large body of preclinical work suggests that HDACi could exert beneficial effects not only in various neurological conditions but also in depression and other psychiatric illnesses.[96,97] To find out whether these encouraging findings from animal model systems will indeed hold up in future clinical trials will be one of the most important tasks ahead for experimental and clinical psychopharmacologists alike.

Acknowledgments

Work in the authors' laboratory is supported by the National Institutes of Health. The authors report no financial conflicts of interest.

References

1. Dehaene S, Changeux JP. Experimental and theoretical approaches to conscious processing. *Neuron*. 2011;70:200–227.
2. Palva JM, Palva S, Kaila K. Phase synchrony among neuronal oscillations in the human cortex. *J Neurosci*. 2005;25:3962–3972.
3. Woon FL, Sood S, Hedges DW. Hippocampal volume deficits associated with exposure to psychological trauma and posttraumatic stress disorder in adults: a meta-analysis. *Prog Neuropsychopharmacol Biol Psychiatry*. 2010;34:1181–1188.
4. Nestler EJ, Hyman SE. Animal models of neuropsychiatric disorders. *Nat Neurosci*. 2010;13:1161–1169.
5. Lehman AF, Lieberman JA, Dixon LB, et al. Practice guideline for the treatment of patients with schizophrenia, second edition. *Am J Psychiatry*. 2004;161:1–56.
6. Krishnan V, Nestler EJ. Linking molecules to mood: new insight into the biology of depression. *Am J Psychiatry*. 2010;167:1305–1320.
7. Kriaucionis S, Heintz N. The nuclear DNA base 5-hydroxymethylcytosine is present in Purkinje neurons and the brain. *Science*. 2009;324:929–930.
8. Jin SG, Wu X, Li AX, Pfeifer GP. Genomic mapping of 5-hydroxymethylcytosine in the human brain. *Nucleic Acids Res*. 2011;39:5015–5024.
9. Song CX, Szulwach KE, Fu Y, et al. Selective chemical labeling reveals the genome-wide distribution of 5-hydroxymethylcytosine. *Nat Biotechnol*. 2011;29:68–72.
10. Maunakea AK, Nagarajan RP, Bilenky M, et al. Conserved role of intragenic DNA methylation in regulating alternative promoters. *Nature*. 2010;466:253–257.
11. Tan M, Luo H, Lee S, et al. Identification of 67 histone marks and histone lysine crotonylation as a new type of histone modification. *Cell*. 2011;146:1016–1028.
12. Kouzarides T. Chromatin modifications and their function. *Cell*. 2007;128:693–705.
13. Taverna SD, Li H, Ruthenburg AJ, Allis CD, Patel DJ. How chromatin-binding modules interpret histone modifications: lessons from professional pocket pickers. *Nat Struct Mol Biol*. 2007;14:1025–1040.
14. Zhou VW, Goren A, Bernstein BE. Charting histone modifications and the functional organization of mammalian genomes. *Nat Rev Genet*. 2011;12:7–18.
15. Woodcock CL. Chromatin architecture. *Curr Opin Struct Biol*. 2006;16:213–220.
16. Jin C, Felsenfeld G. Nucleosome stability mediated by histone variants H3.3 and H2A.Z. *Genes Dev*. 2007;21:1519–1529.
17. Woodcock CL, Skoultchi AI, Fan Y. Role of linker histone in chromatin structure and function: H1 stoichiometry and nucleosome repeat length. *Chromosome Res*. 2006;14:17–25.
18. Pearson EC, Bates DL, Prospero TD, Thomas JO. Neuronal nuclei and glial nuclei from mammalian cerebral cortex. Nucleosome repeat lengths, DNA contents and H1 contents. *Eur J Biochem*. 1984;144:353–360.
19. Day JJ, Sweatt JD. Epigenetic mechanisms in cognition. *Neuron*. 2011;70:813–829.

20. Robison AJ, Nestler EJ. Transcriptional and epigenetic mechanisms of addiction. *Nat Rev Neurosci*. 2011;12:623–637.
21. Bhutani N, Burns DM, Blau HM. DNA demethylation dynamics. *Cell*. 2011;146:866–872.
22. Guo JU, Su Y, Zhong C, Ming GL, Song H. Hydroxylation of 5-methylcytosine by TET1 promotes active DNA demethylation in the adult brain. *Cell*. 2011;145:423–434.
23. Copeland RA, Solomon ME, Richon VM. Protein methyltransferases as a target class for drug discovery. *Nat Rev Drug Discov*. 2009;8:724–732.
24. Rotili D, Mai A. Targeting histone demethylases: a new avenue for the fight against cancer. *Genes Cancer*. 2011;2:663–679.
25. Vermeulen M, Eberl HC, Matarese F, et al. Quantitative interaction proteomics and genome-wide profiling of epigenetic histone marks and their readers. *Cell*. 2010;142:967–980.
26. Haggarty SJ, Tsai LH. Probing the role of HDACs and mechanisms of chromatin-mediated neuroplasticity. *Neurobiol Learn Mem*. 2011;96:41–52.
27. Yan J, Zhang F, Brundage E, et al. Genomic duplication resulting in increased copy number of genes encoding the sister chromatid cohesion complex conveys clinical consequences distinct from Cornelia de Lange. *J Med Genet*. 2009;46:626–634.
28. Nestler EJ. Epigenetic mechanisms in psychiatry. *Biol Psychiatry*. 2009;65:189–190.
29. Maze I, Feng J, Wilkinson MB, Sun H, Shen L, Nestler EJ. Cocaine dynamically regulates heterochromatin and repetitive element unsilencing in nucleus accumbens. *Proc Natl Acad Sci USA*. 2011;108:3035–3040.
30. Li G, Reinberg D. Chromatin higher-order structures and gene regulation. *Curr Opin Genet Dev*. 2011;21:175–186.
31. Rodriguez-Paredes M, Esteller M. Cancer epigenetics reaches mainstream oncology. *Nat Med*. 2011;17:330–339.
32. Huang HS, Matevossian A, Jiang Y, Akbarian S. Chromatin immunoprecipitation in postmortem brain. *J Neurosci Methods*. 2006;156:284–292.
33. Malhotra D, Sebat J. CNVs: harbingers of a rare variant revolution in psychiatric genetics. *Cell*. 2012;148:1223–1241.
34. Cooper GM, Coe BP, Girirajan S, et al. A copy number variation morbidity map of developmental delay. *Nat Genet*. 2011;43:838–846.
35. O'Roak BJ, Deriziotis P, Lee C, et al. Exome sequencing in sporadic autism spectrum disorders identifies severe de novo mutations. *Nat Genet*. 2011;43:585–589.
36. Oberle I, Rousseau F, Heitz D, et al. Instability of a 550-base pair DNA segment and abnormal methylation in fragile X syndrome. *Science*. 1991;252:1097–1102.
37. Al-Mahdawi S, Pinto RM, Ismail O, et al. The Friedreich ataxia GAA repeat expansion mutation induces comparable epigenetic changes in human and transgenic mouse brain and heart tissues. *Hum Mol Genet*. 2008;17:735–746.
38. Tassone F, Hagerman RJ, Gane LW, Taylor AK. Strong similarities of the FMR1 mutation in multiple tissues: postmortem studies of a male with a full mutation and a male carrier of a premutation. *Am J Med Genet*. 1999;84:240–244.
39. De S, Michor F. DNA secondary structures and epigenetic determinants of cancer genome evolution. *Nat Struct Mol Biol*. 2011;18:950–955.
40. Ramocki MB, Peters SU, Tavyev YJ, et al. Autism and other neuropsychiatric symptoms are prevalent in individuals with MeCP2 duplication syndrome. *Ann Neurol*. 2009;66:771–782.
41. Kirov G, Pocklington AJ, Holmans P, et al. De novo CNV analysis implicates specific abnormalities of postsynaptic signalling complexes in the pathogenesis of schizophrenia. *Mol Psychiatry*. 2012;17:142–153.
42. Verhoeven WM, Egger JI, Vermeulen K, van de Warrenburg BP, Kleefstra T. Kleefstra syndrome in three adult patients: further delineation of the behavioral and neurological phenotype shows aspects of a neurodegenerative course. *Am J Med Genet A*. 2011;155A:2409–2415.
43. Klein CJ, Botuyan MV, Wu Y, et al. Mutations in DNMT1 cause hereditary sensory neuropathy with dementia and hearing loss. *Nat Genet*. 2011;43:595–600.
44. de Greef JC, Wang J, Balog J, et al. Mutations in ZBTB24 are associated with immunodeficiency, centromeric instability, and facial anomalies syndrome type 2. *Am J Hum Genet*. 2011;88:796–804.
45. Ehrlich M, Sanchez C, Shao C, et al. ICF, an immunodeficiency syndrome: DNA methyltransferase 3B involvement, chromosome anomalies, and gene dysregulation. *Autoimmunity*. 2008;41:253–271.

46. Akbarian S, Huang HS. Molecular and cellular mechanisms of altered GAD1/GAD67 expression in schizophrenia and related disorders. *Brain Res Rev*. 2006;52:293–304.
47. Aston C, Jiang L, Sokolov BP. Microarray analysis of postmortem temporal cortex from patients with schizophrenia. *J Neurosci Res*. 2004;77:858–866.
48. Benes FM. Amygdalocortical circuitry in schizophrenia: from circuits to molecules. *Neuropsychopharmacology*. 2010;35:239–257.
49. Charych EI, Liu F, Moss SJ, Brandon NJ. GABA(A) receptors and their associated proteins: implications in the etiology and treatment of schizophrenia and related disorders. *Neuropharmacology*. 2009;57:481–495.
50. Dracheva S, Elhakem SL, McGurk SR, Davis KL, Haroutunian V. GAD67 and GAD65 mRNA and protein expression in cerebrocortical regions of elderly patients with schizophrenia. *J Neurosci Res*. 2004;76:581–592.
51. Fillman SG, Duncan CE, Webster MJ, Elashoff M, Weickert CS. Developmental co-regulation of the beta and gamma GABAA receptor subunits with distinct alpha subunits in the human dorsolateral prefrontal cortex. *Int J Dev Neurosci*. 2010;28:513–519.
52. Guidotti A, Auta J, Davis JM, et al. GABAergic dysfunction in schizophrenia: new treatment strategies on the horizon. *Psychopharmacology (Berl)*. 2005;180:191–205.
53. Hakak Y, Walker JR, Li C, et al. Genome-wide expression analysis reveals dysregulation of myelination-related genes in chronic schizophrenia. *Proc Natl Acad Sci USA*. 2001;98:4746–4751.
54. Hashimoto T, Bazmi HH, Mirnics K, Wu Q, Sampson AR, Lewis DA. Conserved regional patterns of GABA-related transcript expression in the neocortex of subjects with schizophrenia. *Am J Psychiatry*. 2008;165: 479–489.
55. Katsel P, Davis KL, Haroutunian V. Variations in myelin and oligodendrocyte-related gene expression across multiple brain regions in schizophrenia: a gene ontology study. *Schizophr Res*. 2005;79:157–173.
56. Martins-de-Souza D, Gattaz WF, Schmitt A, et al. Alterations in oligodendrocyte proteins, calcium homeostasis and new potential markers in schizophrenia anterior temporal lobe are revealed by shotgun proteome analysis. *J Neural Transm*. 2009;116:275–289.
57. Regenold WT, Phatak P, Marano CM, Gearhart L, Viens CH, Hisley KC. Myelin staining of deep white matter in the dorsolateral prefrontal cortex in schizophrenia, bipolar disorder, and unipolar major depression. *Psychiatry Res*. 2007;151:179–188.
58. Sibille E, Wang Y, Joeyen-Waldorf J, et al. A molecular signature of depression in the amygdala. *Am J Psychiatry*. 2009;166:1011–1024.
59. Tkachev D, Mimmack ML, Ryan MM, et al. Oligodendrocyte dysfunction in schizophrenia and bipolar disorder. *Lancet*. 2003;362:798–805.
60. Woo TU, Kim AM, Viscidi E. Disease-specific alterations in glutamatergic neurotransmission on inhibitory interneurons in the prefrontal cortex in schizophrenia. *Brain Res*. 2008;1218:267–277.
61. Benes FM, Lim B, Matzilevich D, Walsh JP, Subburaju S, Minns M. Regulation of the GABA cell phenotype in hippocampus of schizophrenics and bipolars. *Proc Natl Acad Sci USA*. 2007;104:10164–10169.
62. Blatt GJ, Fatemi SH. Alterations in GABAergic biomarkers in the autism brain: research findings and clinical implications. *Anat Rec (Hoboken)*. 2011;294:1646–1652.
63. Curley AA, Arion D, Volk DW, et al. Cortical deficits of glutamic acid decarboxylase 67 expression in schizophrenia: clinical, protein, and cell type-specific features. *Am J Psychiatry*. 2011;168:921–929.
64. Guidotti A, Auta J, Davis JM, et al. Decrease in reelin and glutamic acid decarboxylase67 (GAD67) expression in schizophrenia and bipolar disorder: a postmortem brain study. *Arch Gen Psychiatry*. 2000;57:1061–1069.
65. Hyde TM, Lipska BK, Ali T, et al. Expression of GABA signaling molecules KCC2, NKCC1, and GAD1 in cortical development and schizophrenia. *J Neurosci*. 2011;31:11088–11095.
66. Lisman JE, Coyle JT, Green RW, et al. Circuit-based framework for understanding neurotransmitter and risk gene interactions in schizophrenia. *Trends Neurosci*. 2008;31:234–242.
67. Akbarian S, Kim JJ, Potkin SG, et al. Gene expression for glutamic acid decarboxylase is reduced without loss of neurons in prefrontal cortex of schizophrenics. *Arch Gen Psychiatry*. 1995;52:258–266.
68. Hager GL, McNally JG, Misteli T. Transcription dynamics. *Mol Cell*. 2009;35:741–753.
69. Berger SL. The complex language of chromatin regulation during transcription. *Nature*. 2007;447:407–412.
70. Abel T, Zukin RS. Epigenetic targets of HDAC inhibition in neurodegenerative and psychiatric disorders. *Curr Opin Pharmacol*. 2008;8:57–64.
71. Buckley NJ. Analysis of transcription, chromatin dynamics and epigenetic changes in neural genes. *Prog Neurobiol*. 2007;83:195–210.

72. Duman RS, Newton SS. Epigenetic marking and neuronal plasticity. *Biol Psychiatry*. 2007;62:1–3.
73. Graff J, Mansuy IM. Epigenetic codes in cognition and behaviour. *Behav Brain Res*. 2008;192:70–87.
74. Renthal W, Nestler EJ. Epigenetic mechanisms in drug addiction. *Trends Mol Med*. 2008;14:341–350.
75. Tsankova N, Renthal W, Kumar A, Nestler EJ. Epigenetic regulation in psychiatric disorders. *Nat Rev Neurosci*. 2007;8:355–367.
76. Abdolmaleky HM, Cheng KH, Russo A, et al. Hypermethylation of the reelin (RELN) promoter in the brain of schizophrenic patients: a preliminary report. *Am J Med Genet B Neuropsychiatr Genet*. 2005;134B:60–66.
77. Grayson DR, Jia X, Chen Y, et al. Reelin promoter hypermethylation in schizophrenia. *Proc Natl Acad Sci USA*. 2005;102:9341–9346.
78. Huang HS, Akbarian S. GAD1 mRNA expression and DNA methylation in prefrontal cortex of subjects with schizophrenia. *PLoS One*. 2007;2:e809.
79. Huang HS, Matevossian A, Whittle C, et al. Prefrontal dysfunction in schizophrenia involves mixed-lineage leukemia 1-regulated histone methylation at GABAergic gene promoters. *J Neurosci*. 2007;27:11254–11262.
80. Iwamoto K, Bundo M, Yamada K, et al. DNA methylation status of SOX10 correlates with its downregulation and oligodendrocyte dysfunction in schizophrenia. *J Neurosci*. 2005;25:5376–5381.
81. Shulha HP, Cheung I, Whittle C, et al. Epigenetic signatures of autism: trimethylated H3K4 landscapes in prefrontal neurons. *Arch Gen Psychiatry*. 2012;69:314–324.
82. Lander ES. Initial impact of the sequencing of the human genome. *Nature*. 2011;470:187–197.
83. Petronis A. Epigenetics as a unifying principle in the aetiology of complex traits and diseases. *Nature*. 2010;465:721–727.
84. Wykes SM, Krawetz SA. The structural organization of sperm chromatin. *J Biol Chem*. 2003;278:29471–29477.
85. Balhorn R, Brewer L, Corzett M. DNA condensation by protamine and arginine-rich peptides: analysis of toroid stability using single DNA molecules. *Mol Reprod Dev*. 2000;56:230–234.
86. Hammoud SS, Nix DA, Zhang H, Purwar J, Carrell DT, Cairns BR. Distinctive chromatin in human sperm packages genes for embryo development. *Nature*. 2009;460:473–478.
87. Okada Y, Yamagata K, Hong K, Wakayama T, Zhang Y. A role for the elongator complex in zygotic paternal genome demethylation. *Nature*. 2010;463:554–558.
88. Chong S, Youngson NA, Whitelaw E. Heritable germline epimutation is not the same as transgenerational epigenetic inheritance. *Nat Genet*. 2007;39:574–575. [author reply 575–576].
89. Crepin M, Dieu MC, Lejeune S, et al. Evidence of constitutional MLH1 epimutation associated to transgenerational inheritance of cancer susceptibility. *Hum Mutat*. 2012;33:180–188.
90. Nelson VR, Nadeau JH. Transgenerational genetic effects. *Epigenomics*. 2010;2:797–806.
91. Whitelaw NC, Whitelaw E. Transgenerational epigenetic inheritance in health and disease. *Curr Opin Genet Dev*. 2008;18:273–279.
92. Baltan S, Murphy SP, Danilov CA, Bachleda A, Morrison RS. Histone deacetylase inhibitors preserve white matter structure and function during ischemia by conserving ATP and reducing excitotoxicity. *J Neurosci*. 2011;31:3990–3999.
93. Chuang DM, Leng Y, Marinova Z, Kim HJ, Chiu CT. Multiple roles of HDAC inhibition in neurodegenerative conditions. *Trends Neurosci*. 2009;32:591–601.
94. Fischer A, Sananbenesi F, Mungenast A, Tsai LH. Targeting the correct HDAC(s) to treat cognitive disorders. *Trends Pharmacol Sci*. 2010;31:605–617.
95. Tsou AY, Friedman LS, Wilson RB, Lynch DR. Pharmacotherapy for Friedreich ataxia. *CNS Drugs*. 2009;23:213–223.
96. Covington 3rd HE, Maze I, LaPlant QC, et al. Antidepressant actions of histone deacetylase inhibitors. *J Neurosci*. 2009;29:11451–11460.
97. Morris MJ, Karra AS, Monteggia LM. Histone deacetylases govern cellular mechanisms underlying behavioral and synaptic plasticity in the developing and adult brain. *Behav Pharmacol*. 2010;21:409–419.
98. Schroeder FA, Lin CL, Crusio WE, Akbarian S. Antidepressant-like effects of the histone deacetylase inhibitor, sodium butyrate, in the mouse. *Biol Psychiatry*. 2007;62:55–64.
99. Dong E, Agis-Balboa RC, Simonini MV, Grayson DR, Costa E, Guidotti A. Reelin and glutamic acid decarboxylase67 promoter remodeling in an epigenetic methionine-induced mouse model of schizophrenia. *Proc Natl Acad Sci USA*. 2005;102:12578–12583.
100. Peter CJ, Akbarian S. Balancing histone methylation activities in psychiatric disorders. *Trends Mol Med*. 2011;17:372–379.

101. Kubicek S, O'Sullivan RJ, August EM, et al. Reversal of H3K9me2 by a small-molecule inhibitor for the G9a histone methyltransferase. *Mol Cell*. 2007;25:473–481.
102. Maze I, Covington 3rd HE, Dietz DM, et al. Essential role of the histone methyltransferase G9a in cocaine-induced plasticity. *Science*. 2010;327:213–216.
103. Kelly TK, De Carvalho DD, Jones PA. Epigenetic modifications as therapeutic targets. *Nat Biotechnol*. 2010;28:1069–1078.
104. Han J, Li Y, Wang D, Wei C, Yang X, Sui N. Effect of 5-aza-2-deoxycytidine microinjecting into hippocampus and prelimbic cortex on acquisition and retrieval of cocaine-induced place preference in C57BL/6 mice. *Eur J Pharmacol*. 2010;642:93–98.
105. LaPlant Q, Vialou V, Covington 3rd HE, et al. Dnmt3a regulates emotional behavior and spine plasticity in the nucleus accumbens. *Nat Neurosci*. 2010;13:1137–1143.
106. Levenson JM, Roth TL, Lubin FD, et al. Evidence that DNA (cytosine-5) methyltransferase regulates synaptic plasticity in the hippocampus. *J Biol Chem*. 2006;281:15763–15773.
107. Miller CA, Sweatt JD. Covalent modification of DNA regulates memory formation. *Neuron*. 2007;53:857–869.
108. Salerno S, Da Settimo F, Taliani S, et al. Recent advances in the development of dual topoisomerase I and II inhibitors as anticancer drugs. *Curr Med Chem*. 2010;17:4270–4290.
109. Huang HS, Allen JA, Mabb AM, et al. Topoisomerase inhibitors unsilence the dormant allele of Ube3a in neurons. *Nature*. 2011;481:185–189.
110. Bressler J, Tsai TF, Wu MY, et al. The SNRPN promoter is not required for genomic imprinting of the Prader-Willi/Angelman domain in mice. *Nat Genet*. 2001;28:232–240.
111. Reik W. Stability and flexibility of epigenetic gene regulation in mammalian development. *Nature*. 2007;447:425–432.
112. Jarskog LF, Miyamoto S, Lieberman JA. Schizophrenia: new pathological insights and therapies. *Annu Rev Med*. 2007;58:49–61.
113. Ibrahim HM, Tamminga CA. Schizophrenia: treatment targets beyond monoamine systems. *Annu Rev Pharmacol Toxicol*. 2011;51:189–209.
114. Canitano R, Scandurra V. Psychopharmacology in autism: an update. *Prog Neuropsychopharmacol Biol Psychiatry*. 2011;35:18–28.
115. Cichon S, Craddock N, Daly M, et al. Genomewide association studies: history, rationale, and prospects for psychiatric disorders. *Am J Psychiatry*. 2009;166:540–556.
116. Sebat J, Levy DL, McCarthy SE. Rare structural variants in schizophrenia: one disorder, multiple mutations; one mutation, multiple disorders. *Trends Genet*. 2009;25:528–535.
117. Weiss LA. Autism genetics: emerging data from genome-wide copy-number and single nucleotide polymorphism scans. *Expert Rev Mol Diagn*. 2009;9:795–803.
118. Brown AS. Exposure to prenatal infection and risk of schizophrenia. *Front Psychiatry*. 2011;2:63.
119. Brown AS, Derkits EJ. Prenatal infection and schizophrenia: a review of epidemiologic and translational studies. *Am J Psychiatry*. 2010;167:261–280.
120. Patterson PH. Maternal infection and immune involvement in autism. *Trends Mol Med*. 2011;17:389–394.
121. Patterson PH. Maternal infection: window on neuroimmune interactions in fetal brain development and mental illness. *Curr Opin Neurobiol*. 2002;12:115–118.
122. Hallmayer J, Cleveland S, Torres A, et al. Genetic heritability and shared environmental factors among twin pairs with autism. *Arch Gen Psychiatry*. 2011;68:1095–1102.
123. Akbarian S, Vinuela A, Kim JJ, Potkin SG, Bunney Jr WE, Jones EG. Distorted distribution of nicotinamide-adenine dinucleotide phosphate-diaphorase neurons in temporal lobe of schizophrenics implies anomalous cortical development. *Arch Gen Psychiatry*. 1993;50:178–187.
124. Weinberger DR. Implications of normal brain development for the pathogenesis of schizophrenia. *Arch Gen Psychiatry*. 1987;44:660–669.
125. Siegmund KD, Connor CM, Campan M, et al. DNA methylation in the human cerebral cortex is dynamically regulated throughout the life span and involves differentiated neurons. *PLoS One*. 2007;2:e895.
126. Tang B, Dean B, Thomas EA. Disease- and age-related changes in histone acetylation at gene promoters in psychiatric disorders. *Translational Psychiatry*. 2011;1:e64.
127. Tamura Y, Kunugi H, Ohashi J, Hohjoh H. Epigenetic aberration of the human REELIN gene in psychiatric disorders. *Mol Psychiatry*. 2007;12(519):593–600.
128. Marutha Ravindran CR, Ticku MK. Changes in methylation pattern of NMDA receptor NR2B gene in cortical neurons after chronic ethanol treatment in mice. *Brain Res Mol Brain Res*. 2004;121:19–27.

129. Satta R, Maloku E, Costa E, Guidotti A. Stimulation of brain nicotinic acetylcholine receptors (nAChRs) decreases DNA methyltransferase 1 (DNMT1) expression in cortical and hippocampal GABAergic neurons of Swiss albino mice. *Society for Neuroscience Abstract*. 2007.
130. Satta R, Maloku E, Zhubi A, et al. Nicotine decreases DNA methyltransferase 1 expression and glutamic acid decarboxylase 67 promoter methylation in GABAergic interneurons. *Proc Natl Acad Sci USA*. 2008;105:16356–16361.
131. Kumar A, Choi KH, Renthal W, et al. Chromatin remodeling is a key mechanism underlying cocaine-induced plasticity in striatum. *Neuron*. 2005;48:303–314.
132. Numachi Y, Shen H, Yoshida S, et al. Methamphetamine alters expression of DNA methyltransferase 1 mRNA in rat brain. *Neurosci Lett*. 2007;414:213–217.
133. Numachi Y, Yoshida S, Yamashita M, et al. Psychostimulant alters expression of DNA methyltransferase mRNA in the rat brain. *Ann N Y Acad Sci*. 2004;1025:102–109.
134. Cheng MC, Liao DL, Hsiung CA, Chen CY, Liao YC, Chen CH. Chronic treatment with aripiprazole induces differential gene expression in the rat frontal cortex. *Int J Neuropsychopharmacol*. 2008;11:207–216.
135. Li J, Guo Y, Schroeder FA, et al. Dopamine D2-like antagonists induce chromatin remodeling in striatal neurons through cyclic AMP-protein kinase A and NMDA receptor signaling. *J Neurochem*. 2004;90:1117–1131.
136. Mill J, Tang T, Kaminsky Z, et al. Epigenomic profiling reveals DNA-methylation changes associated with major psychosis. *Am J Hum Genet*. 2008;82:696–711.
137. Shimabukuro M, Jinno Y, Fuke C, Okazaki Y. Haloperidol treatment induces tissue- and sex-specific changes in DNA methylation: a control study using rats. *Behav Brain Funct*. 2006;2:37.
138. Zhang TY, Hellstrom IC, Bagot RC, Wen X, Diorio J, Meaney MJ. Maternal care and DNA methylation of a glutamic acid decarboxylase 1 promoter in rat hippocampus. *J Neurosci*. 2010;30:13130–13137.
139. Alaminos M, Davalos V, Ropero S, et al. EMP3, a myelin-related gene located in the critical 19q13.3 region, is epigenetically silenced and exhibits features of a candidate tumor suppressor in glioma and neuroblastoma. *Cancer Res*. 2005;65:2565–2571.
140. Debinski W, Gibo D, Mintz A. Epigenetics in high-grade astrocytomas: opportunities for prevention and detection of brain tumors. *Ann N Y Acad Sci*. 2003;983:232–242.
141. Felsberg J, Yan PS, Huang TH, et al. DNA methylation and allelic losses on chromosome arm 14q in oligodendroglial tumours. *Neuropathol Appl Neurobiol*. 2006;32:517–524.
142. Uhlmann K, Rohde K, Zeller C, et al. Distinct methylation profiles of glioma subtypes. *Int J Cancer*. 2003;106:52–59.
143. Inda MM, Castresana JS. RASSF1A promoter is highly methylated in primitive neuroectodermal tumors of the central nervous system. *Neuropathology*. 2007;27:341–346.
144. Lindsey JC, Lusher ME, Anderton JA, Gilbertson RJ, Ellison DW, Clifford SC. Epigenetic deregulation of multiple S100 gene family members by differential hypomethylation and hypermethylation events in medulloblastoma. *Br J Cancer*. 2007;97:267–274.
145. Gertz J, Varley KE, Reddy TE, et al. Analysis of DNA methylation in a three-generation family reveals widespread genetic influence on epigenetic regulation. *PLoS Genet*. 2011;7:e1002228.
146. Numata S, Ye T, Hyde TM, et al. DNA methylation signatures in development and aging of the human prefrontal cortex. *Am J Hum Genet*. 2012;90:260–272.
147. Zhang D, Cheng L, Badner JA, et al. Genetic control of individual differences in gene-specific methylation in human brain. *Am J Hum Genet*. 2010;86:411–419.
148. Colantuoni C, Lipska BK, Ye T, et al. Temporal dynamics and genetic control of transcription in the human prefrontal cortex. *Nature*. 2011;478:519–523.
149. Costa E, Davis JM, Dong E, et al. GABAergic cortical deficit dominates schizophrenia pathophysiology. *Crit Rev Neurobiol*. 2004;16:1–23.
150. Costa E, Grayson DR, Mitchell CP, Tremolizzo L, Veldic M, Guidotti A. GABAergic cortical neuron chromatin as a putative target to treat schizophrenia vulnerability. *Crit Rev Neurobiol*. 2003;15:121–142.
151. Mirnics K, Levitt P, Lewis DA. Critical appraisal of DNA microarrays in psychiatric genomics. *Biol Psychiatry*. 2006;60:163–176.
152. Thompson Ray M, Weickert CS, Wyatt E, Webster MJ. Decreased BDNF, trkB-TK+ and GAD67 mRNA expression in the hippocampus of individuals with schizophrenia and mood disorders. *J Psychiatry Neurosci*. 2011;36:195–203.
153. Wong CG, Bottiglieri T, Snead 3rd OC. GABA, gamma-hydroxybutyric acid, and neurological disease. *Ann Neurol*. 2003;54(Suppl 6):S3–12.

154. Wang Y, Fischle W, Cheung W, Jacobs S, Khorasanizadeh S, Allis CD. Beyond the double helix: writing and reading the histone code. *Novartis Found Symp*. 2004;259:3–17. [discussion 17–21, 163–169].
155. Addington AM, Gornick M, Duckworth J, et al. GAD1 (2q31.1), which encodes glutamic acid decarboxylase (GAD67), is associated with childhood-onset schizophrenia and cortical gray matter volume loss. *Mol Psychiatry*. 2005;10:581–588.
156. Straub RE, Lipska BK, Egan MF, Goldberg TE, Kleinman JE, Weinberger DR. Allelic variation in GAD1 (GAD67) is associated with schizophrenia and influences cortical function and gene expression. *Mol Psychiatry*. 2007;12:854–869.
157. Marenco S, Savostyanova AA, van der Veen JW, et al. Genetic modulation of GABA levels in the anterior cingulate cortex by GAD1 and COMT. *Neuropsychopharmacology*. 2010;35:1708–1717.
158. Cukier HN, Rabionet R, Konidari I, et al. Novel variants identified in methyl-CpG-binding domain genes in autistic individuals. *Neurogenetics*. 2010;11:291–303.
159. Li H, Yamagata T, Mori M, Yasuhara A, Momoi MY. Mutation analysis of methyl-CpG binding protein family genes in autistic patients. *Brain Dev*. 2005;27:321–325.
160. Schanen NC. Epigenetics of autism spectrum disorders. *Hum Mol Genet*. 2006;15(Spec No 2):R138–R150.
161. Williams SR, Aldred MA, Der Kaloustian VM, et al. Haploinsufficiency of HDAC4 causes brachydactyly mental retardation syndrome, with brachydactyly type E, developmental delays, and behavioral problems. *Am J Hum Genet*. 2010;87:219–228.
162. Balemans MC, Huibers MM, Eikelenboom NW, et al. Reduced exploration, increased anxiety, and altered social behavior: autistic-like features of euchromatin histone methyltransferase 1 heterozygous knockout mice. *Behav Brain Res*. 2010;208:47–55.
163. Adegbola A, Gao H, Sommer S, Browning M. A novel mutation in JARID1C/SMCX in a patient with autism spectrum disorder (ASD). *Am J Med Genet A*. 2008;146A:505–511.
164. Lim DA, Huang YC, Swigut T, et al. Chromatin remodelling factor Mll1 is essential for neurogenesis from postnatal neural stem cells. *Nature*. 2009;458:529–533.
165. Wynder C, Hakimi MA, Epstein JA, Shilatifard A, Shiekhattar R. Recruitment of MLL by HMG-domain protein iBRAF promotes neural differentiation. *Nat Cell Biol*. 2005;7:1113–1117.
166. Gupta S, Kim SY, Artis S, et al. Histone methylation regulates memory formation. *J Neurosci*. 2010;30:3589–3599.
167. Kim SY, Levenson JM, Korsmeyer S, Sweatt JD, Schumacher A. Developmental regulation of Eed complex composition governs a switch in global histone modification in brain. *J Biol Chem*. 2007;282:9962–9972.
168. Nakamine A, Ouchanov L, Jimenez P, et al. Duplication of 17(p11.2p11.2) in a male child with autism and severe language delay. *Am J Med Genet A*. 2008;146A:636–643.
169. Castermans D, Vermeesch JR, Fryns JP, et al. Identification and characterization of the TRIP8 and REEP3 genes on chromosome 10q21.3 as novel candidate genes for autism. *Eur J Hum Genet*. 2007;15:422–431.
170. Glessner JT, Wang K, Cai G, et al. Autism genome-wide copy number variation reveals ubiquitin and neuronal genes. *Nature*. 2009;459:569–573.
171. Bonora E, Lamb JA, Barnby G, et al. Mutation screening and association analysis of six candidate genes for autism on chromosome 7q. *Eur J Hum Genet*. 2005;13:198–207.
172. Petek E, Windpassinger C, Vincent JB, et al. Disruption of a novel gene (IMMP2L) by a breakpoint in 7q31 associated with Tourette syndrome. *Am J Hum Genet*. 2001;68:848–858.
173. Sakurai T, Ramoz N, Reichert JG, et al. Association analysis of the NrCAM gene in autism and in subsets of families with severe obsessive-compulsive or self-stimulatory behaviors. *Psychiatr Genet*. 2006;16:251–257
174. Melin M, Carlsson B, Anckarsater H, et al. Constitutional downregulation of SEMA5A expression in autism. *Neuropsychobiology*. 2006;54:64–69.
175. Weiss LA, Arking DE, Daly MJ, Chakravarti A. A genome-wide linkage and association scan reveals novel loci for autism. *Nature*. 2009;461:802–808.
176. Scheuerle A, Wilson K. PARK2 copy number aberrations in two children presenting with autism spectrum disorder: further support of an association and possible evidence for a new microdeletion/microduplication syndrome. *Am J Med Genet B Neuropsychiatr Genet*. 2011;156:413–420.
177. Kessler RC, Berglund P, Demler O, et al. The epidemiology of major depressive disorder: results from the national comorbidity survey replication (NCS-R). *J Am Med Assoc*. 2003;289:3095–3105.
178. Frazer A. Antidepressants. *J Clin Psychiatry*. 1997;58(Suppl 6):9–25.

179. Wong ML, Licinio J. From monoamines to genomic targets: a paradigm shift for drug discovery in depression. *Nat Rev Drug Discov*. 2004;3:136–151.
180. Fava M. Diagnosis and definition of treatment-resistant depression. *Biol Psychiatry*. 2003;53:649–659.
181. Trivedi MH. Treatment-resistant depression: new therapies on the horizon. *Ann Clin Psychiatry*. 2003;15: 59–70.
182. Breen G, Webb BT, Butler AW, et al. A genome-wide significant linkage for severe depression on chromosome 3: the depression network study. *Am J Psychiatry*. 2011;168:840–847.
183. Muglia P, Tozzi F, Galwey NW, et al. Genome-wide association study of recurrent major depressive disorder in two European case-control cohorts. *Mol Psychiatry*. 2010;15:589–601.
184. Green T, Gothelf D, Glaser B, et al. Psychiatric disorders and intellectual functioning throughout development in velocardiofacial (22q11.2 deletion) syndrome. *J Am Acad Child Adolesc Psychiatry*. 2009;48:1060–1068.
185. Porteous DJ, Thomson P, Brandon NJ, Millar JK. The genetics and biology of DISC1–an emerging role in psychosis and cognition. *Biol Psychiatry*. 2006;60:123–131.
186. Sherin JE, Nemeroff CB. Post-traumatic stress disorder: the neurobiological impact of psychological trauma. *Dialogues Clin Neurosci*. 2011;13:263–278.
187. Feder A, Nestler EJ, Charney DS. Psychobiology and molecular genetics of resilience. *Nat Rev Neurosci*. 2009;10:446–457.
188. Rusiecki JA, Chen L, Srikantan V, et al. DNA methylation in repetitive elements and post-traumatic stress disorder: a case-control study of US military service members. *Epigenomics*. 2012;4:29–40.
189. Smith AK, Conneely KN, Kilaru V, et al. Differential immune system DNA methylation and cytokine regulation in post-traumatic stress disorder. *Am J Med Genet B Neuropsychiatr Genet*. 2011;156B:700–708.
190. Uddin M, Aiello AE, Wildman DE, et al. Epigenetic and immune function profiles associated with posttraumatic stress disorder. *Proc Natl Acad Sci USA*. 2010;107:9470–9475.
191. McGowan PO, Sasaki A, D'Alessio AC, et al. Epigenetic regulation of the glucocorticoid receptor in human brain associates with childhood abuse. *Nat Neurosci*. 2009;12:342–348.
192. McGowan PO, Sasaki A, Huang TC, et al. Promoter-wide hypermethylation of the ribosomal RNA gene promoter in the suicide brain. *PLoS One*. 2008;3:e2085.
193. Ernst C, Chen ES, Turecki G. Histone methylation and decreased expression of TrkB.T1 in orbital frontal cortex of suicide completers. *Mol Psychiatry*. 2009;14:830–832.
194. Fiori LM, Turecki G. Genetic and epigenetic influences on expression of spermine synthase and spermine oxidase in suicide completers. *Int J Neuropsychopharmacol*. 2010;13:725–736.
195. Weaver IC, Cervoni N, Champagne FA, et al. Epigenetic programming by maternal behavior. *Nat Neurosci*. 2004;7:847–854.
196. Hollis F, Wang H, Dietz D, Gunjan A, Kabbaj M. The effects of repeated social defeat on long-term depressive-like behavior and short-term histone modifications in the hippocampus in male Sprague-Dawley rats. *Psychopharmacology (Berl)*. 2010;211:69–77.
197. Lin H, Geng X, Dang W, et al. Molecular mechanisms associated with the antidepressant effects of the class I histone deacetylase inhibitor MS-275 in the rat ventrolateral orbital cortex. *Brain Res*. 2012;1447:119–125.
198. Tsankova NM, Berton O, Renthal W, Kumar A, Neve RL, Nestler EJ. Sustained hippocampal chromatin regulation in a mouse model of depression and antidepressant action. *Nat Neurosci*. 2006;9:519–525.
199. Zhu H, Huang Q, Xu H, Niu L, Zhou JN. Antidepressant-like effects of sodium butyrate in combination with estrogen in rat forced swimming test: involvement of 5-HT(1A) receptors. *Behav Brain Res*. 2009;196:200–206.
200. Gundersen BB, Blendy JA. Effects of the histone deacetylase inhibitor sodium butyrate in models of depression and anxiety. *Neuropharmacology*. 2009;57:67–74.
201. Ferrante RJ, Kubilus JK, Lee J, et al. Histone deacetylase inhibition by sodium butyrate chemotherapy ameliorates the neurodegenerative phenotype in Huntington's disease mice. *J Neurosci*. 2003;23:9418–9427.
202. Kilgore M, Miller CA, Fass DM, et al. Inhibitors of class 1 histone deacetylases reverse contextual memory deficits in a mouse model of Alzheimer's disease. *Neuropsychopharmacology*. 2010;35:870–880.
203. Rane P, Shields J, Heffernan M, Guo Y, Akbarian S, King JA. The histone deacetylase inhibitor, sodium butyrate, alleviates cognitive deficits in pre-motor stage PD. *Neuropharmacology*. 2012;62:2409–2412.
204. Miskiewicz K, Jose LE, Bento-Abreu A, et al. ELP3 controls active zone morphology by acetylating the ELKS family member Bruchpilot. *Neuron*. 2011;72:776–788.
205. Dokmanovic M, Marks PA. Prospects: histone deacetylase inhibitors. *J Cell Biochem*. 2005;96:293–304.

206. Zhao W, Kruse JP, Tang Y, Jung SY, Qin J, Gu W. Negative regulation of the deacetylase SIRT1 by DBC1. *Nature*. 2008;451:587–590.
207. Libert S, Pointer K, Bell EL, et al. SIRT1 activates MAO-A in the brain to mediate anxiety and exploratory drive. *Cell*. 2011;147:1459–1472.
208. Cheung I, Shulha HP, Jiang Y, et al. Developmental regulation and individual differences of neuronal H3K4me3 epigenomes in the prefrontal cortex. *Proc Natl Acad Sci USA*. 2010;107:8824–8829.
209. Hernandez DG, Nalls MA, Gibbs JR, et al. Distinct DNA methylation changes highly correlated with chronological age in the human brain. *Hum Mol Genet*. 2011;20:1164–1172.
210. Akbarian S. Epigenetics of schizophrenia. *Curr Top Behav Neurosci*. 2010;4:611–628.
211. Gamazon ER, Badner JA, Cheng L, et al. Enrichment of cis-regulatory gene expression SNPs and methylation quantitative trait loci among bipolar disorder susceptibility variants. *Mol Psychiatry*. 2012. doi:10.1038/mp.2011.174 [Epub ahead of print].
212. Taqi MM, Bazov I, Watanabe H, et al. Prodynorphin CpG-SNPs associated with alcohol dependence: elevated methylation in the brain of human alcoholics. *Addict Biol*. 2011;16:499–509.

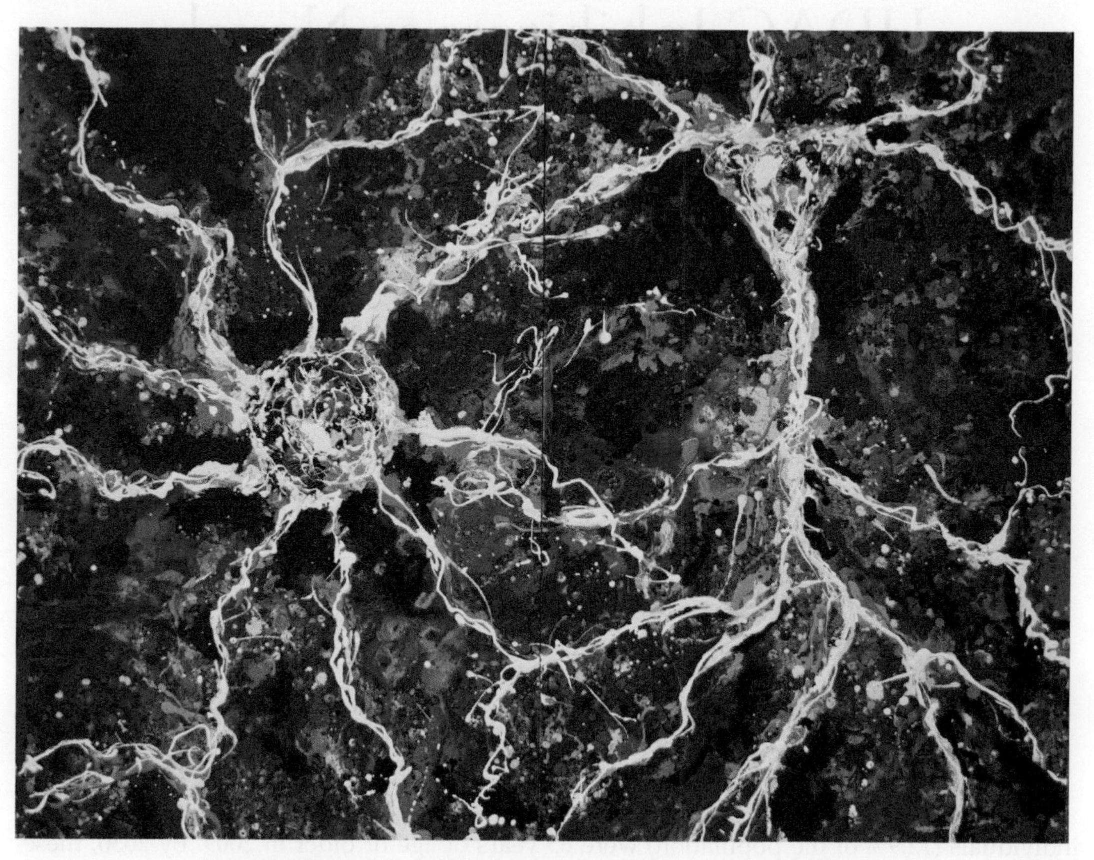

"HDAC Inhibitors and Novel Therapeutics – Granule Cell and Pyramidal Neuron"

J. David Sweatt, acrylic on wood panel (diptych of two 24 x 48 panels), 2011–2012

CHAPTER

8

HDAC Inhibitors as Novel Therapeutics in Aging and Alzheimer's Disease

Alexi Nott,[1] *Daniel M. Fass,*[2,3] *Stephen J. Haggarty*[2,3] *and Li-Huei Tsai*[1,3]

[1]Picower Institute for Learning and Memory, Massachusetts Institute of Technology, Howard Hughes Medical Institute, Cambridge, Massachusetts, USA

[2]Center for Human Genetic Research, Massachusetts General Hospital, Departments of Neurology and Psychiatry, Harvard Medical School, Boston, Massachusetts, USA

[3]Broad Institute of Harvard and MIT, 7 Cambridge Center, Cambridge, Massachusetts, USA

AGING, ALZHEIMER'S DISEASE, AND COGNITIVE DECLINE

The world's population is living longer than ever before. In the USA alone, 39.5 million individuals (12.9% of the population) were 65 years of age or older in 2009. By 2030, these numbers are predicted to increase to 71.2 million people, or 19% of the population (Statistics from Administration of Aging). A natural consequence of this increase in the number of elderly individuals will be an increase in the prevalence of age-associated cognitive decline and neurological disease.

The effects of aging on memory can vary greatly in severity. Mild impairment, termed age-associated memory impairment (AAMI), is considered "normal" in aged individuals.[1] AAMI is generally not debilitating, although it is often worrisome to affected individuals and their families. However, 13.9% of individuals aged 71 years or older will be affected by some form of dementia that will dramatically impair their cognitive abilities. The most severely affected individuals will suffer from profound neurodegeneration, one of the best-characterized and common forms of which is Alzheimer's disease.

Epigenetic Regulation in the Nervous System
DOI: http://dx.doi.org/10.1016/B978-0-12-391494-1.00008-2

Studies on humans have demonstrated that aged individuals with AAMI have cognitive impairments in working and spatial memory, a delayed recall of verbal information, and impaired short-term memory.[2–5] Age-associated cognitive decline has also been observed in other species, including non-human primates and rodents.[6] Imaging studies have indicated that, while the aged brain is structurally intact, the functional activities of the prefrontal cortex and hippocampus are altered.[7,8]

Alzheimer's disease (AD) is currently the most common cause of age-associated dementia, accounting for 70% of all dementia in Americans aged 71 years or older. Statistics for 2011 from the Alzheimer's Association estimate that 5.4 million people in the USA have AD, and this number is expected to swell to 7.7 million by 2030. AD is characterized histologically by extracellular β-amyloid protein plaque formation and intracellular neurofibrillary tangles composed of tau protein. These pathologies are accompanied by neuroinflammation of nearby glial cells and neuronal loss in the hippocampus, cortex, and other brain regions. The most severe clinical characteristic of AD is a progressive decline in declarative memory; the disease is ultimately fatal. Despite intensive studies, the pathogenesis of AD remains to be elucidated and effective therapies still await discovery. For these reasons, exploration of new approaches to understand both the cause and to develop effective new treatments for AD and related dementias are critically needed. In this context, studies on the basic molecular mechanisms of memory involving the epigenetic regulation of gene expression are beginning to suggest new avenues for reversing and potentially preventing memory loss and neurodegeneration.

GENE EXPRESSION DYSREGULATION ASSOCIATED WITH AGE-DEPENDENT MEMORY IMPAIRMENT

The formation of long-term memory is critically dependent upon activity-induced de novo gene expression.[9–13] Using inhibitors of transcription and translation, researchers have demonstrated that, immediately following an experience, a time window exists during which protein synthesis is required for the formation of long-term memory.[9] Since these initial studies, it has been demonstrated that neuronal activity-regulated gene transcription is important for both synapse development and cognitive function (for a recent review see[14]). Likewise, dysregulation of gene transcription has become associated with impairments in age-associated learning and memory.

Transcriptional profiling of cortical regions in humans indicates that aging-related alterations in gene transcription first become apparent after 40 years of age, and are most pronounced during the sixth and seventh decades of life.[15,16] Studies in both humans and primates suggest that only a select number of genes are downregulated in an age-dependent manner, and these changes in gene expression are specific to the cortex.[15,17] Genes downregulated in the aging human cortex are important for synaptic functions implicated in learning and memory, as well as in vesicle transport and mitochondrial functions.[15] Conversely, genes upregulated during aging are commonly associated with stress response pathways.[15] Interestingly, a more recent study on the aging human cortex has demonstrated an additional set of genes that exhibit sexually dimorphic changes in gene expression, which may have been obscured in previous studies. Males have a more robust downregulation of genes for energy production and protein synthesis/transport, whereas females have a more pronounced upregulation of immune-response genes.[16] Similar studies examining aging-related changes in gene transcription in mice and rats have demonstrated a downregulation of genes associated with synaptic and mitochondrial function, as well as with the DNA damage response.[18–20] Behavioral studies in aged rodents have correlated aging-dependent cognitive deficits in the hippocampus with changes in gene expression, including reduced levels of Arc in the CA1 region of the hippocampus.[21,22] These studies suggest that cognitive impairment in the aging mammalian brain is due to dysregulated expression of genes belonging to specific functional groups, such as synaptic proteins, thus pharmacological manipulation of gene transcription could provide a potential therapeutic strategy.

Several microarray studies using hippocampus and entorhinal cortex from the brains of AD patients demonstrated that AD pathology also correlates with changes in gene expression of surviving neurons.[23–25] These included downregulation of genes related to energy pathways and protein folding and, interestingly, upregulation of genes associated with glial transcription factors, epigenetic factors, and tumor suppressor genes.[23] Interestingly, the CK-p25 bi-transgenic mouse, which is used as a model of AD, exhibits a downregulation of genes important for neuroplasticity, echoing the microarray studies looking at age-associated cognitive decline. The CK-p25 mice express p25 under the control of the CaMKII promoter which has been implicated in various neurodegenerative diseases including AD. Expression of p25 can be induced by removing the mice from a doxycycline diet; following a 6-week induction of p25, severe synaptic and neuronal loss in the forebrain is observed together with impaired learning, and downregulation of learning and memory genes.[26,27] At 2 weeks of p25 induction in the CK-p25 mice, neuronal loss is not apparent and memory is not impaired; however, these mice have extensive DNA damage in the cortex and the hippocampus.[28] A microarray study performed at this 2-week induction revealed an upregulation of genes involved in cell-cycle entry and DNA damage response in the forebrain.[28] This implies that during the early stages of AD, when clinical symptoms may not yet be apparent, changes in gene transcription are already evident that are distinct from the dysregulated genes observed at later stages in AD progression. Future pharmacological therapies for aging and AD aimed at manipulating gene expression should therefore aim to enhance genes associated with both neuroplasticity and neuroprotection.

These studies suggest that, during aging and age-associated neurodegeneration, the brain exhibits changes in transcriptional profiles in a region-specific manner. In the next section, we will explore the evidence that epigenetic mechanisms may be responsible for dysregulated gene expression during aging and AD progression.

HISTONE ACETYLATION MARKS ASSOCIATED WITH MEMORY

The epigenetic regulation of gene expression can be mediated by the direct modification of DNA via the methylation of cytosine residues, or through modifications of the histone proteins that package DNA into chromatin. Post-translational modifications of histone proteins include acetylation, methylation, phosphorylation, ubiquitination, sumoylation, and ribosylation.[29] The acetylation of lysine residues (Lys; K) on histone proteins will be the main focus of this chapter. Acetylation of histones can regulate gene transcription by neutralizing the positive charge of the lysine residue, thereby reducing the interaction of the histone with the negatively-charged DNA and leading to chromatin relaxation. Alternatively, acetylated histone lysines can act as docking sites for bromo-domain containing and other chromatin-binding proteins, which allow the recruitment of complexes that promote further epigenetic modifications and regulation of gene transcription.

The concept that learning and memory lead to transient increases in histone acetylation has become well established in the past decade. Learning in rodents results in increased acetylation of histone H3 in the hippocampus,[30,31] prefrontal cortex,[32,33] and lateral amygdala.[34,35] Mice exposed to a novel enriched environment exhibit marked enhancements in associative and spatial learning that have been positively correlated with increased acetylation of histones H3 and H4 in the hippocampus and cortex.[36,37]

Aged mice show impairments in learning-induced histone acetylation increases in the brain. In a study comparing young (3-month-old) to aged (16-month-old) mice, no significant basal differences in histone acetylation were evident. However, following contextual fear-dependent learning, acetylation of H4K12 was elevated in the young, but not the aged, mouse brain.[38] This diminished learning-induced H4K12 acetylation in aged mice correlated with hippocampus-associated memory impairments.[38] A similar study was performed in the APP-PS1 transgenic mouse model of AD, which is characterized by the presence of highly fibrillogenic Aβ_{42} in the brain.[39,40] There were no differences in H4 acetylation between naïve APP-PS1 mice and wild-type controls; however, similar to the results from aged mice, induction of H4 acetylation was reduced following fear-dependent learning in the APP-PS1 mice.[41] Naïve aged Fischer 344 rats were shown to have lower levels of acetylated H3K9 and H4K14 compared to young controls, and this reduction in histone acetylation correlated with a decline in activity-dependent synaptic plasticity in the hippocampus, as measured by long-term potentiation (LTP) induction in brain slices.[42]

In addition to aging, decreases in histone acetylation have also been associated with cognitive impairment in a number of mouse models of age-dependent neurodegeneration, including Parkinson's disease,[43] Huntington disease[44] and AD.[23,36,45,46] An in vitro model of AD, in which cultured cortical neurons overexpress APP, shows decreased acetylation of histone H3.[47] Changes in histone acetylation have also been observed in vivo using two different transgenic mouse models of AD. The Tg2576 mouse expresses human APP containing a double mutation initially found in a Swedish family that increases accumulation of Aβ content in the brain after 6 months. This mouse exhibits spatial cognitive deficits that correlate with decreased levels of acetylated H4 in the cortex and hippocampus.[45] The APP-PS1 transgenic mouse also exhibits impaired cognition, together with decreased acetylation of H3K9, H3K14, and H4K5 (hippocampus and cortex), and H4K12 and H4K16 (hippocampus).[46] Discrepancies in the basal levels of acetylation in aged and transgenic mice may reflect differences in the age and the genetic background of the mice and may be specific for the acetylated histone marks being investigated in these studies. Despite strain- and/or model-dependent differences, overall, the data strongly indicate that aging and neurodegeneration-induced memory impairments correlate with reduced histone acetylation in the brain.

Interestingly, two studies have associated changes in certain histone acetylation marks with the activity of specific histone deacetylase (HDAC) enzymes within the hippocampus. An increase in acetylated H4K5 and H4K12 is linked with HDAC2 loss of function.[48]

Additionally, increased H4K8 acetylation occurs following the focal deletion of HDAC3.[49] These studies highlight the specificity of individual HDACs for histone targets within the epigenetic landscape in the brain, and further underscore the need for additional research to delineate the roles of individual HDACs in the context of histone acetylation and cognition.

HDAC PROTEINS AND COGNITION

The major enzymatic regulators of histone acetylation are the histone acetyltransferases (HATs), which add acetyl groups to lysine residues of histones; and the HDACs, which remove these acetyl groups and that will be the focus of this chapter. HDACs are grouped into four classes based on sequence homology and catalytic mechanism: class I (HDACs 1–3 and 8); class II (HDACs 4–7, 9 and 10); and class IV (HDAC 11) are zinc-dependent enzymes, whereas class III HDACs (SirT 1–7) require the co-factor NAD^+. General HDAC inhibitors against the class I, II and IV zinc-dependent HDAC family members were found to ameliorate effects associated with age-related cognitive decline and provided the first clue that HDAC activity may negatively regulate learning (for recent reviews see[50–53]).

Using both pharmacological and transgenic approaches, it was demonstrated that HDAC2 is a dominant HDAC isoform responsible for negatively regulating cognition.[27,54] Transgenic mice that overexpress HDAC2 had impaired memory formation, whereas HDAC2 knockout mice demonstrated enhanced learning in both contextual and tone-dependent fear conditioning learning paradigms.[54] Mechanistically, HDAC2 was found bound to chromatin at the promoters of memory-related genes, and to regulate negatively synapse formation as observed by changes in spine density and synaptophysin levels.[54] This effect was specific to HDAC2 and was not observed following overexpression of the highly homologous family member, HDAC1.[54]

HDAC2 has also been demonstrated to play a role in impairment of hippocampus-dependent cognition in aging. Expression of HDAC2 protein was elevated in the hippocampi of aged Fischer 344 rats.[42] This corresponded with a decrease in expression of CREB-binding protein (CBP), a HAT protein, reduced acetylation of histone H3 and H4, and a decline in hippocampal LTP.[42] This age-dependent increase in HDAC function was specific for HDAC2, as HDAC1 levels did not increase with aging. These data indicate the importance of HDAC2 as a negative regulator of cognition in the aged brain.[42]

The expression of HDAC2 is also dysregulated in both human AD patient brain tissue and the p25 transgenic mouse model of AD. In both cases, HDAC2, but not HDAC1 or HDAC3, is specifically upregulated in the CA1 region of the hippocampus. RNA interference (RNAi)-mediated reduction of HDAC2 levels in the hippocampus of p25 transgenic mice increased the expression of neuroplasticity genes and alleviated memory deficits associated with p25 activation.[27] These studies provide a strong indication that dysregulated overexpression of HDAC2 is a major epigenetic factor responsible for cognitive deficits in aging and neurodegenerative disease. These also have significant therapeutic implications for treating cognitive disorders as discussed below.

A recent publication has suggested that another class I HDAC member is important for spatial-learning. Focal deletion of HDAC3 in the dorsal hippocampus leads to enhanced novel object recognition memory and object localization memory, and was specific for long-term memory.[49] Interestingly, HDAC3 expression has been reported as being normal in the hippocampus of an AD-mouse model and in the brains of human AD patients.[27] However, it is possible that HDAC3 function may be dysregulated by some other mechanism in aging and AD, and therefore its role in age- and AD-related cognitive decline will need further study in the future. The notion though that multiple HDACs may have a function in regulating memory and neuroplasticity in a manner that is dependent upon the particular type of memory or cognitive impairment again has implications for therapeutic development.

In contrast to the memory-suppressing effects of HDAC2 and HDAC3, HDAC1 and a class III NAD^+-dependent sirtuin, SirT1, may support healthy cognition by virtue of their neuroprotective functions. In the CK-p25 mouse, overexpressed p25 is bound to HDAC1, decreasing its deacetylase activity. HDAC1 inhibition by p25 was associated with increased DNA damage and aberrant cell-cycle activity.[28] Similarly, focal overexpression of SirT1 in the hippocampal CA1 region, or treatment with the SirT1 activator resveratrol, reduced p25-induced neurotoxicity.[55] Thus, by preserving the integrity of neurons, HDAC1 and SirT1 may counteract cognitive decline in AD.

In addition to their neuroprotective functions, HDAC1 and SirT1 may also play direct roles in cognition. A recent publication demonstrated that HDAC1 has an important function in memory extinction.[56] Memory extinction is important for suppressing deleterious responses to stressful memories; a failure of memory extinction can lead to post-traumatic stress disorder. Overexpression of HDAC1 enhances the extinction of fear memories, whereas knock-down of HDAC1 perturbed fear extinction.[55] Interestingly, SirT1 has been recently shown to play a role in cognition by suppressing expression of the brain-specific microRNA miR-134, preventing miR-134-mediated downregulation of cAMP response binding protein (CREB).[57] Loss of function of SirT1 leads to impaired hippocampus-dependent memory, as well as reduced synaptic complexity.[57] These data indicate that the activities of HDAC1 and SirT1 are important for normal cognition.

Given that cognition is suppressed by some HDACs, and facilitated by others, it is clear that future pharmacological strategies for targeting HDACs for memory enhancement will ideally involve isoform-selective inhibitors. It is even possible to imagine a dual therapeutic approach in which inhibitors of memory-suppressing HDACs are administered in combination with activators of memory-facilitating HDACs.

CLASSES OF HDAC INHIBITORS

To date, small molecule HDAC inhibitors that have been studied for memory enhancing effects fall into three chemical structure classes: (1) carboxylic acids; (2) hydroxamic acids; and (3) *ortho*-amino anilides (Figure 8.1). Carboxylic acids, such as butyric acid, phenylbutyric acid (PBA), valproic acid (VPA) and pentyl-4n-valproic acid (PVA), generally consist of one or two branched alkyl chains terminating in a carboxylic acid. Crystallographic studies have shown that the hydroxamic acids, such as suberoylanilide hydroxamic acid (SAHA) and trichostatin A (TSA), consist of a cap region that interacts with the HDAC enzyme surface adjacent to the active site, a linker that extends into the active site, and a hydroxamic acid moiety that chelates the active site zinc.[58,59] *Ortho*-amino anilides, such as RGFP136 and MS-275, share this "cap-linker-biasing element" pharmacophore model,[60] and

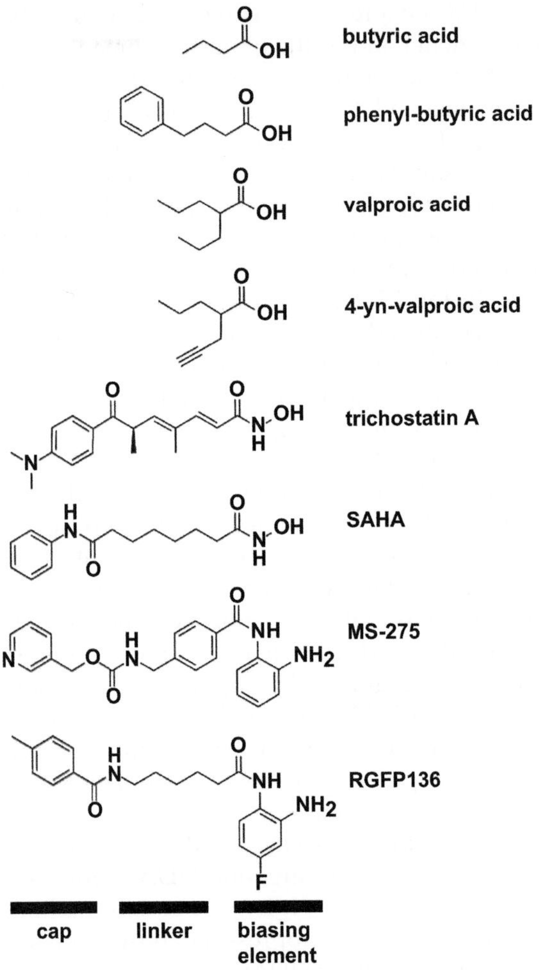

FIGURE 8.1 **Chemical structures of HDAC inhibitors.** Structures of carboxylic acids (butyric acid, phenyl-butyric acid, valproic acid, and 4-yn-valproic acid), hydroxamic acids (trichostatin A and SAHA), and *ortho*-amino anilides (MS-275 and RGFP136) are aligned with the HDAC catalytic site zinc-interacting biasing element on the right, adjacent to the linker and surface-interacting cap elements (if present) to the left.

they utilize an *ortho* position substituted aminophenyl group and anilide moiety to interact with the active site zinc and surrounding amino acid residues through hydrogen bonding.[61]

HDAC INHIBITOR SPECIFICITIES

All memory enhancing HDAC inhibitors target class I isoforms; hydroxamic acids additionally inhibit the class IIb enzyme HDAC6 and, at lower potency, the class I HDAC8 and the class IIa isoforms. Table 8.1 lists the potencies for inhibition of HDAC isoforms by each

TABLE 8.1 HDAC Inhibitor in vitro Potencies with Isoform Specific Substrates

			Class I			Class IIa		Class IIb
Type	**Compound**		**HDAC1**	**HDAC2**	**HDAC3**	**HDAC8**	**HDAC5**	**HDAC6**
Carboxylic acid	Butyric acid[62]	IC_{50}	8.3	7.0	4.8	10.4	>2000	>2000
	PBA[63]	IC_{50}	64	65	260	93	>2000	240
	Valproic acid[62]	IC_{50}	35.5	59.3	218.5	97.1	>2000	>2000
	PVA	–	nd	nd	nd	nd	nd	nd
Hydroxamic acid	SAHA[62]	IC_{50}	0.002	0.003	0.006	0.7	19.3	0.004
	TSA[64]	Ki	0.0002	0.0007	0.0005	0.045	0.26	0.001
Ortho-amino anilide	RGFP136[49]	IC_{50}	5.2	3.0	0.4	ni	ni	ni
	MS-275[64]	Ki	0.022	0.065	0.36	–	–	–

PBA = 4-phenyl-butyric acid; PVA = pentyl-4n-valproic acid; SAHA = suberoylanilide hydroxamic acid; TSA trichostatin A; IC50 = half maximal inhibitory concentration; Ki = dissociation constant; nd = not determined, ni = no inhibition.

inhibitor as determined using in vitro assays with recombinant enzymes. In general, carboxylic acids have low potency but only target class I enzymes; hydroxamic acids have the highest potency but the lowest isoform selectivity; *ortho*-amino anilides fall into an intermediate category with class I isoform selectivity and moderate potency. Additionally, *ortho*-amino anilides show slow-on/slow-off binding kinetics with their interaction with class I HDACs.[65] This property complicates efforts to determine their class I isoform selectivity; on the other hand, longer residence time with their targets could be an advantageous property, allowing for prolonged inhibition of HDAC-mediated deacetylation. Under conditions designed to promote binding equilibrium, RGFP136 is the only inhibitor currently disclosed that shows a tendency towards single HDAC isoform selectivity with preferential inhibition of HDAC3.[49]

HDAC INHIBITOR BRAIN PHARMACOKINETICS

Despite the growing number of behavioral studies, there is only a modest amount of published data reporting HDAC inhibitor brain pharmacokinetic properties. VPA and PVA do enter the brain after systemic administration, although robust mechanisms mediate their efflux and metabolism, limiting their concentration and residence time in the brain.[60,66,67] While butyric acid, SAHA, and TSA have not been measured directly in the brain after systemic delivery, they do induce histone acetylation increases in the brain.[68–70] Two studies report brain pharmacokinetic data for *ortho*-amino anilides. In a positron emission tomography study using carbon-11-labeled MS-275, Hooker and colleagues (2010) found that MS-275 has very poor brain penetration in both rodents and non-human primates.[71] McQuown and colleagues (2011) reported that RGFP136 enters the mouse brain following subcutaneous systemic delivery.[49] Further studies will be required fully to describe and quantify the time courses and dose–response relationships for HDAC inhibitor entry and residence in the brain following systemic administration.

As a consequence of the relative paucity of information regarding brain penetration of HDAC inhibitors (HDACis), some investigators carrying out memory studies in rodents administered with HDACis by direct brain infusion (for example[72]). This approach does allow precise localized administration of inhibitors to specific brain regions. However, the ability to use systemic administration, with confidence in inhibitor brain entry as guided by careful pharmacokinetic studies, would open up new options for experimental design, strengthen the interpretation of inhibitor effects as direct actions in the brain (as opposed to secondary effects mediated by actions on peripheral body tissues), and would provide a physiological condition that more closely mimics typical systemic therapeutic drug administration in humans.

HDAC INHIBITORS AND HISTONE ACETYLATION IN NEURONS

In the last decade, the administration of HDAC inhibitors in vivo has become a well-established paradigm for increasing histone acetylation levels in the brain (for references see Table 8.2). The hydroxamic acid inhibitors, TSA and SAHA, display general inhibition against class I and IIb HDACs, and administration to rodents leads to increased acetylation of histones H3K9, H3K14, H4K5 and H4K12 in the hippocampus, cortex, nucleus accumbens, and lateral amygdala (Table 8.2). SAHA has also been demonstrated to increase H2B acetylation in the hippocampus.[81] Administration of the class I-specific carboxylic acid-based HDACis, butyric acid, PBA and VPA lead to an increased acetylation of the same lysine residues of histone H3 and H4 in the hippocampus, prefrontal cortex and nucleus accumbens (Table 8.2). An increased acetylation of these specific histone residues has been correlated with increased gene expression, and indicates the importance of HDACs, in particular of class I HDACs, on gene transcription (for a recent review see[7]).

Interestingly, administration of the *ortho*-amino anilide, RGFP136, which preferentially inhibits HDAC3, caused an increase in acetylation of H4K8 in the brain. H4K8 has previously been demonstrated to be regulated by HDAC3 activity.[7] However, RGFP136 treatment also reduced the protein levels of HDAC4 (HDAC2 and HDAC3 levels were unchanged); thus, it is unclear whether HDAC3 inhibition, or HDAC4 downregulation, led to increased AcH4K8. In addition, the effects of RGFP136 on other acetylation marks remain to be examined.

HDAC INHIBITORS AND GENE TRANSCRIPTION IN NEURONS

Given that HDAC inhibitors enhance memory by altering gene transcription, an important goal has been the identification of the genes that are regulated by HDAC inhibitors in the brain. Ultimately, to characterize fully the mechanism of memory enhancement by HDAC inhibitors, it will be necessary to understand their effects on transcription, genome-wide, in all brain regions and cell types that mediate cognition. To date, several studies have begun the process of identifying gene expression changes induced by HDAC inhibitors in the brain; these are described below.

TABLE 8.2 Histone Lysine Residues Acetylated in the Presence of HDAC Inhibitors

	Acetylation Mark	Brain Region	Species	Ref
CARBOXYLIC ACIDS				
Butyric acid	H3K14; H4H5	Hippocampus	C57Bl6	36
	H3	Hippocampus/cortex	Swiss Mice post-ORM	31
	H3; H4	Hippocampus	C57Bl6	73
	H3K14	NAcc	C57Bl6/CPP	74
	H3K14	Cortex	Sprague-Dawley rats	75
	H4K12	Hippocampus slices	Chronic stress – Wistar rats	76
	H3K14; H4K12; H4K5	Hippocampus	C57Bl6/AP-PPS1	46
	H3K9	Prefrontal cortex	ICR (PCP-injected) mice	37
PBA	H3K14	Prefrontal cortex	C57Bl6/PQBP1 KD	33
	H3; H4	Cortex and hippocampus	C57Bl6 Tg2576 mice	45
	H3K9	NAcc, dorsal striatum, PFC	Sprague-Dawley rats/CPP	77
VPA	H3; H4	Hippocampus	Sprague-Dawley rats	78
	H4K5/K8/K12/K16	Prefrontal cortex	C57Bl6/WT	32
HYDROXAMIC ACID				
TSA	H3K14; H4K5/K8/K12/K16	Hippocampus slices	Sprague-Dawley rats	30
	H3	Hippocampus	C57Bl6	70
	H3	Hippocampus	C57Bl6	79
	H3	Hippocampus/cortex	Swiss mice post-ORM	31
	H4	Hippocampus	APP-PS1 mice	41
	H3K9; H3K14	Hippocampal culture	Sprague-Dawley rat	80
	H3K9; H4K12	Hippocampus	Aged Fischer 344 rat	42
	H3K9K14; H4K9	Lateral amygdala	Sprague-Dawley rat	34,35
	H3	Hippocampus	C57Bl6/ $Hdh^{Q7/Q111}$	44
SAHA	H2B	Hippocampus	C57Bl6	81
	H3K14	NAcc	C57Bl6	82
	H4K5	Hippocampus	C57Bl6	54
	H4	Hippocampus	B6C3F1J/APP-PS1	62
	H3; H4	Hippocampus	Sprague-Dawley rats	78
	H4K12	Hippocampus	Aged C57Bl6 after FC	38
OAA				
RGFP136	H4K8	Hippocampus	C57Bl6	49
MS-275	H3K14	NAcc	C57Bl6	82

OAA = ortho-amino anilide; PBA = 4-phenyl-butyric acid; VPA = valproic acid; PVA = pentyl-4n-valproic acid; SAHA = suberoylanilide hydroxamic acid; TSA = trichostatin A; NAcc = nucleus accumbens; PFC = prefrontal cortex; ORM = object recognition memory; CPP = conditioned place preference; WT = wild type; ICR = imprinting control region; PCP = phencyclidine; $Hdh^{Q7/Q111}$ = heterozygous Huntington's disease knock-in mice; FC = fear-conditioning; Tg2576 = amyloid precursor protein transgenic mice.

Administration of the carboxylic acid PBA results in increased histone acetylation and potentiated expression of the plasticity markers cFos, GluR1, PSD95, NMDAR1, and NF-κB p50.[33,45] An increase in the expression of learning-induced genes in the hippocampus has been demonstrated using another carboxylic acid, the class I-specific HDACi, butyric acid.[38,46] This includes an increase in Myst4, Fmn2, MarcksL1, Gsk3alpha, GluR1, Snap25, Shank3 and PrkcA gene expression and corresponds with global increases in histone H3 and H4 acetylation; with specific increases in acetylation of H4K12 demonstrated at the promoters of Fmn2 and PrkcA.[38] These studies using carboxylic acid inhibitors with specificity against class I HDACs indicate that attenuation of this subgroup of HDACs is sufficient and may be solely responsible for changes seen in gene expression related to learning and memory. Some light has already been shed on this question in a study using transgenic mice with either HDAC1 or HDAC2 knock-down and overexpression, together with the class I-specific inhibitor butyric acid, indicating that a predominant HDAC that negatively regulates transcription of genes important for learning and memory is HDAC2.[54]

Interestingly, direct infusion of MS-275, an *ortho*-amino anilide inhibitor with in vitro specificity for HDAC1, HDAC2, and HDAC3, increased H3K14 acetylation in the nucleus accumbens. Microarray analysis revealed that HDAC inhibition mediated by MS-275 regulates expression of genes important for cellular morphology (*SLIT*, *TGFalpha*, *p38*, *JNK*, and *Rho*), which the authors imply may be important for dendritic restructuring.[82] Pathways important for regulating gene transcription are also regulated by MS-275, including the transcription factors *CREB*, *ICER*, *REST*, *Co-REST*, *STAT* and *NFAT*.[82] This may also contribute to further changes in histone acetylation and gene expression.

In another key study, PBA increased the expression of the transcription-regulating factor, MeCP2;[77] a protein linked with Rett syndrome, Angelman syndrome, X-linked intellectual disability and autism, all of which are associated with cognitive impairments. A key gene negatively regulated by MeCP2 in a neuronal-activity dependent manner is brain-derived neurotrophic factor (BDNF) and overexpression of *BDNF* has been shown to rescue locomotor deficits in MeCP2 mutant mice. *BDNF* is associated with neuronal plasticity, and learning and memory. The HDACis TSA and VPA increased the activity of *BDNF*-regulated signaling pathways, including an increased expression of BDNF protein.[42] This corresponds with increased acetylation at the level of the *BDNF* promoter of H3K9 and H4K12 in the hippocampus and histone H4 in the prefrontal cortex.[32,42] *BDNF* is one of many immediate-early genes that are rapidly upregulated upon learning and are often used in studies as markers of neuronal plasticity.[7] It is therefore not surprising that an increase in the expression of the immediate-early genes, *cFos*, *Nr4a1* and *Nr4a2*, occurs following administration of HDACis to rodents and that this was coincidental with increased histone acetylation.[31,33,44,79] This indicates that HDACis are an effective mechanism for increasing the transcription of activity-dependent genes that may play an important role in memory formation.

The majority of studies to date have identified the affects of HDACis on histone acetylation and gene transcription in the hippocampus. More work needs to be done to identify transcriptional changes induced by HDACis in other brain regions relevant to memory, such as the cortex. These studies also focus on genes upregulated following HDACi administration. HDACis are known to downregulate about as many genes as they upregulate; future studies should therefore explore whether some key memory-suppressing genes are also downregulated by HDAC inhibitors.

HDAC INHIBITORS AND MEMORY FORMATION

The first observation that pharmacological inhibition of HDAC function had a beneficial effect on memory was performed using VPA, at a time when the ability of this drug to inhibit HDACs had not yet been discovered.[67] In the past decade, a plethora of data has been published documenting improvements in cognition with HDAC inhibitor treatment. The use of several different classes of HDAC is together with multiple memory-testing behavioral paradigms and animal models has established that HDAC activity plays an important role in negatively regulating memory formation (for a summary of the literature see Tables 8.3 and 8.4).

Fear conditioning is a well-defined behavioral paradigm frequently used for studying learning and memory in rodents. Contextual and cued fear conditioning tasks measure the ability of a rodent to learn an association of an aversive experience with environmental cues. The rodent is trained by pairing an auditory cue with exposure to a mild foot shock in an enclosed shock chamber, which causes it to cease movement ("freeze"). Twenty-four hours later, the rodent's ability to learn is measured using time spent freezing as a read out, when placed either into the same cage used to deliver the shock (contextual), or in a novel environment but using the same auditory stimulus given during training the day before (cued). The amygdala is the major brain region that mediates cued fear conditioned memory; the hippocampus, subiculum cortex, anterior cortex, prefrontal cortex, perirhinal cortex, sensory cortex, and medial temporal lobe have been demonstrated to mediate contextual fear conditioning (see review[83]). Improvements in both contextual and cued fear conditioning have been observed in rodents administered with the general hydroxamic acid-based HDACis, TSA and SAHA,[35,36,54,79,95] and using the class I selective HDACis, butyric acid and VPA.[30,54,84,92] Deficits in contextual and cued fear conditioned learning have been observed in APP-PS1, CK-p25, and Tg2576 mouse models of AD; these deficits can be ameliorated by administration of both general and class I-selective HDACis (see Tables 8.3 and 8.4,[36,41,46,62,96]). Studies have also demonstrated that HDAC inhibitors can reverse age-dependent cognitive decline in fear conditioned memory performance in wild-type mice.[38,96] Taken together, these studies demonstrate that HDACis may prove to be a useful therapeutic for a broad spectrum of age-associated decline in learning under both normal and neurodegenerative conditions.

Following fear conditioning, rodents can later be tested for their ability to extinguish bad memories, termed fear extinction. This is observed in mice as a reduction in freezing when the conditioned stimulus, such as context or tone, is repeatedly presented in the absence of the foot shock stimulus with which it has been previously paired. An improvement in fear extinction learning is therefore manifested as a decrease in time spent freezing. Wild-type mice demonstrate improved fear extinction following administration of HDACis, butyric acid, VPA and TSA.[32,85,92,93] Interestingly, MS-275, an HDACi with in vitro selectivity for the class I HDACs 1, 2 and 3, when administered at a high dose, has a negative effect on the ability of mice to extinguish fear memories.[56] A specificity for HDAC1 activity facilitating fear extinction is shown by focal overexpression and knock-down of HDAC1 in the hippocampus.[56] This study is an example of the differing role of class I HDACs on memory as also demonstrated by the negative regulation of HDAC2, and not HDAC1 on hippocampal-dependent fear conditioning,[54] although a suppressive role for HDAC2 in fear extinction has not yet been tested.

TABLE 8.3 Effects of Carboxylic Acid-Based HDAC Inhibitors on Cognition in Rodents

	Behavior	Strain	Drug Delivery	Phenotype	Ref
CARBOXYLIC ACID					
Butyric acid	FC (cont)	Sprague-Dawley rat	1.2 g/kg; ip 1 h pre-training	Increased freezing 24 h post-training	30
	FC (cont)	C57BL6 CK-p25/AD	1.2 g/kg; ip for 28 d pre-training	Increased freezing 24 h and 10 weeks post-training	36
	FC (cont)	Sprague-Dawley rat	1.2 g/kg; ihc 30 m pre-training	Increased freezing of 5-Aza treated rat post-training	84
	FC (cont)	C57Bl6/WT	1.2 g/kg; ip for 21 d pre-training	Increased freezing 24 h post-training	54
	FC (cont)	B6C3F1J APP-PS1/AD	1.2 g/kg; ip for 21 d pre-training	Increased freezing 24 h and 15 d post-training in Tg mice	62
	FC (cue)	C57Bl6/TBI	1.2 g/kg; ip for 7 d pre-training	Increased freezing	73
	FC (cont; cue)	C57Bl6 APP-PS1/AD	1.2 g/kg; ip for 42 d pre-training	Increased freezing in Tg mice 24 h post-training	46
	FE (cue)	C57Bl6/WT	1.2 g/kg; ip pre-training	Potentiated extinction of FC	85
	FE (cue)	C57Bl6/WT	1 g/kg; ip 2 h pre-testing	Potentiated extinction of FC	32
	ORM	Swiss-Webster/WT	250 mg/kg; ip 30 m pre-training	Enhanced object discrimination 24 h post-training	31
	ORM	C57Bl6/WT; CBP^{KIX}/RTS	1.2 g/kg; post-training	Enhanced object discrimination 24 h and 7 d post-training	86
	ORM	Sprague-Dawley rat	10 µg; 0.5 µl/side; iic post-training	Enhanced object discrimination 24 h post-training	75
	ORM	ICR (PCP-injected)	1 g/kg; ip daily 28 d pre-training	Enhanced object discrimination in PCP-injected mice	37
	ORM	C57Bl6/WT; CBP^{KIX}/RTS	1.2 g/kg; ip post-training	Enhanced object discrimination 24 h post-training	87
	ORM	Wistar rats (>8 m)/aging	1.2 g/kg; ip post-training	Enhanced object discrimination 24 h post-training	88
	ORM	SAMP-8/aging	250 mg/kg; ip 30 m pre-training	Enhanced object discrimination 1 h and 24 h post-training	31

(Continued)

TABLE 8.3 (Continued)

	Behavior	Strain	Drug Delivery	Phenotype	Ref
	OLM	C57Bl6/WT	1.2 g/kg; ip post-training	Enhanced object discrimination 24 h post-training	87
	WM	C57BL6 CK-p25/AD	1.2 g/kg; ip daily for 28 d	Reduced escape latency 24 h later	36
	WM	Wistar rat (STZ)/diabetes mellitus	100 mg/kg; ip daily for 26 d	Reduced escape latency of STZ treated mice	89
PBA	FC (cont; cue)	C57Bl6 Tg2576/AD; aging	200 mg/kg; ip daily for 35 d	Increased freezing 24 h post-training	90
	WM	C57Bl6 Tg2576/AD	200 mg/kg; ip daily for 35 d	Reduced escape latency; tested for 9 d	45
	WM	C57Bl6 Tg2576/AD	200 mg/kg; ip daily for 6 months	Reduced escape latency	91
Valproic acid	FC (cont)	B6C3F1J APP-PS1/AD	100 mg/kg; ip for 14 d pre-testing	Increased freezing in Tg mice 24 h post-training	62
	FC (cont)	C57Bl6/WT	100 mg/kg; ip 2 h pre-training	Increased freezing (FC); Enhanced reconsolidation	92
	FE	C57Bl6/WT	100 mg/kg; ip 2 h pre-training	Potentiated extinction of FC 24 h and 7 d post-training	92
	FE (cont)	ICR mice/WT	200 mg/kg; ip daily for 26 d	Increased freezing 2 d, 7 d and 14 d post-FE	93
	FE (cue)	C57Bl6/WT	100 mg/kg; ip 2 h pre-testing	Potentiated extinction of FC	32
	WM	Sprague-Dawley rat/TBI	400 mg/kg; ip daily post-TBI; 5 d	Reduced latency 24 h later in rats with TBI	78
PVA	WM	Wistar rat	16.8 mg/kg daily from P40 to P80	Reduced escape latency as measured on fifth trial	67
	WM	Wistar rat	84 mg/kg; ip 20 m pre-training (4 d)	Reduced escape latency during training; 72 h post-training	94
	WM	Wistar rat (>1.5 y)/aging	84 mg/kg; ip daily for 40 d	Reduced escape latency	109

PBA = 4-phenyl-butyric acid; PVA = pentyl-4n-valproic acid; FC = fear-conditioning; FE = fear extinction; cont = contextual; ORM = object recognition memory; OLM = object location memory; WM = water maze; AD = Alzheimer's disease; WT = wild type; RTS = Rubinstein–Taybi syndrome; PCP = phencyclidine; CBP = CREB binding protein; Tg2576 = amyloid precursor protein transgenic mice; TBI = traumatic brain injury; ip = intraperitoneal injection; icc = intra-cortical cannula; Tg = transgenic mice; 5-Aza = Azacitidine DNA methyltransferase inhibitor; STZ = streptozotocin.

TABLE 8.4 Effects of Hydroxamic Acid- and *Ortho*-Amino Anilide-Based HDAC Inhibitors on Cognition in Rodents

	Behavior	Strain	Drug Delivery	Phenotype	Ref
HYDROXAMIC ACID					
TSA	FC (cont)	C57Bl6/WT	50 ng side; icv immediately post-training	Increased freezing 24h post-training	36
	FC (cont)	C57Bl6/WT	16.5 mM, 1 μl per side; ihc post-training	Increased freezing 24 h post-training	79
	FC (cont)	APP-PS1/AD	2 mg/kg; ip 2 h pre-training	Increased freezing in Tg mice 24 h post-training	41
	FC (cont)	Sprague-Dawley rat	400 nM, 0.8 μl/side; iAM 30 m pre-training	Increased freezing 24 h later	95
	FC (cue)	Sprague-Dawley rat	1 μg; 0.5 μl/side; iLA 1 h post-training	Increased freezing 24 h post-training	35
	FE (cue)	C57Bl6/WT	22 mM, 0.5 μl/side; ihc pre-training	Potentiated extinction of FC	85
	FR (cue)	Sprague-Dawley rat	1 μg; 0.5 μl/side; iLA 1 h post-FR	Increased freezing 24 h post-FR	34
	ORM	C57Bl6 CBP^{HAT-}	2 mg/kg; ip 2 h pre-training	Enhanced object recognition 24 h post-training	70
	ORM	Swiss-Webster/WT	1 mg/kg; ip 30 m pre-training	Increased object discrimination 24 h post-training	31
	ORM	SAMP-8/aging	1 mg/kg; ip 30 m pre-training	Increased object discrimination 1 h and 24 h post-training	31
	ORM	C57Bl6 $Hdh^{Q7/Q111}$/HD	2 mg/kg; ip 2 h pre-training	Enhanced object discrimination 24 h post-training	44
	OLM	C57Bl6/WT	16.5 mM; 0.5 μl/side; ihc post-training	Enhanced object discrimination 24 h post-training	72
	EBCC	Swiss-Webster/WT	1 mg/kg; ip 30 m pre-training	Increased eyelid response	31

(*Continued*)

TABLE 8.4 (Continued)

	Behavior	Strain	Drug Delivery	Phenotype	Ref
SAHA	FC (cont)	C57Bl6 HDAC2OE	25 mg/kg; ip daily for 10 d pre-training	Increased freezing 24 h post-training. *Freezing time of untreated HDAC2OE mice is lower than WT littermates	54
	FC (cont)	B6C3F1J APP-PS1/AD	50 mg/kg; ip daily for 19 d pre-testing	Increased freezing 24 h post-training	62
	FC (cont)	C57Bl6 (16 m)/aging	10 μg/side; ihc 1 h pre-training	Increased freezing 24 h post-training	38
	FC (cont; cue)	C57Bl6/BALBc CBP+/−/RTS	10 μg/μl; 1 μl/side; icv 3 h pre-training	Increased freezing 24 h post-training	81
ORTHO-AMINO ANILIDE					
RGFP136	ORM	C57Bl6/WT	30 mg/kg; sc post-training	Enhanced discrimination 24 h and 7 d post-training	49
	OLM	C57Bl6/WT	30 mg/kg; sc post-training	Enhanced discrimination 24 h and 7 d post-training. *Effect of RGFP136 is abolished in CBP^{KIX} mice	49
MS-275	FE (cont)	C57Bl6/WT	750 ng/μl; 0.5 μl/side; ihc; post-FE trial	Prevents extinction of FC	56
	OLM	C57Bl6/WT	1 mM; 0.5 μl/side; ihc post-training	Enhanced object discrimination 24 h post-training	72

OAA = ortho-amino anilide; SAHA = suberoylanilide hydroxamic acid; TSA = trichostatin A; FC = fear-conditioning; FE = fear extinction; FR = fear reconsolidation; cont = contextual; ORM = object recognition memory; OLM = object location memory; EBCC = eye blink classical conditioning; AD = Alzheimer's disease; WT = wild type; RTS = Rubinstein–Taybi syndrome; HD = Huntington's disease; PCP = phencyclidine; CBP = CREB binding protein; icv = intracerebroventricular; ip = intraperitoneal injection; ihc = intrahippocampal cannula; sc = subcutaneous injection; iAM = intra-amygdala cannula; iLA = intra-lateral amygdala cannula; Tg = transgenic mice.

A deficit in fear extinction memory is exhibited in two different transgenic mouse models of AD, APP-PS1 and TASTPM. A deficit in fear extinction in these mice precedes other cognitive abnormalities and occurs prior to the deposition of amyloid plaques.[97–99] The authors have suggested that these findings parallel observations in patients with mild AD and may provide a diagnostic test for early stages of AD progression.[99] It will be of great interest to test whether HDAC inhibitors can reverse deficits in fear extinction in mouse models of AD.

The Water Maze task is a spatial learning navigation test in which mice are placed in a tank of opaque water, and are required to swim to a hidden platform using distant visual cues in the environment.[100,101] The most consistent finding is that lesions to the hippocampus lead to impaired memory acquisition in the Water Maze task; however, lesions to the medial septum and cortex have also been associated with impaired learning with this task (see review[83]). A dramatic impairment in Water Maze learning has been demonstrated in aged rats,[102–104] and also in amyloid overexpressing transgenic mouse models of AD.[105–108] Administration of the HDACi VPA has been shown to reverse impairments in Water Maze performance in 1.5-year-aged Wistar rats.[109] Similarly, improvements were observed in Tg2576 and CK-p25 transgenic mouse models of AD following administration of the HDACis, butyric acid and PBA.[36,45,91] HDACis have also been demonstrated to improve spatial cognition in rodent models of diabetes mellitus and following traumatic brain injury.[78,89] In sum, HDACi can ameliorate spatial learning deficits in rodent models due to aging, neurodegeneration, and other medical conditions affecting learning.

Novel object recognition (NOR) is a test of episodic memory based on the premise that rodents will explore a novel object more than a familiar one, but only if they remember the familiar one. In this test, animals are first habituated to two or more objects. Following habituation, two test protocols are possible: in the first protocol, one familiar object is replaced with a new visually distinct object to test object recognition memory (ORM); in the second protocol, one of the familiar objects is moved to a new location to test object location memory (OLM). Treatment with the HDACis TSA and SAHA enhances both ORM and OLM in wild-type rodents.[31,72,86,87] Similar observations were reported with the HDAC1 and HDAC2 specific inhibitor, MS-275, and the HDAC3-selective HDACi RGFP136,[49,72] suggesting that the memory formation process necessary for NOR is antagonized by class I HDAC activity.

Transgenic mice overexpressing SAMP-8, which display generalized accelerated aging, including cognitive impairments associated with the accumulation of protein aggregates,[110] have a deficit in ORM. This deficit can be ameliorated by administration of either butyric acid or TSA.[31] Butyric acid was also able to reverse aging-related deficits in ORM observed using 18-month aged Wistar rats.[88] TSA alleviated an ORM deficit in a Huntington's mouse model of neurodegeneration.[44] However, it remains to be seen whether the same positive effects can be achieved with transgenic models of AD.

HDACis can also reverse ORM impairments in several different knock-in transgenic mouse strains which feature loss of function of the histone acetyl transferase, CREB *b*inding *p*rotein (CBP).[70,86,87] Converse to HDAC function, CBP is recruited by DNA-binding transcription factors, such as CREB, to promote localized increases in acetylation of histones and nuclear factors culminating in gene transcription regulation. This suggests that some of the genes important for learning and memory that are activated by HDACis could be CREB/CBP target genes. This is corroborated by HDACi studies previously mentioned that have demonstrated an increase in histone acetylation at promoters of the *Bdnf* and *PrkcA* genes, both of which contain CREB binding sites,[32,38,42] as well as transcriptional upregulation of a number of CREB-regulated genes, including *Bdnf* and *cFos*.

The development of inhibitors with higher specificity and selectivity for individual HDACs will provide further insight into the contribution of HDAC isoforms on cognition. Some light has already been shed on this question through in vivo studies with knock-down and overexpression of individual HDACs, in combination with pharmacological approaches.[49,54,56] These studies indicate that the predominant HDAC that negatively regulates learning and memory is HDAC2, while HDAC3 is demonstrated negatively to regulate spatial memory. Conversely, the activity of the highly homologous HDAC1 may have beneficial effects on memory extinction.[54]

NON-HISTONE TARGETS OF HDAC INHIBITORS

The focus of this chapter has been on the epigenetic effects of HDACi on cognition in aging and AD. However, HDACs are known to deacetylate proteins other than histones, and the possibility of regulation of non-histone protein acetylation by HDACis should be considered both from a mechanistic point of view as well as for efforts aiming to identify unique biomarkers of HDAC inhibition. A prominent example of this is HDAC6, which is mainly cytoplasmic. HDAC6 expression is dysregulated in the brains of AD patients, with significantly higher protein levels in both the cortex and hippocampus.[111] Phospho-tau is a major component of neurofibrillary tangles in AD, and HDAC6 has been demonstrated to bind tau.[111] Administration of the HDAC6-specific inhibitor, tubacin, was able to attenuate the phosphorylation of tau and therefore potentially modulate its accumulation.[111] HDAC6 has been demonstrated to deacetylate Hsp90, augmenting the refolding of Hsp90 client proteins that includes tau. Loss of HDAC6 activity, using either HDAC6 siRNA or an analog of TSA, augments binding of Hsp90 with Hsp90 inhibitors and potentiates the degradation of tau in neurons.[112] The authors suggest that HDAC6 inhibition may provide an alternative approach for treating tauopathies such as AD; however, this question still remains to be addressed. The HDAC inhibitors TSA and SAHA both inhibit HDAC6, thus potentially regulating tau and other cytoplasmic targets. In addition, the effects of HDAC inhibitors on the acetylation of non-histone proteins is not limited to HDAC6 substrates; class I HDACs have been demonstrated to deacetylate a number of transcription factors and nuclear receptors (see review[113]). Non-histone targets of HDACis should therefore be considered in future studies assessing effects of HDACis on cognition.

FUTURE HDAC INHIBITOR DRUG DESIGN

An important future goal for novel HDAC inhibitor discovery will be to develop inhibitors with true single isoform specificity. All currently known HDAC inhibitors bind to the catalytic sites in these enzymes and are substrate competitive inhibitors. Due to the high degree of homology between the catalytic sites in the different isoforms (for example[114]), further efforts to optimize the selectivity of these inhibitors will likely achieve only limited, incremental improvements. This obstacle will likely be greatest for HDACs 1 and 2, which are almost completely homologous in their catalytic domains: human HDAC1 and 2 catalytic domains are 93% identical at the primary amino acid sequence level. Thus, to discover true single isoform-selective HDAC inhibitors, it will likely be necessary to find compounds capable of allosteric inhibition.

An alternative strategy for discovery of selective HDAC inhibitors could involve targeting HDAC complexes, rather than specific HDAC isoforms. HDAC complexes mediate specific functions through coordinated interactions with distinct effectors, with emerging evidence that different tissues may exhibit unique complexes. Thus, targeting specific HDAC complexes with small molecule inhibitors could allow for selective modulation of distinct chromatin-mediated functions in the brain. Small molecule inhibitors could target specific HDAC complexes in several different ways. First, the catalytic sites of HDAC isoforms might exist in unique, complex-dependent conformations due to the specific sets of binding interactions that mediate assembly of HDACs into each complex. Indeed, Bantscheff and colleagues have shown that catalytic site-binding HDAC inhibitors have different affinities for each HDAC complex.[115] Second, small molecule inhibitors could potentially inhibit HDAC complex function by disrupting complex integrity, rather than (and/or in addition to) blocking enzymatic deacetylase activity. The potential for discovery of this type of inhibitor effect was demonstrated by Smith and colleagues, who showed that SAHA specifically disrupts the interaction between ING2 and the SIN3 complex.[116] To discover novel HDAC inhibitors that function by targeting specific HDAC complexes, it will be necessary to utilize HDAC complexes, rather than recombinant free HDAC enzymes, in compound screening assays. Such HDAC complex assays could also be used to define further the functions of known catalytic site HDAC inhibitors in a more physiological context.

An indication of HDAC complex composition being regulated in neurons has been demonstrated with a physiological signaling molecule, nitric oxide. A study has demonstrated that HDAC2 can undergo post-translational modification by nitric oxide at two specific cysteine residues, termed nitrosylation. Nitrosylation of HDAC2 leads to its dissociation from CREB-regulated gene promoters, an elevation in acetylation of local histones, and increased gene transcription. Nitrosylation leads to a loss of function of HDAC2 without affecting its enzymatic activity. An attractive hypothesis is that nitric oxide disrupts the association of HDAC2 with repressive complexes found at these gene promoters. Interestingly, a recent study has demonstrated that knockout mice lacking neuronal nitric oxide synthase (nNOS), the major source of nitric oxide in neurons, have deficits in contextual fear-dependent learning. Furthermore, this deficit in learning can be rescued by administration of butyric acid.[117] It would therefore be desirable to find a small molecule inhibitor that acts like nitric oxide; for example disrupting specific HDAC2-containing complexes that are associated with regulating learning and memory genes, without completely obliterating the function of the protein. Therefore, future research will need to focus on defining which HDAC complexes are found within individual cell types to a resolution of gene-specific promoters, with the hope of one day developing therapeutics that will allow HDAC-containing complexes to be targeted that specifically modulate the enhancement of cognition and have beneficial effects in treating aging-related cognitive decline and neurodegeneration.

Acknowledgments

We would like to thank members of the Tsai and Haggarty laboratories for their helpful discussions regarding the role of epigenetic mechanisms in the brain. LHT is supported through funding from the NIH (R01DA028301, R01NS051874, NS078839), the Neurodegeneration Consortium, and the Howard Hughes Medical Institute. SJH is supported through funding from the NIH (R01DA028301, R01DA030321, NS078839), the Stanley Medical Research Institute, and the Tau Consortium.

References

1. Penner MR, Roth TL, Barnes CA, Sweatt JD. An epigenetic hypothesis of aging-related cognitive dysfunction. *Front Aging Neurosci*. 2010;2:9.
2. Albert M, Duffy FH, Naeser M. Nonlinear changes in cognition with age and their neuropsychologic correlates. *Can J Psychol*. 1987;41(2):141–157.
3. Craik FI, Moscovitch M, McDowd JM. Contributions of surface and conceptual information to performance on implicit and explicit memory tasks. *J Exp Psychol Learn Mem Cogn*. 1994;20(4):864–875.
4. Zelinski EM, Burnight KP. Sixteen-year longitudinal and time lag changes in memory and cognition in older adults. *Psychol Aging*. 1997;12(3):503–513.
5. Petersen RC, Smith G, Kokmen E, Ivnik RJ, Tangalos EG. Memory function in normal aging. *Neurology*. 1992;42(2):396–401.
6. Loerch PM, Lu T, Dakin KA, et al. Evolution of the aging brain transcriptome and synaptic regulation. *PLoS One*. 2008;3(10):e3329.
7. Wang D, Xia X, Weiss RE, Refetoff S, Yen PM. Distinct and histone-specific modifications mediate positive versus negative transcriptional regulation of TSHalpha promoter. *PLoS One*. 2010;5(3):e9853.
8. Burke SN, Barnes CA. Neural plasticity in the ageing brain. *Nat Rev Neurosci*. 2006;7(1):30–40.
9. Davis HP, Squire LR. Protein synthesis and memory: a review. *Psychol Bull*. 1984;96(3):518–559.
10. Frey U, Krug M, Reymann KG, Matthies H. Anisomycin an inhibitor of protein synthesis, blocks late phases of LTP phenomena in the hippocampal CA1 region in vitro. *Brain Res*. 1988;452(1–2):57–65.
11. Kandel ER. The molecular biology of memory storage: a dialog between genes and synapses. *Biosci Rep*. 2001;21(5):565–611.
12. Barco A, Lopez de Armentia M, Alarcon JM. Synapse-specific stabilization of plasticity processes: the synaptic tagging and capture hypothesis revisited 10 years later. *Neurosci Biobehav Rev*. 2008;32(4):831–851.
13. Gold PE. Protein synthesis inhibition and memory: formation vs amnesia. *Neurobiol Learn Mem*. 2008;89(3):201–211.
14. West AE, Greenberg ME. Neuronal activity-regulated gene transcription in synapse development and cognitive function. *Cold Spring Harb Perspect Biol*. 2011;3(6):1–21.
15. Lu T, Pan Y, Kao SY, et al. Gene regulation and DNA damage in the ageing human brain. *Nature*. 2004;429(6994):883–891.
16. Berchtold NC, Cribbs DH, Coleman PD, et al. Gene expression changes in the course of normal brain aging are sexually dimorphic. *Proc Natl Acad Sci USA*. 2008;105(40):15605–15610.
17. Fraser HB, Khaitovich P, Plotkin JB, Paabo S, Eisen MB. Aging and gene expression in the primate brain. *PLoS Biol*. 2005;3(9):e274.
18. Jiang CH, Tsien JZ, Schultz PG, Hu Y. The effects of aging on gene expression in the hypothalamus and cortex of mice. *Proc Natl Acad Sci USA*. 2001;98(4):1930–1934.
19. Blalock EM, Chen KC, Sharrow K, et al. Gene microarrays in hippocampal aging: statistical profiling identifies novel processes correlated with cognitive impairment. *J Neurosci*. 2003;23(9):3807–3819.
20. Xu X, Zhan M, Duan W, et al. Gene expression atlas of the mouse central nervous system: impact and interactions of age, energy intake and gender. *Genome Biol*. 2007;8(11):R234.
21. Verbitsky M, Yonan AL, Malleret G, Kandel ER, Gilliam TC, Pavlidis P. Altered hippocampal transcript profile accompanies an age-related spatial memory deficit in mice. *Learn Mem*. 2004;11(3):253–260.
22. Penner MR, Roth TL, Chawla MK, et al. Age-related changes in Arc transcription and DNA methylation within the hippocampus. *Neurobiol Aging*. 2011;32(12):2198–2210.

23. Blalock EM, Geddes JW, Chen KC, Porter NM, Markesbery WR, Landfield PW. Incipient Alzheimer's disease: microarray correlation analyses reveal major transcriptional and tumor suppressor responses. *Proc Natl Acad Sci USA*. 2004;101(7):2173–2178.
24. Small SA, Kent K, Pierce A, et al. Model-guided microarray implicates the retromer complex in Alzheimer's disease. *Ann Neurol*. 2005;58(6):909–919.
25. Blalock EM, Buechel HM, Popovic J, Geddes JW, Landfield PW. Microarray analyses of laser-captured hippocampus reveal distinct gray and white matter signatures associated with incipient Alzheimer's disease. *J Chem Neuroanat*. 2011;42(2):118–126.
26. Fischer A, Sananbenesi F, Pang PT, Lu B, Tsai LH. Opposing roles of transient and prolonged expression of p25 in synaptic plasticity and hippocampus-dependent memory. *Neuron*. 2005;48(5): 825–838.
27. Graff J, Rei D, Guan JS, et al. An epigenetic blockade of cognitive functions in the neurodegenerating brain. *Nature*. 2012;483(7388):222–226.
28. Kim D, Frank CL, Dobbin MM, et al. Deregulation of HDAC1 by p25/Cdk5 in Neurotoxicity. *Neuron*. 2008;60(5):803–817.
29. Suganuma T, Workman JL. Signals and combinatorial functions of histone modifications. *Annu Rev Biochem*. 2011;80:473–499.
30. Levenson JM, O'Riordan KJ, Brown KD, Trinh MA, Molfese DL, Sweatt JD. Regulation of histone acetylation during memory formation in the hippocampus. *J Biol Chem*. 2004;279(39):40545–40559.
31. Fontan-Lozano A, Romero-Granados R, Troncoso J, Munera A, Delgado-Garcia JM, Carrion AM. Histone deacetylase inhibitors improve learning consolidation in young and in KA-induced-neurodegeneration and SAMP-8-mutant mice. *Mol Cell Neurosci*. 2008;39(2):193–201.
32. Bredy TW, Wu H, Crego C, Zellhoefer J, Sun YE, Barad M. Histone modifications around individual BDNF gene promoters in prefrontal cortex are associated with extinction of conditioned fear. *Learn Mem*. 2007;14(4):268–276.
33. Ito H, Yoshimura N, Kurosawa M, Ishii S, Nukina N, Okazawa H. Knock-down of PQBP1 impairs anxiety-related cognition in mouse. *Hum Mol Genet*. 2009;18(22):4239–4254.
34. Maddox SA, Schafe GE. Epigenetic alterations in the lateral amygdala are required for reconsolidation of a Pavlovian fear memory. *Learn Mem*. 2011;18(9):579–593.
35. Monsey MS, Ota KT, Akingbade IF, Hong ES, Schafe GE. Epigenetic alterations are critical for fear memory consolidation and synaptic plasticity in the lateral amygdala. *PLoS One*. 2011;6(5):e19958.
36. Fischer A, Sananbenesi F, Wang X, Dobbin M, Tsai L-H. Recovery of learning and memory is associated with chromatin remodelling. *Nature*. 2007;447(7141):178–182.
37. Koseki T, Mouri A, Mamiya T, et al. Exposure to enriched environments during adolescence prevents abnormal behaviours associated with histone deacetylation in phencyclidine-treated mice. *Int J Neuropsychopharmacol*. 2011:1–13.
38. Peleg S, Sananbenesi F, Zovoilis A, et al. Altered histone acetylation is associated with age-dependent memory impairment in mice. *Science*. 2010;328(5979):753–756.
39. Radde R, Bolmont T, Kaeser SA, et al. Abeta42-driven cerebral amyloidosis in transgenic mice reveals early and robust pathology. *EMBO Rep*. 2006;7(9):940–946.
40. Gengler S, Hamilton A, Holscher C. Synaptic plasticity in the hippocampus of a APP/PS1 mouse model of Alzheimer's disease is impaired in old but not young mice. *PLoS One*. 2010;5(3):e9764.
41. Francis YI, Fa M, Ashraf H, et al. Dysregulation of histone acetylation in the APP/PS1 mouse model of Alzheimer's disease. *J Alzheimers Dis*. 2009;18(1):131–139.
42. Zeng Y, Tan M, Kohyama J, et al. Epigenetic enhancement of BDNF signaling rescues synaptic plasticity in aging. *J Neurosci*. 2011;31(49):17800–17810.
43. Kontopoulos E, Parvin JD, Feany MB. Alpha-synuclein acts in the nucleus to inhibit histone acetylation and promote neurotoxicity. *Hum Mol Genet*. 2006;15(20):3012–3023.
44. Giralt A, Puigdellivol M, Carreton O, et al. Long-term memory deficits in Huntington's disease are associated with reduced CBP histone acetylase activity. *Hum Mol Genet*. 2012;21(6):1203–1216.
45. Ricobaraza A, Cuadrado-Tejedor M, Perez-Mediavilla A, Frechilla D, Del Rio J, Garcia-Osta A. Phenylbutyrate ameliorates cognitive deficit and reduces tau pathology in an Alzheimer's disease mouse model. *Neuropsychopharmacology*. 2009;34(7):1721–1732.

46. Govindarajan N, Agis-Balboa RC, Walter J, Sananbenesi F, Fischer A. Sodium butyrate improves memory function in an Alzheimer's disease mouse model when administered at an advanced stage of disease progression. *J Alzheimers Dis*. 2011;26(1):187–197.
47. Rouaux C, Jokic N, Mbebi C, Boutillier S, Loeffler JP, Boutillier AL. Critical loss of CBP/p300 histone acetylase activity by caspase-6 during neurodegeneration. *EMBO J*. 2003;22(24):6537–6549.
48. Guan JS, Haggarty SJ, Giacometti E, et al. HDAC2 negatively regulates memory formation and synaptic plasticity. *Nature*. 2009;459(7243):55–60.
49. McQuown SC, Barrett RM, Matheos DP, et al. HDAC3 is a critical negative regulator of long-term memory formation. *J Neurosci*. 2011;31(2):764–774.
50. Fischer A, Sananbenesi F, Mungenast A, Tsai LH. Targeting the correct HDAC(s) to treat cognitive disorders. *Trends Pharmacol Sci*. 2010;31(12):605–617.
51. Haggarty SJ, Tsai LH. Probing the role of HDACs and mechanisms of chromatin-mediated neuroplasticity. *Neurobiol Learn Mem*. 2011;96(1):41–52.
52. Graff J, Kim D, Dobbin MM, Tsai LH. Epigenetic regulation of gene expression in physiological and pathological brain processes. *Physiol Rev*. 2011;91(2):603–649.
53. Stilling RM, Fischer A. The role of histone acetylation in age-associated memory impairment and Alzheimer's disease. *Neurobiol Learn Mem*. 2011;96(1):19–26.
54. Guan JS, Haggarty SJ, Giacometti E, et al. HDAC2 negatively regulates memory formation and synaptic plasticity. *Nature*. 2009;459(7243):55–60.
55. Kim D, Nguyen MD, Dobbin MM, et al. SIRT1 deacetylase protects against neurodegeneration in models for Alzheimer's disease and amyotrophic lateral sclerosis. *EMBO J*. 2007;26(13):3169–79.
56. Bahari-Javan S, Maddalena A, Kerimoglu C, et al. HDAC1 regulates fear extinction in mice. *J Neurosci*. 2012;32(15):5062–5073.
57. Gao J, Wang WY, Mao YW, et al. A novel pathway regulates memory and plasticity via SIRT1 and miR-134. *Nature*. 2010;466(7310):1105–1109.
58. Finnin MS, Donigian JR, Cohen A, et al. Structures of a histone deacetylase homologue bound to the TSA and SAHA inhibitors. *Nature*. 1999;401(6749):188–193.
59. Vannini A, Volpari C, Filocamo G, et al. Crystal structure of a eukaryotic zinc-dependent histone deacetylase, human HDAC8, complexed with a hydroxamic acid inhibitor. *Proc Natl Acad Sci USA*. 2004;101(42):15064–15069.
60. Sternson SM, Wong JC, Grozinger CM, Schreiber SL. Synthesis of 7200 small molecules based on a substructural analysis of the histone deacetylase inhibitors trichostatin and trapoxin. *Organic letters*. 2001;3(26):4239–4242.
61. Bressi JC, Jennings AJ, Skene R, et al. Exploration of the HDAC2 foot pocket: synthesis and SAR of substituted N-(2-aminophenyl)benzamides. *Bioorg Med Chem Lett*. 2010;20(10):3142–3145.
62. Kilgore M, Miller CA, Fass DM, et al. Inhibitors of class 1 histone deacetylases reverse contextual memory deficits in a mouse model of Alzheimer's disease. *Neuropsychopharmacology*. 2010;35(4):870–880.
63. Fass DM, Shah R, Ghosh B, et al. Effect of inhibiting histone deacetylase with short-chain carboxylic acids and their hydroxamic acid analogs on vertebrate development and neuronal chromatin. *ACS Med Chem Lett*. 2010;2(1):39–42.
64. Bradner JE, Mak R, Tanguturi SK, et al. Chemical genetic strategy identifies histone deacetylase 1 (HDAC1) and HDAC2 as therapeutic targets in sickle cell disease. *Proc Natl Acad Sci USA*. 2010;107(28):12617–12622.
65. Chou CJ, Herman D, Gottesfeld JM. Pimelic diphenylamide 106 is a slow, tight-binding inhibitor of class I histone deacetylases. *J Biol Chem*. 2008;283(51):35402–35409.
66. Cornford EM, Diep CP, Pardridge WM. Blood-brain barrier transport of valproic acid. *J Neurochem*. 1985;44(5):1541–1550.
67. Murphy KJ, Fox GB, Foley AG, et al. Pentyl-4-yn-valproic acid enhances both spatial and avoidance learning, and attenuates age-related NCAM-mediated neuroplastic decline within the rat medial temporal lobe. *J Neurochem*. 2001;78(4):704–714.
68. Ferrante RJ, Kubilus JK, Lee J, et al. Histone deacetylase inhibition by sodium butyrate chemotherapy ameliorates the neurodegenerative phenotype in Huntington's disease mice. *J Neurosci*. 2003;23(28):9418–9427.
69. Hockly E, Richon VM, Woodman B, et al. Suberoylanilide hydroxamic acid, a histone deacetylase inhibitor, ameliorates motor deficits in a mouse model of Huntington's disease. *Proc Natl Acad Sci USA*. 2003;100(4):2041–2046.

70. Korzus E, Rosenfeld MG, Mayford M. CBP histone acetyltransferase activity is a critical component of memory consolidation. *Neuron*. 2004;42(6):961–972.
71. Hooker JM, Kim SW, Alexoff D, et al. Histone deacetylase inhibitor, MS-275, exhibits poor brain penetration: PK studies of [C]MS-275 using Positron emission tomography. *ACS Chem Neurosci*. 2010;1(1): 65–73.
72. Hawk JD, Florian C, Abel T. Post-training intrahippocampal inhibition of class I histone deacetylases enhances long-term object-location memory. *Learn Mem*. 2011;18(6):367–370.
73. Dash PK, Orsi SA, Moore AN. Histone deactylase inhibition combined with behavioral therapy enhances learning and memory following traumatic brain injury. *Neuroscience*. 2009;163(1):1–8.
74. Malvaez M, Sanchis-Segura C, Vo D, Lattal KM, Wood MA. Modulation of chromatin modification facilitates extinction of cocaine-induced conditioned place preference. *Biol Psychiatry*. 2010;67(1):36–43.
75. Roozendaal B, Hernandez A, Cabrera SM, et al. Membrane-associated glucocorticoid activity is necessary for modulation of long-term memory via chromatin modification. *J Neurosci*. 2010;30(14):5037–5046.
76. Ferland CL, Schrader LA. Regulation of histone acetylation in the hippocampus of chronically stressed rats: a potential role of sirtuins. *Neuroscience*. 2011;174:104–114.
77. Pastor V, Host L, Zwiller J, Bernabeu R. Histone deacetylase inhibition decreases preference without affecting aversion for nicotine. *J Neurochem*. 2011;116(4):636–645.
78. Dash PK, Orsi SA, Zhang M, et al. Valproate administered after traumatic brain injury provides neuroprotection and improves cognitive function in rats. *PLoS One*. 2010;5(6):e11383.
79. Vecsey CG, Hawk JD, Lattal KM, et al. Histone deacetylase inhibitors enhance memory and synaptic plasticity via CREB:CBP-dependent transcriptional activation. *J Neurosci*. 2007;27(23):6128–6140.
80. Tian F, Marini AM, Lipsky RH. Effects of histone deacetylase inhibitor Trichostatin A on epigenetic changes and transcriptional activation of Bdnf promoter 1 by rat hippocampal neurons. *Ann N Y Acad Sci*. 2010;1199:186–193.
81. Alarcon JM, Malleret G, Touzani K, et al. Chromatin acetylation, memory, and LTP are impaired in CBP+/- mice: a model for the cognitive deficit in Rubinstein-Taybi syndrome and its amelioration. *Neuron*. 2004;42(6):947–959.
82. Covington III HE, Maze I, LaPlant QC, et al. Antidepressant actions of histone deacetylase inhibitors. *J Neurosci*. 2009;29(37):11451–11460.
83. Crawley JN. *What's Wrong with my Mouse? Behavioral Phenotyping of Transgenic and Knockout Mice*, 2nd ed. Hoboken, NJ: Wiley-Interscience; 2007.
84. Miller CA, Campbell SL, Sweatt JD. DNA methylation and histone acetylation work in concert to regulate memory formation and synaptic plasticity. *Neurobiol Learn Mem*. 2008;89(4):599–603.
85. Lattal KM, Barrett RM, Wood MA. Systemic or intrahippocampal delivery of histone deacetylase inhibitors facilitates fear extinction. *Behav Neurosci*. 2007;121(5):1125–1131.
86. Stefanko DP, Barrett RM, Ly AR, Reolon GK, Wood MA. Modulation of long-term memory for object recognition via HDAC inhibition. *Proc Natl Acad Sci USA*. 2009;106(23):9447–9452.
87. Haettig J, Stefanko DP, Multani ML, Figueroa DX, McQuown SC, Wood MA. HDAC inhibition modulates hippocampus-dependent long-term memory for object location in a CBP-dependent manner. *Learn Mem*. 2011;18(2):71–79.
88. Reolon GK, Maurmann N, Werenicz A, et al. Posttraining systemic administration of the histone deacetylase inhibitor sodium butyrate ameliorates aging-related memory decline in rats. *Behav Brain Res*. 2011;221(1):329–332.
89. Sharma B, Singh N. Attenuation of vascular dementia by sodium butyrate in streptozotocin diabetic rats. *Psychopharmacology (Berl)*. 2011;215(4):677–687.
90. Ricobaraza A, Cuadrado-Tejedor M, Marco S, Pérez-Otaño I, García-Osta A. Phenylbutyrate rescues dendritic spine loss associated with memory deficits in a mouse model of Alzheimer disease. *Hippocampus*. 2012;22(5):1040–1050.
91. Ricobaraza A, Cuadrado-Tejedor M, Garcia-Osta A. Long-term phenylbutyrate administration prevents memory deficits in Tg2576 mice by decreasing Abeta. *Front Biosci (Elite Ed)*. 2011;3:1375–1384.
92. Bredy TW, Barad M. The histone deacetylase inhibitor valproic acid enhances acquisition, extinction, and reconsolidation of conditioned fear. *Learn Mem*. 2008;15(1):39–45.
93. Li S, Murakami Y, Wang M, Maeda K, Matsumoto K. The effects of chronic valproate and diazepam in a mouse model of posttraumatic stress disorder. *Pharmacol Biochem Behav*. 2006;85(2):324–331.

94. O'Loinsigh ED, Gherardini LM, Gallagher HC, Foley AG, Murphy KJ, Regan CM. Differential enantioselective effects of pentyl-4-yn-valproate on spatial learning in the rat, and neurite outgrowth and cyclin D3 expression in vitro. *J Neurochem*. 2004;88(2):370–379.
95. Yeh SH, Lin CH, Gean PW. Acetylation of nuclear factor-kappaB in rat amygdala improves long-term but not short-term retention of fear memory. *Mol Pharmacol*. 2004;65(5):1286–1292.
96. Ricobaraza A, Cuadrado-Tejedor M, Marco S, Perez-Otano I, Garcia-Osta A. Phenylbutyrate rescues dendritic spine loss associated with memory deficits in a mouse model of Alzheimer disease. *Hippocampus*. 2012;22(5):1040–1050.
97. Rattray I, Scullion GA, Soulby A, Kendall DA, Pardon MC. The occurrence of a deficit in contextual fear extinction in adult amyloid-over-expressing TASTPM mice is independent of the strength of conditioning but can be prevented by mild novel cage stress. *Behav Brain Res*. 2009;200(1):83–90.
98. Rattray I, Pitiot A, Lowe J, et al. Novel cage stress alters remote contextual fear extinction and regional T2 magnetic resonance relaxation times in TASTPM mice overexpressing amyloid. *J Alzheimers Dis*. 2010;20(4):1049–1068.
99. Bonardi C, de Pulford F, Jennings D, Pardon MC. A detailed analysis of the early context extinction deficits seen in APPswe/PS1dE9 female mice and their relevance to preclinical Alzheimer's disease. *Behav Brain Res*. 2011;222(1):89–97.
100. Morris RGM. Spatial localization does not require the presence of local cues. *Learn Motivat*. 1981;12(2):239–260.
101. Morris R. Developments of a water-maze procedure for studying spatial learning in the rat. *J Neurosci Methods*. 1984;11(1):47–60.
102. Gallagher M, Burwell R, Burchinal M. Severity of spatial learning impairment in aging: development of a learning index for performance in the Morris water maze. *Behav Neurosci*. 1993;107(4):618–626.
103. Chouinard ML, Gallagher M, Yasuda RP, Wolfe BB, McKinney M. Hippocampal muscarinic receptor function in spatial learning-impaired aged rats. *Neurobiol Aging*. 1995;16(6):955–963.
104. Geinisman Y, Ganeshina O, Yoshida R, Berry RW, Disterhoft JF, Gallagher M. Aging, spatial learning, and total synapse number in the rat CA1 stratum radiatum. *Neurobiol Aging*. 2004;25(3):407–416.
105. Moran PM, Higgins LS, Cordell B, Moser PC. Age-related learning deficits in transgenic mice expressing the 751-amino acid isoform of human beta-amyloid precursor protein. *Proc Natl Acad Sci USA*. 1995;92(12):5341–5345.
106. D'Hooge R, Nagels G, Westland CE, Mucke L, De Deyn PP. Spatial learning deficit in mice expressing human 751-amino acid beta-amyloid precursor protein. *Neuroreport*. 1996;7(15–17):2807–2811.
107. Hsiao K, Chapman P, Nilsen S, et al. Correlative memory deficits, Abeta elevation, and amyloid plaques in transgenic mice. *Science*. 1996;274(5284):99–102.
108. Hsiao KK, Borchelt DR, Olson K, et al. Age-related CNS disorder and early death in transgenic FVB/N mice overexpressing Alzheimer amyloid precursor proteins. *Neuron*. 1995;15(5):1203–1218.
109. Foley AG, Gallagher HC, Murphy KJ, Regan CM. Pentyl-4-yn-valproic acid reverses age-associated memory impairment in the Wistar rat. *Neurobiol Aging*. 2004;25(4):539–546.
110. Butterfield DA, Poon HF. The senescence-accelerated prone mouse (SAMP8): a model of age-related cognitive decline with relevance to alterations of the gene expression and protein abnormalities in Alzheimer's disease. *Exp Gerontol*. 2005;40(10):774–783.
111. Ding H, Dolan PJ, Johnson GV. Histone deacetylase 6 interacts with the microtubule-associated protein tau. *J Neurochem*. 2008;106(5):2119–2130.
112. Cook C, Gendron TF, Scheffel K, et al. Loss of HDAC6, a novel CHIP substrate, alleviates abnormal tau accumulation. *Hum Mol Genet*. 2012;21(13):2936–45.
113. Glozak MA, Sengupta N, Zhang X, Seto E. Acetylation and deacetylation of non-histone proteins. *Gene*. 2005;363:15–23.
114. Lahm A, Paolini C, Pallaoro M, et al. Unraveling the hidden catalytic activity of vertebrate class IIa histone deacetylases. *Proc Natl Acad Sci USA*. 2007;104(44):17335–17340.
115. Bantscheff M, Hopf C, Savitski MM, et al. Chemoproteomics profiling of HDAC inhibitors reveals selective targeting of HDAC complexes. *Nat Biotechnol*. 2011;29(3):255–265.
116. Smith KT, Martin-Brown SA, Florens L, Washburn MP, Workman JL. Deacetylase inhibitors dissociate the histone-targeting ING2 subunit from the Sin3 complex. *Chem Biol*. 2010;17(1):65–74.
117. Itzhak Y, Anderson KL, Kelley JB, Petkov M. Histone acetylation rescues contextual fear conditioning in nNOS KO mice and accelerates extinction of cued fear conditioning in wild type mice. *Neurobiol Learn Mem*. 2012;97:409–417.

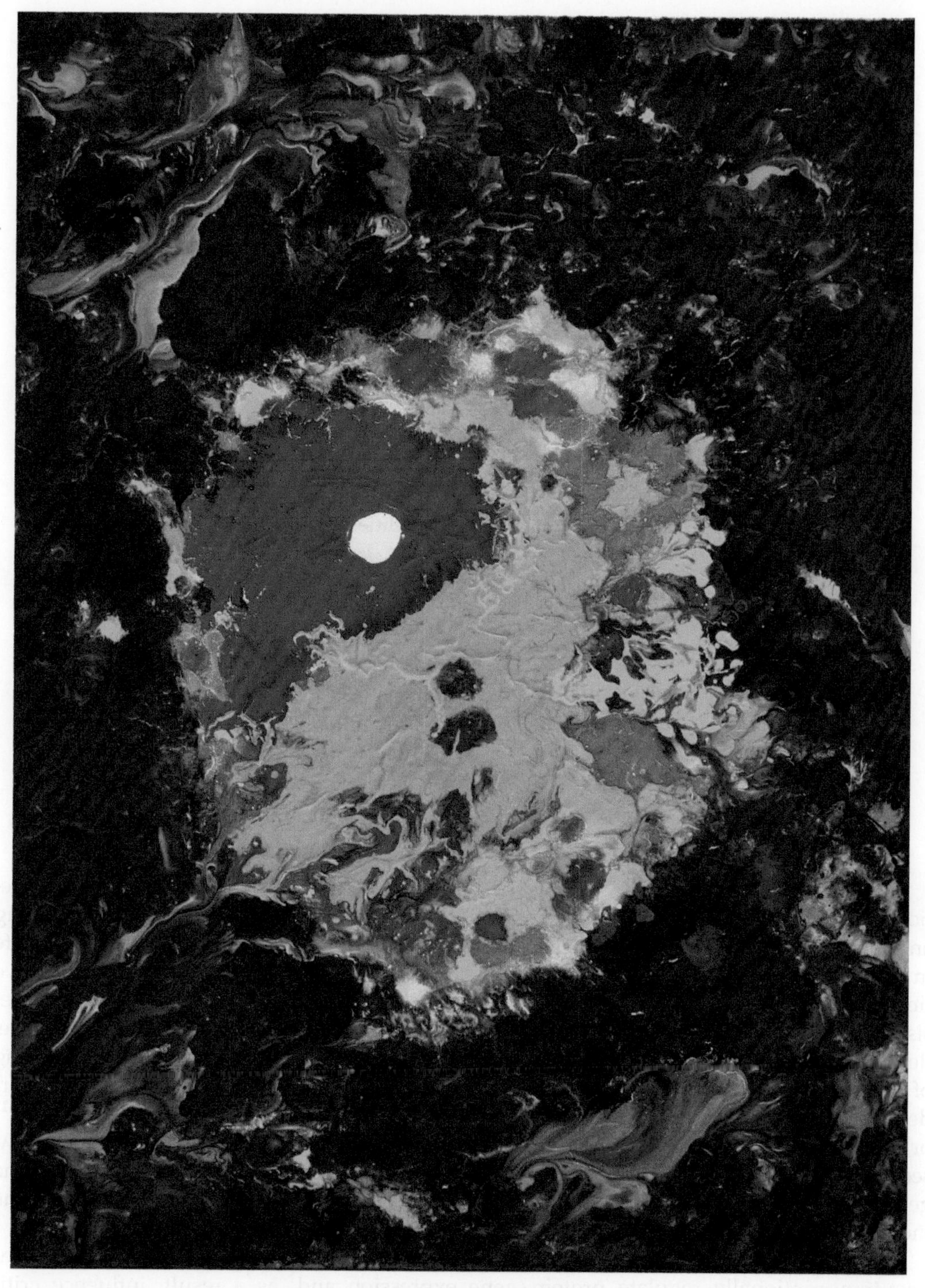

"Neuronal Fate Determination"
J. David Sweatt, acrylic on canvas (40 x 30), 2012

CHAPTER

9

miRNAs and Neurodevelopmental Disorders

Quan Lin[1,2] *and Yi E. Sun*[1,2,3,4]

[1]Department of Psychiatry and Behavioral Sciences
[2]Intellectual Development and Disabilities Research Center, David Geffen School of Medicine
[3]Department of Molecular and Medical Pharmacology, UCLA, Los Angeles, California, USA
[4]Stem Cell Translational Research Center, Shanghai Tongji Hospital, Department of Regenerative Medicine, Tongji University School of Medicine, Shanghai, China

INTRODUCTION

Abnormalities in early brain development can lead to a broad spectrum of brain dysfunction including impairment in sensory and motor activities, speech and language, learning and memory, emotion, as well as social interactions, which severely affects the quality of life in adulthood. Neurodevelopmental disorders can be triggered by changes during sensitive time-windows in development, potentially involving alterations in neuro- and glio- genesis, cell migration, neuronal maturation processes including axon formation and guidance, dendritic morphogenesis, spine and synaptic formation, which are responsible for wiring of neural circuitry and hence cognitive functions later in life. Neurodevelopmental disorders can be elicited by genetic abnormalities[1–10] as well as environmental interferences,[11–13] or a combination of multiple genetic and environmental etiologies (for an excellent review, see[14]). Although substantial progress has been made towards a better understanding of the genetic and environmental etiologies of these devastating disorders, the causes and the molecular nature of many psychiatric disorders remain largely obscure.[14–17]

MicroRNAs (miRNAs) comprise a family of endogenous, non-coding small RNAs that post-transcriptionally regulate protein/gene expression and, as a result, influence critical developmental processes in not only plants, invertebrates, but also mammals.[19] It is

Epigenetic Regulation in the Nervous System
DOI: http://dx.doi.org/10.1016/B978-0-12-391494-1.00009-4

becoming evident that altered expression of this class of small, non-coding RNAs is linked to neurodevelopmental diseases, which can either be causal, in the case of schizophrenia-related miRNA polymorphism,[21–24] or still correlative at this stage.[20,25–28] Recently, a large-scale high-power genome-wide study (GWAS) by Ripke and colleagues demonstrated a causal role of miR-137 in schizophrenia[22]; however, the detailed molecular etiology of this miRNA in schizophrenia is yet to be elucidated. The correlative roles of miRNA are largely revealed through studies of post-mortem brains and animal disease models such as William syndrome, schizophrenia, bipolar disorder, Rett syndrome, Fragile X syndrome, epilepsy, and fetal alcohol disorder,[17,20,25–39,41–43,45–51,53–54,56–62] suggesting that miRNA could be widely involved in a broad spectrum of neurological/neurodevelopmental disorders.

Among all possible causes for neurodevelopmental disorders, environmental factors are particularly interesting, because environmental insults during early development can sometimes elicit changes, which may be manifested later in life even if the insult is no longer there. Normal cognitive and emotional functions are dependent on proper brain development, including a series of critical events in the developing brain throughout gestation and early postnatal stages, such as neural progenitor differentiation, cell migration, dendrite and axon morphogenesis, and spine and synaptic formation and maturation. In the past few years, several lines of studies showed that, in the developing mammalian brain, miRNAs regulated neural stem cell fate specification, neuronal migration, dendritic morphogenesis, axon complexity, spine formation, synaptic function, and memory,[40,44,63–77] raising the possibility that miRNAs, via regulating these early neural developmental processes, elicit a significant impact on cognitive function later in life and leading to abnormalities in behavior. In this review, we will mainly discuss the role of miRNAs in brain development (i.e. cell fate specification, dendrite and axon morphogenesis, dendritic spine and synaptic formation), as well as the environmental and/or genetic regulation of miRNA expression, which can lead to profound changes in behavior and hence is linked to neurological disorders.

miRNA BIOGENESIS

miRNAs were first identified to control the timing of larval development in *Caenorhabditis elegans*.[78,79] Since then, the number of confidently identified miRNAs has exceeded one hundred in *C. elegans* and *Drosophila melanogaster*, and close to one thousand in humans and mice.[80–83] The biologically active, mature miRNAs are processed through three steps: (1) generation of a primary miRNA transcript (pri-miRNA); (2) the pri-miRNA is then processed to create a precursor miRNA (pre-miRNA); and (3) in the cytoplasm, the pre-miRNA becomes a mature miRNA duplex. Most mammalian miRNAs are transcribed by RNA polymerase II (Pol II).[84] Pol II generates pri-miRNA that consists of one or more imperfect hairpin structures. The pri-miRNA is then trimmed into an ≈70 nucleotide (nt) pre-miRNA by the microprocessor complex. The two key components of this microprocessor include an RNase III enzyme, Dorsha, and the RNA-binding protein DiGeorge critical region 8 (DGCR8) (DGCR8 is also known as Pasha [Partner of Dorsha] in *Drosophila* and *C. elegans*). DGCR8 recognizes the hairpin in pri-miRNA and Dorsha cleaves both strands ≈11 base pairs (bp) away from the base of the stem.[85] The released pre-miRNA is then exported to the cytoplasm via the Exportin-5 pathway.[86,87] In the cytoplasm, the pre-miRNA is chopped into ≈22bp by a cytoplasmic RNase III type protein, Dicer. The resulting miRNA duplex comprises of ≈22bp double-stranded from each arm of the original hairpin. One of the strands associates with the Argonaute-containing RNA induced silencing complex (RISC) and this strand is then stabilized. In contrast, the other miRNA strand, known as the passenger strand, is degraded. Once mature miRNAs are loaded onto the Argonaute (AGO) protein of the silencing complex, miRNAs often pair with 3′untranslated regions (3′UTR) of target mRNAs to inhibit directly gene expression at post-transcriptional level by translational repression, mRNA destabilization, or a combination of the two.[78,88–92]

miRNA REGULATION OF CELL FATES AND MIGRATION IN THE DEVELOPING BRAIN

Recent evidence shows that abnormal cortical development appears to contribute to neurodevelopmental disorders such as autism spectrum disorder (ASD).[93–96] Understanding mechanisms that govern critical time-windows and events of brain development may greatly help in revealing the bases of various neurodevelopmental disorders. miRNAs are highly expressed in the mammalian central nervous system (CNS) and regulate neural stem cell fate specification and neuronal migration. The essential regulatory role of miRNA in development emerged from the original study of embryonic stem (ES) cells deficient for *dicer*. Ablation of *dicer* disturbs ES cell division and proliferation, causing entire loss of pluripotent stem cells.[97–99] Consistent with *dicer* knockout, knockout of DGCR8, an RNA-binding protein that assists the RNase III enzyme Drosha to process pri-miRNAs, also blocks stem cell differentiation in mice.[100] Thus, miRNAs can regulate cell fates at the level of both proliferation and differentiation.

Among the various microRNAs, miR-9 is perhaps the most studied member in the context of the CNS. In fish, miR-9 has been shown to refine the mid-hind brain boundary.[101] In frogs, miR-9 is required for proper neurogenesis during brain development.[71] In rodents, miR-9 controls neural progenitor cell proliferation and differentiation in the developing cerebral cortex.[73,102] Shibata and colleagues showed that miR-9 is critical for proper reelin-positive Cajal–Retzius cells to differentiate terminally during early cortical development.[73,102] Cajal–Reztius cells are known to provide a critical instructive cue for proper layer formation in the cortex. In conventional miR-9-double-mutant mice, cortical layers and the ventricular zone are remarkably reduced with expanded lateral ventricles.[73] Perturbation of miR-9 affects neuronal differentiation and migration in mice[67,73] and in human ES cell-derived neural progenitors.[68] Moreover, miR-9 also influences proliferation of neural progenitor cells in the developing mouse cortex.[73] In addition to miR-9, miR-9*, miR-124, miR-34, and miR-125b have been implicated to promote neurogenesis in mice[73,76,103,104,145] as well as human neuroblastoma cells.[66] The robust effect of miRNAs in controlling cell fate can be shown by two recent studies. Overexpression of a brain-specific miRNA, miR-124 in HeLa cells induces HeLa cell gene expression toward neuron-specific molecular profiles.[89] miR-9* and miR-124 can instruct human fibroblasts to express neuronal specific genes, and convert them to neuronal-like cells.[105] Moreover, these miRNAs have also been shown to be associated with neuropsychiatric disorders such as schizophrenia, autism spectrum disorders, Fragile X syndrome, and epilepsy (Table 9.1). In sum, accumulating evidence has demonstrated essential roles of miRNAs in regulating proper neural differentiation in development, suggesting that

Disease	miRNA-SNP	miRNA Related to Disease	Function Related Gene	Sample Source	References
Schizophrenia and bipolar disorders					
	miR-206, -198				Hensen et al. (2007)[21]
	miR-502, -510, -890, -892b, -934				Sun et al. (2009)[24]
	Let-7f-2, miR-188-3p, -510-3p, -660, -325-3p, -509-3p, pre-miR-18b, pre-miR-502, pre-miR-505				Feng et al. (2009)[23]
	miR-137				Ripke et al. (2011)[22]
		miR-26b, -30a/b/d/e, -29a/b/c, -195, -92, -20b, -212, -7, -24, -9-3p, miR-106b		Prefrontal cortex of schizophrenia subjects	Perkins et al. (2007)[49]
		miR-181b	VSNL1, GRIA2	Superior temporal gyrus	Beveridge et al. (2008)[106]
		miR-195	BDNF	Prefrontal cortex of schizophrenia subjects	Mellios et al. (2009)[45]
		miR-34a, -132, -132*, -212, -544, -7, -154*		Prefrontal cortex of schizophrenia subjects	Kim et al. (2010)[107]
		miR-519c, -409-3p, 652, -382, -532, -199a*, -17-5p, -542-3p, -199b, -592, -495, -487a, -425-5p, -152, -148b, -134, -150, -105, -187, -154, -767-5p, -548b, -590, -502, -452*, -25, -328, -92b, -433, -222, -512-3p, 423, -193a		Dorsolateral prefrontal cortex of schizophrenia subjects	Santarelli et al. (2011)[51]
		miR-17	NPAS3		Wong et al. (2012)[57]
		miR-132	DNMT3A, GATA2, and DPYSL3	Dorsolateral prefrontal cortex of schizophrenia or bipolar disorder subjects	Miller et al. (2012)[46]

	miR-29, -147, -199, -297				Glinsky et al. (2008)[108]
		miR-504, -454*, -29a, -520c-3p, -140-3p, -145*, -767-5p, -22*, -145, -874, -133b, -154*, -32, hsa- -573, -889		Dorsolateral prefrontal cortex of bipolar disorder subjects	Kim et al. (2010)[107]
		miR-330, -33, -193b, -545, -138, -151, -210, -324-3p, -22, -425, -181a, -106b, -193a, -192, -301, -27b, -148b, -338, -639, -15a, -186, -99a, -190, -339		Schizophrenia and bipolar disorder subjects	Moreau et al. (2011)[47]
Autism spectrum disorders					
	bipolar disorders	miR-184	MeCP2	Mouse E16 cortical neurons	Nomura et al. (2008)[26]
		miR-212	MeCP2	Rat striatum	Im et al. (2010)[42]
		miR-29b, -329, -199b, -382, -296, -221, -92, -146a/b, -130, -122a, -342, -409	MeCP2	RTT mouse brain	Urdinguio et al. (2010)[27]
		miR-206, -874, -299*, -674, -137, -455, -543, -7a*, -495, -377, -744*, -29a*, -15a*, -34b-3p, -666-5p; -29c, -let-7a*, -140	MeCP2	RTT mouse cerebellum	Wu et al. (2010)[28]
		miR-199b-5p, -548o, -577, -486-3p, -455-3p, -338-3p, -199a-5p, -650, -486-5p, -125b, -10a, -196a		Autism spectrum disorder subjects	Ghahramani Seno et al. (2011)[38]
Fragile X syndrome					
		miR-19a/b, -142, -302b*, -323-3p	FMRP	Human HEK-293 cells	Yi et al. (2010)[61]
Epilepsy		miR-132, -125a/b, -128, -143, -100, -127, -138, -9,	FMRP	FMR1 knock out mouse brain	Edbauer et al. (2010)[35]

TABLE 9.1 (Continued)

Disease	miRNA-SNP	miRNA Related to Disease	Function Related Gene	Sample Source	References
		miR-213, -132, -30c, -26a, -375, -99a, -24, -124a, -22, -34a, -125a, -101-1, -29b, -125b, -199a, -196b, -150, -151, -145, -29a, -181c, -215, -181b, -25, -10b, -21		Rat hippocampus	Hu et al. (2011)[41]
		miR-10b, -21, -27a, -29a, -30e, -101, -103, -107, -125a, -127, -132, -133b, -134, -139, -145, -146b, -148b, -153, -181c, -199a, -200a, -219, -323, -326, - 328, -330, -374, -375, -381, -422b, -425, -451, -487b, -497, 507, -509, -518b/c/d, -520b/c/g, -532, -629, -657		Mouse hippocampus	Jimenez-Mateos et al. (2011)[43]
		miR-21	NT-3	Rat hippocampus	Risbud et al. (2011)[50]
		miR-98, -352, let-7e/d, -185, -99a, -134, -127, -379, -137, -324-5P, -27b, -383, -132, -24, -29a, -139-5P, -9*, -23a/b, -146a, -140*, -126		Rat hippocampus	Song et al. (2011)[53]
Environmental factors					
		miR-21, -335, -9, and -153		Mouse cerebral cortical neural precursors	Sathyan et al. (2007)[109]
		miR-10a/b, -9, -145, -30a-3p -152, -200a, -496, -296, -30e-5p, -362, -339, -29c, -154		Mouse brain	Wang et al. (2009)[25]

NPAS3: Neuronal Per-Arnt-Sim(PAS) domain protein 3; DNMT3A: DNA (cytosine-5)-methyltransferase 3A; GATA2: GATA binding protein 2; DPYSL3: dihydropyrimidinase-like 3; BDNF: brain-derived neurotrophic factor; VSNL1: visinin-like 1; GRIA2: ionotropic AMPA glutamate receptor subunit; MeCP2: methyl CpG binding protein 2; FMRP: fragile X mental retardation 1; NT-3: neurotrophin-3.

dysregulation of these miRNAs may affect normal brain formation which, in turn, triggers neurological disorders later in life.

miRNA REGULATION OF NEURAL PLASTICITY

The perinatal period represents a critical time-window opening to neural plasticity changes. This period also renders individuals most vulnerable to influences from external disturbances. Perturbations of normal series of developmental events during this time period can lead to quite profound functional consequences that manifest later in life.[110–113] This critical period is associated with neural plasticity (i.e. dendritogenesis, spine and synaptic formation, and ion channel compositions), a major salient feature of the nervous system during brain development. Neurons form a single axon and multiple dendrites, which govern the directional flow of information in the CNS. Dendrites integrate synaptic inputs, triggering the generation of action potentials at the level of the soma, which then propagate along the axon, making presynaptic contacts onto the targeted brain regions. In the past 30 years, a number of key studies using Golgi staining of post-mortem materials from children with developmental disorders demonstrated that changes in dendritic branching and dendritic spine formation were associated with autism spectrum disorder and mental retardation.[114–122] For instance, dendrites and dendritic spines consolidate synaptic inputs and are the first stage for signal processing.[18,123–125] The branching pattern of dendritic trees and dendritic spines greatly affects the flow of synaptic signals to the soma and, therefore, influences overall neuronal information processing, which ultimately contributes to learning, memory, and other cognitive functions. An increasing amount of evidence has suggested that miRNAs are involved in modeling dendrite morphology, dendritic spine density and structure, synaptic strength, and axon guidance, elongation, and branching (Table 9.2).

The role of miRNA in regulating neural plasticity and cognitive functions has been revealed in animal models. Universally eliminating miRNAs in mature neurons in the adult mouse forebrain results in improved learning and memory, while increasing long filopodia spines and altering synaptic transmission.[136] Studies by Gao and colleagues and Yang and colleagues showed that miR-134 and miR-124 are correlatively upregulated in *sirtuin 1* (SirT1) and Rap guanine nucleotide exchange factor (GEF) 3 (RapGEF3, EPAC) mutant mice, respectively. Both studies claim that the impaired synaptic plasticity and cognitive functions are due to the dysregulation of the expression of these miRNAs, since knock-down of these miRNAs in mutant adult brains rescues the phenotype.[69,137]

There is accumulating evidence suggesting that exposure to isolated events (such as stress) during critical periods of development profoundly influences late-onset biobehavioral traits during an individual's adult life, including metabolic alterations and neuropsychiatric conditions such as anxiety, depression, schizophrenia, as well as autism spectrum disorder.[138–143] The mechanisms underlying this intriguing phenomenon or the biological substrates for such "dormant, yet long-lasting, biological memory" have fascinated scientists for decades. For example, Gross and colleagues showed that ablation of serotonon1A receptor function in the forebrain during postnatal development, but not in

TABLE 9.2 miRNA Regulation of Dendrite, Dendritic Spine, and Axon Morphology in Mammalian CNS

miRNA ID	Functions	Upstream Regulator	Target	References
miR-132	Dendrite[63,35,126], spine[35,40,133,135] and synaptic physiology,[35,135] and cognitive function[40]	CREB[63]	p250GAP[63,126]	Vo et al. (2005),[63] Wayman et al. (2008),[126] Impey et al. (2010),[133] Edbauer et al. (2010),[35] Hansen et al. (2010),[40] Mellios et al. (2011)[135]
miR-125b	Spine and synaptic physiology[35]		NR2A[35]	Edbauer et al. (2010)[35]
Let-7	Spine			Edbauer et al. (2010)[35]
miR-22	Spine			Edbauer et al. (2010)[35]
miR-134	Dendrite[127] and spine[64]	BDNF[64], Mef2[127]	LimK1[64], Pum2[127]	Schratt et al. (2006),[64] Fiore et al. (2009)[127]
miR-124	Dendrite[65,76], Spine[35]	REST[76]	BAF53a[76]	Yoo et al. (2009),[76] Yu et al. (2008)[65]
miR-9*	Dendrite[76]	REST[76]	BAF53a[76]	Yoo et al. (2009)[76]
miR-138	Spine and synaptic physiology		APT1	Siegel et al. (2009)[77]
miR-375	Dendrite		HuD	Abdelmohsen et al. (2010)[128]
miR-485	Spine and synaptic physiology		SV2A	Cohen et al. (2011)[132]
miR-29a/b	Spine		Arpc3	Lippi et al. (2011)[134]
miR-34a	Dendrite, spine, and synaptic physiology	TAp73	Syt-1, stx-1A	Agostini et al. (2011)[130,131]
miR-9	Dendrite, spine, and cognitive function		Diap1, Pfn2	Lin et al. unpublished observation
miR-9	Axon guidance and branching		Map1b[75]	Shibata et al. (2011),[73] Dajas-Bailador et al. (2012)[75]

CREB: cAMP-response element binding protein; p250GAP: GTPase-activating protein; NR2A: a NMDA receptor subunit; BDNF: brain derived neurotrophic factor; LimK1: LIM domain kinase 1; Pum2: translational repressor Pumilio2; BAF: mammalian SWI/SNF; REST: repressor element-1 silencing transcription factor; APT1: acyl protein thioesterase 1; HuD: a member of the embryonic-lethal abnormal vision (elav)/Hu protein family; SV2A: a presynaptic vesicle protein; Arpc3: a subunit of the ARP2/3 actin nucleation complex; Syt-1: synaptotagmin-1; stx-1A: syntaxin-1A; TAp73: a p53-family member; Daip1: diaphanous homolog 1; Pfn2: profillin 2; Map1b: microtubule-associated protein 1b.

the adult, causes an anxiety-related behavioral phenotype in mice.[141] Another study shows that briefly disrupting Disrupted-in-schizophrenia 1 (DISC1) during postnatal development in a mouse model of schizophrenia, but not in the adult, is adequate to induce phenotypes linked to this disorder.[143] The development and refinement of neuronal circuitry are regulated by both cell intrinsic and neuronal activity-dependent means. Studies by Wayman and colleagues and Fiore and colleagues showed that the expression level of miR-132 and miR-134 could be upregulated by BDNF treatment and KCl depolarization in immature neurons to promote dendritic morphogenesis.[126,127] Thus, this evidence indicates that dysregulation of endogenous miRNA expression by extrinsic cues could contribute to morphological abnormality in dendritic arborization which, in turn, might influence normal cognition.[116,118,120] Ethanol treatment induced major fetal teratogenesis in mice and caused mental retardation in their offspring, namely lower locomotor activity and impaired task acquisition. Sathyan and colleagues found that miR-9, miR-21, miR-153, and miR-335 were downregulated by ethanol in a fetal mouse cerebral cortex-derived neurosphere culture model.[109] In adult mouse brain, alcohol upregulates miR-9 expression and mediates calcium- and voltage-activated potassium channel mRNA splice variants. In addition, miR-132 has been shown to be upregulated upon monocular deprivation or dark rearing, disruption of its expression leads to altered ocular dominance plasticity.[135] Although increasing evidence implies the possibility that cognitive abnormalities can result from perturbed miRNA expression/activity, it appears that all studies focused either on miRNA regulation at a cellular level in cell migration, dendritic morphology, and synaptic physiology, or on the passive dysregulation of miRNA expression in post-mortem human samples such as schizophrenia or animal models of neurodevelopmental disorders such as Rett syndrome. To elucidate definitively the role of miRNA during the critical developmental time-window that influences cognitive function later in life, our group employed a Tet-ON inducible system, which allows the expression of a brain-enriched miRNA, miR-9, to be restricted within a precise period of embryonic development (E14-P1) in mice. Strikingly, we found that perturbation of miR-9 expression in the prelimbic frontal cortex (PLPFC) during the critical time-window of dendritogenesis[144] altered dendrite morphogenesis and spine formation which, in turn, affected fear-related learning and memory in the adult (Lin et al., unpublished observation). In search of the underlying mechanism for this effect, we found that miR-9 strongly influenced dendritic morphology as well as spine formation by directly regulating organizers of the actin cytoskeleton. Manipulations of these miR-9 targets in vivo mimicked the effect of miR-9 on dendritic complexity and fear learning, further substantiating a causal influence of developmental regulation by miR-9 on memory capacity across the lifespan. Interestingly, two recent elegant studies by Shibata and colleagues and Dajas-Bailador and colleagues show that miR-9 regulates axon guidance, elongation, and branching, suggesting that dysregulated axon guidance may also contribute to altered fear learning.[73,75] Our findings on overexpression of miR-9 in the PLPFC demonstrated that miR-9 led to enhanced cognition, suggesting a critical role for small non-coding RNAs in fine-tuning neural plasticity during brain development which, in turn, modulate cognitive behavior in the adult. In summary, these initial findings could have revealed only the tip of the iceberg in the role of dysregulated miRNAs in cognitive abnormalities.

CONCLUDING REMARKS

miRNAs are a group of newly discovered endogenous genes, not encoding proteins, but which have a profound effect on gene expression at post-transcriptional levels, and thus heavily influence protein synthesis. In addition, miRNA expression is highly enriched in the

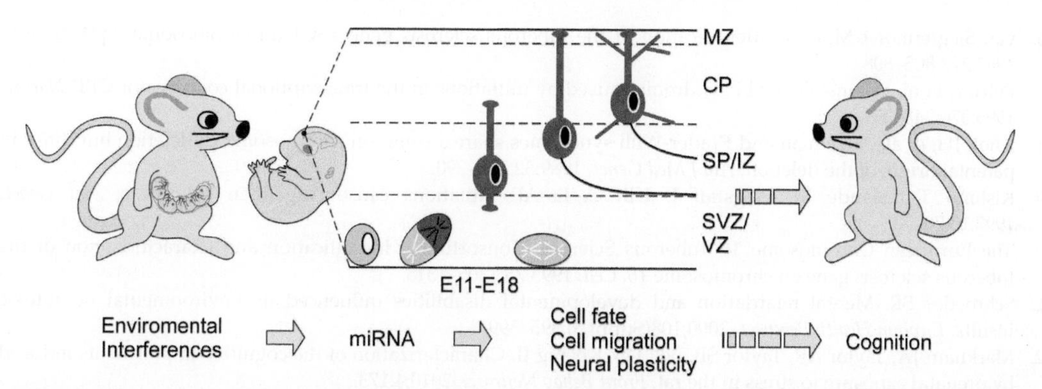

FIGURE 9.1 Dysregulation of miRNA by environmental interference during the critical time-window of brain development is involved in alternations in cell fates, cell migration, dendritic morphogenesis, dendritic spine formation and axon guidance and branching that are potentially responsible for cognitive deficits in the adult. MZ: marginal zone; CP: cortical plate; IZ: intermediate zone; SP: subplate zones; SVZ: subventricular zone; VZ: ventricular zone.

mammalian brain. The broad spectrum of miRNA functions including its role in regulating brain development has been uncovered at an accelerated rate. In this review, we intend to include the majority of the publications on miRNA regulation of brain development without emphasizing their direct roles in regulating neural plasticity and brain circuitry function in the adult brain. The other focus of the review is the link between early developmental regulations and the long-lasting effect that manifests later in life, which is a classic feature of "epigenetic mechanisms". Here, the long-lasting effect of miRNAs could be evoked via alterations in structures of neuronal cells, i.e. neuronal morphology, and hence brain circuitry wiring patterns. We provide a precise example on how spatially and temporally restricted miR-9 overexpression during embryonic stages changes neuronal dendritic morphology in a long-lasting manner and, subsequently, influences fear-learning circuitry, which can be manifested much later in adult life (Figure 9.1). Such mechanisms, though not directly influencing chromatin structures, have to also be considered as epigenetic. Finally, given miRNA duplexes, both the sense and antisense versions, can potentially be developed into drugs, understanding the interaction between miRNA and genetic and/or environmental cues can provide the basis for future targeting miRNAs as novel therapeutic interventions for reversing cognitive deficits and neurological impairments.

References

1. Verkerk AJ, et al. Identification of a gene (FMR-1) containing a CGG repeat coincident with a breakpoint cluster region exhibiting length variation in fragile X syndrome. *Cell*. 1991;65:905–914.
2. Amir RE, et al. Rett syndrome is caused by mutations in X-linked MECP2, encoding methyl-CpG-binding protein 2. *Nat Genet*. 1999;23:185–188.
3. Francke U. Williams-Beuren syndrome: genes and mechanisms. *Hum Mol Genet*. 1999;8:1947–1954.
4. Megarbane A, et al. The 50th anniversary of the discovery of trisomy 21: the past, present, and future of research and treatment of Down syndrome. *Genet Medicine*. 2009;11:611–616.
5. Ramocki MB, Tavyev YJ, Peters SU. The MECP2 duplication syndrome. *Am J Med Genet A*. 2010;152A:1079–1088.

6. van Slegtenhorst M, et al. Identification of the tuberous sclerosis gene TSC1 on chromosome 9q34. *Science*. 1997;277:805–808.
7. Petrij F, et al. Rubinstein-Taybi syndrome caused by mutations in the transcriptional co-activator CBP. *Nature*. 1995;376:348–351.
8. Knoll JH, et al. Angelman and Prader-Willi syndromes share a common chromosome 15 deletion but differ in parental origin of the deletion. *Am J Med Genet*. 1989;32:285–290.
9. Kishino T, Lalande M, Wagstaff J. UBE3A/E6-AP mutations cause Angelman syndrome. *Nat Genet*. 1997;15:70–73.
10. The European Chromosome 16 Tuberous Sclerosis Consortium. Identification and characterization of the tuberous sclerosis gene on chromosome 16. *Cell*. 1993;75:1305–1315.
11. Schroeder SR. Mental retardation and developmental disabilities influenced by environmental neurotoxic insults. *Environ Health Perspect*. 2000;108(Suppl 3):395–399.
12. Markham JA, Taylor AR, Taylor SB, Bell DB, Koenig JI. Characterization of the cognitive impairments induced by prenatal exposure to stress in the rat. *Front Behav Neurosci*. 2010;4:173.
13. Khashan AS, et al. Higher risk of offspring schizophrenia following antenatal maternal exposure to severe adverse life events. *Arch Gen Psychiatry*. 2008;65:146–152.
14. van Loo KM, Martens GJ. Genetic and environmental factors in complex neurodevelopmental disorders. *Curr Genomics*. 2007;8:429–444.
15. Betancur C. Etiological heterogeneity in autism spectrum disorders: more than 100 genetic and genomic disorders and still counting. *Brain Res*. 2011;1380:42–77.
16. Gejman PV, Sanders AR, Kendler KS. Genetics of schizophrenia: new findings and challenges. *Annu Rev Genomics Hum Genet*. 2011;12:121–144.
17. Beveridge NJ, Cairns MJ. MicroRNA dysregulation in schizophrenia. *Neurobiol Dis*. 2012;46:263–271.
18. Segev I, London M. Untangling dendrites with quantitative models. *Science*. 2000;290:744–750.
19. Bartel DP. MicroRNAs: genomics, biogenesis, mechanism, and function. *Cell*. 2004;116:281–297.
20 Burmistrova OA, et al. MicroRNA in schizophrenia: genetic and expression analysis of miR-130b (22q11). *Biochemistry (Mosc)*. 2007;72:578–582.
21. Hansen T, et al. Brain expressed microRNAs implicated in schizophrenia etiology. *PLoS One*. 2007;2:e873.
22. Ripke S, et al. Genome-wide association study identifies five new schizophrenia loci. *Nat Genet*. 2011;43:969–976.
23. Feng J, et al. Evidence for X-chromosomal schizophrenia associated with microRNA alterations. *PLoS One*. 2009;4:e6121.
24. Sun G, et al. SNPs in human miRNA genes affect biogenesis and function. *RNA*. 2009;15:1640–1651.
25. Wang LL, et al. Ethanol exposure induces differential microRNA and target gene expression and teratogenic effects which can be suppressed by folic acid supplementation. *Hum Reprod*. 2009;24:562–579.
26. Nomura T, et al. MeCP2-dependent repression of an imprinted miR-184 released by depolarization. *Hum Mol Genet*. 2008;17:1192–1199.
27. Urdinguio RG, et al. Disrupted microRNA expression caused by Mecp2 loss in a mouse model of Rett syndrome. *Epigenetics*. 2010;5:656–663.
28. Wu H, et al. Genome-wide analysis reveals methyl-CpG-binding protein 2-dependent regulation of microRNAs in a mouse model of Rett syndrome. *Proc Natl Acad Sci USA*. 2010;107:18161–18166.
29. Abu-Elneel K, et al. Heterogeneous dysregulation of microRNAs across the autism spectrum. *Neurogenetics*. 2008;9:153–161.
30. Aronica E, et al. Expression pattern of miR-146a, an inflammation-associated microRNA, in experimental and human temporal lobe epilepsy. *Eur J Neurosci*. 2010;31:1100–1107.
31. Beveridge NJ, Gardiner E, Carroll AP, Tooney PA, Cairns MJ. Schizophrenia is associated with an increase in cortical microRNA biogenesis. *Mol Psychiatry*. 2010;15:1176–1189.
32. Cheever A, Blackwell E, Ceman S. Fragile X protein family member FXR1P is regulated by microRNAs. *RNA*. 2010;16:1530–1539.
33. Cheever A, Ceman S. Phosphorylation of FMRP inhibits association with dicer. *RNA*. 2009;15:362–366.
34. de Leon-Guerrero SD, Pedraza-Alva G, Perez-Martinez L. In sickness and in health: the role of methyl-CpG binding protein 2 in the central nervous system. *Eur J Neurosci*. 2011;33:1563–1574.
35. Edbauer D, et al. Regulation of synaptic structure and function by FMRP-associated microRNAs miR-125b and miR-132. *Neuron*. 2010;65:373–384.

36. Fung LK, Quintin EM, Haas BW, Reiss AL. Conceptualizing neurodevelopmental disorders through a mechanistic understanding of fragile X syndrome and Williams syndrome. *Curr Opin Neurol*. 2012;25:112–124.
37. Garbett KA, et al. Novel animal models for studying complex brain disorders: BAC-driven miRNA-mediated in vivo silencing of gene expression. *Mol Psychiatry*. 2010;15:987–995.
38. Ghahramani Seno MM, et al. Gene and miRNA expression profiles in autism spectrum disorders. *Brain Res*. 2011;1380:85–97.
39. Guo AY, Sun J, Jia P, Zhao Z. A novel microRNA and transcription factor mediated regulatory network in schizophrenia. *BMC Syst Biol*. 2010;4:10.
40. Hansen KF, Sakamoto K, Wayman GA, Impey S, Obrietan K. Transgenic miR132 alters neuronal spine density and impairs novel object recognition memory. *PLoS One*. 2010;5:e15497.
41. Hu K, et al. Expression profile of microRNAs in rat hippocampus following lithium-pilocarpine-induced status epilepticus. *Neurosci Lett*. 2011;488:252–257.
42. Im HI, Hollander JA, Bali P, Kenny PJ. MeCP2 controls BDNF expression and cocaine intake through homeostatic interactions with microRNA-212. *Nat Neurosci*. 2010;13:1120–1127.
43. Jimenez-Mateos EM, et al. miRNA Expression profile after status epilepticus and hippocampal neuroprotection by targeting miR-132. *Am J Pathol*. 2011;179:2519–2532.
44. Kocerha J, et al. MicroRNA-219 modulates NMDA receptor-mediated neurobehavioral dysfunction. *Proc Natl Acad Sci USA*. 2009;106:3507–3512. doi:10.1073/pnas.0805854106.
45. Mellios N, et al. Molecular determinants of dysregulated GABAergic gene expression in the prefrontal cortex of subjects with schizophrenia. *Biol Psychiatry*. 2009;65:1006–1014.
46. Miller BH, et al. MicroRNA-132 dysregulation in schizophrenia has implications for both neurodevelopment and adult brain function. *Proc Natl Acad Sci USA*. 2012;109:3125–3130.
47. Moreau MP, Bruse SE, David-Rus R, Buyske S, Brzustowicz LM. Altered microRNA expression profiles in postmortem brain samples from individuals with schizophrenia and bipolar disorder. *Biol Psychiatry*. 2011;69:188–193.
48. Muinos-Gimeno M, et al. Allele variants in functional MicroRNA target sites of the neurotrophin-3 receptor gene (NTRK3) as susceptibility factors for anxiety disorders. *Hum Mutat*. 2009;30:1062–1071.
49. Perkins DO, et al. MicroRNA expression in the prefrontal cortex of individuals with schizophrenia and schizoaffective disorder. *Genome Biol*. 2007;8:R27.
50. Risbud RM, Lee C, Porter BE. Neurotrophin-3 mRNA a putative target of miR21 following status epilepticus. *Brain Res*. 2011;1424:53–59.
51. Santarelli DM, Beveridge NJ, Tooney PA, Cairns MJ. Upregulation of dicer and microRNA expression in the dorsolateral prefrontal cortex Brodmann area 46 in schizophrenia. *Biol Psychiatry*. 2011;69:180–187.
52. Soares AR, et al. Ethanol exposure induces up-regulation of specific microRNAs in zebrafish embryos. *Toxicol Sci*. 2012. doi:10.1093/toxsci/kfs068.
53. Song YJ, et al. Temporal lobe epilepsy induces differential expression of hippocampal miRNAs including let-7e and miR-23a/b. *Brain Res*. 2011;1387:134–140.
54. Tabares-Seisdedos R, Rubenstein JL. Chromosome 8p as a potential hub for developmental neuropsychiatric disorders: implications for schizophrenia, autism and cancer. *Mol Psychiatry*. 2009;14:563–589.
55. Uchida S, et al. Characterization of the vulnerability to repeated stress in Fischer 344 rats: possible involvement of microRNA-mediated down-regulation of the glucocorticoid receptor. *Eur J Neurosci*. 2008;27:2250–2261.
56. Voineskos AN, et al. Neurexin-1 and frontal lobe white matter: an overlapping intermediate phenotype for schizophrenia and autism spectrum disorders. *PLoS One*. 2011;6:e20982.
57. Wong J, et al. Expression of NPAS3 in the human cortex and evidence of its posttranscriptional regulation by miR-17 during development, with implications for schizophrenia. *Schizophr Bull*. 2012. doi:10.1093/schbul/sbr177.
58. Wu L, et al. A novel function of microRNA let-7d in regulation of galectin-3 expression in attention deficit hyperactivity disorder rat brain. *Brain Pathol*. 2010;20:1042–1054.
59. Xu XL, et al. FXR1P but not FMRP regulates the levels of mammalian brain-specific microRNA-9 and microRNA-124. *J Neurosci*. 2011;31:13705–13709.
60. Yang Y, et al. The bantam microRNA is associated with drosophila fragile X mental retardation protein and regulates the fate of germline stem cells. *PLoS Genet*. 2009;5:e1000444.
61. Yi YH, et al. Experimental identification of microRNA targets on the 3' untranslated region of human FMR1 gene. *J Neurosci Methods*. 2010;190:34–38.

62. Zhu Y, Kalbfleisch T, Brennan MD, Li Y. A MicroRNA gene is hosted in an intron of a schizophrenia-susceptibility gene. *Schizophr Res*. 2009;109:86–89.
63. Vo N, et al. A cAMP-response element binding protein-induced microRNA regulates neuronal morphogenesis. *Proc Natl Acad Sci USA*. 2005;102:16426–16431.
64. Schratt GM, et al. A brain-specific microRNA regulates dendritic spine development. *Nature*. 2006;439:283–289.
65. Yu JY, Chung KH, Deo M, Thompson RC, Turner DL. MicroRNA miR-124 regulates neurite outgrowth during neuronal differentiation. *Exp Cell Res*. 2008;314:2618–2633.
66. Le MT, et al. MicroRNA-125b promotes neuronal differentiation in human cells by repressing multiple targets. *Mol Cell Biol*. 2009;29:5290–5305.
67. Zhao C, Sun G, Li S, Shi Y. A feedback regulatory loop involving microRNA-9 and nuclear receptor TLX in neural stem cell fate determination. *Nat Struct Mol Biol*. 2009;16:365–371.
68. Delaloy C, et al. MicroRNA-9 coordinates proliferation and migration of human embryonic stem cell-derived neural progenitors. *Cell Stem Cell*. 2010;6:323–335.
69. Gao J, et al. A novel pathway regulates memory and plasticity via SIRT1 and miR-134. *Nature*. 2010;466:1105–1109.
70. Magill ST, et al. MicroRNA-132 regulates dendritic growth and arborization of newborn neurons in the adult hippocampus. *Proc Natl Acad Sci USA*. 2010;107:20382–20387.
71. Bonev B, Pisco A, Papalopulu N. MicroRNA-9 reveals regional diversity of neural progenitors along the anterior-posterior axis. *Dev Cell*. 2011;20:19–32.
72. Otaegi G, Pollock A, Hong J, Sun T. MicroRNA miR-9 modifies motor neuron columns by a tuning regulation of foxp1 levels in developing spinal cords. *J Neurosci*. 2011;31:809–818.
73. Shibata M, Nakao H, Kiyonari H, Abe T, Aizawa S. MicroRNA-9 regulates neurogenesis in mouse telencephalon by targeting multiple transcription factors. *J Neurosci*. 2011;31:3407–3422.
74. Han R, et al. MiR-9 promotes the neural differentiation of mouse bone marrow mesenchymal stem cells via targeting zinc finger protein 521. *Neurosci Lett*. 2012;515:147–152.
75. Dajas-Bailador F, et al. MicroRNA-9 regulates axon extension and branching by targeting Map1b in mouse cortical neurons. *Nat Neurosci*. 2012. doi:10.1038/nn.3082.
76. Yoo AS, Staahl BT, Chen L, Crabtree GR. MicroRNA-mediated switching of chromatin-remodelling complexes in neural development. *Nature*. 2009;460:642–646.
77. Siegel G, et al. A functional screen implicates microRNA-138-dependent regulation of the depalmitoylation enzyme APT1 in dendritic spine morphogenesis. *Nat Cell Biol*. 2009;11:705–716.
78. Lee RC, Feinbaum RL, Ambros V. The C. elegans heterochronic gene lin-4 encodes small RNAs with antisense complementarity to lin-14. *Cell*. 1993;75:843–854.
79. Reinhart BJ, et al. The 21-nucleotide let-7 RNA regulates developmental timing in Caenorhabditis elegans. *Nature*. 2000;403:901–906.
80. Ruby JG, et al. Large-scale sequencing reveals 21U-RNAs and additional microRNAs and endogenous siRNAs in C. elegans. *Cell*. 2006;127:1193–1207.
81. Ruby JG, et al. Evolution, biogenesis, expression, and target predictions of a substantially expanded set of Drosophila microRNAs. *Genome Res*. 2007;17:1850–1864.
82. Landgraf P, et al. A mammalian microRNA expression atlas based on small RNA library sequencing. *Cell*. 2007;129:1401–1414.
83. Chiang HR, et al. Mammalian microRNAs: experimental evaluation of novel and previously annotated genes. *Genes Dev*. 2010;24:992–1009.
84. Lee Y, et al. MicroRNA genes are transcribed by RNA polymerase II. *EMBO J*. 2004;23:4051–4060.
85. Han J, et al. Molecular basis for the recognition of primary microRNAs by the Drosha-DGCR8 complex. *Cell*. 2006;125:887–901.
86. Yi R, Qin Y, Macara IG, Cullen BR. Exportin-5 mediates the nuclear export of pre-microRNAs and short hairpin RNAs. *Genes Dev*. 2003;17:3011–3016.
87. Lund E, Guttinger S, Calado A, Dahlberg JE, Kutay U. Nuclear export of microRNA precursors. *Science*. 2004;303:95–98.
88. Guo H, Ingolia NT, Weissman JS, Bartel DP. Mammalian microRNAs predominantly act to decrease target mRNA levels. *Nature*. 2010;466:835–840.
89. Lim LP, et al. Microarray analysis shows that some microRNAs downregulate large numbers of target mRNAs. *Nature*. 2005;433:769–773.

90. Filipowicz W, Bhattacharyya SN, Sonenberg N. Mechanisms of post-transcriptional regulation by microRNAs: are the answers in sight? *Nat Rev Genet*. 2008;9:102–114.
91. Hausser J, Landthaler M, Jaskiewicz L, Gaidatzis D, Zavolan M. Relative contribution of sequence and structure features to the mRNA binding of Argonaute/EIF2C-miRNA complexes and the degradation of miRNA targets. *Genome Res*. 2009;19:2009–2020.
92. Hendrickson DG, et al. Concordant regulation of translation and mRNA abundance for hundreds of targets of a human microRNA. *PLoS Biol*. 2009;7:e1000238.
93. Aylward EH, Minshew NJ, Field K, Sparks BF, Singh N. Effects of age on brain volume and head circumference in autism. *Neurology*. 2002;59:175–183.
94. Piven J, et al. An MRI study of brain size in autism. *Am J Psychiatry*. 1995;152:1145–1149.
95. Amaral DG, Schumann CM, Nordahl CW. Neuroanatomy of autism. *Trends Neurosci*. 2008;31:137–145.
96. Courchesne E, et al. Unusual brain growth patterns in early life in patients with autistic disorder: an MRI study. *Neurology*. 2001;57:245–254.
97. Murchison EP, Partridge JF, Tam OH, Cheloufi S, Hannon GJ. Characterization of Dicer-deficient murine embryonic stem cells. *Proc Natl Acad Sci USA*. 2005;102:12135–12140.
98. Bernstein E, et al. Dicer is essential for mouse development. *Nat Genet*. 2003;35:215–217.
99. Kanellopoulou C, et al. Dicer-deficient mouse embryonic stem cells are defective in differentiation and centromeric silencing. *Genes Dev*. 2005;19:489–501.
100. Wang Y, Medvid R, Melton C, Jaenisch R, Blelloch R. DGCR8 is essential for microRNA biogenesis and silencing of embryonic stem cell self-renewal. *Nat Genet*. 2007;39:380–385.
101. Leucht C, et al. MicroRNA-9 directs late organizer activity of the midbrain-hindbrain boundary. *Nat Neurosci*. 2008;11:641–648.
102. Shibata M, Kurokawa D, Nakao H, Ohmura T, Aizawa S. MicroRNA-9 modulates Cajal-Retzius cell differentiation by suppressing Foxg1 expression in mouse medial pallium. *J Neurosci*. 2008;28:10415–10421.
103. Makeyev EV, Zhang J, Carrasco MA, Maniatis T. The MicroRNA miR-124 promotes neuronal differentiation by triggering brain-specific alternative pre-mRNA splicing. *Mol Cell*. 2007;27:435–448.
104. Maiorano NA, Mallamaci A. Promotion of embryonic cortico-cerebral neuronogenesis by miR-124. *Neural Dev*. 2009;4:40.
105. Yoo AS, et al. MicroRNA-mediated conversion of human fibroblasts to neurons. *Nature*. 2011;476:228–231.
106. Beveridge NJ, et al. Dysregulation of miRNA 181b in the temporal cortex in schizophrenia. *Hum Mol Genet*. 2008;17:1156–1168.
107. Kim AH, et al. MicroRNA expression profiling in the prefrontal cortex of individuals affected with schizophrenia and bipolar disorders. *Schizophr Res*. 2010;124:183–191.
108. Glinsky GV. An SNP-guided microRNA map of fifteen common human disorders identifies a consensus disease phenocode aiming at principal components of the nuclear import pathway. *Cell Cycle*. 2008;7:2570–2583.
109. Sathyan P, Golden HB, Miranda RC. Competing interactions between micro-RNAs determine neural progenitor survival and proliferation after ethanol exposure: evidence from an ex vivo model of the fetal cerebral cortical neuroepithelium. *J Neurosci*. 2007;27:8546–8557.
110. Rapoport JL, Addington AM, Frangou S, Psych MR. The neurodevelopmental model of schizophrenia: update 2005. *Mol Psychiatry*. 2005;10:434–449.
111. Li J, et al. A nationwide study on the risk of autism after prenatal stress exposure to maternal bereavement. *Pediatrics*. 2009;123:1102–1107.
112. Kinney DK, Munir KM, Crowley DJ, Miller AM. Prenatal stress and risk for autism. *Neurosci Biobehav Rev*. 2008;32:1519–1532.
113. Lupien SJ, McEwen BS, Gunnar MR, Heim C. Effects of stress throughout the lifespan on the brain, behaviour and cognition. *Nat Rev Neurosci*. 2009;10:434–445.
114. Huttenlocher PR. Synaptic and dendritic development and mental defect. *UCLA Forum Med Sci*. 1975:123–140.
115. Huttenlocher PR. Dendritic development in neocortex of children with mental defect and infantile spasms. *Neurology*. 1974;24:203–210.
116. Huttenlocher PR. Dendritic development and mental defect. *Neurology*. 1970;20:381.
117. Huttenlocher PR. Dendritic and synaptic pathology in mental retardation. *Pediatr Neurol*. 1991;7:79–85.
118. Kaufmann WE, MacDonald SM, Altamura CR. Dendritic cytoskeletal protein expression in mental retardation: an immunohistochemical study of the neocortex in Rett syndrome. *Cereb Cortex*. 2000;10:992–1004.

119. Kaufmann WE, Moser HW. Dendritic anomalies in disorders associated with mental retardation. *Cereb Cortex*. 2000;10:981–991.
120. Marin-Padilla M. Structural abnormalities of the cerebral cortex in human chromosomal aberrations: a Golgi study. *Brain Res*. 1972;44:625–629.
121. Purpura DP. Dendritic differentiation in human cerebral cortex: normal and aberrant developmental patterns. *Adv Neurol*. 1975;12:91–134.
122. Purpura DP. Normal and aberrant neuronal development in the cerebral cortex of human fetus and young infant. *UCLA Forum Med Sci*. 1975:141–169.
123. Reyes A. Influence of dendritic conductances on the input-output properties of neurons. *Annu Rev Neurosci*. 2001;24:653–675.
124. Mehta MR. Cooperative LTP can map memory sequences on dendritic branches. *Trends Neurosci*. 2004;27:69–72.
125. Williams SR, Wozny C, Mitchell SJ. The back and forth of dendritic plasticity. *Neuron*. 2007;56:947–953.
126. Wayman GA, et al. An activity-regulated microRNA controls dendritic plasticity by down-regulating p250GAP. *Proc Natl Acad Sci USA*. 2008;105:9093–9098.
127. Fiore R, et al. Mef2-mediated transcription of the miR379-410 cluster regulates activity-dependent dendritogenesis by fine-tuning Pumilio2 protein levels. *EMBO J*. 2009;28:697–710.
128. Abdelmohsen K, et al. MiR-375 inhibits differentiation of neurites by lowering HuD levels. *Mol Cell Biol*. 2010;30:4197–4210.
129. Christensen M, Larsen LA, Kauppinen S, Schratt G. Recombinant adeno-associated virus-mediated microRNA delivery into the postnatal mouse brain reveals a role for miR-134 in dendritogenesis in vivo. *Front Neural Circuits*. 2010;3:16.
130. Agostini M, et al. microRNA-34a regulates neurite outgrowth, spinal morphology, and function. *Proc Natl Acad Sci USA*. 2011;108:21099–21104.
131. Agostini M, et al. Neuronal differentiation by TAp73 is mediated by microRNA-34a regulation of synaptic protein targets. *Proc Natl Acad Sci USA*. 2011;108:21093–21098.
132. Cohen JE, Lee PR, Chen S, Li W, Fields RD. MicroRNA regulation of homeostatic synaptic plasticity. *Proc Natl Acad Sci USA*. 2011;108:11650–11655.
133. Impey S, et al. An activity-induced microRNA controls dendritic spine formation by regulating Rac1-PAK signaling. *Mol Cell Neurosci*. 2010;43:146–156.
134. Lippi G, et al. Targeting of the Arpc3 actin nucleation factor by miR-29a/b regulates dendritic spine morphology. *J Cell Biol*. 2011;194:889–904.
135. Mellios N, et al. miR-132, an experience-dependent microRNA, is essential for visual cortex plasticity. *Nat Neurosci*. 2011;14:1240–1242.
136. Konopka W, et al. MicroRNA loss enhances learning and memory in mice. *J Neurosci*. 2010;30:14835–14842.
137. Yang Y, et al. EPAC null mutation impairs learning and social interactions via aberrant regulation of miR-124 and Zif268 translation *Neuron*. 2012;73:774–788.
138. Liu D, et al. Maternal care, hippocampal glucocorticoid receptors, and hypothalamic-pituitary-adrenal responses to stress. *Science*. 1997;277:1659–1662.
139. Caldji C, et al. Maternal care during infancy regulates the development of neural systems mediating the expression of fearfulness in the rat. *Proc Natl Acad Sci USA*. 1998;95:5335–5340.
140. Francis D, Diorio J, Liu D, Meaney MJ. Nongenomic transmission across generations of maternal behavior and stress responses in the rat. *Science*. 1999;286:1155–1158.
141. Gross C, et al. Serotonin1A receptor acts during development to establish normal anxiety-like behaviour in the adult. *Nature*. 2002;416:396–400.
142. Weaver IC, et al. Epigenetic programming by maternal behavior. *Nat Neurosci*. 2004;7:847–854.
143. Li W, et al. Specific developmental disruption of disrupted-in-schizophrenia-1 function results in schizophrenia-related phenotypes in mice. *Proc Natl Acad Sci USA*. 2007;104:18280–18285.
144. Barnes AP, Polleux F. Establishment of axon-dendrite polarity in developing neurons. *Annu Rev Neurosci*. 2009;32:347–381.
145. Aranha MM, Santos DM, Sola S, Steer CJ, Rodrigues CM. miR-34a regulates mouse neural stem cell differentiation. *PLoS One*. 2011;6(8):e21396. Epub 2011/08/23. doi: 10.1371/journal.pone.0021396. PubMed PMID: 21857907; PubMed Central PMCID: PMC3153928.

"Epigenetics in Neurodevelopmental Disorders"

J. David Sweatt, acrylic on wood panel (24 x 48), 2012

CHAPTER

10

Imprinting in the CNS and Neurodevelopmental Disorders

Weston T. Powell and Janine M. LaSalle

Medical Microbiology and Immunology, Genome Center, Medical Institute of Neurodevelopmental Disorders, University of California, Davis School of Medicine, Davis, California, USA

IMPRINTING AND NEURODEVELOPMENTAL DISORDERS

Proper development of the human brain into a fully functional, robust organ depends on an intricate web of genetically encoded developmental transcription networks, cell-to-cell communication via signaling pathways, and environmental cues. Central to all these developmental programs are various epigenetic mechanisms, such as DNA methylation, histone modifications, and nuclear organization. Development of the mammalian brain is exceptional in the diverse roles epigenetic mechanisms play. During early neurodevelopment epigenetic processes serve to guide intrinsic transcriptional programs to specify neural precursor cells (NPCs) that will differentiate into neurons that migrate to populate the entire brain. In fetal life, neurons begin to mature and continue to differentiate into distinct neuronal subpopulations that can be classified based on function (i.e. γ-aminobutyric acid-(GABA)ergic vs glutamatergic; excitatory vs inhibitory), microarchitecture (cortical layering), or organ architecture (brain region). While these pathways are governed largely by intrinsic transcription networks and developmental epigenetic programs, developing neurons are also able to incorporate external environmental cues that can alter developmental trajectories and synaptic plasticity later in life. The mammalian brain, and the human brain in particular, continues to develop postnatally, and epigenetic processes have been shown to play a role in regulating neuronal synaptic plasticity, activity-dependent maturation, and dendritic pruning. Since neurodevelopment relies on such complex genetic and environmental interactions, there are many possible sources of disruption that can lead to a variety of neurodevelopmental disorders, particularly those with complex genetic bases. Epigenetic

Epigenetic Regulation in the Nervous System
DOI: http://dx.doi.org/10.1016/B978-0-12-391494-1.00010-0

regulation is hypothesized to play a central role in the pathogenesis of complex neurobehavioral disorders such as autism and schizophrenia.[1,2]

Neurodevelopmental disorders involving genomic imprinting, such as Prader–Willi syndrome (PWS) and Angelman syndrome (AS), offered the first clinical examples of an epigenetic process (imprinting) that could be disrupted during neurodevelopment and lead to a phenotype.[3] Similarly, the finding that Rett syndrome was caused by mutations in the protein methyl-CpG binding protein 2 (MeCP2) showed that proteins important in interpreting epigenetic marks such as DNA methylation were important in neurodevelopment.[4] Recently, researchers have focused on possible dysregulation of epigenetic processes in neurodevelopmental diseases such as X-linked mental retardation, autism, schizophrenia, and attention deficit hyperactivity disorder/attention deficit disorder (ADHD/ADD).[5–7] The role of proper control of epigenetic processes is seen in the spectrum of neurodevelopmental disorders and phenotypes caused by dysregulation of proper genomic imprinting in humans and in animal models.

GENOMIC IMPRINTING

Genomic imprinting is the regulation of gene expression in a parent-of-origin-specific manner through epigenetic mechanisms. Here we will refer to imprinting as allele-specific silencing based on parent of origin, so if a gene is "maternally-imprinted" then we mean that it is silenced on the allele that was maternally inherited through the oocyte. Most imprinted loci use differentially methylated regions (DMRs) that act to control the expression pattern of a cluster of imprinted genes (Figure 10.1).[8,9] Many imprinted genes show high levels of expression in placenta and/or brain, and some also show regional and temporal changes of expression during neurodevelopment.[10] The importance of imprinting for proper brain and placenta formation may be a result of the need to synchronize development of the two organs because the brain is a late-developing organ with high energy and metabolic requirements.[11] While imprinted genes are rare, with only approximately 100 genes completely silenced in a parental allele-specific manner, many more transcripts showed allelic skewing in their expression levels in brain.[12,13] However, recent reports have shown that such apparent allelic biases of expression could be due to improper interpretation of deep sequencing of brain RNA from mouse species crosses.[14] Therefore, imprinted loci can be conservatively defined by both the presence of at least one differentially methylated region and the presence of at least one gene exhibiting complete silencing of one parental allele. Nevertheless, some genes that are not imprinted, but are located close to imprinted loci, have shown allelic biases in expression in neurodevelopmental disorders.[15,16] Further studies will be necessary to show the true extent of allelic bias in transcription and the functional relevance of parentally biased transcription during neurodevelopment.

MECHANISMS OF IMPRINTING

DNA Methylation

Regions that exhibit allele-specific differences in methylation, called differentially methylated regions (DMRs), serve as important imprinting control regions (ICRs) for imprinted

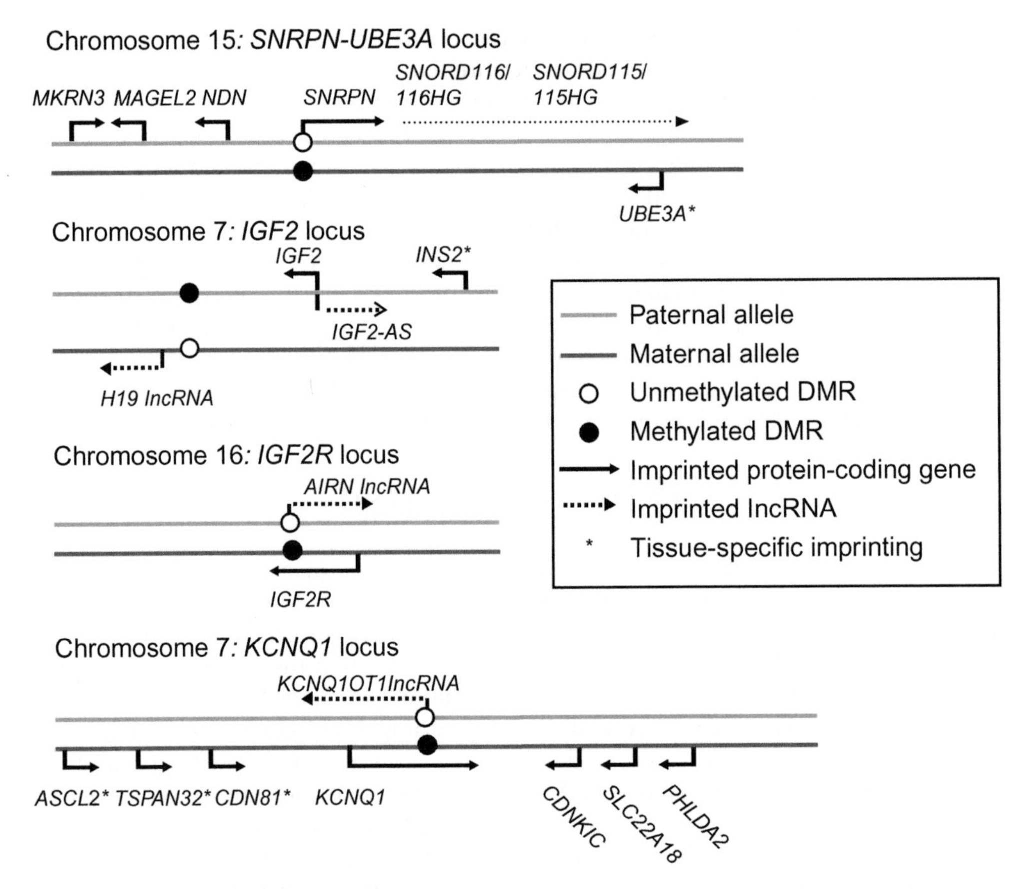

FIGURE 10.1 **Map of imprinted loci.** Schematic of four imprinted loci showing direction and organization of imprinted genes, differential methylated regions (DMR) and presence of long, non-coding RNAs (lncRNA). Figure is not to scale.

clusters of genes.[8] In some cases, DMRs overlap with CpG island promoters, such as at the *PWS-ICR*, which is heavily methylated on the silent maternal allele and unmethylated on the active paternal allele (see Figure 10.1). For other allele-specific DMRs, such as the *H19/Igf2* locus, methylation prevents the binding of the chromatin insulator CTCF and thereby disrupts enhancer–promoter interactions on one allele. In both cases, the active unmethylated allele of ICRs is critical to long-range regulation of gene expression in the imprinted loci.[17,18]

The allele-specific pattern of methylation is established early in development, soon after fertilization, and the exact mechanisms that cause methylation on one allele, or protection from methylation on the other allele, have not yet been fully elucidated. One mechanism that protects CpG island promoters from DNA methylation is the formation of RNA:DNA hybrids called R-loops. R-loops have been demonstrated at two imprinted loci, *Snrpn* and *Airn*.[19] R-loop formation in an allele-specific manner may function to block DNA methylation on one allele of a CpG island, thus creating a DMR. The role of DMRs in specifying parentally imprinted alleles seems to be a general mechanism of imprinting, but how cells can then demonstrate tissue- or developmental stage-specific patterns of imprinting is less well understood. Most DMRs serve as imprinting control regions in multiple tissues and persist throughout development, suggesting that mechanisms other than DMRs act to control tissue and developmental stage-specific imprinting. Some imprinted genes, such as *UBE3A*, are biallelically expressed in some tissues, but imprinted in other tissues without a corresponding differential methylation mark at the promoter. Instead, the finding that nearly all imprinted loci in the mammalian genome contain long, non-coding RNAs (lncRNAs) raises the possibility that lncRNAs could play a more important role in tissue-specific imprinting than DMRs.

lncRNA

The past few years have seen an explosion of research into the role of lncRNAs in directing epigenetic mechanisms of transcriptional control. While knowledge and appreciation of non-coding transcripts has blossomed with the advent of next-generation sequencing, some of the first examples of functional lncRNAs were from imprinted loci. Best characterized is the lncRNA *Xist*, which serves to silence the inactive X-chromosome by creating a nuclear RNA cloud and recruiting silencing complexes to coat the inactive X-chromosome.[20–24] While X-inactivation is random in most tissues, in mice the paternal X-chromosome is imprinted and preferentially silenced in early development.[25] Nearly all imprinted loci express lncRNAs, often in the opposite direction from the silenced allele of an imprinted protein-coding gene (see Figure 10.1).[8,26]

One appeal of lncRNAs acting to control expression of imprinted genes is their potential ability to regulate spatiotemporal lability in imprinting status. The lncRNA *Kcnq1ot1* provides one example of differential imprinting regulation. By increasing the size of the nuclear RNA cloud formed by *Kcnq1ot1* in placenta, a larger number of genes localize within the cloud and become silenced on the paternal allele.[27] In contrast to differentially methylated regions, which maintain the covalent methylation mark in all tissues and timepoints, the lncRNA can be processed, degraded, and modified. For example, the length of the *SNRPN-UBE3A-antisense* transcript determines *UBE3A* imprinting in neurons, while the differential methylation pattern marking the maternal and paternal alleles remains constant in all tissues.[28] A portion of the *SNRPN-UBE3A-antisense* transcript arising from the *SNORD116* region (the critical region deleted in PWS) has been shown to share some characteristics with other lncRNA.[29]

Most investigations of the PWS critical region have focused on the function of the repeated small, nucleolar RNAs (snoRNAs) arising from processed introns of the *SNRPN-UBE3A-antisense* transcript; however, the function of the snoRNAs derived from this locus is currently unclear. While C/D box snoRNAs from other loci serve to guide modification of ribosomal RNA in the nucleolus, those encoded by the *SNORD115* and *SNORD116* loci are mostly considered "orphans" with minimal rRNA homology. *SNORD115* has been implicated in regulating alternative splicing of a serotonin receptor gene;[30] however, the implications of this finding to the PWS phenotype are unclear, since it is the *SNORD116* cluster responsible for PWS. Both *SNORD115* and *SNORD116* encoded snoRNAs localize primarily to the nucleolus of mature postnatal neurons, and their deficiency coincides with significantly smaller nucleoli in Purkinje neurons of PWS patients, suggesting a possible role in nucleolar maturation.[31]

The portions of the *SNRPN-UBE3A-antisense* transcript that have received much less attention are the exons flanking the snoRNA-containing intron that are spliced to form a non-coding RNA estimated to be 4–6 kb in length called *116HG* and *115HG*.[29] RNA fluorescence in situ hybridization of primary neurons showed that *116HG* and *115HG* localize to the nucleus and form non-overlapping RNA clouds, reminiscent of distinct nuclear localizations of other lncRNA such as *Xist* and *Kcnq1ot1*.[27,29] The function of *116HG* has not yet been investigated, but this lncRNA at the heart of the PWS locus could play a central role in the disease pathogenesis and potentially function in regulating the imprinting of *UBE3A* on the paternal allele in mature neurons. Recent studies have shown that lncRNAs exhibit regional and temporal specific patterns of expression in the brain, raising the possibility that the PWS imprinted lncRNA could play an important role in brain function.[32–34] PWS has been the first example of a human developmental disease caused by loss of a non-coding RNA, but the question remains: Is it due to the loss of the snoRNA, *SNORD116*, or the lncRNA, *116HG*?

MATERNAL AND PATERNAL GENOMES IN DEVELOPMENT

The reproductive necessity for balanced contributions by the maternal and paternal genomes was first seen from the lack of viability of complete androgenetic (paternal-derived) or parthenogenetic/gynogenetic (maternal-derived) embryos. Parthenogenetic embryos failed primarily because they lacked a functional placenta, most likely due to dysregulation of *H19*, *IGF2* and *IGF2R*, which have been shown to function in placental development.[35,36] However, in the early 1990s, a series of experiments was performed in which normal cells were injected into androgenetic (AG) or parthenogenetic (PG) blastocysts in order to create AG-chimeras or PG-chimeras that were viable beyond early embryonic development.[37,38] The chimeras that survived had low levels of AG or PG contributions (<40%), but analysis revealed that AG cells preferentially contributed to the development of the limbic areas of the brain, whereas PG-derived cells appeared most often in cortical areas, the striatum, and the hippocampus. Individual imprinted loci critical for proper development of the AG-derived or PG-derived brain regions have not yet been identified, although there does appear to be a bias of some imprinted genes for either cortical regions or for limbic regions.[10] One caveat of the AG and PG chimera studies is that survival of AG or PG cells in certain brain regions may be due to a small number of genes, or possibly to non-cell-autonomous interactions with neighboring normal cells that do not have the AG or PG genome. The finding that full AG and PG embryos are not viable also raises the possibility that some of the phenotypes seen in individuals with imprinted loci deleted may result from failure to form fully functional non-neuronal tissues. In particular, the placenta has a clear role in proper neurodevelopment because of its endocrine function, importance for metabolism, waste removal, and function in translating environmental cues to the developing embryo. However, in the past few years, work has begun to shed light on the role of imprinted loci in neurodevelopment and the dysregulation of imprinted genes in neurodevelopment.

NEURODEVELOPMENTAL DISORDERS OF IMPRINTED LOCI

Prader–Willi Syndrome and Angelman Syndrome

Proper expression of the imprinted 15q11-13 locus is important for neurodevelopment, as loss of the region leads to Prader–Willi syndrome or Angelman syndrome, depending on which parental allele is lost. PWS presents in infancy as hypotonia and failure to thrive before progressing at about 2 years of age to hyperphagia, mental retardation, breathing abnormalities, short stature, and obsessive–compulsive behaviors.[39,40] In contrast, AS is characterized by stereotyped hand movements, severe mental retardation and developmental delay, lack of speech, ataxia, a happy disposition, and severe seizures.[41] Both result from abnormalities in chromosome 15q11-13, including epigenetic dysregulation of the region. PWS results from a deficiency of the paternal allele, through deletions (65–75%), maternal uniparental disomy (20–30%), or methylation abnormalities (1–3%).[42] AS is caused by deficiency of the maternally-inherited allele resulting from deletions (70%), paternal uniparental disomy (2–5%), loss of methylation on the maternal allele (7–9%), or point mutations of the maternal allele of the gene *UBE3A* (15%).[43] While no specific point mutations are observed in PWS, three patients with PWS were shown to carry microdeletions of a cluster of tandemly repeated non-coding RNAs, including snoRNAs called *SNORD116* and intervening exons called *116HG*.[44–46]

Transcriptional control of *SNRPN* and *UBE3A* are intertwined in the pathogenesis of PWS and AS, as they rely on complex, overlapping layers of epigenetic regulation. In mature neurons, *UBE3A* is transcribed only from the maternal allele, but silencing of the paternal allele is determined by the nearby *SNRPN* promoter, also called the Prader–Willi syndrome imprinting control region (PWS–ICR).[47] Silencing of paternal *UBE3A* occurs by transcription of an antisense transcript under the control of the *SNRPN* promoter, and methylation of the *SNRPN* promoter prevents antisense transcription, thereby permitting maternal expression of *UBE3A*.[47] The chromatin on the paternal allele undergoes a large scale decondensation in neurons that coincides with upregulation of the antisense transcript, suggesting that the correct chromatin environment is also needed for paternal silencing of *UBE3A*.[31] The antisense transcript is the 3′ end of a long (>600 kb) transcript originating at the *SNRPN* promoter. *SNRPN* is expressed ubiquitously and, in non-neuronal tissues, transcription initiating at the *SNRPN* promoter terminates either just upstream of the *Snord116* region in mouse tissues or after the *SNORD116* region in human tissues.[28,29,48–51] However, in neurons, transcription proceeds through the cluster of repeated snoRNAs and eventually terminates after generating the *UBE3A-antisense* transcript.[47,51,52] The mechanism by which transcription switches from terminating in the *SNORD116* region to terminating in the *UBE3A-antisense* transcript is not known, but is thought to relate to the large chromatin decondensation seen in the transcribed region.[31] Recent work showed that chromosome loops formed by the PWS–ICR change in a neuronal cell model of differentiation, suggesting that long-range chromatin interactions may also play a role in transcriptional control in the region.[53,54] A recent study showed *Ube3a-antisense* transcription decreased and paternal *Ube3a* expression increased with treatment by topoisomerase inhibitors. These results suggest that treatments that change the chromatin structure and looping may be useful in future therapies for Angelman syndrome.[55]

Ube3a knockout mice recapitulate many of the phenotypes seen in human patients, such as motor abnormalities and learning deficits.[56] The precise functions of the protein UBE3A in neurodevelopment are still being investigated, although it appears to have role in regulating the activity-dependent turnover of Arc, a protein with synaptic function, and in mediating Ephexin5-directed excitatory synaptic development.[57,58] In addition to regulating cytoplasmic turnover of proteins, UBE3A may also have a nuclear role in regulating transcription in neurons because concomitant with paternal silencing is a change in cellular localization of UBE3A to the nucleus.[59]

Mouse models of PWS have begun to unlock the pathogenesis of the syndrome, but some human–mouse differences in presentation persist. Large deletions of mouse chromosome 7q spanning the region orthologous to 15q11-13 that affect multiple imprinted genes show a much more severe phenotype (early neonatal lethality) than is seen in human patients with large paternal deletions in 15q11-13.[60] Differences in early life care may explain why children with 15q11-13 deletions survive the neonatal period while mice pups do not. However, mouse models mimicking the smallest PWS-causing deletions spanning the *Snord116* repeat region do not become obese, although they do have increased metabolic rates and are hyperphagic.[61] The difference in presentation could result from the differences in expression of *SNORD116/Snord116* between mice and humans. Humans have detectable levels of *SNORD116* in most non-neuronal tissues, whereas in mice *Snord116* expression is limited to neurons.[29,48–50,62]

15q11-q13 and Complex Neurodevelopmental Disorders

Genetic studies have associated polymorphisms and rearrangements of 15q11-q13 with autism, schizophrenia, bipolar disorder, and epilepsy.[7,63,64] The most common cytogenetic abnormality seen in autism, accounting for 1–3% of cases, is duplication of maternal 15q11-q13.[65,66] Patients with duplications of chromosome 15q11-q13 have autistic traits, such as reduced speech, reduced social behaviors, seizures, sensory abnormalities, and stereotypic behaviors.[67] Since maternal duplication most often causes autism, whereas paternal duplication results in developmental delay, but not autism, studies have hypothesized that aberrant transcriptional control of imprinted genes plays a central role in autism and similar neurodevelopmental disorders.

Analysis of gene expression in 15q11-q13 in cortical samples with extra copies of 15q11-q13 revealed changes that did not correlate with copy number, as *SNRPN* and *GABRB3* expression actually were decreased.[68] *UBE3A* expression would be expected to increase with maternal duplication, but expression levels were lower than would be predicted by gene copy number.[68] Changes in *UBE3A* transcript levels are also seen in brain samples from Rett syndrome patients with mutations in *MECP2*, encoding methyl CpG binding protein 2.[69] Interestingly, the level of methylation of the maternal *PWS–ICR* positively correlated with *UBE3A* levels in 15q11-q13 duplication syndrome, reinforcing that idea that MeCP2 binding to the maternal methylated allele enhances *UBE3A* levels in mature neurons.[68] The phenotypic overlap of Rett syndrome, Angelman syndrome, and autism, along with decreased *UBE3A* levels in neurons from afflicted individuals has led to studies seeking to link *UBE3A* function to developmental pathways and behaviors.

Part of the difficulty in identifying specific pathways affected by altered epigenetic regulation may come from the different mechanisms human cells and mouse cells use to control transcription in the region. Previous studies have shown a role for MeCP2 function in direct homologous pairing of 15q11-q13 in human neurons, but pairing of the orthologous locus does not occur in mouse neurons.[70] A human neuronal cell line containing an extra maternal copy of 15q11-13 showed altered pairing and decreased transcription of the paternal allele.[53] In contrast, a mouse model with duplication of the orthologous 7q region showed increased transcription of all the imprinted genes in the locus upon both maternal and paternal inheritance, with the exception of *Ube3a*, which showed no change when the extra copy was inherited paternally.[71] The mouse duplication model showed some social deficits, but only upon paternal inheritance of the duplication. As noted above, maternal duplication results in autism in humans, and individuals carrying paternal duplications have developmental delay or develop normally.[65,72,73] Supporting maternal duplication in the pathogenesis of autism is a mouse model with increased gene dosage of *Ube3a* that shows impaired social behavior.[74] Human–mouse differences in development, phenotypes, and regulation are important to keep in mind when considering the role of imprinted loci in neurodevelopment.

OTHER IMPRINTED GENES AND CNS FUNCTION

While 15q11-13 is the only imprinted human locus to date that has been proven to play a role in the pathogenesis of human neurodevelopmental and behavioral diseases, studies in mice of imprinted loci have shown a role for other imprinted genes in brain development and function. Imprinted expression of genes has been hypothesized to be part of the evolution of a "social brain" based on deficits in maternal care in mouse knockout studies.[75] However, a direct link between imprinted genes and evolution of social behaviors has not been definitively shown.

Peg3

Mice that carry deletions on the paternal allele of *Peg3* have fewer oxytocin-producing hypothalamic cells, and the reduction may occur via the p53 mediated apoptotic pathway.[76,77] A behavioral consequence of the deletion is decreased learning of sexual activity, in particular a decreased preference in male mice for female urine; female mice showed decreased maternal care of pups.[78,79] *Peg3* also appears to regulate genes in both the placenta and in the developing hypothalamus, and is upregulated in the hypothalamus in response to starvation.[80] Co-regulation of imprinted genes in placental and brain development is hypothesized to occur because the fetal brain has high energy demands and is particularly sensitive to insults such as starvation. An essential dual role of *Peg3* in placental development and early brain development could explain why no human neurodevelopmental disorders have yet been linked to *Peg3* dysfunction. If the placental defect caused by *Peg3* were more severe in humans, then embryos carrying a *Peg3* mutation would abort early in development and no disease would present. In support of this, *Peg3* mutations in cattle have been shown to cause spontaneous abortions.[81]

Grb10

Growth factor receptor-bound protein 10 (*Grb10*) is paternally imprinted, maternally expressed in most somatic tissues in the mouse, but switches to a maternal imprint and paternal expression in certain brain regions.[82] In a mouse knockout model, loss of the paternal allele resulted in a mild behavioral phenotype where deletion mice were more dominant by a tube test as compared to wild-type littermates.[82] In humans, *GRB10* is maternally imprinted in the brain, although it is biallelically expressed in most other tissues, except for trophoblast cells in the placenta.[83] Conservation suggests that *GRB10* may play a role in human social behavior, but the link to human behavior has not been investigated.

CONCLUSIONS

Human disorders related to abnormalities on chromosome 15q11-q13 provide the clearest link between imprinted gene expression and normal brain function. However, multiple other imprinted genes appear to cause behavioral phenotypes in mouse models, suggesting that imprinted gene expression may play a more general role in proper brain development and function. Connecting mouse studies, such as those done with *Peg3* and *Grb10* deficient mice, to studies done on human subjects remains a critical avenue for future studies. While many genes conserve imprinting and tissue-specific patterns of expression from mouse to human, some notable differences exist.[84] Given that only ≈150 imprinted genes have been found in mice and half that number in humans, at this point it is difficult to decipher the significance of species-specific patterns of imprinting.[75] Similarly, various hypotheses have been posited that genomic imprinting is driven by intragenomic conflict between parental genomes, by the development of a social brain, or by generational pressures.[75,85] But, thus far, concrete evidence supporting these hypotheses has been lacking. The role of imprinted genes in driving behavioral phenotypes remains appealing because epigenetic regulation is hypothesized to play a central role in complex disorders, such as autism and schizophrenia, and imprinted genes offer the clearest example of epigenetic regulation of transcription. Similarly, bipolar disorder shows a maternal influence in inherited susceptibility, although a specific imprinted locus has not been linked to the maternal risk.[86] Future investigations of imprinted genes and loci that may be dysregulated in neuropsychiatric disorders may therefore be warranted. Furthermore, therapeutic strategies to counteract the epigenetic marks at imprinted loci could be potentially beneficial for a wider range of neurodevelopmental disorders beyond Angelman and Prader–Willi syndromes.

References

1. LaSalle JM. A genomic point-of-view on environmental factors influencing the human brain methylome. *Epigenetics*. 2011;6(7):862–869.
2. St Clair D. Copy number variation and schizophrenia. *Schizophr Bull*. 2009;35(1):9–12.
3. Williams CA, Zori RT, Stone JW, Gray BA, Cantu ES, Ostrer H. Maternal origin of 15q11-13 deletions in Angelman syndrome suggests a role for genomic imprinting [see comments]. *Am J Med Genet*. 1990;35(3):350–353.
4. Amir RE, Van den Veyver IB, Wan M, Tran CQ, Francke U, Zoghbi HY. Rett syndrome is caused by mutations in X-linked MECP2, encoding methyl- CpG-binding protein 2. *Nat Gen*. 1999;23(2):185–188.

5. Cook Jr EH, Scherer SW. Copy-number variations associated with neuropsychiatric conditions. *Nature*. 2008;455(7215):919–923.
6. Leung KN, Chamberlain SJ, Lalande M, LaSalle JM. Neuronal chromatin dynamics of imprinting in development and disease. *J Cell Biochem*. 2011;112(2):365–373.
7. Stefansson H, Rujescu D, Cichon S, et al. Large recurrent microdeletions associated with schizophrenia. *Nature*. 2008;455(7210):232–236.
8. Lewis A, Reik W. How imprinting centres work. *Cytogenet Genome Res*. 2006;113(1–4):81–89.
9. Xie W, Barr CL, Kim A, et al. Base-resolution analyses of sequence and parent-of-origin dependent DNA methylation in the mouse genome. *Cell*. 2012;148(4):816–831.
10. Wilkinson LS, Davies W, Isles AR. Genomic imprinting effects on brain development and function. *Nat Rev Neurosci*. 2007;8(11):832–843.
11. Keverne EB, Curley JP. Epigenetics, brain evolution and behaviour. *Front Neuroendocrinol*. 2008;29(3): 398–412.
12. Gregg C, Zhang J, Butler JE, Haig D, Dulac C. Sex-specific parent-of-origin allelic expression in the mouse brain. *Science*. 2010;329(5992):682–685.
13. Gregg C, Zhang J, Weissbourd B, et al. High-resolution analysis of parent-of-origin allelic expression in the mouse brain. *Science*. 2010;329(5992):643–648.
14. DeVeale B, van der Kooy D, Babak T. Critical evaluation of imprinted gene expression by RNA-Seq: a new perspective. *PLoS Genet*. 2012;8(3):e1002600.
15. Hogart A, Nagarajan RP, Patzel KA, Yasui DH, Lasalle JM. 15q11-13 GABAA receptor genes are normally biallelically expressed in brain yet are subject to epigenetic dysregulation in autism-spectrum disorders. *Hum Mol Genet*. 2007;16(6):691–703.
16. Hogart A, Patzel KA, Lasalle JM. Gender influences monoallelic expression of ATP10A in human brain. *Hum Genet*. 2008;124(3):235–242.
17. Reik W, Walter J. Genomic imprinting: parental influence on the genome. *Nat Rev Genet*. 2001;2(1):21–32.
18. Wan LB, Bartolomei MS. Regulation of imprinting in clusters: noncoding RNAs versus insulators. *Adv Genet*. 2008;61:207–223.
19. Ginno PA, Lott PL, Christensen HC, Korf I, Chedin F. R-loop formation is a distinctive characteristic of unmethylated human CpG island promoters. *Mol Cell*. 2012;45(6):814–825.
20. Brockdorff N, Ashworth A, Kay GF, et al. The product of the mouse Xist gene is a 15kb inactive X-specific transcript containing no conserved ORF and located in the nucleus. *Cell*. 1992;71(3):515–526.
21. Brown CJ, Hendrich BD, Rupert JL, et al. The human XIST gene: analysis of a 17kb inactive X-specific RNA that contains conserved repeats and is highly localized within the nucleus. *Cell*. 1992;71(3):527–542.
22. Kay GF, Penny GD, Patel D, Ashworth A, Brockdorff N, Rastan S. Expression of Xist during mouse development suggests a role in the initiation of X chromosome inactivation. *Cell*. 1993;72(2):171–182.
23. Norris DP, Patel D, Kay GF, et al. Evidence that random and imprinted Xist expression is controlled by preemptive methylation. *Cell*. 1994;77(1):41–51.
24. Zhao J, Sun BK, Erwin JA, Song JJ, Lee JT. Polycomb proteins targeted by a short repeat RNA to the mouse X chromosome. *Science*. 2008;322(5902):750–756.
25. Takagi N, Sasaki M. Preferential inactivation of the paternally derived X chromosome in the extraembryonic membranes of the mouse. *Nature*. 1975;256(5519):640–642.
26. Ponting CP, Oliver PL, Reik W. Evolution and functions of long noncoding RNAs. *Cell*. 2009;136(4):629–641.
27. Redrup L, Branco MR, Perdeaux ER, et al. The long noncoding RNA Kcnq1ot1 organises a lineage-specific nuclear domain for epigenetic gene silencing. *Development*. 2009;136(4):525–530.
28. Runte M, Huttenhofer A, Gross S, Kiefmann M, Horsthemke B, Buiting K. The IC-SNURF-SNRPN transcript serves as a host for multiple small nucleolar RNA species and as an antisense RNA for UBE3A. *Hum Mol Genet*. 2001;10(23):2687–2700.
29. Vitali P, Royo H, Marty V, Bortolin-Cavaille ML, Cavaille J. Long nuclear-retained non-coding RNAs and allele-specific higher-order chromatin organization at imprinted snoRNA gene arrays. *J Cell Sci*. 2010; 123(Pt 1):70–83.
30. Kishore S, Stamm S. The snoRNA HBII-52 regulates alternative splicing of the serotonin receptor 2C. *Science*. 2006;311(5758):230–232.
31. Leung KN, Vallero RO, DuBose AJ, Resnick JL, LaSalle JM. Imprinting regulates mammalian snoRNA-encoding chromatin decondensation and neuronal nucleolar size. *Hum Mol Genet*. 2009;18(22):4227–4238.

32. Mercer TR, Dinger ME, Mariani J, Kosik KS, Mehler MF, Mattick JS. Noncoding RNAs in long-term memory formation. *Neuroscientist*. 2008;14(5):434–445.
33. Mercer TR, Dinger ME, Sunkin SM, Mehler MF, Mattick JS. Specific expression of long noncoding RNAs in the mouse brain. *Proc Natl Acad Sci USA*. 2008;105(2):716–721.
34. Mercer TR, Qureshi IA, Gokhan S, et al. Long noncoding RNAs in neuronal-glial fate specification and oligodendrocyte lineage maturation. *BMC Neurosci*. 2010;11:14.
35. Barton SC, Surani MA, Norris ML. Role of paternal and maternal genomes in mouse development. *Nature*. 1984;311(5984):374–376.
36. Surani MA, Barton SC, Norris ML. Development of reconstituted mouse eggs suggests imprinting of the genome during gametogenesis. *Nature*. 1984;308(5959):548–550.
37. Allen ND, Logan K, Lally G, Drage DJ, Norris ML, Keverne EB. Distribution of parthenogenetic cells in the mouse brain and their influence on brain development and behavior. *Proc Natl Acad Sci USA*. 1995;92(23):10782–10786.
38. Keverne EB, Fundele R, Narasimha M, Barton SC, Surani MA. Genomic imprinting and the differential roles of parental genomes in brain development. *Brain Res Dev Brain Res*. 1996;92(1):91–100.
39. Butler JV, Whittington JE, Holland AJ, Boer H, Clarke D, Webb T. Prevalence of, and risk factors for, physical ill-health in people with Prader-Willi syndrome: a population-based study. *Dev Med Child Neurol*. 2002;44(4):248–255.
40. Webb T, Clarke D, Hardy CA, Kilpatrick MW, Corbett J, Dahlitz M. A clinical, cytogenetic, and molecular study of 40 adults with the Prader-Willi syndrome. *J Med Genet*. 1995;32(3):181–185.
41. Clayton-Smith J, Pembrey ME. Angelman syndrome. *J Med Genet*. 1992;29(6):412–415.
42. Cassidy SB, Schwartz S, Miller JL, Driscoll DJ. Prader-Willi syndrome. *Genet Med*. 2012;14(1):10–26.
43. Chamberlain SJ, Lalande M. Angelman syndrome, a genomic imprinting disorder of the brain. *J Neurosci*. 2010;30(30):9958–9963.
44. de Smith AJ, Purmann C, Walters RG, et al. A deletion of the HBII-85 class of small nucleolar RNAs (snoRNAs) is associated with hyperphagia, obesity and hypogonadism. *Hum Mol Genet*. 2009;18(17):3257–3265.
45. Duker AL, Ballif BC, Bawle EV, et al. Paternally inherited microdeletion at 15q11.2 confirms a significant role for the SNORD116 C/D box snoRNA cluster in Prader-Willi syndrome. *Eur J Hum Genet*. 2010;18(11):1196–1201.
46. Sahoo T, del Gaudio D, German JR, et al. Prader-Willi phenotype caused by paternal deficiency for the HBII-85 C/D box small nucleolar RNA cluster. *Nat Genet*. 2008;40(6):719–721.
47. Chamberlain SJ, Brannan CI. The Prader-Willi syndrome imprinting center activates the paternally expressed murine Ube3a antisense transcript but represses paternal Ube3a. *Genomics*. 2001;73(3):316–322.
48. Cavaille J, Buiting K, Kiefmann M, et al. Identification of brain-specific and imprinted small nucleolar RNA genes exhibiting an unusual genomic organization. *Proc Natl Acad Sci USA*. 2000;97(26):14311–14316.
49. Cavaille J, Vitali P, Basyuk E, Huttenhofer A, Bachellerie JP. A novel brain-specific box C/D small nucleolar RNA processed from tandemly repeated introns of a noncoding RNA gene in rats. *J Biol Chem*. 2001;276(28):26374–26383.
50. de los Santos T, Schweizer J, Rees CA, Francke U. Small evolutionarily conserved RNA, resembling C/D box small nucleolar RNA, is transcribed from PWCR1, a novel imprinted gene in the Prader-Willi deletion region, which is highly expressed in brain. *Am J Hum Genet*. 2000;67(5):1067–1082.
51. Le Meur E, Watrin F, Landers M, Sturny R, Lalande M, Muscatelli F. Dynamic developmental regulation of the large non-coding RNA associated with the mouse 7C imprinted chromosomal region. *Dev Biol*. 2005;286(2):587–600.
52. Rougeulle C, Cardoso C, Fontes M, Colleaux L, Lalande M. An imprinted antisense RNA overlaps UBE3A and a second maternally expressed transcript. *Nat Genet*. 1998;19(1):15–16.
53. Meguro-Horike M, Yasui DH, Powell W, et al. Neuron-specific impairment of inter-chromosomal pairing and transcription in a novel model of human 15q-duplication syndrome. *Hum Mol Genet*. 2011;20(19):3798–3810.
54. Yasui DH, Scoles HA, Horike S, et al. 15q11.2-13.3 chromatin analysis reveals epigenetic regulation of CHRNA7 with deficiencies in Rett and autism brain. *Hum Mol Genet*. 2011;20(22):4311–4323.
55. Huang HS, Allen JA, Mabb AM, et al. Topoisomerase inhibitors unsilence the dormant allele of Ube3a in neurons. *Nature*. 2011;481(7380):185–189.
56. van Woerden GM, Harris KD, Hojjati MR, et al. Rescue of neurological deficits in a mouse model for Angelman syndrome by reduction of alphaCaMKII inhibitory phosphorylation. *Nat Neurosci*. 2007;10(3):280–282.

57. Greer PL, Hanayama R, Bloodgood BL, et al. The Angelman syndrome protein Ube3A regulates synapse development by ubiquitinating arc. *Cell*. 2010;140(5):704–716.
58. Margolis SS, Salogiannis J, Lipton DM, et al. EphB-mediated degradation of the RhoA GEF Ephexin5 relieves a developmental brake on excitatory synapse formation. *Cell*. 2010;143(3):442–455.
59. Dindot SV, Antalffy BA, Bhattacharjee MB, Beaudet AL. The Angelman syndrome ubiquitin ligase localizes to the synapse and nucleus, and maternal deficiency results in abnormal dendritic spine morphology. *Hum Mol Genet*. 2008;17(1):111–118.
60. Tsai TF, Jiang YH, Bressler J, Armstrong D, Beaudet AL. Paternal deletion from Snrpn to Ube3a in the mouse causes hypotonia, growth retardation and partial lethality and provides evidence for a gene contributing to Prader-Willi syndrome. *Hum Mol Genet*. 1999;8(8):1357–1364.
61. Ding F, Li HH, Zhang S, et al. SnoRNA Snord116 (Pwcr1/MBII-85) deletion causes growth deficiency and hyperphagia in mice. *PLoS One*. 2008;3(3):e1709.
62. Runte M, Kroisel PM, Gillessen-Kaesbach G, et al. SNURF-SNRPN and UBE3A transcript levels in patients with Angelman syndrome. *Hum Genet*. 2004;114(6):553–561.
63. Flomen RH, Collier DA, Osborne S, et al. Association study of CHRFAM7A copy number and 2bp deletion polymorphisms with schizophrenia and bipolar affective disorder. *Am J Med Genet B Neuropsychiatr Genet*. 2006;141B(6):571–575.
64. Helbig I, Mefford HC, Sharp AJ, et al. 15q13.3 microdeletions increase risk of idiopathic generalized epilepsy. *Nat Genet*. 2009;41(2):160–162.
65. Cook Jr EH, Lindgren V, Leventhal BL, et al. Autism or atypical autism in maternally but not paternally derived proximal 15q duplication. *Am J Hum Genet*. 1997;60(4):928–934.
66. Schanen NC. Epigenetics of autism spectrum disorders. *Hum Mol Genet*. 2006:R138–R150. [15 Spec No 2].
67. Dennis NR, Veltman MW, Thompson R, Craig E, Bolton PF, Thomas NS. Clinical findings in 33 subjects with large supernumerary marker(15) chromosomes and 3 subjects with triplication of 15q11-q13. *Am J Med Genet A*. 2006;140(5):434–441.
68. Scoles HA, Urraca N, Chadwick SW, Reiter LT, Lasalle JM. Increased copy number for methylated maternal 15q duplications leads to changes in gene and protein expression in human cortical samples. *Mol Autism*. 2011;2(1):19.
69. Samaco RC, Hogart A, LaSalle JM. Epigenetic overlap in autism-spectrum neurodevelopmental disorders: MECP2 deficiency causes reduced expression of UBE3A and GABRB3. *Hum Mol Genet*. 2005;14:483–492.
70. Thatcher K, Peddada S, Yasui D, LaSalle JM. Homologous pairing of 15q11-13 imprinted domains in brain is developmentally regulated but deficient in Rett and autism samples. *Hum Mol Genet*. 2005;14:785–797.
71. Nakatani J, Tamada K, Hatanaka F, et al. Abnormal behavior in a chromosome-engineered mouse model for human 15q11-13 duplication seen in autism. *Cell*. 2009;137(7):1235–1246.
72. Browne CE, Dennis NR, Maher E, et al. Inherited interstitial duplications of proximal 15q: genotype-phenotype correlations. *Am J Hum Genet*. 1997;61(6):1342–1352.
73. Mohandas TK, Park JP, Spellman RA, et al. Paternally derived de novo interstitial duplication of proximal 15q in a patient with developmental delay. *Am J Med Genet*. 1999;82(4):294–300.
74. Smith SE, Zhou YD, Zhang G, Jin Z, Stoppel DC, Anderson MP. Increased gene dosage of Ube3a results in autism traits and decreased glutamate synaptic transmission in mice. *Sci Transl Med*. 2011;3(103) [103ra97].
75. Curley JP. Is there a genomically imprinted social brain? *Bioessays*. 2011;33(9):662–668.
76. Li L, Keverne EB, Aparicio SA, Ishino F, Barton SC, Surani MA. Regulation of maternal behavior and offspring growth by paternally expressed Peg3. *Science*. 1999;284(5412):330–333.
77. Relaix F, Wei X, Li W, et al. Pw1/Peg3 is a potential cell death mediator and cooperates with Siah1a in p53-mediated apoptosis. *Proc Natl Acad Sci USA*. 2000;97(5):2105–2110.
78. Champagne FA, Curley JP, Swaney WT, Hasen NS, Keverne EB. Paternal influence on female behavior: the role of Peg3 in exploration, olfaction, and neuroendocrine regulation of maternal behavior of female mice. *Behav Neurosci*. 2009;123(3):469–480.
79. Swaney WT, Curley JP, Champagne FA, Keverne EB. The paternally expressed gene Peg3 regulates sexual experience-dependent preferences for estrous odors. *Behav Neurosci*. 2008;122(5):963–973.
80. Broad KD, Keverne EB. Placental protection of the fetal brain during short-term food deprivation. *Proc Natl Acad Sci USA*. 2011;108(37):15237–15241.
81. Flisikowski K, Venhoranta H, Nowacka-Woszuk J, et al. A novel mutation in the maternally imprinted PEG3 domain results in a loss of MIMT1 expression and causes abortions and stillbirths in cattle (Bos taurus). *PLoS One*. 2010;5(11):e15116.

82. Garfield AS, Cowley M, Smith FM, et al. Distinct physiological and behavioural functions for parental alleles of imprinted Grb10. *Nature*. 2011;469(7331):534–538.
83. Monk D, Arnaud P, Frost J, et al. Reciprocal imprinting of human GRB10 in placental trophoblast and brain: evolutionary conservation of reversed allelic expression. *Hum Mol Genet*. 2009;18(16):3066–3074.
84. Moore GE, Oakey R. The role of imprinted genes in humans. *Genome Biol*. 2011;12(3):106.
85. Haig D, Graham C. Genomic imprinting and the strange case of the insulin-like growth factor II receptor. *Cell*. 1991;64(6):1045–1046.
86. Nicholls RD. The impact of genomic imprinting for neurobehavioral and developmental disorders. *J Clin Invest*. 2000;105(4):413–418.

"Long INterspersed ElementS – Effect of Epigenetic Background"

J. David Sweatt, acrylic on wood panel (diptych of two 24 x 48 panels), 2011–2012

CHAPTER

11

Neuronal Genomic and Epigenetic Diversity

Michael J. McConnell[1,2,3] *and Fred H. Gage*[1]

[1]Laboratory of Genetics

[2]Crick-Jacobs Center for Theoretical and Computational Biology, Salk Institute for Biological Studies, La Jolla, California, USA

[3]Department of Biochemistry and Molecular Genetics, Center for Brain Immunology and Glia, School of Medicine, University of Virginia, Charlottesville, Virginia, USA

INTRODUCTION

Mobile element activity in eukaryotic genomes was first established through the pioneering work of Barbara McClintock in maize.[1] Mobile elements are now known in virtually every organism's genome and are generally considered to be parasites. Different elements are observed to specialize in different genomes. For example, at least 55 spontaneous mouse mutants are linked to one subset of mobile elements, endogenous retroviruses;[2] more than 60 human diseases have been attributed to other mobile elements, LINE1 and Alu retrotransposons.[3,4] Importantly however, mobile element activity is not strictly deleterious, as several examples of host genomes co-opting mobile element sequences, termed exaptation, have been reported.[5,6]

Genome sequencing technologies are revealing surprising levels of inter-individual variability in human genomes, and retrotransposon polymorphisms are among the most prevalent copy number variations (CNVs).[7,8] Curiously, Alu polymorphisms seem much more prevalent, suggesting elevated activity relative to LINE1s, perhaps owing to their small size and greater abundance, but their prevalence may also reflect a more neutral impact on the host genome and less negative selection in the population.

Mobile elements exhibit meiotic drive; they are transmitted by different rules than the rest of the genome. This quality underlies the notion of mobile elements as selfish genes because they achieve non-Mendelian (>50%) transmission to progeny.[9,10] Thus, mobile

Epigenetic Regulation in the Nervous System
DOI: http://dx.doi.org/10.1016/B978-0-12-391494-1.00011-2

elements are in intragenomic competition with host genomes (i.e. host genes are typically transmitted to progeny at Mendelian frequencies), leading to a natural evolution of host mechanisms that can flatten the playing field.[11]

Host genomes have evolved multiple epigenetic mechanisms to slow mobile element activity. Cytosine methylation and the PIWI system are major means by which epigenetic silencing acts to limit mobile element activity in the host germline. *Bricolage* – the idea that evolution "tinkers" with an existing mechanism, rather than creating a new mechanism from scratch[12] – posits that epigenetic genome regulation has been co-opted from its original roles in mobile element antagonism to pleiotropic roles in gene regulation during development and in the plasticity of mature multicellular systems.[13,14]

Herein, we focus on retrotransposon activity during brain development, with particular emphasis on the autonomous mobile element LINE1. We review retrotransposon biology and epigenetic silencing of mobile elements during germline development. We also review an ever-expanding literature describing multiple dimensions (e.g. aneuploidy, LINE1, and Alu) of genomic mosaicism among neurons. These data support our view that almost every neuron in an individual's brain has a unique version of that individual's genome, and necessarily complicate understanding of how epigenetic mechanisms operate in individual neurons and neural circuits. Finally, we promote the idea that single cell genomic and epigenomic approaches will be essential to understanding how intragenomic competition plays out somatically during brain development and over an individual's lifetime.

MOBILE ELEMENTS IN MAMMALIAN GENOMES

Mobile element-derived sequences account for ≈40% of the mammalian genome sequence.[15] Two main classes of mobile elements are known (Figure 11.1): DNA transposons that mobilize via a DNA intermediate, and retrotransposons that mobilize via an RNA intermediate.[16] DNA transposons use a "cut-and-paste" mechanism and change their location but not their copy number; retrotransposons use a "copy-and-paste" mechanism that amplifies their copy number and locations in the host genome. Although DNA transposons are generally thought to have become extinct in mammalian genomes, there is evidence that some DNA transposons have been active in mammalian genomes (e.g. bat,[17–19] wooly mammoth[20]). Nonetheless, retrotransposons are clearly the most prevalent class of active mobile elements in mammalian genomes.

Retrotransposons are broadly divided into two groups based on their promoters.[21,22] One group – including human endogenous retroviruses (HERV), mouse endogenous retroviruses (MERV) and intracisternal A particles (IAPs) – uses long terminal repeat (LTR)-containing promoters. The second group of retrotransposons includes non-LTR retrotransposons: LINEs (long interspersed nuclear elements) are transcribed by RNA polymerase II,[23] and SINEs (short interspersed nuclear elements) are transcribed by RNA polymerase III.[24] ERVs and LINEs have autonomous mobility; they encode the proteins that chaperone, target, and reverse transcribe their mRNA into a new genomic locus. SINEs, on the other hand, are non-autonomous; these ≈280 bp RNA molecules mobilize by hijacking the LINE machinery.

Bicistronic LINE transcripts encode two proteins: ORF1 and ORF2. After LINE1 mRNA is transported to the cytoplasm, these proteins are translated and preferentially associate in *cis* with the transcript from which they were derived.[25,26] ORF1 is an RNA binding protein that seems to act as

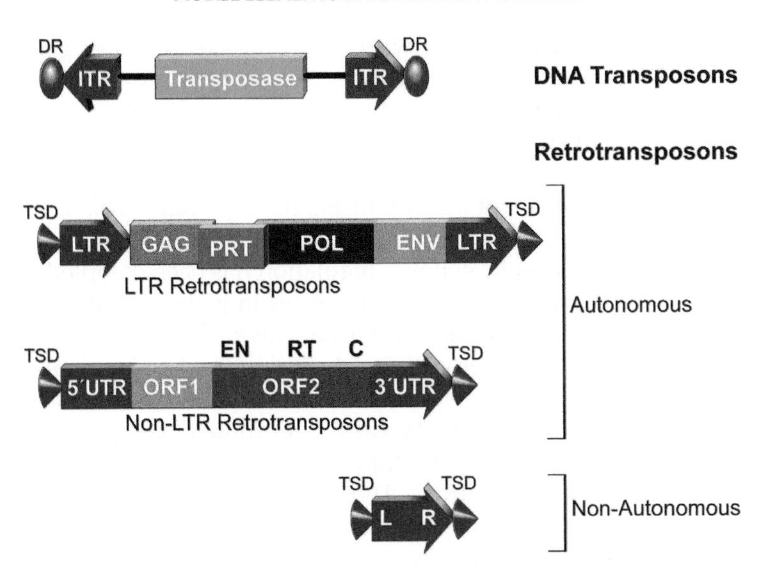

FIGURE 11.1 **The genomic structure of selfish genetic elements.** DNA transposons (e.g. mariner, ≈1.4 kb) encode a transposase flanked by inverted terminal repeats (ITR) and direct repeats (DR). Retrotransposons are broadly divided into those that are flanked by long terminal repeats (LTRs) and those that are not (non-LTR). LTR retrotransposons (e.g. HERV, ≈9.4 kb) are derived from retroviruses and these typically encode similar proteins: a group-specific antigen (GAG), a protease (PRT), a polymerase (POL) with endonuclease (EN) and reverse transcriptase (RT) activities, and a dysfunctional envelope protein (ENV). Non-LTR transposons can be autonomous (e.g. LINE, ≈6 kb) or non-autonomous (e.g. SINE, ≈280 bp), as discussed further in the main text. Retrotransposon insertions are flanked by target site duplications (TSD).

a chaperone,[27] whereas the ORF2 protein has both endonuclease and reverse transcriptase activity.[28,29] The endonuclease activity of ORF2 has limited sequence specificity (5′- TTTT/A -3′, and variants of this),[28,30,31] leaving many genomic loci "available" for new insertions. The predominant model for ORF2-mediated insertion of mRNA is described as target-primed reverse transcription (TPRT),[23] although this is not the only means for pasting LINE1 sequence into the genome.[32] The TPRT model proceeds as follows: after single strand nicking by ORF2, the polyA tail of the LINE1 transcript anneals to the overhanging polyT stretch which primes second strand synthesis. However, reverse transcription is not typically completed, leaving most new insertions truncated without a promoter or intact genes and, therefore, incapable of subsequent mobility.

Much of what is known about LINE1 activity derives from studies using a LINE1 reporter assay developed by John Moran and colleagues.[33,34] The key element of this approach is the incorporation of an antisense reporter gene (e.g. GFP) or selectable marker that is interrupted by an intron (Figure 11.2). Importantly, the intron has splice sites that are in the sense strand of the LINE1 but are antisense to the reporter. Thus, the reporter gene must pass through an RNA intermediate so that it can be spliced into an active form and then inserted into the genome, where an endogenous promoter can drive reporter gene expression. An important control, the same construct with an inactive ORF2, reports very little or no "leaky" reporter gene expression in this system.

SINEs are non-autonomous mobile elements; the predominant SINEs in humans are known as Alu elements.[35] A second group, SVA (named after three main components: SINE-R,

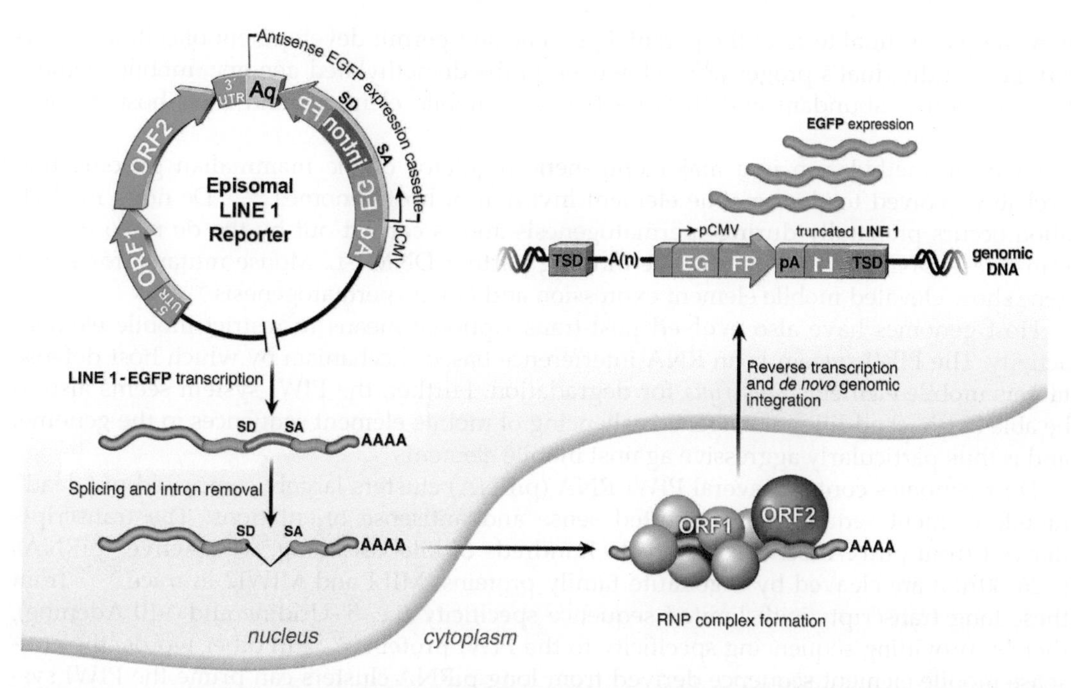

FIGURE 11.2 **Schematic of the LINE1 reporter assay.** In the prototypical assay, the LINE1 (L1) reporter, a full-length retrotransposon encoding active ORF1 and ORF2, is maintained in the nucleus as an episome. An antisense reporter gene expression cassette inserted into the 3′UTR of the retrotransposon is the key element for reporting retrotransposition. The reporter gene (e.g. GFP) is interrupted by an intron; however, the splice donor (SD) and splice acceptor (SA) sites are antisense to the reporter gene. Thus, the reporter transcript must pass through an RNA intermediate to create a functional reporter sequence. Retrotransposition mediated by ORF1 and ORF2 inserts the now-functional reporter expression cassette into the genome. Once retrotransposed, a constitutive promoter (e.g. CMV) drives expression of the reporter in cells that are retrotransposition competent.

VNTR and Alu), is less abundant but still active in human genomes.[36,37] The Alu sequence has evolved from 7SL RNA, a component of the signal recognition particle, and forms an extensive secondary structure. This structure of Alus is thought to favor their proximity to ongoing translation (i.e. association with ribosomes) and enables "hijacking" of newly translated LINE1 proteins. Thus, SINE RNA mobility is directly dependent on LINE1 transcription and translation.

Mobile element insertions have broad potential to alter a cell's transcriptome. De novo insertions into a gene can bring about alleles with either transcriptional or post-transcriptional changes. For example, the mobile element sequence contains transcription factor binding sites, alternative splice sites, antisense promoters, and premature polyadenylation signals.[22,38]

HOST DEFENSE

Near-global demethylation of host genomes during gametogenesis provides an important niche for mobile element survival. Demethylation of primordial germ cell genomes is essential to reset the parental genome and permit development of cellular diversity in an individual's progeny.[39,40] However, in the demethylated genome, mobile element transcription is abundant and the race between mobile elements and their host genome begins.[41]

Cytosine methylation is a major epigenetic regulator of the mammalian genome that probably evolved to slow mobile element invasion of host genomes.[42–45] De novo methylation occurs prenatally during spermatogenesis and is carried out by the de novo methyl transferase DNMT2a and an obligate binding partner DNMT-L. Mouse mutants for either gene show elevated mobile element expression and failed spermatogenesis.[46–48]

Host genomes have also evolved post-transcriptional means to restrict mobile element activity. The PIWI system is an RNA-interference based mechanism by which host defense targets mobile element *transcripts* for degradation. Further, the PIWI system seems also to be able to direct additional epigenetic silencing of mobile element sequences in the genome, and is thus particularly aggressive against mobile elements.[49]

Host genomes contain several PIWI RNA (piRNA) clusters largely comprised of "dead" mobile element sequences in inverted sense and antisense orientations. The transcripts derived from piRNA loci can be tens to hundreds of kilobases long.[50–52] "Active" piRNAs (≈26–30 nts) are cleaved by argonaute family proteins, MILI and MIWI2 in mice,[53–56] from these long transcripts with limited sequence specificity (i.e. 5′-Uridine and +10 Adenine), thereby providing sequencing specificity to the PIWI proteins.[57–59] In other words, the antisense mobile element sequence derived from long piRNA clusters can prime the PIWI system to target "active" (e.g. retrotransposition competent) mobile element transcripts directly and degrade them.[53,60] Furthermore, during transcript degradation, part of the transposon-derived RNA becomes loaded on a second PIWI protein. Now loaded with sense sequence from an actively transcribed mobile element, this second PIWI molecule initiates a so-called "ping pong cycle", where transposon sequence-loaded PIWI complexes can return to the piRNA transcripts and can selectively generate new piRNAs with increasing specificity toward actively transcribed mobile element sequences.

A novel element of the PIWI system is the control of transcriptional repression via cytosine methylation.[61] Although the mechanism is unclear, PIWI protein-deficient mice show clear deficits in cytosine methylation and abundant mobile element transcription.[53–55] Male PIWI knockout mice, like DNMT knockout mice, are sterile due to failed spermatogenesis. One model posits that the RNA sequence loaded on to a PIWI protein guides the PIWI complex to a mobile element sequence in the genome. However, the link between PIWI complexes and cytosine methylation does not appear to be direct, and may involve sequence-directed histone modifications that promote subsequent methylation and transcriptional inactivation of the mobile element sequence.

SOMATIC MOBILE ELEMENT ACTIVITY

Despite multiple host defense mechanisms, mobile elements survive in mammalian genomes;[62–64] current estimates suggest that a new full-length LINE insertion occurs surprisingly frequently (<1/100 human genomes). Mobile element survival is promoted when new insertions contribute to the progeny germline. Moreover, sexual reproduction provides

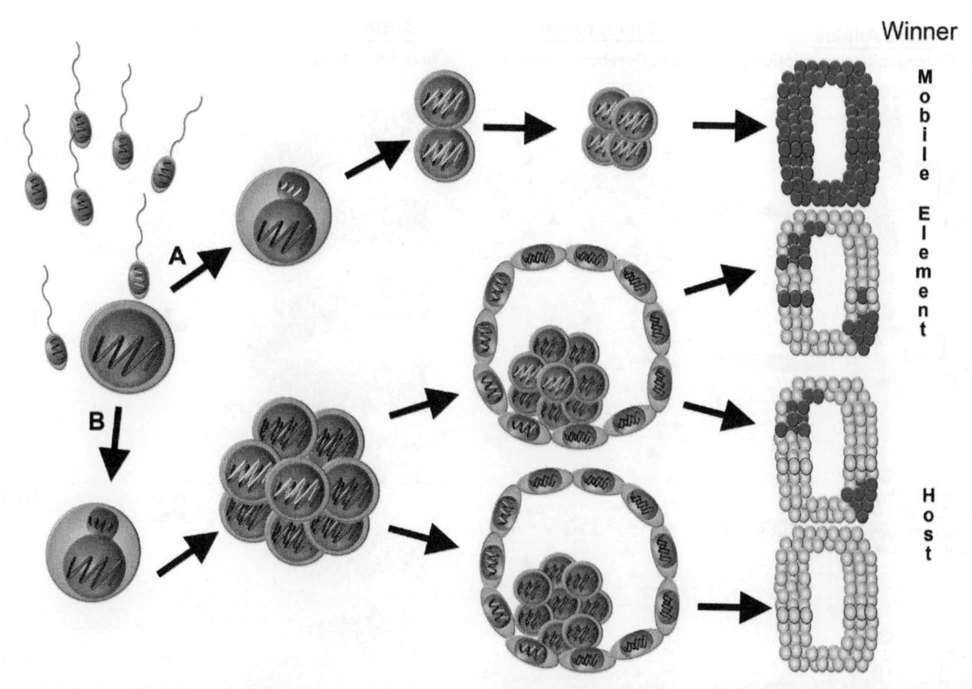

FIGURE 11.3 **Mobile elements race against host defense to reach the progeny germline.** (A) When mobile elements "jump" in the parental germline (red double helix), that gamete succeeds in fertilization, and does not abrogate progeny reproduction; that element wins the race against host genes. (B) When de novo insertions occur after fertilization, the mobile element may or may not contribute a new polymorphism to the progeny germline. For example, a "jump" that is completed during the second cell division after fertilization will be present in 1 of 4 cells and, after the next cell cycle, 2 of 8 cells in the 8-cell embryo. To reach the progeny germline, a cell with a de novo insertion (red cell) must become a primordial germ cell progenitor (dark, outlined cells).

novel habitats for mobile elements to invade: for example, deleterious de novo insertions affecting one allele can be compensated from the unaffected allele in the other parent's genome.[65] Thus, a new insertion in a successful gamete is the most direct way for all cells in the embryo to contain a de novo polymorphism. However, new insertions that occur after fertilization present various possibilities; the contribution of polymorphic genomes to the individual may not always contribute to that individual's primordial germ cell population.

As illustrated in Figure 11.3, insertions that occur during early embryogenesis (e.g. in one of 2, 4, or 8 cells) do not necessarily contribute to the germline. Precedent for segregation of abnormal genomes to extra-embryonic tissues exists. Aneuploidy is observed in the placenta from karyotypically normal human births,[66] and recent single cell genomic analysis of human 8-cell embryos showed a surprisingly high prevalence of mosaic genomic copy number variations,[67] far higher than the occurrence of mosaic abnormalities in the general population (≈1%).[68,69] In addition, mosaicism in the embryonic stem cell (ESC) population does necessarily contribute to the primordial germ cell population (e.g. consider the not insignificant frequency of failed germline transmission during transgenic mouse creation).

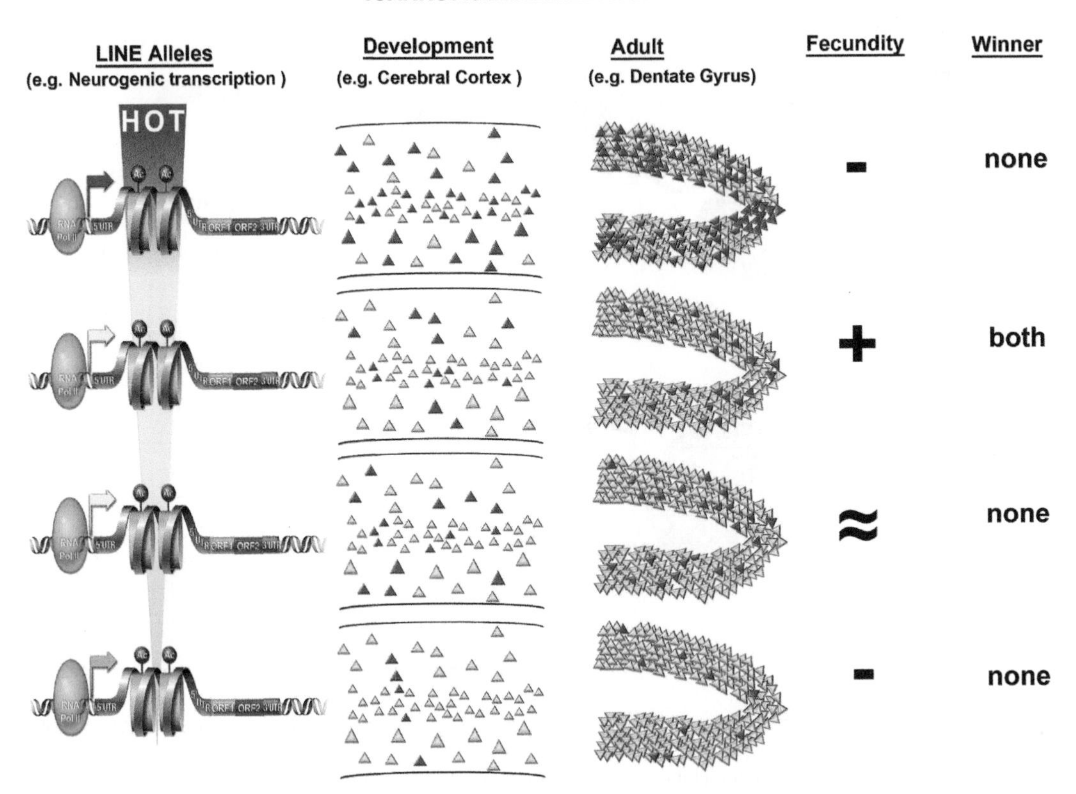

FIGURE 11.4 **Potential for symbiotic somatic mobility.** Speculative scenarios are presented to illustrate how somatic mobile element activity could be beneficial to both host and mobile element survival. Some LINE1 alleles are "hot" while others are less active.[62,75,76] Differential LINE1 activity alters the size of the neuronal population with diverse genomes (red cells). Importantly, red cells represent the diverse population, thus each red cell might be expected to have a unique genome; yellow cells are expected to have the same genome. We suggest that either too much (top row) or too little mosaicism (bottom row) is detrimental to the performance of neural circuits and, thereby, limits reproductive success (i.e. fecundity). However, some "normal" level of mosaicism (brought about by yellow LINE1 alleles) might or might not affect individual behavior in beneficial ways, perhaps due to the specific genomic differences, or specific neuronal cell types, that occur.

It has traditionally been thought that mobile element insertions occur concomitant with their transcription during gametogenesis. However, recent data from the Kazazian laboratory have re-written this view.[70] In a study of transgenic mouse lines containing various genomically encoded versions of the LINE1 reporter described above, the authors report that new insertions typically occurred after fertilization, leading to somatic mosaicism in progeny.[71] Given the stability of LINE1 ribonucleoprotein complexes,[72] it seems possible that these complexes could form during gametogenesis and persist through fertilization.

Contribution to the germline, while direct, is not the only means by which mobile elements can survive.[73,74] Although controversial, the possibility exists that mobile element activity in somatic cells can influence the survival and reproductive fitness of an individual (Figure 11.4). Thus, mobile elements may not always be *selfish* parasites. Alleles with a propensity to jump in somatic tissues can also be successful (i.e. increase their frequency in the population) if their somatic activity leads to increased host survival and reproduction (i.e. fecundity). This idea seems especially attractive given recent studies that show preferential activation of LINE elements in neurons, as an individual's behavior is a direct consequence of neuronal function.

GENOME DIVERSITY AMONG NEURONS

Data accumulated over the past 12 years show that some subsets of neuronal genomes are variants of the individual's genome. This idea was put forward more than 40 years ago when Dreyer, Gray and Hood were considering the problem of target specificity for neuronal axons.[77] Although somatic gene rearrangement was not explicitly implicated, the authors surmised that the mechanisms underlying target choice in the nervous system might be related to the mechanisms that generate antibody diversity in the immune system. After Tonegawa[78] showed that somatic gene rearrangement generates antibody diversity in the immune system, Schatz, Baltimore, and colleagues[79,80] showed that the RAG genes mediate this rearrangement. The possibility of gene rearrangement in the brain gained traction when Chun and colleagues[81] showed expression of one RAG gene, RAG-1, in the brain. Follow-up studies by others, however, showed that lymphocyte-like gene rearrangement (e.g. V(D)J recombination) does not operate in neuronal genomes.[82,83] Later studies further ruled out somatic DNA rearrangement at candidate loci like the protocadherins or olfactory receptor clusters.[84,85] Instead, neuronal genomes seem to diversify via less restrictive means.

The first clear demonstration of genomic diversity among neurons showed that whole chromosome gains and losses (i.e. aneuploidy) occurred in ≈33% of neural progenitor cells (NPCs) during mouse cerebral cortical development[86] and that some aneuploid NPCs survived and went on to become functional neurons.[87] Follow-up studies showed that aneuploidy in neurons was brought about by chromosome missegregation and led to altered gene expression.[88,89] Current estimates are that 10% of human neurons are aneuploid;[90,91] however, euploid neurons also have diverse genomes.

Mobile element activity diversifies neuronal genomes even further.[38] Several lines of in vitro and in vivo evidence show high-level activation of LINE1 retrotransposition during neurogenesis. First, transcription from LINE1 promoters occurs concomitantly with neurogenesis. Transcription from LINE1 promoters is controlled similarly to the pro-neuronogenic transcription factor NeuroD.[94] Hybrid SOX/LEF binding sites in LINE1 promoters are repressed when occupied by the neural stem cell (NSC) transcription factor SOX2. SOX2 levels go down as NSCs differentiate into NPCs and then neurons.[95] This progression is driven, in part, by WNT signaling via β-catenin. A β-catenin:TCF/LEF complex then occupies SOX/LEF sites and transactivates both LINE1 and NeuroD promoters. Endogenous LINE1 transcripts from retrotransposition-competent subfamilies are also detected during the differentiation of human embryonic stem cell (hESC)-derived NPCs into neurons.[95] Second, the gold standard assay for retrotransposition competence reports high levels of activity in many neurogenesis paradigms. In vitro, this is a plasmid-based assay that reports GFP fluorescence only after the plasmid-based LINE1 element progresses through an RNA

BOX 11.1

WHAT DOES GENOMIC DIVERSITY AMONG NEURONS DO?

Answers to this question are necessarily speculative because the technological ability to measure the genomic, epigenomic, transcriptomic, and proteomic states of hundreds of individual neurons is not sufficiently advanced. Nonetheless, two observations provide straightforward hypotheses. For one, during cortical neurogenesis, sister neurons in different cortical layers have been shown preferentially to form synapses with each other.[92] The source of information that directs sister progeny connectivity is still unknown; however, it is clear that these sister neurons are progeny of the same NPC. One hypothesis follows from the eventuality that genomic changes in each NPC will be shared among their subsequent progeny. Perhaps genomic changes provide lineage information that influences the development of neuronal connectivity. A second study reports that neurons in the same olfactory glomerulus have different outputs to identical stimuli.[93] This property is further shown to increase the information content of the neural circuit. Thus, one hypothesis would contend that genomic diversity within the glomerulus could alter the activation properties of each neuron leading to the diverse responses observed. In either scenario, altered levels or types of genomic diversity could directly impact the development and function of neural circuits.

intermediate and integrates into the genome. In vivo evidence of retrotransposition competence in neurons comes from a transgenic mouse with a genomically encoded reporter. These mice show GFP-positive neurons throughout the brain.[96] Importantly, non-neuronal GFP-positive cells were not observed in these brains. Third, analysis of post-mortem human brains shows additional LINE1 DNA in neurons relative to other somatic tissues examined. Using taqman qPCR, with multiple non-mobile control loci, Coufal and colleagues[95] reported higher levels of LINE1 DNA in bulk brain tissue. By spiking these samples with known copies of LINE1 DNA, it was estimated that neurons contain between 80 and 800 additional copies on average. A targeted sequencing approach recently revealed very high levels of mobile element insertions in brain genomes that were consistent with taqman qPCR estimates of intra-individual variability between brain regions.[97]

Perhaps the most interesting aspect of LINE1 activation during neurogenesis is that it indicates that the race between parasitic DNA and host defense plays out again during an individual's development. LINE1 promoters contain methyl CpG islands that are typically methylated and inactivated early during development, but LINE promoters are hypomethylated in the brain.[95] The methylated DNA binding protein MeCP2, together with the histone deacetylase HDAC1, occupies LINE1 CpG islands in NSCs, but much less occupancy is observed in NPCs and neurons. Furthermore, the absence of MeCP2 (which in humans leads to Rett's syndrome, an autism spectrum disorder) leads to additional activation of the LINE1-GFP reporter both in vitro and in vivo, as well as elevated levels of LINE1 DNA in neurons as detected by qPCR.[98,99]

De novo LINE1 insertions in neuronal genomes have broad potential to alter neuronal transcriptomes on a neuron-to-neuron basis.[38,100] For example, a clonal NSC line was found to have a marked (by the retrotransposition reporter gene) insertion into an intron of the neuronal gene PSD-93. Rather than ameliorate expression of this gene, the LINE1 insertion led to elevated expression and a pro-neuronogenic phenotype.[96] Many other such insertions have been mapped, and clonally derived cell lines show both positive and negative effects of LINE1 on neuronal transcriptomes in both human and rodent systems.[95,96] Furthermore, mobile element activity can lead to additional genomic diversity which, in turn, can engage epigenetic mechanisms differently in different neurons.

ADDITIONAL GENOMIC DIVERSITY CAN FOLLOW FROM RETROTRANSPOSON ACTIVITY

The current model for retrotransposon insertion begins with TPRT, as discussed above. However, the details surrounding the resolution of new mobile element insertion are less clear. One clue comes from the formation of target site duplications that flank new insertions. Here, the site of initial single strand nicking by ORF2 is duplicated at the 5′ end of new insertions, which implies that a double strand break (DSB) occurred at some point during cDNA synthesis, leaving the complementary strand (to the nick) to prime second strand synthesis.

DSB formation has been associated with LINE1 activity in a variety of experimental settings. First, transient expression of a LINE1 reporter in HeLa cells leads to phosphorylation of a histone subunit, γ-H2aX, and detection of DSBs as phospho-γH2aX foci.[101] Here, γ-H2aX foci are dependent on both the DNA damage signaling protein, ATM, and active ORF2. Second, an additional marker of DNA DSBs, p53 binding protein, is observed to form ORF2-dependent foci in HeLa and other human cell lines.[102] Third, genetic studies in human and rodent cell lines indicate that the ERCC1/ XPF flap endonuclease, a component of nucleotide excision repair that also participates in the resolution of DSBs, acts to limit retrotransposition.[103] Finally, other genetic studies in chicken DT40 cells also implicate the non-homologous end-joining (NHEJ) DSB repair pathway in limiting retrotransposition.[104]

However, the interplay between DNA repair and retrotransposition is clearly more complicated. Constitutive expression of LINE1 reporters in HeLa cells upregulates DNA repair mechanisms that reduce DSB formation and limit retrotransposition.[105] Yet, in cells deficient for NHEJ components, the retrotransposon sequence seems to facilitate DNA repair, particularly at telomeres.[106] Furthermore, in *Atm*-deficient mice, an elevated prevalence of somatic LINE1 insertions is observed.[107] Moreover, DNA repair-associated histone modifications seem to induce local epigenetic silencing around de novo insertions.[108]

Mobile element activity, and associated DNA damage and repair, can bring about additional changes in a cell's genome. Subchromosomal deletions have been attributed to LINE1 and Alu insertions.[109–111] As well, new mobile element insertions on one chromosome can mediate non-allelic homologous recombination leading to deletions, amplifications, and inversions.[112,113]

NATURE, NURTURE AND NEURONAL GENOMES

The interaction between nature (i.e. genes) and nurture (i.e. environment) plays out directly in neural circuits. Following from the pioneering studies of Hubel and Wiesel,[114,115] ocular dominance plasticity in the visual system illustrates this interaction directly.[116] Very briefly, binocular zones in visual cortex receive environmental input from both eyes, in contrast to surrounding monocular zones that receive input only from the contralateral eye (e.g. left eye to right cortex). When input to one eye is removed, termed monocular deprivation, the ipsilateral (e.g. left eye to left cortex) projection from the open eye is observed to expand into monocular zones that have been deprived of input. Thus, environmental inputs can lead to functional, structural modification of neural circuits that were initially formed by genetic programs during development.[117,118]

Behavior is another environmental input that shapes neural circuits. As described in more detail in other chapters, recreational drug use leads to long-lasting changes in neural circuits that reinforce addictive behavior. These changes are brought about by epigenetic modifications of neuronal genomes that lead to chromatin reorganization in specific neurons and alter gene expression.[119] Expanding from these observations and others, epigenetic modifications have been put forward to explain non-genetic differences between individuals as revealed through studies of monozygotic twins.[120–124] A shortcoming of this view is that it is grounded in the assumption that all neurons in an individual have identical genomes.

In fact, the race between mobile elements and host defense may sometimes be initiated by environmental stimuli. When LINE1 reporter mice are provided a running wheel – an environmental enrichment known to increase neurogenesis in the adult hippocampus[125,126] – elevated LINE1 retrotransposition is observed among newborn dentate granule neurons.[127] Also intriguing is the observation that mobile element transcripts become abundant during epigenetic modification in an addiction model,[128] although the relationship between transcript abundance and mobility in this setting is not yet known. Furthermore, some components of the PIWI system have also been detected in mouse neurons,[129] and the PIWI system was very recently implicated in the plasticity of an invertebrate's neural circuits.[130]

HOW DO WE STUDY EPIGENETIC INTERACTIONS WITH NEURONAL MOSAICISM?

Even without considering neuronal mosaicism, one already expects neuron-to-neuron epigenomic differences in any given brain region or neuronal cell type. These differences are demonstrated at least in part by the observation that DNA methylation levels among neurons are more variable than among non-neuronal genomes in the same brain.[131] However, when the epigenetic changes observed in a particular brain region are altered by an experimental manipulation, like drug administration, the changes reported are derived from a population of neurons, resulting in a population average. An open question is whether the change is occurring in all the neurons or in a fraction of the neurons (e.g. 40% or 90%). Moreover, extending findings from twin studies, the percentage of neurons with particular epigenetically encoded modifications could be expected to vary from individual to individual, despite the common assumption that these individuals have the same genome in every neuron. Genomic diversity among neurons adds significant complexity to this model, perhaps increasing the likelihood of non-genetic variability in individual behavior.

A careful description of genomic diversity is critical for understanding how mosaic neuronal genomes can alter the epigenetic potential of neurons and neural circuits, yet sufficient single cell resolution to obtain this description is lacking. Nonetheless, we are able to glimpse some pathological consequences associated with altered mosaics of neuronal genomes: elevated levels of chromosome 1 mosaicism are associated with schizophrenia,[132] chromosome 21 mosaicism is associated with Alzheimer's disease,[133] and broadly more aneuploidy, as well as mosaic chromosomal translocations, have been associated with Ataxia-telangiectasia.[134,135]

Levels of endogenous retrotransposition are also difficult to ascertain. Rough estimates obtained by spiking LINE1 DNA into bulk samples of non-CNS DNA suggest that human neurons contain between 80 and 800 more L1 copies than non-CNS cells.[95] However, it is unknown whether this estimate represents thousands (or tens of thousands) of insertions in a subset of neurons or ≈500 insertions in most neurons. Further, if most neurons have an average number of insertions, owing to the minimal insertion site specificity of the ORF2 endonuclease, one could expect substantial variability among the insertion sites in each neuron. Moreover, other mobile elements (Alus and SVAs) also diversify neuronal genomes, apparently to different extents in different individuals.[97] Taking these findings together with one report of a large CNV present in one brain region but not another (in the same individual),[136] distinct CNVs in phenotypically discordant monozygotic twins,[137] extensive DNA content variation (many neurons with >100 mb of additional DNA) among human frontal cortical neurons,[138] and there is every indication that neural circuits in the mammalian brain are formed from a multidimensional mosaic of unique neuronal genomes.

We propose that the evolutionary race between mobile elements and host defense plays out during neurogenesis, setting individualized courses for brain development and plasticity. Neurons with distinct genomes should also have different epigenomes. In a straightforward scenario, mobile element insertions into genes that encode chromatin-modifying enzymes could lead to subsets of neurons that have either increased or decreased propensity (relative to neighboring neurons) to carry out epigenetic modifications. Likewise, new LINE insertions seem to instigate local repression through histone modifications, and near full-length LINE insertions will insert de novo CpG islands. Short Alu elements are cytosine-rich; thus, new Alu insertions have the direct ability to alter the potential epigenetic landscape on a neuron-to-neuron basis. Moreover, large changes in DNA copy number (e.g. aneuploidy) could readily engage dosage compensation-type mechanisms of epigenetic control[139] as well as changing the copy number of genes whose epigenetic regulation has been shown to directly alter neural function.

CONCLUDING REMARKS

Single cell approaches will be essential for unraveling the complex mosaic of genomes and epigenomes in mammalian brains, as well as the effect of mosaic genomes on the performance of neural circuits. Science is just beginning to glimpse genomic diversity at the single cell level. Recent single cell genomic studies have found remarkable levels and types of heterogeneity in tumors and in 8-cell human embryos.[67,140] Combined single cell genomic and epigenomic technologies are likely on the horizon. The application of these approaches to neuroscience promises substantial insight into how nature and nurture interact at the level of neural circuits over an individual's lifetime.

Acknowledgments

We thank A. Denli, J. Erwin, J. Han and M. Kagalwala for insightful discussions, J. Simon for assistance with artwork, and M.L. Gage for editorial assistance. We regret that space constraints invariably lead to the omission of some relevant references. This work was supported by a Crick-Jacobs Junior Fellowship to MJM and an NIH Director's Transformative Research Award (R01 MH095741) to FHG.

References

1. McClintock B. The origin and behavior of mutable loci in maize. *Proc Natl Acad Sci USA*. 1950;36(6):344–355.
2. Maksakova IA, Romanish MT, Gagnier L, Dunn CA, van de Lagemaat LN, Mager DL. Retroviral elements and their hosts: insertional mutagenesis in the mouse germ line. *PLoS Genet*. 2006;2(1):e2.
3. Beck CR, Garcia-Perez JL, Badge RM, Moran JV. LINE-1 elements in structural variation and disease. *Annu Rev Genomics Hum Genet*. 2011;12:187–215.
4. Solyom S, Kazazian Jr HH. Mobile elements in the human genome: implications for disease. *Genome Med*. 2012;4(2):12.
5. Bourque G. Transposable elements in gene regulation and in the evolution of vertebrate genomes. *Curr Opin Genet Dev*. 2009;19(6):607–612.
6. Feschotte C. Transposable elements and the evolution of regulatory networks. *Nat Rev Genet*. 2008;9(5):397–405.
7. Lander ES, Linton LM, Birren B, et al. Initial sequencing and analysis of the human genome. *Nature*. 2001;409(6822):860–921.
8. Mills RE, Walter K, Stewart C, et al. Mapping copy number variation by population-scale genome sequencing. *Nature*. 2011;470(7332):59–65.
9. Doolittle WF, Sapienza C. Selfish genes, the phenotype paradigm and genome evolution. *Nature*. 1980;284(5757):601–603.
10. Orgel LE, Crick FH, Sapienza C. Selfish DNA. *Nature*. 1980;288(5792):645–646.
11. Burt A, Trivers R. *Genes in conflict: the biology of selfish genetic elements*. Cambridge, Mass: Belknap Press of Harvard University Press; 2006.
12. Jacob F. Evolution and tinkering. *Science*. 1977;196(4295):1161–1166.
13. Suzuki S, Ono R, Narita T, et al. Retrotransposon silencing by DNA methylation can drive mammalian genomic imprinting. *PLoS Genet*. 2007;3(4):e55.
14. Suzuki S, Shaw G, Kaneko-Ishino T, Ishino F, Renfree MB. The evolution of mammalian genomic imprinting was accompanied by the acquisition of novel CpG islands. *Genome Biol Evol*. 2011;3:1276–1283.
15. Mandal PK, Kazazian Jr HH. SnapShot: vertebrate transposons. *Cell*. 2008;135(1):192. [e1].
16. Craig NL. *Mobile DNA II*. Washington, D.C. ASM Press; 2002.
17. Pritham EJ, Feschotte C. Massive amplification of rolling-circle transposons in the lineage of the bat Myotis lucifugus. *Proc Natl Acad Sci USA*. 2007;104(6):1895–1900.
18. Ray DA, Feschotte C, Pagan HJ, et al. Multiple waves of recent DNA transposon activity in the bat, Myotis lucifugus. *Genome Res*. 2008;18(5):717–728.
19. Ray DA, Pagan HJ, Thompson ML, Stevens RD. Bats with hATs: evidence for recent DNA transposon activity in genus Myotis. *Mol Biol Evol*. 2007;24(3):632–639.
20. Zhao F, Qi J, Schuster SC. Tracking the past: interspersed repeats in an extinct Afrotherian mammal, Mammuthus primigenius. *Genome Res*. 2009;19(8):1384–1392.
21. Deininger PL, Moran JV, Batzer MA, Kazazian Jr HH. Mobile elements and mammalian genome evolution. *Curr Opin Genet Dev*. 2003;13(6):651–658.

22. Goodier JL, Kazazian Jr HH. Retrotransposons revisited: the restraint and rehabilitation of parasites. *Cell.* 2008;135(1):23–35.
23. Ostertag EM, Kazazian Jr HH. Biology of mammalian L1 retrotransposons. *Annu Rev Genet.* 2001;35:501–538.
24. Roy AM, West NC, Rao A, et al. Upstream flanking sequences and transcription of SINEs. *J Mol Biol.* 2000;302(1):17–25.
25. Kulpa DA, Moran JV. Cis-preferential LINE-1 reverse transcriptase activity in ribonucleoprotein particles. *Nat Struct Mol Biol.* 2006;13(7):655–660.
26. Wei W, Gilbert N, Ooi SL, et al. Human L1 retrotransposition: cis preference versus trans complementation. *Mol Cell Biol.* 2001;21(4):1429–1439.
27. Hohjoh H, Singer MF. Sequence-specific single-strand RNA binding protein encoded by the human LINE-1 retrotransposon. *EMBO J.* 1997;16(19):6034–6043.
28. Feng Q, Moran JV, Kazazian Jr HH, Boeke JD. Human L1 retrotransposon encodes a conserved endonuclease required for retrotransposition. *Cell.* 1996;87(5):905–916.
29. Mathias SL, Scott AF, Kazazian Jr HH, Boeke JD, Gabriel A. Reverse transcriptase encoded by a human transposable element. *Science.* 1991;254(5039):1808–1810.
30. Cost GJ, Boeke JD. Targeting of human retrotransposon integration is directed by the specificity of the L1 endonuclease for regions of unusual DNA structure. *Biochemistry.* 1998;37(51):18081–18093.
31. Jurka J. Sequence patterns indicate an enzymatic involvement in integration of mammalian retroposons. *Proc Natl Acad Sci USA.* 1997;94(5):1872–1877.
32. Morrish TA, Gilbert N, Myers JS, et al. DNA repair mediated by endonuclease-independent LINE-1 retrotransposition. *Nat Genet.* 2002;31(2):159–165.
33. Moran JV, Holmes SE, Naas TP, DeBerardinis RJ, Boeke JD, Kazazian Jr HH. High frequency retrotransposition in cultured mammalian cells. *Cell.* 1996;87(5):917–927.
34. Rangwala SH, Kazazian Jr HH. The L1 retrotransposition assay: a retrospective and toolkit. *Methods.* 2009;49(3):219–226.
35. Bennett EA, Keller H, Mills RE, et al. Active Alu retrotransposons in the human genome. *Genome Res.* 2008;18(12):1875–1883.
36. Damert A, Raiz J, Horn AV, et al. 5'-Transducing SVA retrotransposon groups spread efficiently throughout the human genome. *Genome Res.* 2009;19(11):1992–2008.
37. Ostertag EM, Goodier JL, Zhang Y, Kazazian Jr HH. SVA elements are nonautonomous retrotransposons that cause disease in humans. *Am J Hum Genet.* 2003;73(6):1444–1451.
38. Singer T, McConnell MJ, Marchetto MC, Coufal NG, Gage FH. LINE-1 retrotransposons: mediators of somatic variation in neuronal genomes?. *Trends Neurosci.* 2010;33(8):345–354. [Epub 2010 May 12. PubMed PMID: 20471112; PubMed Central PMCID: PMC2916067].
39. Hajkova P, Erhardt S, Lane N, et al. Epigenetic reprogramming in mouse primordial germ cells. *Mech Dev.* 2002;117(1–2):15–23.
40. Surani MA, Hajkova P. Epigenetic reprogramming of mouse germ cells toward totipotency. *Cold Spring Harb Symp Quant Biol.* 2010;75:211–218.
41. Branciforte D, Martin SL. Developmental and cell type specificity of LINE-1 expression in mouse testis: implications for transposition. *Mol Cell Biol.* 1994;14(4):2584–2592.
42. Beauregard A, Curcio MJ, Belfort M. The take and give between retrotransposable elements and their hosts. *Annu Rev Genet.* 2008;42:587–617.
43. Levin HL, Moran JV. Dynamic interactions between transposable elements and their hosts. *Nat Rev Genet* 2011;12(9):615–627.
44. Munoz-Lopez M, Macia A, Garcia-Canadas M, Badge RM, Garcia-Perez JL. An epi [c] genetic battle: LINE-1 retrotransposons and intragenomic conflict in humans. *Mob Genet Elements.* 2011;1(2):122–127.
45. Yoder JA, Walsh CP, Bestor TH. Cytosine methylation and the ecology of intragenomic parasites. *Trends Genet.* 1997;13(8):335–340.
46. Bourc'his D, Bestor TH. Meiotic catastrophe and retrotransposon reactivation in male germ cells lacking Dnmt3L. *Nature.* 2004;431(7004):96–99.
47. Li E, Bestor TH, Jaenisch R. Targeted mutation of the DNA methyltransferase gene results in embryonic lethality. *Cell.* 1992;69(6):915–926.
48. Okano M, Bell DW, Haber DA, Li E. DNA methyltransferases Dnmt3a and Dnmt3b are essential for de novo methylation and mammalian development. *Cell.* 1999;99(3):247–257.

49. Aravin AA, Hannon GJ, Brennecke J. The Piwi-piRNA pathway provides an adaptive defense in the transposon arms race. *Science*. 2007;318(5851):761–764.
50. Aravin A, Gaidatzis D, Pfeffer S, et al. A novel class of small RNAs bind to MILI protein in mouse testes. *Nature*. 2006;442(7099):203–207.
51. Girard A, Sachidanandam R, Hannon GJ, Carmell MA. A germline-specific class of small RNAs binds mammalian Piwi proteins. *Nature*. 2006;442(7099):199–202.
52. Lau NC, Seto AG, Kim J, et al. Characterization of the piRNA complex from rat testes. *Science*. 2006;313(5785):363–367.
53. Carmell MA, Girard A, van de Kant HJ, et al. MIWI2 is essential for spermatogenesis and repression of transposons in the mouse male germline. *Dev Cell*. 2007;12(4):503–514.
54. Deng W, Lin H. miwi, a murine homolog of piwi, encodes a cytoplasmic protein essential for spermatogenesis. *Dev Cell*. 2002;2(6):819–830.
55. Kuramochi-Miyagawa S, Kimura T, Ijiri TW, et al. Mili, a mammalian member of piwi family gene, is essential for spermatogenesis. *Development*. 2004;131(4):839–849.
56. Kuramochi-Miyagawa S, Kimura T, Yomogida K, et al. Two mouse piwi-related genes: miwi and mili. *Mech Dev*. 2001;108(1–2):121–133.
57. Gunawardane LS, Saito K, Nishida KM, et al. A slicer-mediated mechanism for repeat-associated siRNA 5' end formation in Drosophila. *Science*. 2007;315(5818):1587–1590.
58. Saito K, Nishida KM, Mori T, et al. Specific association of Piwi with rasiRNAs derived from retrotransposon and heterochromatic regions in the Drosophila genome. *Genes Dev*. 2006;20(16):2214–2222. [PMCID: 1553205].
59. Vagin VV, Sigova A, Li C, Seitz H, Gvozdev V, Zamore PD. A distinct small RNA pathway silences selfish genetic elements in the germline. *Science*. 2006;313(5785):320–324.
60. Aravin AA, Sachidanandam R, Bourc'his D, et al. A piRNA pathway primed by individual transposons is linked to de novo DNA methylation in mice. *Mol Cell*. 2008;31(6):785–799. [PMCID: 2730041].
61. Aravin AA, Bourc'his D. Small RNA guides for de novo DNA methylation in mammalian germ cells. *Genes Dev*. 2008;22(8):970–975.
62. Beck CR, Collier P, Macfarlane C, et al. LINE-1 retrotransposition activity in human genomes. *Cell*. 2010;141(7):1159–1170.
63. Huang CR, Schneider AM, Lu Y, et al. Mobile interspersed repeats are major structural variants in the human genome. *Cell*. 2010;141(7):1171–1182.
64. Iskow RC, McCabe MT, Mills RE, et al. Natural mutagenesis of human genomes by endogenous retrotransposons. *Cell*. 2010;141(7):1253–1261.
65. Bestor TH. Sex brings transposons and genomes into conflict. *Genetica*. 1999;107(1–3):289–295.
66. Kalousek DK, Dill FJ. Chromosomal mosaicism confined to the placenta in human conceptions. *Science*. 1983;221(4611):665–667.
67. Vanneste E, Voet T, Le Caignec C, et al. Chromosome instability is common in human cleavage-stage embryos. *Nat Med*. 2009;15(5):577–583.
68. Conlin LK, Thiel BD, Bonnemann CG, et al. Mechanisms of mosaicism, chimerism and uniparental disomy identified by single nucleotide polymorphism array analysis. *Hum Mol Genet*. 2010;19(7):1263–1275.
69. Rodriguez-Santiago B, Malats N, Rothman N, et al. Mosaic uniparental disomies and aneuploidies as large structural variants of the human genome. *Am J Hum Genet*. 2010;87(1):129–138.
70. Kazazian Jr HH. Mobile DNA transposition in somatic cells. *BMC Biol*. 2011;9:62.
71. Kano H, Godoy I, Courtney C, et al. L1 retrotransposition occurs mainly in embryogenesis and creates somatic mosaicism. *Genes Dev*. 2009;23(11):1303–1312.
72. Doucet AJ, Hulme AE, Sahinovic E, et al. Characterization of LINE-1 ribonucleoprotein particles. *PLoS Genet*. 2010;6:10.
73. De S. Somatic mosaicism in healthy human tissues. *Trends Genet*. 2011;27(6):217–223. [Epub 2011 Apr 14. Review. PubMed PMID: 21496937].
74. Youssoufian H, Pyeritz RE. Mechanisms and consequences of somatic mosaicism in humans. *Nat Rev Genet*. 2002;3(10):748–758.
75. Brouha B, Schustak J, Badge RM, et al. Hot L1s account for the bulk of retrotransposition in the human population. *Proc Natl Acad Sci USA*. 2003;100(9):5280–5285.
76. Lutz SM, Vincent BJ, Kazazian Jr HH, Batzer MA, Moran JV. Allelic heterogeneity in LINE-1 retrotransposition activity. *Am J Hum Genet*. 2003;73(6):1431–1437.

77. Dreyer WJ, Gray WR, Hood L. The genetic, molecular, and cellular basis of antibody formation: some facts and a unifying hypothesis. *Cold Spring Harb Symp Quant Biol*. 1967;32:353–367.
78. Hozumi N, Tonegawa S. Evidence for somatic rearrangement of immunoglobulin genes coding for variable and constant regions. *Proc Natl Acad Sci USA*. 1976;73(10):3628–3632.
79. Oettinger MA, Schatz DG, Gorka C, Baltimore D. RAG-1 and RAG-2, adjacent genes that synergistically activate V(D)J recombination. *Science*. 1990;248(4962):1517–1523.
80. Schatz DG, Baltimore D. Stable expression of immunoglobulin gene V(D)J recombinase activity by gene transfer into 3T3 fibroblasts. *Cell*. 1988;53(1):107–115.
81. Chun JJ, Schatz DG, Oettinger MA, Jaenisch R, Baltimore D. The recombination activating gene-1 (RAG-1) transcript is present in the murine central nervous system. *Cell*. 1991;64(1):189–200.
82. Abeliovich A, Gerber D, Tanaka O, Katsuki M, Graybiel AM, Tonegawa S. On somatic recombination in the central nervous system of transgenic mice. *Science*. 1992;257(5068):404–410.
83. Schatz DG, Chun JJ. V(D)J recombination and the transgenic brain blues. *New Biol*. 1992;4(3):188–196.
84. Eggan K, Baldwin K, Tackett M, et al. Mice cloned from olfactory sensory neurons. *Nature*. 2004;428(6978):44–49.
85. Tasic B, Nabholz CE, Baldwin KK, et al. Promoter choice determines splice site selection in protocadherin alpha and gamma pre-mRNA splicing. *Mol Cell*. 2002;10(1):21–33.
86. Rehen SK, McConnell MJ, Kaushal D, Kingsbury MA, Yang AH, Chun J. Chromosomal variation in neurons of the developing and adult mammalian nervous system. *Proc Natl Acad Sci USA*. 2001;98(23):13361–13366.
87. Kingsbury MA, Friedman B, McConnell MJ, et al. Aneuploid neurons are functionally active and integrated into brain circuitry. *Proc Natl Acad Sci USA*. 2005;102(17):6143–6147.
88. Kaushal D, Contos JJ, Treuner K, et al. Alteration of gene expression by chromosome loss in the postnatal mouse brain. *J Neurosci*. 2003;23(13):5599–5606.
89. Yang AH, Kaushal D, Rehen SK, et al. Chromosome segregation defects contribute to aneuploidy in normal neural progenitor cells. *J Neurosci*. 2003;23(32):10454–10462.
90. Rehen SK, Yung YC, McCreight MP, et al. Constitutional aneuploidy in the normal human brain. *J Neurosci*. 2005;25(9):2176–2180.
91. Yurov YB, Iourov IY, Vorsanova SG, et al. Aneuploidy and confined chromosomal mosaicism in the developing human brain. *PLoS One*. 2007;2(6):e558.
92. Yu YC, Bultje RS, Wang X, Shi SH. Specific synapses develop preferentially among sister excitatory neurons in the neocortex. *Nature*. 2009;458(7237):501–504.
93. Padmanabhan K, Urban NN. Intrinsic biophysical diversity decorrelates neuronal firing while increasing information content. *Nat Neurosci*. 2010;13(10):1276–1282.
94. Kuwabara T, Hsieh J, Muotri A, Yeo G, et al. Wnt-mediated activation of NeuroD1 and retro-elements during adult neurogenesis. *Nat Neurosci*. 2009;12(9):1097–1105.
95. Coufal NG, Garcia-Perez JL, Peng GE, et al. L1 retrotransposition in human neural progenitor cells. *Nature*. 2009;460(7259):1127–1131.
96. Muotri AR, Chu VT, Marchetto MC, Deng W, Moran JV, Gage FH. Somatic mosaicism in neuronal precursor cells mediated by L1 retrotransposition. *Nature*. 2005;435(7044):903–910.
97. Baillie JK, Barnett MW, Upton KR, et al. Somatic retrotransposition alters the genetic landscape of the human brain. *Nature*. 2011;479(7374):534–537.
98. Marchetto MC, Carromeu C, Acab A, et al. A model for neural development and treatment of Rett syndrome using human induced pluripotent stem cells. *Cell*. 2010;143(4):527–539.
99. Muotri AR, Marchetto MC, Coufal NG, et al. L1 retrotransposition in neurons is modulated by MeCP2. *Nature*. 2010;468(7322):443–446.
100. Martin SL. Developmental biology: Jumping-gene roulette. *Nature*. 2009;460(7259):1087–1088.
101. Gasior SL, Wakeman TP, Xu B, Deininger PL. The human LINE-1 retrotransposon creates DNA double-strand breaks. *J Mol Biol*. 2006;357(5):1383–1393.
102. Belgnaoui SM, Gosden RG, Semmes OJ, Haoudi A. Human LINE-1 retrotransposon induces DNA damage and apoptosis in cancer cells. *Cancer Cell Int*. 2006;6:13.
103. Gasior SL, Roy-Engel AM, Deininger PL. ERCC1/XPF limits L1 retrotransposition. *DNA Repair (Amst)*. 2008;7(6):983–989.
104. Suzuki J, Yamaguchi K, Kajikawa M, et al. Genetic evidence that the non-homologous end-joining repair pathway is involved in LINE retrotransposition. *PLoS Genet*. 2009;5(4):e1000461.

105. Wallace NA, Belancio VP, Faber Z, Deininger P. Feedback inhibition of L1 and alu retrotransposition through altered double strand break repair kinetics. *Mob DNA*. 2010;1(1):22.
106. Morrish TA, Garcia-Perez JL, Stamato TD, Taccioli GE, Sekiguchi J, Moran JV. Endonuclease-independent LINE-1 retrotransposition at mammalian telomeres. *Nature*. 2007;446(7132):208–212.
107. Coufal NG, Garcia-Perez JL, Peng GE, et al. Ataxia telangiectasia mutated (ATM) modulates long interspersed element-1 (L1) retrotransposition in human neural stem cells. *Proc Natl Acad Sci USA*. 2011;108(51):20382–20387.
108. Garcia-Perez JL, Morell M, Scheys JO, et al. Epigenetic silencing of engineered L1 retrotransposition events in human embryonic carcinoma cells. *Nature*. 2010;466(7307):769–773.
109. Callinan PA, Wang J, Herke SW, Garber RK, Liang P, Batzer MA. Alu retrotransposition-mediated deletion. *J Mol Biol*. 2005;348(4):791–800.
110. Gilbert N, Lutz-Prigge S, Moran JV. Genomic deletions created upon LINE-1 retrotransposition. *Cell*. 2002;110(3):315–325.
111. Symer DE, Connelly C, Szak ST, et al. Human l1 retrotransposition is associated with genetic instability in vivo. *Cell*. 2002;110(3):327–338.
112. Hastings PJ, Lupski JR, Rosenberg SM, Ira G. Mechanisms of change in gene copy number. *Nat Rev Genet*. 2009;10(8):551–564.
113. Lupski JR. Retrotransposition and structural variation in the human genome. *Cell*. 2010;141(7):1110–1112.
114. Hubel DH, Wiesel TN. The period of susceptibility to the physiological effects of unilateral eye closure in kittens. *J Physiol*. 1970;206(2):419–436.
115. Wiesel TN, Hubel DH. Effects of visual deprivation on morphology and physiology of cells in the cats lateral geniculate body. *J Neurophysiol*. 1963;26:978–993.
116. Katz LC, Shatz CJ. Synaptic activity and the construction of cortical circuits. *Science*. 1996;274(5290):1133–1138.
117. Huberman AD, Feller MB, Chapman B. Mechanisms underlying development of visual maps and receptive fields. *Annu Rev Neurosci*. 2008;31:479–509.
118. Monahan K, Rudnick ND, Kehayova PD, et al. Role of CCCTC binding factor (CTCF) and cohesin in the generation of single-cell diversity of Protocadherin-alpha gene expression. *Proc Natl Acad Sci USA*. 2012;109:9125–9130.
119. Robison AJ, Nestler EJ. Transcriptional and epigenetic mechanisms of addiction. *Nat Rev Neurosci*. 2011;12(11):623–637.
120. Baranzini SE, Mudge J, van Velkinburgh JC, et al. Genome, epigenome and RNA sequences of monozygotic twins discordant for multiple sclerosis. *Nature*. 2010;464(7293):1351–1356.
121. Coolen MW, Statham AL, Qu W, et al. Impact of the genome on the epigenome is manifested in DNA methylation patterns of imprinted regions in monozygotic and dizygotic twins. *PLoS One*. 2011;6(10):e25590.
122. Dempster EL, Pidsley R, Schalkwyk LC, et al. Disease-associated epigenetic changes in monozygotic twins discordant for schizophrenia and bipolar disorder. *Hum Mol Genet*. 2011;20(24):4786–4796.
123. Haque FN, Gottesman II, Wong AH. Not really identical: epigenetic differences in monozygotic twins and implications for twin studies in psychiatry. *Am J Med Genet C Semin Med Genet*. 2009;151C(2):136–141.
124. Singh SM, Murphy B, O'Reilly R. Epigenetic contributors to the discordance of monozygotic twins. *Clin Genet*. 2002;62(2):97–103.
125. van Praag H, Christie BR, Sejnowski TJ, Gage FH. Running enhances neurogenesis, learning, and long-term potentiation in mice. *Proc Natl Acad Sci USA*. 1999;96(23):13427–13431.
126. van Praag H, Kempermann G, Gage FH. Running increases cell proliferation and neurogenesis in the adult mouse dentate gyrus. *Nat Neurosci*. 1999;2(3):266–270.
127. Muotri AR, Zhao C, Marchetto MC, Gage FH. Environmental influence on L1 retrotransposons in the adult hippocampus. *Hippocampus*. 2009;19(10):1002–1007.
128. Maze I, Feng J, Wilkinson MB, Sun H, Shen L, Nestler EJ. Cocaine dynamically regulates heterochromatin and repetitive element unsilencing in nucleus accumbens. *Proc Natl Acad Sci USA*. 2011;108(7):3035–3040.
129. Lee EJ, Banerjee S, Zhou H, et al. Identification of piRNAs in the central nervous system. *RNA*. 2011;17(6):1090–1099.
130. Rajasethupathy P, Antonov I, Sheridan R, et al. A role for neuronal piRNAs in the epigenetic control of memory-related synaptic plasticity. *Cell*. 2012;149(3):693–707.

131. Iwamoto K, Bundo M, Ueda J, et al. Neurons show distinctive DNA methylation profile and higher interindividual variations compared with non-neurons. *Genome Res*. 2011;21(5):688–696.
132. Yurov YB, Iourov IY, Vorsanova SG, et al. The schizophrenia brain exhibits low-level aneuploidy involving chromosome 1. *Schizophr Res*. 2008;98(1–3):139–147.
133. Geller LN, Potter H. Chromosome missegregation and trisomy 21 mosaicism in Alzheimer's disease. *Neurobiol Dis*. 1999;6(3):167–179.
134. Iourov IY, Vorsanova SG, Liehr T, Kolotii AD, Yurov YB. Increased chromosome instability dramatically disrupts neural genome integrity and mediates cerebellar degeneration in the ataxia-telangiectasia brain. *Hum Mol Genet*. 2009;18(14):2656–2669.
135. McConnell MJ, Kaushal D, Yang AH, et al. Failed clearance of aneuploid embryonic neural progenitor cells leads to excess aneuploidy in the Atm-deficient but not the Trp53-deficient adult cerebral cortex. *J Neurosci*. 2004;24(37):8090–8096.
136. Piotrowski A, Bruder CE, Andersson R, et al. Somatic mosaicism for copy number variation in differentiated human tissues. *Hum Mutat*. 2008;29(9):1118–1124.
137. Bruder CE, Piotrowski A, Gijsbers AA, et al. Phenotypically concordant and discordant monozygotic twins display different DNA copy-number-variation profiles. *Am J Hum Genet*. 2008;82(3):763–771.
138. Westra JW, Rivera RR, Bushman DM, et al. Neuronal DNA content variation (DCV) with regional and individual differences in the human brain. *J Comp Neurol*. 2010;518(19):3981–4000.
139. Veitia RA, Bottani S, Birchler JA. Cellular reactions to gene dosage imbalance: genomic, transcriptomic and proteomic effects. *Trends Genet*. 2008;24(8):390–397.
140. Navin N, Kendall J, Troge J, et al. Tumour evolution inferred by single-cell sequencing. *Nature*. 2011;472(7341):90–94. [Epub 2011 Mar 13. PubMed PMID: 21399628].

"Neural Stem Cell"

J. David Sweatt, acrylic on canvas (48 x 48), 2011

CHAPTER

12

Adult Neurogenesis

Jenny Hsieh[1] *and Hongjun Song*[2]

[1]Department of Molecular Biology, UT Southwestern Medical Center, Dallas, Texas, USA

[2]Institute for Cell Engineering, Departments of Neurology and Neuroscience, Johns Hopkins University School of Medicine, Baltimore, Maryland, USA

INTRODUCTION

Epigenetic control of gene expression provides a molecular mechanism to regulate stem cell plasticity in embryonic and adult tissues, including the mammalian central nervous system (CNS). The ability of epigenetic mechanisms to induce changes in phenotype without altering genotype confers stem cells the ability to self-renew as well as transmit their genetic information to differentiated progeny.[1] During development, neural stem cells (NSCs) progress through sequential stages of neurogenesis and gliogenesis to give rise to embryonic brain and spinal cord structures.[2,3] Postnatally and during adulthood, NSCs in discrete anatomical regions continue to give rise to mature neurons and glia, contributing to repair and regeneration.[4–7] A major effort has been to uncover the transcriptional and epigenetic networks that govern the development of NSCs in the embryonic and adult brain. Ultimately, by fully understanding the molecular mechanisms that control NSC plasticity, we hope to develop more effective strategies to harness their potential for therapy.

Why approach NSC biology through the study of epigenetics? First, NSC self-renewal and differentiation most likely requires the coordinated actions of many genes, and not a single gene. Epigenetic mechanisms involving changes in chromatin structure, including both DNA and histone modifications, or the actions of non-coding RNAs, such as microRNAs (miRNAs), provide means for regulating arrays of genes. Second, epigenetic mechanisms are usually associated with heritable changes in gene expression, which may be critical for lineage-specific differentiation programs and stem-cell maintenance. Third, cellular reprogramming, such as the ability of a cell to dedifferentiate or transdifferentiate, often achieved through transcription-factor overexpression, may be due to the reversible feature of epigenetic mechanisms. It has become increasingly clear that dysregulated epigenetic

Epigenetic Regulation in the Nervous System
DOI: http://dx.doi.org/10.1016/B978-0-12-391494-1.00012-4

states could contribute to inefficient reprogramming. For these reasons, understanding the physiological epigenetic changes associated with NSC proliferation and differentiation may contribute to the restoration of dysregulated stem-cell plasticity by normalizing gene expression in pathological contexts.

In this chapter, we present an overview of NSC biology, highlighting four classes of epigenetic mechanisms in NSC function, followed by a description of epigenetic regulation of adult neurogenesis in vivo. We also discuss emerging work on epigenetics and induced pluripotent stem cells (iPSCs) for neuroregenerative medicine. From a biological standpoint, epigenetic mechanisms offer mechanistic insights on neuronal lineage commitment and terminal differentiation. From a translational perspective, it provides new approaches and possible strategies for restoring neurogenesis after damage or injury.

EPIGENETIC CONTROL OF NSCS: A FEW GUIDING PRINCIPLES

A major advance in the field came from the ability to successfully culture and control the expansion of NSCs and characterize their cellular and molecular properties in vitro. Pioneering studies from Reynolds and Weiss demonstrated that NSCs could be harvested from the adult mammalian brain and maintained in an undifferentiated state for many passages.[8] These cells can be cultured as monolayers on coated substrates (e.g. polyornithine and laminin) or as free-floating clonal aggregates, known as neurospheres.[8,9] Cultured NSCs can be induced to proliferate with mitogenic growth factors (e.g. epidermal growth factor [EGF], fibroblast growth factor 2 [FGF-2]) or undergo lineage-specific differentiation into neurons and glia under defined conditions.[10,11] NSCs from rodent hippocampal subgranular zone (SGZ) and subventricular zone (SVZ) can also be derived into monolayer culture and remain multipotent.[12,13]

A defining feature of NSCs is their ability to undergo self-renewal and differentiate into neurons and glial cells (e.g. astrocytes, and oligodendrocytes) (Figure 12.1). First, extrinsic signaling molecules are provided from the local tissue microenvironment or niche. In turn, niche-derived signals control intrinsic gene expression programs required for proliferation or differentiation. The ability of NSCs to integrate extrinsic and intrinsic signals to control their plasticity can be explained by the regulated expression of transcription factors under the control of epigenetic mechanisms. To control multiple steps of neurogenesis, either during development or in the adult CNS, epigenetic mechanisms play essential roles.

As mentioned in other chapters, the current concept of epigenetic mechanisms refers to potentially heritable changes in gene expression without a change at the DNA sequence level.[14,15] Originally described by Conrad Waddington, the "epigenetic landscape" attempts to explain how identical genotypes could result in a wide variety of phenotypic variation through the process of development.[16] In recent years, epigenetic regulation has emerged as an equally important mechanism for gene regulation compared to regulation by transcription factors. In this chapter, we highlight the function of four main classes of epigenetic mechanisms: histone modifications, DNA methylation, chromatin remodeling, and non-coding RNAs (Figure 12.2).

In this next section, we focus on recent reports highlighting the role of the four types of epigenetic mechanisms in neural fate specification, in a broad range of NSCs from both embryonic and adult stages, before delving deeper into adult neurogenesis.

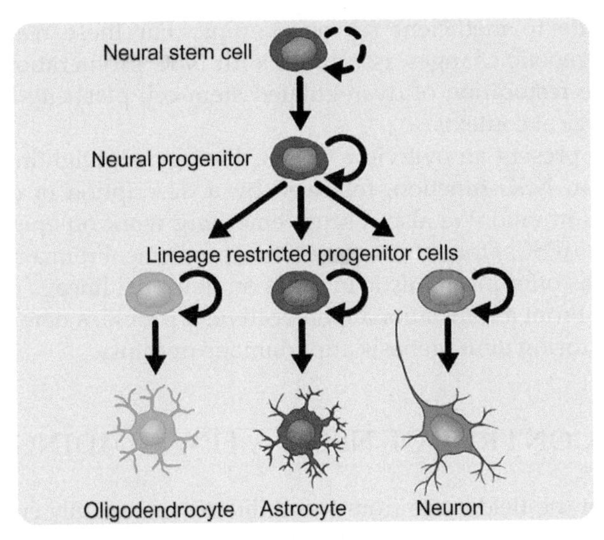

FIGURE 12.1 **Adult neural stem cells undergo self-renewal and differentiation to the three major central nervous system cell types: neurons, oligodendrocytes, and astrocytes.**

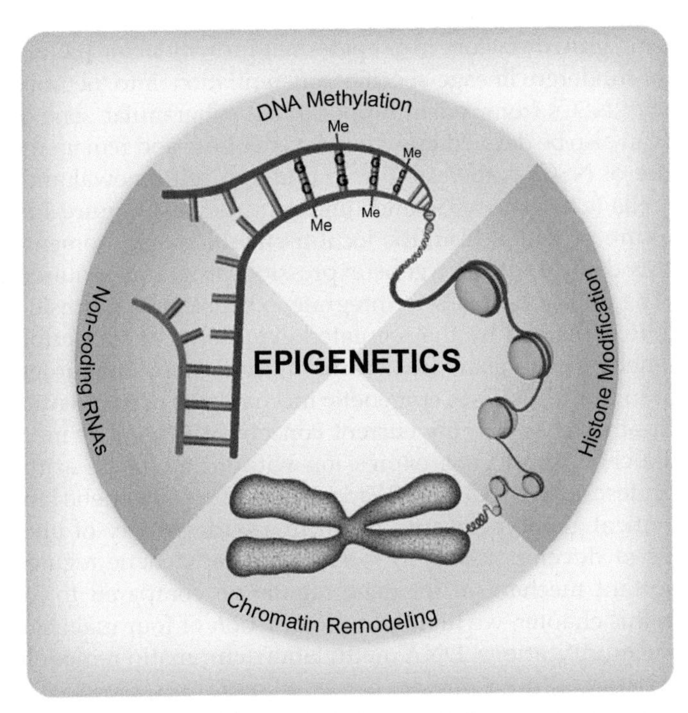

FIGURE 12.2 **The four major categories of epigenetic mechanisms: DNA methylation, histone modification, chromatin remodeling, and non-coding RNAs.** The dynamic nature of epigenetic mechanisms offers an additional layer of gene regulation, allowing precise changes in neuronal production rates in response to environmental flux.

HISTONE MODIFICATION

To begin evaluating the broad importance of histone modification, work from our laboratory and others treated cultured NSCs with histone deacetylase inhibitors (HDACi). Neural progenitors from adult rat hippocampus displayed reduced proliferation and robustly differentiated into neuronal cells in the presence of HDACi, such as trichostatin A (TSA), sodium butyrate (NaB), and valproic acid (VPA).[17] On the other hand, inhibition of class I and II HDACs in developing mouse embryos with TSA resulted in a dramatic reduction in cortical neurogenesis as well as in the ganglionic eminences but a modest increase in neurogenesis in the cortex.[18] These results suggested that, in some contexts, such as in adult stages, histone acetylation could be more permissive for neuronal differentiation. Alternatively, different neural precursors in distinct brain regions could have context-specific responses to HDACi.

Histone methylation has also been demonstrated to be important for NSC proliferation. Knock-down of the lysine-specific demethylase (LSD1) led to decreased NSC proliferation through recruitment of the orphan nuclear hormone receptor TLX.[19] The study also performed intracranial viral transduction of LSD1 small interfering RNAs and intraperitoneal injection of the LSD1 inhibitors pargyline and tranylcypromine and demonstrated a dramatic reduction of neural progenitor proliferation in the hippocampal dentate gyrus, further supporting the important role of LSD1 in NSCs.

Similar to NSC proliferation and neuronal differentiation, histone modification changes have also been demonstrated to be important for glial lineage specification.[20,21] As a major type of glial cell in the CNS, oligodendrocytes generate a myelin sheath for the wrapping and insulation of neighboring axons. Oligodendrocytes arise from oligodendrocyte precursor cells (OPCs) from the embryonic ventricular zone in the developing neural tube[22,23] or dorsal spinal cord at later embryonic stages.[23–25] Interestingly, treatment of OPCs with TSA or NaB blocked differentiation into mature oligodendrocytes. In mice, conditional removal of HDAC1 and HDAC2 resulted in myelin defects and decreased oligodendrocytes, consistent with the critical role of HDACs in oligodendrocyte differentiation.[26]

DNA METHYLATION

There is considerable evidence that DNA methylation plays a critical role in astrocyte differentiation, particularly in the timing of the neuronal-to-astrocyte switch in the developing neocortex.[27] Elegant studies have demonstrated the importance of extracellular signals, including ciliary neurotrophic factor, leukemia inhibitory factor (LIF), and cardiotrophin-1 to activate the JAK-STAT pathway, for promoting astrocytic differentiation.[28–32] Astrocyte differentiation is associated with an upregulation of the glial fibrillary acidic protein (GFAP) gene. Takizawa, Nakashima and colleagues were the first to describe a mechanism in which the signal transducer and activator of transcription 3 (STAT3) activation of GFAP was blocked in early E11.5 telencephalic neuroepithelial cells, as well as in post-mitotic neurons, by CpG methylation of the STAT3 binding site.[27] Through this mechanism, GFAP expression was repressed, even in the presence of LIF signaling, one of the triggers of STAT3 phosphorylation and activation. Interestingly, E14.5 neuroepithelial cells became demethylated at the STAT3 binding site within the GFAP promoter and became responsive to LIF signaling, leading to GFAP activation and astrocyte differentiation.

The bone morphogenetic protein and Notch pathways also cross-talk with the JAK-STAT pathway to induce astrocytic differentiation.[33–36] In an elegant study, forced expression of the Notch intracellular domain in E14.5 neuroepithelial cells to activate Notch signaling promotes demethylation of astrocyte-specific gene promoters, including GFAP and S100B, allowing precocious astrocyte differentiation in response to LIF.[37] Recently, it was shown that the polycomb group complex (PcG) restricts neurogenic competence of NSCs and promotes the transition of NSCs' fate from neurogenic to astrogenic. Inactivation of PcG by knockout of the Ring1B or Ezh2 gene or Eed knock-down prolonged the neurogenic phase of NPCs and delayed the onset of the astrogenic phase. Moreover, PcG was found to repress the promoter of the proneural gene neurogenin1 in a developmental stage-dependent manner. These results demonstrate a role of PcG to control the temporal regulation of NPC fate.[36]

While Hes5 is one of the target genes of the Notch signaling pathway,[38] its expression is still detectable in null mutants of Notch pathway genes at E8.5,[39,40] suggesting alternative mechanisms of controlling its expression in early embryos. Interestingly, one recent study evaluated an epigenetic mechanism for regulation of Hes5 expression in E8.5 embryos. This study demonstrated that mammalian Glial Cells Missing 1 and 2 (GCM1/2) is required for expression of the transcription factor Hes5 by demethylation of the Hes5 promoter in a DNA replication-independent active process to induce NSC properties from embryonic neuroepithelial cells.[41] Consistent with the importance of DNA methylation, embryos cultured in the presence of a DNA methyltransferase inhibitor, 5'aza-2'deoxycytidine, resulted in significantly lower methylation of the Hes5 promoter and induction of Hes5 expression.

REGULATORY NON-CODING RNAS AND CHROMATIN REMODELING

As described elsewhere, regulatory non-coding RNAs function at the post-transcriptional level as a class of epigenetic mechanisms. Several recent studies have described the functions of both small and long forms of non-coding RNAs in neural development and differentiation.[42,43] MiRNAs (or miRs) have been shown to regulate neurogenesis, dendritic development, and synaptic plasticity.[44,45] For some of the brain-enriched miRNAs studied to date, they appear to play major roles to induce NSC differentiation into specific cell types. For example, miR-9*, and miR-124b are shown to repress the expression of the transcriptional regulator BAF53a, a subunit of the SWI/SNF chromatin remodeling complex which, in turn, blocks neural progenitor proliferation.[46] Strikingly, miR-9* and miR-124b are targets of the transcriptional repressor REST/NRSF, which was shown to be required to block precocious neuronal differentiation and maintain NSC self-renewal.[46–48]

In another study, the let-7 miRNA has been shown to be important for regulating NSC proliferation and differentiation.[49] Expression of let-7 is under negative feedback control by the RNA binding protein Lin-28.[50] Overexpression of let-7b resulted in accelerated neuronal differentiation and antisense-mediated knock-down of let-7b led to increased NSC proliferation. Strikingly, let-7b appears to inhibit the nuclear hormone receptor TLX mentioned earlier, which is important to regulate cell cycle progression of NSCs during development,[51] by binding to its 3′ UTR, in addition to the 3′ UTR of cyclin D1. Feedback regulatory loops involving miRs and transcription factors to promote neuronal differentiation are quite common, with another case being the cross-talk between miR-137 and TLX in embryonic NSCs.[52] Together, these experiments demonstrate the interplay between miRNAs and transcription factors to control the dynamics between NSC proliferation and differentiation.

Besides short miRNAs that inhibit gene expression by translational repression, long noncoding RNAs (lncRNAs), defined as greater than 200 nt in length, have also been implicated in neural patterning and development and in the pathogenesis of neuropsychiatric diseases.[45,53] There are examples of both sense and antisense lncRNAs to specific genes that have apparent roles in neural differentiation. For example, Nkx2.2 antisense (Nkx2.2as) is an antisense lncRNA to the Nkx2.2 gene, which is expressed in the developing mammalian forebrain and plays an essential role during oligodendrocyte development.[54] Forced expression of Nkx2.2as in cultured NSCs promoted oligodendrocyte differentiation through an upregulation of the Nkx2.2 mRNA, although the exact mechanism by which this lncRNA controls Nkx2.2 expression remains to be elucidated.[55]

Another lncRNA worthy of mention is Evfl, a 2.7 kb lncRNA transcribed upstream from the homeobox-containing transcription factor Dlx-6.[56] Evfl has an alternatively spliced transcript called Evf2, which has been demonstrated to function as a transcriptional co-activator of Dlx-3 important for the Dlx5/6 enhancer during forebrain development.[57] Loss of Evf2 results in reduced GABAergic interneurons and deficits in synaptic inhibition in the developing hippocampus.[56] At this point, the detailed function of lncRNAs is unknown, although bioinformatic studies of genomic locations and comparison of expression patterns of lncRNAs and their adjacent protein coding transcripts posits a *cis*-regulation mechanism of lncRNAs in controlling gene expression.[53]

EPIGENETIC REGULATION OF ADULT NEUROGENESIS

Within the adult brain, one region that continues to generate new neurons is the SGZ in the dentate gyrus of the hippocampus (Figure 12.3). In the SGZ niche, newly generated neurons structurally and functionally mature from resident stem and progenitor cells in about six to eight weeks.[58,59] Quiescent NSCs, known as Type-1 cells, possess a single radial process ending in a tree-like tuft, and express markers such as glial fibrillary acidic protein (GFAP) and Nestin.[60–63] Recently, a second class of Type-1 cell, characterized by short, horizontal processes, was identified.[64] Horizontal Type-1 cells appear to divide more quickly and can be generated from radial Type-1 cells.[65] Quiescent NSCs proliferate to generate Type-2A and Type-2B cells (that look like small cells with short tangential processes and are often found in clusters in the SGZ) as well as Type-3 neuroblasts.[61,64–67] The major difference between Type-2A, -2B and -3 cells is that Type-2B cells were initially identified in Nestin-GFP reporter mice and expression of the immature neuronal marker doublecortin (DCX) in Nestin-GFP+ cells defines the transition between Type-2A and Type-2B, whereas Nestin-negative Type-3 cells express DCX only.[68] In addition, the SRY box transcription factor 2 (Sox2) identifies Type-2A cells (as well as Type-1 cells), which transition to Type-2B and Type-3 cells.[64,69,70] Ultimately, Type-3 cells give rise to immature and mature granule neurons, which are associated with downregulation of DCX and upregulation of calretinin and NeuN.

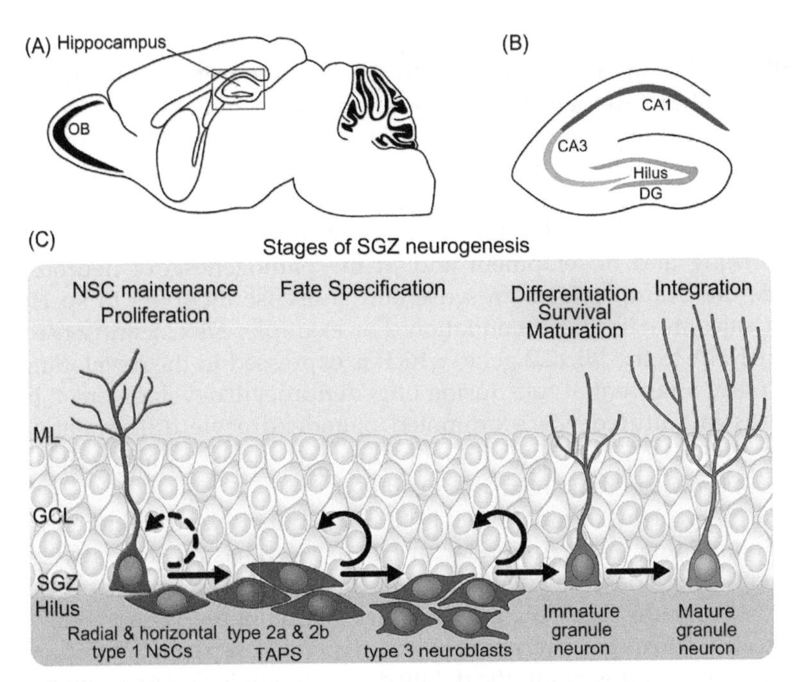

FIGURE 12.3 **Adult neurogenesis in the hippocampal dentate gyrus.** (A) Sagittal view of the rodent brain with the boxed region outlining hippocampal formation. OB: olfactory bulb. (B) Schematic of the hippocampus with CA1, CA3, DG and hilus regions. (C) The SGZ niche is comprised of radial and horizontal Type-1 neural stem cells (NSCs), early-stage Type-2a transit-amplifying progenitors (TAPs) and late-stage Type-3 TAPs, immature granule neurons and mature granule neurons. The progression from Type-1 NSCs to mature granule neurons in adult SGZ is a multistep process with distinct stages (labeled on top). ML: molecular layer. GCL: granule cell layer.

The second region where adult neurogenesis occurs is within the SVZ of the lateral ventricle (Figure 12.4). Over the course of three weeks, a diverse array of neurons, including deep granule interneurons and calbindin+ periglomerular cells (from ventral/medial SVZ progenitors) and superficial granule interneurons and calretinin+ periglomerular cells (from dorsal/anterior SVZ progenitors), are generated from resident SVZ NSCs.[71,72] Similar to the SGZ, quiescent NSCs called Type B1 cells express astroglia-associated markers such as the astrocyte-specific glutamate transporter (GLAST), brain-lipid-binding protein (BLBP), GFAP, and Nestin.[71,73–77] Interestingly, Type B1 cells contain a non-motile primary cilium that extends into the ventricular cerebrospinal fluid, suggesting a signaling role of the cilium in the regulation of proliferation and differentiation.[73,78,79] In contrast, Type B2 cells have astrocytic features, but do not contact the ventricle. Type C cells, the progeny of Type B cells, rapidly proliferate and are often found in clusters nearby blood vessels.[80] Similar to Type-2 TAPs in the SGZ, Type C cells upregulate Ascl1, Pax6, and the transcription factor Dlx2.[81] Finally, Type C cells give rise to Type A neuroblasts, which form migratory chains encased within astrocyte tubes.[82] Most of the olfactory bulb neurons are γ-aminobutyric acid- (GABA)ergic interneurons; however, a recent fate mapping study suggested the existence of a small number of glutamatergic juxtaglomerular neurons,[83] leaving open the question how SVZ NSCs contribute to distinct neuronal subtypes.

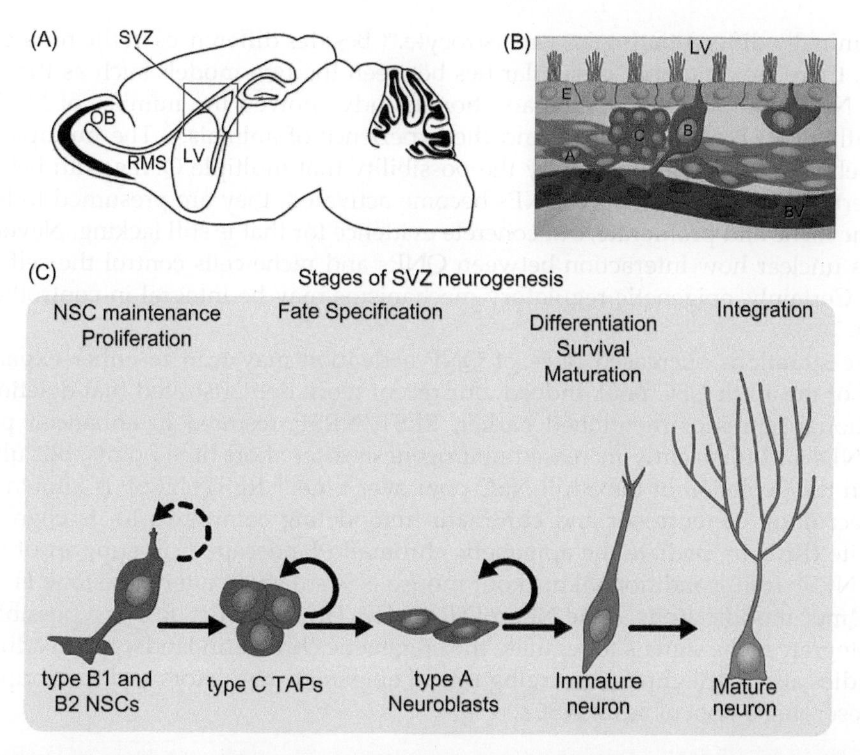

FIGURE 12.4 **Adult neurogenesis in the subventricular zone (SVZ) and rostral migratory stream (RMS).** (A) Sagittal view of the rodent brain with the boxed region outlining the SVZ region next to the lateral ventricle (LV). OB: olfactory bulb. (B) Schematic of the SVZ with ependymal cells (E), blood vessel cells (BV) and distinct stem/progenitor cell types (Type B, C, and A). (C) The SVZ niche is comprised of astrocyte-like Type B1 and B2 neural stem cells (NSCs), Type C transit-amplifying progenitors (TAPs), Type A neuroblasts, immature neurons and mature neurons. The progression from Type B NSCs to mature neurons in adult SVZ is a multistep process with distinct stages (labeled on top).

In this next section, we will discuss recent studies that highlight how epigenetic mechanisms can function cell-autonomously to control each stage of adult neurogenesis, with an emphasis on adult SGZ NSCs.[84,85] However, it is important to point out we cannot directly extrapolate results from the SGZ to the SVZ because it is likely there are region-specific differences. Also, epigenetic mechanisms controlling adult neurogenesis may be exerted non-cell autonomously.[86]

NSC SELF-RENEWAL AND MAINTENANCE OF QUIESCENCE

As mentioned above, adult NSCs have a slow proliferation rate and are defined as quiescent neural progenitors (QNPs).[87] Currently, there are two models of NSC self-renewal in the adult SGZ. In one model, QNPs divide symmetrically to generate two identical daughter stem cells or asymmetrically to produce one daughter stem cell and one restricted progenitor cell.[65] In a second model, QNPs only divide asymmetrically a finite number of times before terminally differentiating into an astrocyte.[67] Besides differences in the timing of NSC depletion, there are a number of similarities between the two models such as the differentiation of NSCs into astrocytes. One additional study showed the number of NSCs could also be influenced by their location and the experience of animals.[88] The multiple models of QNP behavior can be reconciled by the possibility that multiple QNPs with heterogeneous properties are present.[89] Once QNPs become activated, they are presumed to leave the neurogenic niche and proliferate, but concrete evidence for that is still lacking. Nevertheless, it remains unclear how interaction between QNPs and niche cells control the self-renewal of QNPs. Certainly, epigenetic regulatory mechanisms may be integral in controlling QNP activation.

In some situations, increased rates of QNP activation may lead to either expansion or depletion of the adult NSC pool. Indeed, our recent work demonstrated that deletion of the transcriptional repressor mentioned earlier, REST/NRSF, resulted in enhanced proliferation of QNPs and transiently increased neurogenesis after short time points, but ultimately resulted in the depletion of the adult NSC pool over time.[48] REST/NRSF is known to function by recruiting co-repressor and chromatin remodeling complexes to its cognate DNA binding site (RE1) to mediate the epigenetic chromatin landscape.[90] In support of this, loss of REST/NRSF from conditional knockout mouse NSCs display altered histone H3K27me3 and H3K4me2 modifications at the NeuroD RE1 site. Thus, REST/NRSF is a possible candidate to integrate niche signals to regulate the epigenetic chromatin landscape in adult NSCs. These studies also highlight the emerging role of epigenetic regulators and transcription factors in preserving a pool of adult NSCs.

PROLIFERATION OF TRANSIT-AMPLIFYING PROGENITORS

After QNPs become activated, they generate TAPs, which can undergo several rounds of additional proliferation to generate neuroblasts, immature and mature neurons. Both QNPs and TAPs are capable of proliferation, however, the underlying mechanisms may not be identical. For example, a member of the polycomb group (PcG) repressor complex Bmi1, which catalyzes trimethylation of H3K27, was shown to induce proliferation of QNPs in adult SVZ, rather than TAPs by inhibiting cyclin-dependent kinase inhibitors, p16 and p19.[91] In contrast, the kinase inhibitory protein 1 (Kip) family member, cell cycle inhibitor $p27^{Kip1}$ can specifically control the proliferation of TAPs in the adult SVZ, without affecting QNPs.[92] These results suggest that there may be separate cell cycle genes being targeted in different cells depending on the stage of neurogenesis, which may be influenced by the intrinsic epigenetic state.

The proliferation of TAPs (and possibly the activation of QNPs) in adult neurogenic regions also requires TLX, possibly due to its restricted expression and activity in adult QNPs and TAPs, and consistent with its role in vitro.[93–95] Recently, many studies have taken advantage of conditional and/or inducible mice in adult neurogenesis research.[68] In this system, a tamoxifen-dependent $CreER^{T2}$ recombinase driven by a cell-specific promoter/enhancer[96] allele is crossed with a floxed allele of the gene of interest (for loss-of-function studies) and/or a reporter gene (e.g. ROSA26RloxP-stop-loxP lacZ or GFP) to allow fate mapping of Cre-recombined reporter-positive cells in adult SGZ. Using inducible genetic lineage tracing, it was demonstrated that TLX-expressing cells contribute to both proliferating and quiescent Type-1 SGZ NSCs.[95] In addition, genome-wide gene expression analyses in TLX-deficient NSCs suggest that many genes involved in cell proliferation show significant changes.[95,97]

How do multiple regulators, such as REST/NRSF and TLX, coordinately regulate NSC quiescence activation? One possibility is that TLX and REST/NRSF may function in an interconnected transcriptional/epigenetic network involving miRNA genes. In addition to regulating coding genes, REST/NRSF has also been shown to regulate neurogenesis, in part through repressing the actions of miRNAs, such as miR-124a and miR-9.[47] Since this first study, the list of miRNAs that are predicted or regulated by REST/NRSF has increased.[98–100] Based on software prediction and in vitro expression pattern analysis, the two miRNAs let-7b and miR-9 were confirmed to target the TLX 3′UTR to repress its expression.[49,101] Consistent with expression levels of TLX being important in NSC proliferation, overexpression of let-7b and miR-9 in cultured NSCs resulted in reduced TLX expression and diminished proliferation. These conclusions are still controversial as several recent studies suggest that miR-9 increases, not decreases, the expression of TLX.[102,103] Together, these results indicate extensive cross-talk mechanisms exist between transcription factors and epigenetic regulators, such as miRNAs.

NEURONAL VS GLIAL LINEAGE SPECIFICATION OF ADULT NSCS

As TAPs in the adult SGZ commit to neuronal or glial lineages and further differentiate into glutamatergic granule neurons or astrocytes, there is a switch in the transcription factor program to control the later stages of neurogenesis. Compared to embryonic development, less is known regarding how epigenetic mechanisms control neuronal vs glial specification during adult stages. For example, during development, deletion of the DNA methyltransferase 1 (DNMT1) in Nestin-positive neural progenitor cells resulted in an increase of mature astrocytes.[104] Loss of DNMT1 mediated passive demethylation of CpG islands in the astrocyte gene GFAP promoter as well as in the promoter of STAT3 resulted in an upregulation of astrocyte genes by binding to their promoters together with CBP/p300.[104] Interestingly, there was no observable phenotype when DNMT1 was deleted from adult neurons. Since DNMT3b is not expressed in adult stages,[105] it is thus speculated that DNMT3a and DNMT1 may play redundant roles in adult neurons.[106] Supporting this notion, conditional deletion of DNMT1 and 3a using a post-mitotic neuron-specific Cre recombinase (Camk2α) led to impaired synaptic plasticity, learning and memory through de-repression of immune genes, such as the major histocompatibility complex 1.[107]

Methylated DNA can also be bound by the family of methyl-CpG binding (MBD) proteins, which is comprised of MBDs 1–4 and MeCP2.[108] In contrast to DNMT1 knockout mice, MBD1 knockout mice did not display any dramatic developmental defects. However, the generation and survival of adult-generated neurons is severely impaired while astrogliogenesis is unaffected in adult mice lacking MBD1.[109] Moreover, MBD1 knockout mice also showed deficits in spatial learning and long-term potentiation. However, the underlying epigenetic regulatory mechanisms are unknown. Recent work provides insight into this question by identifying two downstream targets.[110,111] One is a commonly used growth factor in NSC culture, FGF-2, whose overexpression in vitro can suppress differentiation in the presence of neurogenic signals. Loss of MBD1 is associated with hypomethylation of the FGF-2 promoter, which resulted in increased expression, while global methylation is unaffected. Another target miR-184 was identified through miRNA array analysis of MBD1 knockout cells. Removal of MBD1 from the miR-184 promoter led to increased miR-184 levels. Furthermore, this microRNA was shown to bind to the 3′UTR of Numblike (Numbl) mRNA to suppress its expression. Finally, exogenous expression of Numbl rescued neuronal differentiation deficits found in MBD1 knockout mice, thus forming an elegant regulatory network between two epigenetic mechanisms, DNA methylation and miRNA regulation.

Another member of the MBD family of epigenetic regulators, MeCP2 is also involved in NSC fate specification, apparently through distinct mechanisms from MBD1. Ectopic expression of MeCP2 in neural progenitor cells in vitro induced neuronal differentiation even in conditions that favor astrocyte differentiation, while glial differentiation was blocked.[106] The Ras-ERK pathway is responsible for the MeCP2-dependent neuronal differentiation after detailed analyses, however, the exact mechanisms by which MeCP2, a putative suppressor, functions to activate the ERK pathway is still unknown. Thus, these studies reinforce the concept that epigenetic regulatory mechanisms work in concert with transcription factors to control the essential fate choice between neuronal vs glial lineages during development and in the adult brain.

Finally, large-scale chromatin changes may also control neuronal cell fate choice. One study demonstrated that under oxidative conditions in vitro and in vivo, embryonic cortical NPCs adopt an astrocytic lineage at the expense of a neuronal lineage. This process is mediated by the binding of the transcription factor Hes1 together with Sirt1, a mammalian NAD+ dependent HDAC, to the promoter of pro-neuronal genes such as Ascl1 (Mash1) to repress its expression.[112] However, upon differentiation conditions, Sirt1 translocates into the nucleus to promote neuronal differentiation possibly through repression of Hes1.[113] Interestingly, a previous study suggested that the Ascl1 locus undergoes large-scale chromatin reorganization from the periphery to the interior of the nucleus upon differentiation induction.[114] It will be interesting to determine whether large-scale chromatin changes may be fundamentally important to control neuronal lineage specification or control neuronal fate specification in adult NSCs.

DIFFERENTIATION, SURVIVAL, AND MATURATION OF ADULT-GENERATED NEURONS

Transcription factors working together with epigenetic regulators also control the survival and maturation of neuroblasts and immature neurons as they ultimately transition into functionally mature neurons. As previously mentioned, HDACi are powerful chemical biological tools to interrogate epigenetic mechanisms in NSCs. HDAC1 and HDAC2 play redundant roles in many different systems, such as adipogenesis, oligodendrocyte differentiation, and progression of neural precursors to neurons during brain development.[26,115,116] In these studies, the authors did not observe a dramatic phenotype with single deletion of HDAC1 or HDAC2, but HDAC1:HDAC2 double knockout mice showed severe defects. Considering that both HDACs 1 and 2 are members of the Class I HDACs, they are speculated to play redundant roles in the developing and adult brain. Nevertheless, one study showed that overexpression of HDAC2, but not HDAC1 in neurons, resulted in defects in hippocampus-dependent memory formation by Pavlovian fear conditioning and Morris Water maze tests.[117] Another study demonstrated that HDAC2 deletion, both globally as well as conditionally using a GLAST-CreERT2, had a severe impact on the maturation and survival of adult-generated neurons in the SGZ and SVZ.[118] However, these studies do not exclude a role for HDAC1 in adult neurogenesis.

Members of the PcG and Trithorax (TrxG) group of proteins comprise antagonistic chromatin complexes. The mixed-lineage leukaemia 1 (Mll1) gene, a TrxG member that encodes an H3K4 methyltransferase, was shown to be required for neuronal differentiation in the adult SVZ.[119] Upon closer inspection, these authors demonstrated neurogenesis was severely impaired in Mll1 knockout mice because of the downregulation of the Mll1 target gene Dlx2, which was previously described to function in interneuron development.[120] Interestingly, loss of Mll1 did not affect the level of H3K4 methylation at the Dlx2 promoter, but H3K27 methylation was increased. Given that JmjC domain-containing H3K27 demethylases, UTX and JMJD3, can form complexes with H3K4 methyltransferases,[121,122] it would be interesting to see if the phenotype observed in the Mll1 knockout is due to a failure to recruit JMJD3, thus resulting in enhanced H3K27 methylation. The many types of histone modifications, which can occur on different histone residues, plus the identification of many histone methyltransferases and demethylases suggest an intricate molecular mechanism regulating neuronal differentiation, survival and maturation.

So far, we have discussed the effects of DNA methylation, histone modification marks, chromatin remodeling, and non-coding RNAs in the process of adult neurogenesis under basal conditions. As emphasized in this review, the nature of epigenetic phenomena lends itself to dynamic regulation during development and in the adult, in a tissue- and cell type-specific manner that is highly dependent on the niche. Indeed, one example where environmental changes during disease pathogenesis may potentially lead to alterations of the epigenome comes from drug addiction research. For example, repeated exposure to drugs of abuse (e.g. cocaine) was shown to induce changes in DNA methylation and histone acetylation, and treatment with HDAC inhibitors potentiated the behavioral sensitivity to cocaine.[123] This and other examples have given rise to the emerging concept of "behavioral epigenetics".[124] Compared to this well established connection between epigenetics and pathophysiology of the nervous system, there is little known regarding the potential of targeting epigenetic mechanisms underlying adult neurogenesis in translational applications. Next, we focus on an applied perspective and ask how might fundamental knowledge of epigenetics underlying adult neurogenesis be harnessed to treat neurological disease? In this next section, we discuss potential therapeutic approaches with chromatin-modifying drugs, most notably, HDACi which is already licensed for clinical use in oncology – relatively little attention has focused on their potential application in neuroregenerative medicine.

EPIGENETICS, iPSCs, AND STRATEGIES FOR NEURAL REPAIR

NSCs have the potential to treat a variety of human neurological conditions, including acute spinal cord injury, neurodegeneration, stroke, and epilepsy.[125–127] Two strategies currently employed for stem cell therapy are exogenous stem cell transplantation or endogenous stem cell fate conversion.[125,128] Indeed, the first successful bone marrow transplantation was conducted decades ago for effective treatment of myeloma and leukemia.[129] Currently, there are two types of stem cells that have been proposed to be used for transplantation. One is pluripotent human embryonic stem (ES) cells, but they have their own limitations, including heterogeneity and ethical issues.[130,131] The pioneering work of Yamanaka and colleagues[132] has illustrated that reprogramming of differentiated cell types to stem-like cells, or alternatively, from one lineage cell type directly into a different lineage cell type, is achievable, which is one potential way to bypass the ethical concerns of human ES cells.[132] Despite the risk of genomic insertion and capability of forming tumors, iPSCs still have become powerful tools to interrogate basic mechanisms of disease.[133]

Recently, the conversion of fibroblasts from Rett syndrome, schizophrenia, and ALS patients into human iPSCs and/or directly into neurons has provided novel ways to study underlying disease pathogenesis.[134–136] Moreover, iPSCs derived from patients may provide new platforms for high-throughput drug screening. Despite these advantages, iPSCs are not completely equal to ES cells in several ways, such as in their gene expression signatures, copy number variation, and coding mutations.[137–139] Indeed, genome-wide DNA methylome analysis has suggested differences in DNA methylation between ES cells and iPSCs, which may have an impact on gene expression and genomic stability.[140] Certainly, these important differences need to be taken into consideration when applying iPSCs to the clinical setting.

A complementary source for stem cell transplantation is multipotent adult NSCs, which can differentiate into the three major neural cell types. Massive neuronal cell loss is characteristic of many neurological diseases, such as stroke and spinal cord injury (SCI).[141,142] Neuronal replacement is one anticipated strategy to restore function after damage or injury in the nervous system. However, under pathological contexts, transplanted NSCs predominantly adopt glial fates, instead of generating neurons.[143] Thus, combining chemical compounds that regulate cell fate specification with stem cell therapy may provide a promising strategy for neuronal replacement.[144] In a recent study involving SCI, the HDACi valproic acid (VPA) was used to coax transplanted NSCs to undergo neuronal fates, resulting in restoration of hindlimb function in mice after SCI.[145] Despite hopes that stem cell transplantation is a promising way for treating neurological disease, confounding issues such as heterogeneity, donor cell rejection, and tumorigenesis are considered major disadvantages.[146] Thus, understanding and enhancing endogenous NSC pools may be more efficacious for disease prevention or treatment.

In fact, increased SGZ and SVZ neurogenesis after ischemic insults have been suggested to be an example of an intrinsic repair mechanism.[147] Stroke results from ischemia, the occlusion of the cerebral artery by thrombus or embolism, or from hemorrhage.[148] Recent work has shown that in both rodents and human brain after stroke, neuroblasts generated from NSCs residing in the SVZ could migrate into the injured sites and partly compensate for local cell death.[149] However, the effect of this endogenous response after stroke is limited for functional regeneration. As discussed above, epigenetics can control multiple stages of adult neurogenesis, therefore, targeting epigenetic regulators by small molecules, or "epi-drugs", may provide a powerful method to modulate endogenous NSCs and enhance neurogenesis after stroke. Indeed, HDACi such as sodium butyrate (NaB) have been highlighted in stroke recovery for their roles in stimulating neurogenesis, reducing the infarct volume, and significantly ameliorating neuronal injury.[150–151] The beneficial effects of NaB have been associated with hyperacetylation of histone H3 and upregulation of several non-histone proteins involved in neuroprotection, such as B-cell lymphoma-2 and heat shock protein 70. While these proof-of-concept examples are starting points to translate epidrugs to the clinical setting, there are still major hurdles to overcome such as issues of specificity, delivery, and deciding which epigenetic regulatory mechanism(s) is best suited for drug discovery.

CONCLUSIONS AND FUTURE PERSPECTIVES

In summary, ongoing neurogenesis in the adult brain is governed by three general mechanisms: (1) cell-extrinsic factors from the niche that signal in an autocrine and/or paracrine fashion; (2) integration of niche signals by transcription factors to control NSC fate; and (3) fine-tuning of gene expression by epigenetic regulators that form a regulatory circuit with transcription factors. Each of these mechanisms and their associated networks are used to varying degrees during different stages of adult NSC self-renewal and differentiation. Despite the ability of the brain to activate select pools of adult NSCs, these collective endogenous mechanisms are still inadequate to fully restore neuronal production and function to the adult brain or spinal cord following damage or injury. We envision that more restoration of function to the nervous system after injury or during disease will benefit from a full understanding of the epigenetic control mechanisms (Figure 12.5). A set of therapeutic targets will involve strategies for preserving the NSC pool, directing NSCs and transit-amplifying cells to adopt a neuronal fate, promoting differentiation, integration, and survival of immature neurons, or preventing the death of terminally differentiated neurons. While this chapter describes a subset of epigenetic factors controlling neurogenesis and discusses possible therapeutic strategies involving adult NSCs, there is still abundant information to be gained.

Towards the future, additional studies focusing on the basic anatomy of the adult neurogenic niche, the intrinsic properties of adult SGZ and SVZ NSCs, and the application of discoveries in model organisms to our knowledge of postnatal and adult neurogenesis in humans will guide our efforts to harness the potential of adult NSCs for therapy. While we have solved the genetic code and continue to annotate the human genome, there is still a long way to go before we "crack" the epigenetic code. It has become increasingly apparent that epigenetic mechanisms play an essential role in controlling stem cells and the undifferentiated state. In the clinical arena, if brain disease is a consequence of altered or dysregulated gene expression, and if changes in gene expression can be stabilized through epigenetic mechanisms, in principle, drugs targeting DNA methylation or histone-modifying enzymes could normalize gene expression in various disease settings and represent a potential strategy of reversing the effects of the disease. However, there is still great debate surrounding how changes in epigenetic mechanisms affect the brain and behavior, and the cause-and-effect relationship underlying the pathophysiology of nervous system disorders. A long-term goal is to understand fully how changes in epigenetics affect adult neurogenesis and behavior, and how behavior can change the epigenetic status of NSCs and their progeny.

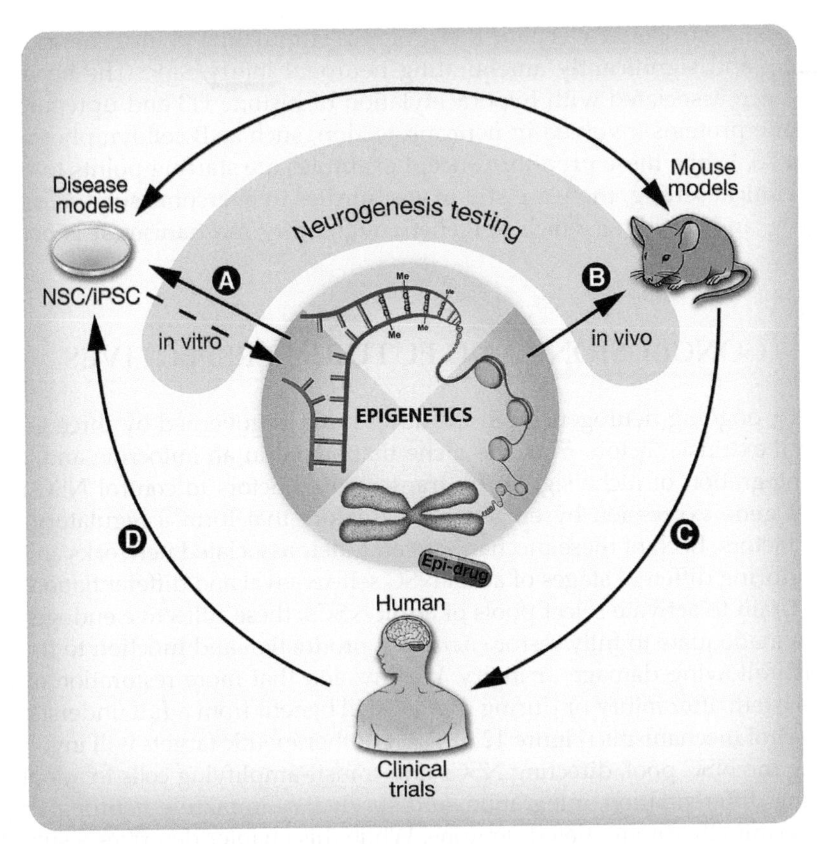

FIGURE 12.5 **Harnessing the potential of epigenetic gene regulatory mechanisms to translate neurogenesis research to the clinic.** Epigenetic regulation of adult neurogenesis can be investigated using two complementary approaches. (A) In one approach, neural stem/progenitor cells are established and grown in culture to assay potential epigenetic effects on neurogenesis. (B) Insights from in vitro models can be further tested in vivo using genetically engineered (knockout or transgenic) mice or vice versa. (C) An eventual goal is to develop effective therapeutic treatments (e.g. stem cell transplantation) or drugs (e.g. inhibitors of histone modifying enzymes, so-called "epi-drugs") for human clinical trials. (D) In parallel, patient-derived somatic cells can be reprogrammed to ES-like cells or directly converted to somatic cell types (e.g. neurons and glia) for modeling disease-in-a-dish and/or high-throughput drug screening.

Until now, only a handful of epi-drugs targeting HDACs and DNMTs are proven for clinical use. However, their specificity and the number of epigenetic mechanisms they target are still limited, considering the complexity of human nervous system disorders, such as stroke, spinal cord injury, and Parkinson's disease. Further understanding of each epigenetic regulatory mechanism will elucidate the detailed regulatory network underlying adult neurogenesis, especially points of convergence that integrate multiple damage/stress signals to the NSC genome. While the regeneration of neurons seems a lofty goal, we speculate that by targeting key nodal points, coupled with emerging tools to study epigenetics in vivo, better therapeutics will be developed to restore nervous system function following damage or in the setting of neurodegenerative diseases.

References

1. Buszczak M, Spradling AC. Searching chromatin for stem cell identity. *Cell*. 2006;125:233–236.
2. Guillemot F. Cell fate specification in the mammalian telencephalon. *Prog Neurobiol*. 2007;83:37–52.
3. Okano H, Temple S. Cell types to order: temporal specification of CNS stem cells. *Curr Opin Neurobiol*. 2009;19:112–119.
4. Altman J, Das GD. Autoradiographic and histological evidence of postnatal hippocampal neurogenesis in rats. *J Comp Neurol*. 1965;124:319–335.
5. Gage FH. Mammalian neural stem cells. *Science*. 2000;287:1433–1438.
6. Alvarez-Buylla A, Garcia-Verdugo JM. Neurogenesis in adult subventricular zone. *J Neurosci*. 2002;22:629–634.
7. Ming GL, Song H. Adult neurogenesis in the mammalian central nervous system. *Annu Rev Neurosci*. 2005;28:223–250.
8. Reynolds BA, Weiss S. Generation of neurons and astrocytes from isolated cells of the adult mammalian central nervous system. *Science*. 1992;255:1707–1710.
9. Reynolds BA, Weiss S. Clonal and population analyses demonstrate that an EGF-responsive mammalian embryonic CNS precursor is a stem cell. *Dev Biol*. 1996;175:1–13.
10. Takahashi J, Palmer TD, Gage FH. Retinoic acid and neurotrophins collaborate to regulate neurogenesis in adult-derived neural stem cell cultures. *J Neurobiol*. 1999;38:65–81.
11. Hsieh J, Aimone JB, Kaspar BK, Kuwabara T, Nakashima K, Gage FH. IGF-I instructs multipotent adult neural progenitor cells to become oligodendrocytes. *J Cell Biol*. 2004;164:111–122.
12. Palmer TD, Takahashi J, Gage FH. The adult rat hippocampus contains primordial neural stem cells. *Mol Cell Neurosci*. 1997;8:389–404.
13. Ray J, Gage FH. Differential properties of adult rat and mouse brain-derived neural stem/progenitor cells. *Mol Cell Neurosci*. 2006;31:560–573.
14. Jaenisch R, Bird A. Epigenetic regulation of gene expression: how the genome integrates intrinsic and environmental signals. *Nat Genet*. 2003;33(suppl):245–254.
15. Goldberg AD, Allis CD, Bernstein E. Epigenetics: a landscape takes shape. *Cell*. 2007;128:635–638.
16. Waddington CH. *The Strategy of the Genes*. London: Geoge Allen & Unwin; 1957.
17. Hsieh J, Nakashima K, Kuwabara T, Mejia E, Gage FH. Histone deacetylase inhibition-mediated neuronal differentiation of multipotent adult neural progenitor cells. *Proc Natl Acad Sci U S A*. 2004;101:16659–16664.
18. Shaked M, Weissmuller K, Svoboda H, et al. Histone deacetylases control neurogenesis in embryonic brain by inhibition of BMP2/4 signaling. *PLoS One*. 2008;3:e2668.
19. Sun G, Alzayady K, Stewart R, et al. Histone demethylase LSD1 regulates neural stem cell proliferation. *Mol Cell Biol*. 2010;30:1997–2005.
20. Liu J, Casaccia P. Epigenetic regulation of oligodendrocyte identity. *Trends Neurosci*. 2010;33:193–201.
21. Liu J, Sandoval J, Doh ST, Cai L, Lopez-Rodas G, Casaccia P. Epigenetic modifiers are necessary but not sufficient for reprogramming non-myelinating cells into myelin gene-expressing cells. *PLoS One*. 2010;5:e13023.
22. Miller RH. Oligodendrocyte origins. *Trends Neurosci*. 1996;19:92–96.
23. Cai J, Qi Y, Hu X, et al. Generation of oligodendrocyte precursor cells from mouse dorsal spinal cord independent of Nkx6 regulation and Shh signaling. *Neuron*. 2005;45:41–53.
24. Fogarty M, Richardson WD, Kessaris N. A subset of oligodendrocytes generated from radial glia in the dorsal spinal cord. *Development*. 2005;132:1951–1959.
25. Vallstedt A, Klos JM, Ericson J. Multiple dorsoventral origins of oligodendrocyte generation in the spinal cord and hindbrain. *Neuron*. 2005;45:55–67.
26. Ye F, Chen Y, Hoang T, et al. HDAC1 and HDAC2 regulate oligodendrocyte differentiation by disrupting the beta-catenin-TCF interaction. *Nat Neurosci*. 2009;12:829–838.
27. Takizawa T, Nakashima K, Namihira M, et al. DNA methylation is a critical cell-intrinsic determinant of astrocyte differentiation in the fetal brain. *Dev Cell*. 2001;1:749–758.
28. Barnabe-Heider F, Wasylnka JA, Fernandes KJ, et al. Evidence that embryonic neurons regulate the onset of cortical gliogenesis via cardiotrophin-1. *Neuron*. 2005;48:253–265.
29. Johe KK, Hazel TG, Muller T, Dugich-Djordjevic MM, McKay RD. Single factors direct the differentiation of stem cells from the fetal and adult central nervous system. *Genes Dev*. 1996;10:3129–3140.
30. Bonni A, Sun Y, Nadal-Vicens M, et al. Regulation of gliogenesis in the central nervous system by the JAK-STAT signaling pathway. *Science*. 1997;278:477–483.

31. Rajan P, McKay RD. Multiple routes to astrocytic differentiation in the CNS. *J Neurosci*. 1998;18:3620–3629.
32. He F, Ge W, Martinowich K, et al. A positive autoregulatory loop of Jak-STAT signaling controls the onset of astrogliogenesis. *Nat Neurosci*. 2005;8:616–625.
33. Nakashima K, Yanagisawa M, Arakawa H, et al. Synergistic signaling in fetal brain by STAT3-Smad1 complex bridged by p300. *Science*. 1999;284:479–482.
34. Gaiano N, Fishell G. The role of notch in promoting glial and neural stem cell fates. *Annu Rev Neurosci*. 2002;25:471–490.
35. Kamakura S, Oishi K, Yoshimatsu T, Nakafuku M, Masuyama N, Gotoh Y. Hes binding to STAT3 mediates crosstalk between Notch and JAK-STAT signalling. *Nat Cell Biol*. 2004;6:547–554.
36. Hirabayashi Y, Suzki N, Tsuboi M, et al. Polycomb limits the neurogenic competence of neural precursor cells to promote astrogenic fate transition. *Neuron*. 2009;63:600–613.
37. Namihira M, Kohyama J, Semi K, et al. Committed neuronal precursors confer astrocytic potential on residual neural precursor cells. *Dev Cell*. 2009;16:245–255.
38. Artavanis-Tsakonas S, Rand MD, Lake RJ. Notch signaling: cell fate control and signal integration in development. *Science*. 1999;284:770–776.
39. de la Pompa JL, Wakeham A, Correia KM, et al. Conservation of the Notch signalling pathway in mammalian neurogenesis. *Development*. 1997;124:1139–1148.
40. Donoviel DB, Hadjantonakis AK, Ikeda M, Zheng H, Hyslop PS, Bernstein A. Mice lacking both presenilin genes exhibit early embryonic patterning defects. *Genes Dev*. 1999;13:2801–2810.
41. Hitoshi S, Ishino Y, Kumar A, et al. Mammalian Gcm genes induce Hes5 expression by active DNA demethylation and induce neural stem cells. *Nat Neurosci*. 2011;14:957–964.
42. Liu C, Zhao X. MicroRNAs in adult and embryonic neurogenesis. *Neuromolecular Med*. 2009;11:141–152.
43. Shi Y, Zhao X, Hsieh J, et al. MicroRNA regulation of neural stem cells and neurogenesis. *J Neurosci*. 2010;30:14931–14936.
44. Kosik KS. The neuronal microRNA system. *Nat Rev Neurosci*. 2006;7:911–920.
45. Mehler MF, Mattick JS. Noncoding RNAs and RNA editing in brain development, functional diversification, and neurological disease. *Physiol Rev*. 2007;87:799–823.
46. Yoo AS, Staahl BT, Chen L, Crabtree GR. MicroRNA-mediated switching of chromatin-remodelling complexes in neural development. *Nature*. 2009;460:642–646.
47. Conaco C, Otto S, Han JJ, Mandel G. Reciprocal actions of REST and a microRNA promote neuronal identity. *Proc Natl Acad Sci USA*. 2006;103:2422–2427.
48. Gao Z, Ure K, Ding P, et al. The master negative regulator REST/NRSF controls adult neurogenesis by restraining the neurogenic program in quiescent stem cells. *J Neurosci*. 2011;31:9772–9786.
49. Zhao C, Sun G, Li S, et al. MicroRNA let-7b regulates neural stem cell proliferation and differentiation by targeting nuclear receptor TLX signaling. *Proc Natl Acad Sci USA*. 2010;107:1876–1881.
50. Rybak A, Fuchs H, Smirnova L, et al. A feedback loop comprising lin-28 and let-7 controls pre-let-7 maturation during neural stem-cell commitment. *Nat Cell Biol*. 2008;10:987–993.
51. Li W, Sun G, Yang S, Qu Q, Nakashima K, Shi Y. Nuclear receptor TLX regulates cell cycle progression in neural stem cells of the developing brain. *Mol Endocrinol*. 2008;22:56–64.
52. Sun J, Ming GL, Song H. Epigenetic regulation of neurogenesis in the adult mammalian brain. *Eur J Neurosci*. 2011;33:1087–1093.
53. Ponjavic J, Oliver PL, Lunter G, Ponting CP. Genomic and transcriptional co-localization of protein-coding and long non-coding RNA pairs in the developing brain. *PLoS Genet*. 2009;5:e1000617.
54. Price M, Lazzaro D, Pohl T, et al. Regional expression of the homeobox gene Nkx-2.2 in the developing mammalian forebrain. *Neuron*. 1992;8:241–255.
55. Tochitani S, Hayashizaki Y. Nkx2.2 antisense RNA overexpression enhanced oligodendrocytic differentiation. *Biochem Biophys Res Commun*. 2008;372:691–696.
56. Kohtz JD, Fishell G. Developmental regulation of EVF-1, a novel non-coding RNA transcribed upstream of the mouse Dlx6 gene. *Gene Expr Patterns*. 2004;4:407–412.
57. Feng J, Bi C, Clark BS, Mady R, Shah P, Kohtz JD. The Evf-2 noncoding RNA is transcribed from the Dlx-5/6 ultraconserved region and functions as a Dlx-2 transcriptional coactivator. *Genes Dev*. 2006;20:1470–1484.
58. van Praag H, Schinder AF, Christie BR, Toni N, Palmer TD, Gage FH. Functional neurogenesis in the adult hippocampus. *Nature*. 2002;415:1030–1034.

59. Zhao C, Teng EM, Summers Jr RG, Ming GL, Gage FH. Distinct morphological stages of dentate granule neuron maturation in the adult mouse hippocampus. *J Neurosci*. 2006;26:3–11.
60. Seri B, Garcia-Verdugo JM, McEwen BS, Alvarez-Buylla A. Astrocytes give rise to new neurons in the adult mammalian hippocampus. *J Neurosci*. 2001;21:7153–7160.
61. Kempermann G, Jessberger S, Steiner B, Kronenberg G. Milestones of neuronal development in the adult hippocampus. *Trends Neurosci*. 2004;27:447–452.
62. Ables JL, Decarolis NA, Johnson MA, et al. Notch1 is required for maintenance of the reservoir of adult hippocampal stem cells. *J Neurosci*. 2010;30:10484–10492.
63. Mira H, Andreu Z, Suh H, et al. Signaling through BMPR-IA regulates quiescence and long-term activity of neural stem cells in the adult hippocampus. *Cell Stem Cell*. 2010;7:78–89.
64. Lugert S, Basak O, Knuckles P, et al. Quiescent and active hippocampal neural stem cells with distinct morphologies respond selectively to physiological and pathological stimuli and aging. *Cell Stem Cell*. 2010;6:445–456.
65. Bonaguidi MA, Wheeler MA, Shapiro JS, et al. In vivo clonal analysis reveals self-renewing and multipotent adult neural stem cell characteristics. *Cell*. 2011;145:1142–1155.
66. Kronenberg G, Reuter K, Steiner B, et al. Subpopulations of proliferating cells of the adult hippocampus respond differently to physiologic neurogenic stimuli. *J Comp Neurol*. 2003;467:455–463.
67. Encinas JM, Michurina TV, Peunova N, et al. Division-coupled astrocytic differentiation and age-related depletion of neural stem cells in the adult hippocampus. *Cell Stem Cell*. 2011;8:566–579.
68. Dhaliwal J, Lagace DC. Visualization and genetic manipulation of adult neurogenesis using transgenic mice. *Eur J Neurosci*. 2011;33:1025–1036.
69. Encinas JM, Vaahtokari A, Enikolopov G. Fluoxetine targets early progenitor cells in the adult brain. *Proc Natl Acad Sci USA*. 2006;103:8233–8238.
70. Suh H, Consiglio A, Ray J, Sawai T, D'Amour KA, Gage FH. In vivo fate analysis reveals the multipotent and self-renewal capacities of Sox2+ neural stem cells in the adult hippocampus. *Cell Stem Cell*. 2007;1:515–528.
71. Merkle FT, Mirzadeh Z, Alvarez-Buylla A. Mosaic organization of neural stem cells in the adult brain. *Science*. 2007;317:381–384.
72. Lledo PM, Merkle FT, Alvarez-Buylla A. Origin and function of olfactory bulb interneuron diversity. *Trends Neurosci*. 2008;31:392–400.
73. Doetsch F, Caille I, Lim DA, Garcia-Verdugo JM, Alvarez-Buylla A. Subventricular zone astrocytes are neural stem cells in the adult mammalian brain. *Cell*. 1999;97:703–716.
74. Merkle FT, Tramontin AD, Garcia-Verdugo JM, Alvarez-Buylla A. Radial glia give rise to adult neural stem cells in the subventricular zone. *Proc Natl Acad Sci USA*. 2004;101:17528–17532.
75. Spassky N, Merkle FT, Flames N, Tramontin AD, Garcia-Verdugo JM, Alvarez-Buylla A. Adult ependymal cells are postmitotic and are derived from radial glial cells during embryogenesis. *J Neurosci*. 2005;25:10–18.
76. Liu X, Bolteus AJ, Balkin DM, Henschel O, Bordey A. GFAP-expressing cells in the postnatal subventricular zone display a unique glial phenotype intermediate between radial glia and astrocytes. *Glia*. 2006;54:394–410.
77. Nomura T, Goritz C, Catchpole T, Henkemeyer M, Frisen J. EphB signaling controls lineage plasticity of adult neural stem cell niche cells. *Cell Stem Cell*. 2010;7:730–743.
78. Mirzadeh Z, Merkle FT, Soriano-Navarro M, Garcia-Verdugo JM, Alvarez-Buylla A. Neural stem cells confer unique pinwheel architecture to the ventricular surface in neurogenic regions of the adult brain. *Cell Stem Cell*. 2008;3:265–278.
79. Shen Q, Wang Y, Kokovay E, et al. Adult SVZ stem cells lie in a vascular niche: a quantitative analysis of niche cell-cell interactions. *Cell Stem Cell*. 2008;3:289–300.
80. Kriegstein A, Alvarez-Buylla A. The glial nature of embryonic and adult neural stem cells. *Annu Rev Neurosci*. 2009;32:149–184.
81. Brill MS, Snapyan M, Wohlfrom H, et al. A dlx2- and pax6-dependent transcriptional code for periglomerular neuron specification in the adult olfactory bulb. *J Neurosci*. 2008;28:6439–6452.
82. Goergen EM, Bagay LA, Rehm K, Benton JL, Beltz BS. Circadian control of neurogenesis. *J Neurobiol*. 2002;53:90–95.
83. Brill MS, Ninkovic J, Winpenny E, et al. Adult generation of glutamatergic olfactory bulb interneurons. *Nat Neurosci*. 2009;12:1524–1533.
84. Ma DK, Marchetto MC, Guo JU, Ming GL, Gage FH, Song H. Epigenetic choreographers of neurogenesis in the adult mammalian brain. *Nat Neurosci*. 2010;13:1338–1344.

85. Jiang YHJ. Harnessing adult neurogenesis by cracking the epigenetic code. *Future Neurology*. 2012;7:65–79.
86. Ma DK, Jang MH, Guo JU, et al. Neuronal activity-induced Gadd45b promotes epigenetic DNA demethylation and adult neurogenesis. *Science*. 2009;323:1074–1077.
87. Orford KW, Scadden DT. Deconstructing stem cell self-renewal: genetic insights into cell-cycle regulation. *Nat Rev Genet*. 2008;9:115–128.
88. Dranovsky A, Picchini AM, Moadel T, et al. Experience dictates stem cell fate in the adult hippocampus. *Neuron*. 2011;70:908–923.
89. Bonaguidi MA, Song J, Ming GL, Song H. A unifying hypothesis on mammalian neural stem cell properties in the adult hippocampus. *Curr Opin Neurobiol*. 2012;22:1–8.
90. Ballas N, Grunseich C, Lu DD, Speh JC, Mandel G. REST and its corepressors mediate plasticity of neuronal gene chromatin throughout neurogenesis. *Cell*. 2005;121:645–657.
91. Molofsky AV, Pardal R, Iwashita T, Park IK, Clarke MF, Morrison SJ. Bmi-1 dependence distinguishes neural stem cell self-renewal from progenitor proliferation. *Nature*. 2003;425:962–967.
92. Doetsch F, Verdugo JM, Caille I, Alvarez-Buylla A, Chao MV, Casaccia-Bonnefil P. Lack of the cell-cycle inhibitor p27Kip1 results in selective increase of transit-amplifying cells for adult neurogenesis. *J Neurosci*. 2002;22:2255–2264.
93. Shi Y, Chichung Lie D, Taupin P, et al. Expression and function of orphan nuclear receptor TLX in adult neural stem cells. *Nature*. 2004;427:78–83.
94. Zhang CL, Zou Y, He W, Gage FH, Evans RM. A role for adult TLX-positive neural stem cells in learning and behaviour. *Nature*. 2008;451:1004–1007.
95. Niu W, Zou Y, Shen C, Zhang CL. Activation of postnatal neural stem cells requires nuclear receptor TLX. *J Neurosci*. 2011;31:13816–13828.
96. Weber P, Metzger D, Chambon P. Temporally controlled targeted somatic mutagenesis in the mouse brain. *Eur J Neurosci*. 2001;14:1777–1783.
97. Renault VM, Rafalski VA, Morgan AA, et al. FoxO3 regulates neural stem cell homeostasis. *Cell Stem Cell*. 2009;5:527–539.
98. Wu J, Xie X. Comparative sequence analysis reveals an intricate network among REST, CREB and miRNA in mediating neuronal gene expression. *Genome Biol*. 2006;7:R85.
99. Otto SJ, McCorkle SR, Hover J, et al. A new binding motif for the transcriptional repressor REST uncovers large gene networks devoted to neuronal functions. *J Neurosci*. 2007;27:6729–6739.
100. Gao Z, Ding P, Hsieh J. Profiling of REST-dependent microRNAs reveals dynamic modes of expression. *Front Neurosci*. 2012;6:67.
101. Zhao C, Sun G, Li S, Shi Y. A feedback regulatory loop involving microRNA-9 and nuclear receptor TLX in neural stem cell fate determination. *Nat Struct Mol Biol*. 2009;16:365–371.
102. Shibata M, Kurokawa D, Nakao H, Ohmura T, Aizawa S. MicroRNA-9 modulates Cajal-Retzius cell differentiation by suppressing Foxg1 expression in mouse medial pallium. *J Neurosci*. 2008;28:10415–10421.
103. Shibata M, Nakao H, Kiyonari H, Abe T, Aizawa S. MicroRNA-9 regulates neurogenesis in mouse telencephalon by targeting multiple transcription factors. *J Neurosci*. 2011;31:3407–3422.
104. Fan G, Martinowich K, Chin MH, et al. DNA methylation controls the timing of astrogliogenesis through regulation of JAK-STAT signaling. *Development*. 2005;132:3345–3356.
105. Feng J, Chang H, Li E, Fan G. Dynamic expression of de novo DNA methyltransferases Dnmt3a and Dnmt3b in the central nervous system. *J Neurosci Res*. 2005;79:734–746.
106. Tsujimura K, Abematsu M, Kohyama J, Namihira M, Nakashima K. Neuronal differentiation of neural precursor cells is promoted by the methyl-CpG-binding protein MeCP2. *Exp Neurol*. 2009;219:104–111.
107. Feng J, Zhou Y, Campbell SL, et al. Dnmt1 and Dnmt3a maintain DNA methylation and regulate synaptic function in adult forebrain neurons. *Nat Neurosci*. 2010;13:423–430.
108. Ballestar E, Wolffe AP. Methyl-CpG-binding proteins. Targeting specific gene repression. *Eur J Biochem*. 2001;268:1–6.
109. Zhao X, Ueba T, Christie BR, et al. Mice lacking methyl-CpG binding protein 1 have deficits in adult neurogenesis and hippocampal function. *Proc Natl Acad Sci USA*. 2003;100:6777–6782.
110. Li X, Barkho BZ, Luo Y, et al. Epigenetic regulation of the stem cell mitogen Fgf-2 by Mbd1 in adult neural stem/progenitor cells. *J Biol Chem*. 2008;283:27644–27652.
111. Liu C, Teng ZQ, Santistevan NJ, et al. Epigenetic regulation of miR-184 by MBD1 governs neural stem cell proliferation and differentiation. *Cell Stem Cell*. 2010;6:433–444.

112. Prozorovski T, Schulze-Topphoff U, Glumm R, et al. Sirt1 contributes critically to the redox-dependent fate of neural progenitors. *Nat Cell Biol*. 2008;10:385–394.
113. Hisahara S, Chiba S, Matsumoto H, et al. Histone deacetylase SIRT1 modulates neuronal differentiation by its nuclear translocation. *Proc Natl Acad Sci USA*. 2008;105:15599–15604.
114. Williams RR, Azuara V, Perry P, et al. Neural induction promotes large-scale chromatin reorganisation of the Mash1 locus. *J Cell Sci*. 2006;119:132–140.
115. Montgomery RL, Hsieh J, Barbosa AC, Richardson JA, Olson EN. Histone deacetylases 1 and 2 control the progression of neural precursors to neurons during brain development. *Proc Natl Acad Sci USA*. 2009;106:7876–7881.
116. Haberland M, Carrer M, Mokalled MH, Montgomery RL, Olson EN. Redundant control of adipogenesis by histone deacetylases 1 and 2. *J Biol Chem*. 2010;285:14663–14670.
117. Guan JS, Haggarty SJ, Giacometti E, et al. HDAC2 negatively regulates memory formation and synaptic plasticity. *Nature*. 2009;459:55–60.
118. Jawerka M, Colak D, Dimou L, et al. The specific role of histone deacetylase 2 in adult neurogenesis. *Neuron Glia Biol*. 2010;6:93–107.
119. Lim DA, Huang YC, Swigut T, et al. Chromatin remodelling factor Mll1 is essential for neurogenesis from postnatal neural stem cells. *Nature*. 2009;458:529–533.
120. Long JE, Garel S, Alvarez-Dolado M, et al. Dlx-dependent and -independent regulation of olfactory bulb interneuron differentiation. *J Neurosci*. 2007;27:3230–3243.
121. Hong S, Cho YW, Yu LR, Yu H, Veenstra TD, Ge K. Identification of JmjC domain-containing UTX and JMJD3 as histone H3 lysine 27 demethylases. *Proc Natl Acad Sci USA*. 2007;104:18439–18444.
122. Issaeva I, Zonis Y, Rozovskaia T, et al. Knockdown of ALR (MLL2) reveals ALR target genes and leads to alterations in cell adhesion and growth. *Mol Cell Biol*. 2007;27:1889–1903.
123. Renthal W, Nestler EJ. Epigenetic mechanisms in drug addiction. *Trends Mol Med*. 2008;14:341–350.
124. Lester BM, Tronick E, Nestler E, et al. Behavioral epigenetics. *Ann N Y Acad Sci*. 2011;1226:14–33.
125. Lindvall O, Kokaia Z, Martinez-Serrano A. Stem cell therapy for human neurodegenerative disorders-how to make it work. *Nat Med*. 2004;10(suppl):S42–S50.
126. Parent JM, Jessberger S, Gage FH, Gong C. Is neurogenesis reparative after status epilepticus? *Epilepsia*. 2007;48(suppl 8):69–71.
127. Kernie SG, Parent JM. Forebrain neurogenesis after focal Ischemic and traumatic brain injury. *Neurobiol Dis*. 2010;37:267–274.
128. Goldman S. Stem and progenitor cell-based therapy of the human central nervous system. *Nat Biotechnol*. 2005;23:862–871.
129. Bensinger WI. The current status of hematopoietic stem cell transplantation for multiple myeloma. *Clin Adv Hematol Oncol*. 2004;2:46–52.
130. McLaren A. Important differences between sources of embryonic stem cells. *Nature*. 2000;408:513.
131. McLaren A. Ethical and social considerations of stem cell research. *Nature*. 2001;414:129–131.
132. Takahashi K, Yamanaka S. Induction of pluripotent stem cells from mouse embryonic and adult fibroblast cultures by defined factors. *Cell*. 2006;126:663–676.
133. Saha K, Jaenisch R. Technical challenges in using human induced pluripotent stem cells to model disease. *Cell Stem Cell*. 2009;5:584–595.
134. Marchetto MC, Carromeu C, Acab A, et al. A model for neural development and treatment of Rett syndrome using human induced pluripotent stem cells. *Cell*. 2010;143:527–539.
135. Brennand KJ, Simone A, Jou J, et al. Modelling schizophrenia using human induced pluripotent stem cells. *Nature*. 2011;473:221–225.
136. Son EY, Ichida JK, Wainger BJ, et al. Conversion of mouse and human fibroblasts into functional spinal motor neurons. *Cell Stem Cell*. 2011;9:205–218.
137. Chin MH, Mason MJ, Xie W, et al. Induced pluripotent stem cells and embryonic stem cells are distinguished by gene expression signatures. *Cell Stem Cell*. 2009;5:111–123.
138. Gore A, Li Z, Fung HL, et al. Somatic coding mutations in human induced pluripotent stem cells. *Nature*. 2011;471:63–67.
139. Hussein SM, Batada NN, Vuoristo S, et al. Copy number variation and selection during reprogramming to pluripotency. *Nature*. 2011;471:58–62.

140. Lister R, Pelizzola M, Kida YS, et al. Hotspots of aberrant epigenomic reprogramming in human induced pluripotent stem cells. *Nature*. 2011;471:68–73.
141. Colbourne F, Li H, Buchan AM, Clemens JA. Continuing postischemic neuronal death in CA1: influence of ischemia duration and cytoprotective doses of NBQX and SNX-111 in rats. *Stroke*. 1999;30:662–668.
142. Beattie MS, Hermann GE, Rogers RC, Bresnahan JC. Cell death in models of spinal cord injury. *Prog Brain Res*. 2002;137:37–47.
143. Picard-Riera N, Decker L, Delarasse C, et al. Experimental autoimmune encephalomyelitis mobilizes neural progenitors from the subventricular zone to undergo oligodendrogenesis in adult mice. *Proc Natl Acad Sci USA*. 2002;99:13211–13216.
144. Emre N, Coleman R, Ding S. A chemical approach to stem cell biology. *Curr Opin Chem Biol*. 2007;11:252–258.
145. Abematsu M, Tsujimura K, Yamano M, et al. Neurons derived from transplanted neural stem cells restore disrupted neuronal circuitry in a mouse model of spinal cord injury. *J Clin Invest*. 2010;120:3255–3266.
146. Amariglio N, Hirshberg A, Scheithauer BW, et al. Donor-derived brain tumor following neural stem cell transplantation in an ataxia telangiectasia patient. *PLoS Med*. 2009;6:e1000029.
147. Jin K, Mao XO, Batteur SP, McEachron E, Leahy A, Greenberg DA. Caspase-3 and the regulation of hypoxic neuronal death by vascular endothelial growth factor. *Neuroscience*. 2001;108:351–358.
148. Doyle KP, Simon RP, Stenzel-Poore MP. Mechanisms of ischemic brain damage. *Neuropharmacology*. 2008;55:310–318.
149. Yamashita T, Ninomiya M, Hernandez Acosta P, et al. Subventricular zone-derived neuroblasts migrate and differentiate into mature neurons in the post-stroke adult striatum. *J Neurosci*. 2006;26:6627–6636.
150. Kim HJ, Leeds P, Chuang DM. The HDAC inhibitor, sodium butyrate, stimulates neurogenesis in the ischemic brain. *J Neurochem*. 2009;110:1226–1240.
151. Langley B, Brochier C, Rivieccio MA. Targeting histone deacetylases as a multifaceted approach to treat the diverse outcomes of stroke. *Stroke*. 2009;40:2899–2905.

"Transgenerational Inheritance"
J. David Sweatt, acrylic on wood panel (24 x 48), 2011–2012

CHAPTER

13

Transgenerational Inheritance in Mammals

Isabelle M. Mansuy,[1] Rahia Mashoodh[2] and Frances A. Champagne[2]

[1]Medical Faculty of the University of Zurich and Department of Health Science and Technology, Swiss Federal Institute of Technology Zurich, Switzerland

[2]Columbia University, Department of Psychology, New York, USA

INTRODUCTION

The concept of inheritance is integral to evolutionary theory and to the understanding of the origins of individual differences in phenotypic traits. Though, historically, this concept has been broadly used, since the discovery of DNA, "inheritance" has come to be linked almost exclusively to the transmission of genetic variation across generations. Thus, we are similar to our ancestors and descendants due to similarities in DNA sequence. However, when we consider the pathways leading to individual differences in traits, such as personality, intelligence, or health, it is clear that genetic factors are not the only contributors, but that environmental factors also play an essential role. The debate regarding the relative influence of nature (genetics) and nurture (environment) has given way to the notion that these factors interact to produce phenotypic variability.[1,2] Given the importance of within-generation interplay between genes and environment in the emergence of individual characteristics, there has been increasing interest in whether this interplay can likewise lead to inheritance of acquired traits across generations to produce transgenerational effects. Though transgenerational effects of environmental factors are well documented in humans and animals, the mechanisms that permit such inheritance have yet to be elucidated. Advances in the understanding of the pathways that dynamically regulate gene expression may provide insight into these mechanisms.

Epigenetic regulation of gene expression involves molecular mechanisms which can modulate gene transcription without altering the DNA sequence.[3] These mechanisms are

Epigenetic Regulation in the Nervous System
DOI: http://dx.doi.org/10.1016/B978-0-12-391494-1.00013-6

critical during development and adulthood, and regulate fundamental processes such as cellular differentiation and functioning by activating or silencing genes.[4] Their role in cellular differentiation is particularly critical because it contributes to cellular phenotypes. To maintain these phenotypes when cells divide however, the epigenetic "marks" they are associated with must persist and be transferred to daughter cells during mitosis. Such transfer suggests the possibility that an epigenetic profile may also be transmitted across generations. However, because the genome undergoes genome-wide epigenetic reprogramming post-fertilization,[5] it has been assumed that epigenetic marks in gametes are erased during the formation of the embryo, and that all cellular traces of the "epigenetic history" of parents is lost. The discovery of imprinted genes, a class of genes that are expressed from either the maternal or the paternal allele, challenged this view and led to the recognition that parental information present in germ cells can be epigenetically maintained across meiosis. The finding that epigenetic pathways are plastic and change in response to a wide range of environmental exposures, particularly during fetal and postnatal development,[6,7] coupled with the accumulating evidence that, in some cases, this epigenetic variation can be passed from parent to offspring, strengthens the idea that epigenetic mechanisms participate in the inheritance of acquired phenotypes.

In this chapter, we highlight findings illustrating the transgenerational effects of environmental experiences and the role of epigenetic processes in mediating this inheritance. We explore in particular both germline- and experience-dependent inheritance of epigenetic variation (see Figure 13.1), which represent divergent pathways through which epigenetic information can be transmitted across generations. Though divergent, these pathways are not mutually exclusive and their interplay in inheritance is discussed. Broadening the concept of inheritance to include epigenetics and other processes independent of the DNA sequence can be challenging from a mechanistic perspective. However, it is essential that this concept be seriously explored as it should ultimately provide new insight into the dynamics of evolutionary processes and a better understanding of the genotype–phenotype relationship.[8]

TRANSGENERATIONAL EFFECTS: EPIDEMIOLOGICAL AND LABORATORY STUDIES

Maternal Influence

The life experiences of an individual can lead to enduring physiological, behavioral or psychological changes. Further, there is increasing evidence that these experiences can also influence subsequent generations, wherein the experience of grandparents (F0) can affect phenotypic outcomes in both their offspring (F1) and grand-offspring (F2). Such transmission can occur through matriline or patriline mechanisms and is likely to depend upon the unique interactions of mothers and fathers with offspring. For transmission through mothers, there is evidence that maternal care plays a critical role and significantly influences the maternal behavior of female offspring, thus perpetuating maternal effects into the following generation. In humans, mother–infant attachment traits (secure, anxious/resistant, avoidant, disorganized)[9,10] as well as the level of parental bonding[11] are often similar when examined in successive generations. Likewise, in rhesus and pigtail macaques, the frequency of postpartum maternal contact or, in contrast, the rate of maternal rejection and infant abuse, can also be transmitted across the matriline.[12–14] This transmission is robust and cross-fostering studies in abusive and non-abusive macaque females have indicated that transmission of abusive behaviors from mother to daughter depends on the experience of abuse in postnatal life.[15] Females born to abusive mothers and fostered to a non-abusive mother do not show any infant abuse, suggesting that this trait is behaviorally transmitted.

Matrilineal transmission of maternal behavior is also evident in laboratory rodents. In rats, maternal behavior varies among mothers during the first week postpartum. Individual differences in pup licking/grooming (LG) differentiate lactating dams into low LG and high LG,[16] and this distinction in the F0 generation is maintained in F1 and F2 generation females.[16,17] Thus, in stable environmental conditions, the offspring and grand-offspring of low LG females provide low LG to their offspring, whereas offspring and grand-offspring of high LG females provide high levels of LG to their progeny. Similar to the transgenerational effects of abuse in macaques, cross-fostering studies have demonstrated that transmission of maternal LG from mother to female offspring depends on the level of maternal LG received in infancy,[16,18] whether provided by the biological mother or any other lactating foster mother. Importantly, there is evidence that maternal behaviors are sensitive to environmental conditions. For example, in rodents, maternal behaviors are perturbed by chronic exposure to stress,[19] manipulation of the juvenile environment,[17] maternal diet,[20] or disruption of cage bedding.[21] Such manipulations interfere with maternal care, and modify the predicted inheritance of multiple behaviors.

Social enrichment can also have transgenerational implications. Though standard laboratory rearing of rodents typically consists of a single lactating dam and her pups, under naturalistic conditions a more common rearing strategy consists of multiple females caring for pups in a communal nest.[22] Adult females exposed to communal nesting engage in higher pup-directed behaviors and LG during the postpartum period.[23] As adults, female offspring of communally-reared females display higher levels of maternal care toward their pups, even in non-communal conditions.[23] Likewise, exposure to social/physical environmental enrichment during the juvenile period can have transgenerational benefits. In wild-type mice, or in mice with a deficit in long-term potentiation due to a deficiency in the signaling molecule Ras-GRF, exposure to an enriched environment improved synaptic plasticity and enhanced memory. This effect is transmitted to the offspring through females, but not males. Further, the improvements in memory persist even after postnatal cross-fostering to non-enriched foster mothers, suggesting a transgenerational inheritance via prenatal or female gametic mechanisms.[24]

Paternal Influence

In most mammals, the early rearing environment is characterized by intense prenatal and postnatal mother–infant interactions, with limited or no interaction with the father. However, despite such a limited parental contribution, paternal influence on offspring development that persists into the F2 generation and beyond has been observed. In rodents, exposure of males to alcohol prior to mating is associated with reduced offspring litter size and birth weight, increased mortality, and numerous cognitive and behavioral abnormalities.[25–30] Likewise, the offspring of cocaine-exposed males perform poorly on tests of attention, spatial working memory and spontaneous alternation, and have reduced cerebral volume.[31,32] In addition, males' exposure to poor housing conditions with low oxygen and high carbon dioxide prior to mating, increases blood hemoglobin in the female offspring.[33] In utero exposure to the glucocorticoid receptor agonist, dexamethasone, increases glucose intolerance in male offspring and grand-offspring, even if mothers of these offspring are not exposed.[34] The metabolic effect of prenatal dexamethasone exposure however does not persist beyond the F2 generation.

Variation in the dietary regimen of fathers can also have transmissible effects in the offspring. In humans, analysis of archival records from Sweden, in which crop success (used as a proxy for food intake) can be related to longevity across successive generations, has suggested that the level of nutrition during the slow growth period that precedes puberty is associated with diabetes and mortality from cardiovascular disease in grand-offspring.[35,36] This association is sex-specific, with paternal grandfather nutrition predicting grandson mortality and paternal grandmother nutrition predicting grand-daughter longevity.[37] A transgenerational impact of nutrition also occurs in laboratory rodents. A 24-hour complete fast two weeks prior to mating in males reduces serum glucose and alters the level of corticosterone and insulin-like growth factor 1 in the offspring.[38] Prenatal protein restriction can also affect growth and metabolism in offspring and grand-offspring.[39] Likewise, during the juvenile period, male mice exposed to a low-protein diet have offspring with detectable changes in the expression of genes involved in lipid and cholesterol function.[40] In female mice, exposure to caloric restriction during late gestation also impairs glucose tolerance in the grand-offspring, even if the direct offspring (parents of the grand-offspring) do not undergo any restriction and are provided with ad libitum food. High fat diet is also associated with phenotypic effects in the offspring that can persist across multiple generations. In rats, female offspring of males fed a high-fat diet have impaired insulin secretion and glucose tolerance, and β-cell dysfunction in the pancreas.[41] When exposed in utero, offspring have increased body length and insulin sensitivity, and transmit these phenotypes to the following two generations (F2 and F3) through the patriline.[42,43] Transmission can also occur through the matriline, but these phenotypes only persist until the second generation (F2). In humans, paternal consumption of betel nuts (containing nitrosamines which are carcinogenic) increases the risk for metabolic syndrome dose-dependently in the offspring. Similarly in mice, 2–6 days of betel nut consumption by males is associated with increased glucose intolerance in the offspring through three consecutive generations.[44,45] Though these studies suggest a transgenerational inheritance of phenotype (metabolic dysfunction), it is unclear whether the underlying mechanism of these phenotypic outcomes is likewise transgenerational.[46]

In addition to specific external exposures, endogenous factors (likely epigenetic) in males can also be a determinant of offspring phenotype. In Balb/c isogenic mice, in which individuals show natural variability in open field activity, the level of activity in males determines activity in the female offspring, even if males and daughters never interact.[47] In humans, transmission of higher risk of autism and schizophrenia as a function of increased paternal age has been demonstrated.[48–50] Likewise, in genetically identical rodents, paternal age has a significant effect on the offspring. The progeny of "old" fathers have a shorter lifespan and poor learning and memory performance,[51,52] even in the absence of any postnatal contact with fathers.

Overall, it is apparent that both maternal and paternal experiences can have consequences for offspring development and, in some instances, the effects persist for several subsequent generations. Such transgenerational inheritance of phenotype suggests the role of the germline, however, in most cases, distinguishing the contribution of germline mechanisms from developmental factors (during embryogenesis and postnatal life) is difficult, particularly in the matriline. The most critical question relevant to both parental pathways is regarding the specific molecular processes that mediate these transgenerational effects.

EPIGENETIC MODIFICATIONS AND THE INHERITANCE OF SPECIFIC TRAITS

Epigenetic Mechanisms

Regulation of gene expression through epigenetic modifications provides a potential mechanistic route through which environmental experiences can lead to persistent changes in cellular phenotypes. Epigenetic modifications are dynamic. As such, they confer some plasticity to the genome and allow genes to respond to changing environmental factors. However, some epigenetic modifications can be stable, and support the notion that variations in gene expression are persistent and heritable.[53] There are multiple epigenetic mechanisms, but three mechanisms in particular, DNA methylation, post-translational modifications (PTMs) of histone proteins, and microRNAs, have been associated with environmentally-induced changes in the epigenome. DNA methylation is a process of gene silencing (but can also induce gene activation; see Chapter 1) mediated by conversion of cytosine into 5-methylcytosine by de novo or maintenance DNA methyltransferases (DNMTs) (i.e. DNMT3 or DNMT1 respectively).[3,53,54] Methylation depends on the presence of methyl donors provided by nutrients such as folic acid, methionine and choline, and affects transcription by recruiting methyl-DNA binding proteins (MBDs) such as MeCP2 to the chromatin.[55] DNA methylation acts in concert with PTMs of histones, proteins associated with the DNA that form the core of nucleosomes, to regulate gene expression. PTMs such as acetylation, phosphorylation, mono-, bi, or trimethylation, and ubiquination of specific amino acids (mostly lysine, serine and arginine) can reversibly alter chromatin structure and the accessibility of DNA to transcriptional machinery.[56,57] The nature and site-specificity of PTMs establish a histone code specific for each gene, with implications for gene transcription. For example, acetylation of lysines (K) in histone H3 is typically associated with transcriptional activation, whereas H3 di- or trimethylation on K9 generally reduces transcription and trimethylation on K4 (H3K4Me3) increases transcription ([58,59], see Chapter 1). In addition to DNA methylation and histone PTMs, microRNAs (miRNAs) also contribute to the epigenetic regulation of gene expression. miRNAs are small, non-coding RNAs that can suppress gene activity by binding to specific mRNAs and preventing their translation, inducing their cleavage, or promote their degradation.[60] Many miRNAs have been identified in various tissues and cells including germ cells, and can specifically target multiple genes.[61,62] Although DNA methylation, histone PTMs, and miRNAs are independent processes, they interact dynamically to regulate gene transcription.

Involvement of the Germline

It is well established that epigenetic mechanisms play a critical role in developmental processes and modulate cellular differentiation and the activity of gene networks.[4,63] Further, there is increasing evidence that epigenetic pathways are highly plastic and respond to experience beyond the early stages of embryogenesis and throughout development.[6,64] However, the role of these mechanisms in transgenerational inheritance of experiences has remained more speculative. Within the study of germline inheritance, a critical issue is whether epigenetic marks can be maintained following the post-fertilization events typical of mammalian reproduction. One of the most important of these events is the genome-wide reprogramming of the epigenome via DNA demethylation, a process meant to reset the parental genome.[5] To be transmitted to the offspring, epigenetic marks induced by environmental factors or stochastic events need to escape this reprogramming. Reprogramming occurs in two waves: a first wave occurs in the zygote, where the paternal genome is actively demethylated shortly after fertilization then remethylated just prior to implantation of the blastocyst.[65] The second wave occurs during embryogenesis following sex determination where DNA is remethylated in germ cells in a sex-specific fashion.[66] Remarkably, certain genetic loci within the germline, in particular imprinted genes and retrotransposable elements, have the unique capacity to retain their methylation profile across multiple generations despite these waves of epigenetic reprogramming. Significantly, these loci are sensitive to environmental factors.[67]

Evidence in mice has shown that variations in methylation status of an intracisternal-A particle (IAP) element, a long terminal repeat retrotransposon, can result in heritable phenotypic variability. When inserted into an exon of the agouti gene (*Avy*), variation in the expression of this IAP results in a range of phenotypic characteristics including nuanced coat color pigmentation and a propensity for obesity.[68] This phenotypic variation can be inherited through the matriline and is thought to be mediated by female gametes since transmission occurs even after embryo transfer, which eliminates any post-fertilization maternal effects. Similarly, an IAP element inserted into the 5′ region of the *AxinFu* allele, a gene responsible for embryonic axis formation, results in the expression of aberrant gene transcripts and a kinked-tail phenotype when methylated.[69] Interestingly, methylation status of the IAP is consistent across tissues (germ and somatic) and correlates with the degree of tail kink indicating that DNA methylation at this region directly controls the expression of the *AxinFu* gene. Both the anatomical phenotype and methylation status can be inherited by the offspring through maternal and paternal lineages. Importantly, the epigenetic state of *Avy* can be modified by environmental factors such as diet (methyl supplementation), environmental toxicants (bisphenol A), or drug exposure (ethanol), suggesting that epigenetic inheritance at these loci can be driven by environmental factors.[68,70–72] This environmentally induced modification to *Avy* has implications for the development of obesity, and may suggest an epigenetic basis for the development and transmission of obesity risk in humans as a consequence of disruption to DNA methylation levels at loci associated with growth and metabolism.[73]

Imprinted genes represent another class of DNA loci that can retain a "memory" of their ancestral epigenetic marks. These genes are expressed in a parent-of-origin manner, by way of the silencing of either the maternal or paternal allele classically via DNA methylation.

These epigenetic marks escape reprogramming following fertilization, and persist in the developing embryo and through adulthood.[74] About 100 imprinted genes have been identified in both rodents and humans, though a recent study suggests that many more alleles may show parent-of-origin specific DNA methylation.[75] Interestingly, gene expression and methylation status of imprinted genes in sperm is plastic and can be altered by environmental factors. In humans, increased demethylation of two paternally imprinted genes that are normally hypermethylated, *H19* and *IG-DMR*, occurs in sperm of fathers associated with chronic alcohol consumption.[76] Likewise in mice, methylation of the paternally imprinted *H19* gene is decreased, but methylation of the maternally imprinted *Snrpn* gene is increased in the sperm of second-generation offspring following superovulation of F0 females, a procedure used for assisted reproduction (consisting of injection of sex hormones). These observations indicate that some epigenetic marks can escape reprogramming and be transgenerationally inherited.[77] Evidence of variation in histone modifications in sperm cells that persist following fertilization suggests that there may be multiple epigenetic pathways that are potentially heritable.[78,79]

TRANSFER OF EPIGENETIC VARIATION ACROSS GENERATIONS

The transmission of epigenetic marks induced by environmental factors across generations may occur through two distinct routes (Figure 13.1). Within the patriline, particularly in species that do not engage in paternal care of offspring, transmission likely occurs via the germline. In this case, paternal or grandparental environmental exposures are thought to induce epigenetic alterations in the gametes or in precursor cells that persist in the absence of continued exposure, which are then passed on to offspring. In the case of transmission via the matriline, the epigenetic effects are less likely to involve the germline, but persist across generations through behavioral/social transfer (e.g. the transfer of maternal behaviors across generations). Such transfer requires that the experience or environmental conditions responsible for the traits be repeated in each generation to re-establish the associated epigenetic profile in the following generations. Distinguishing between germline and behavioral transmission can be difficult experimentally, particularly in the case of prenatal or postnatal exposures in which germ cells and precursors in the F1 offspring, which give rise to F2 individuals, are also exposed to the inducing environmental factors. Though both pathways can lead to stable inheritance of acquired traits, they are divergent from a mechanistic perspective.

GERMLINE TRANSMISSION OF PATERNAL EFFECTS

Potential mechanisms for paternal transmission of the effects of environmental factors have been proposed. One of these mechanisms involves DNA methylation. In humans, monozygotic twin studies show that the epigenetic profile of most cells drift with age, and while young twins have epigenetic similarities, aged twins are more divergent in their DNA methylation profile.[80,81] Further, in rodents, the impact of paternal age has been associated with hypermethylation of ribosomal DNA in sperm and liver cells in aged males

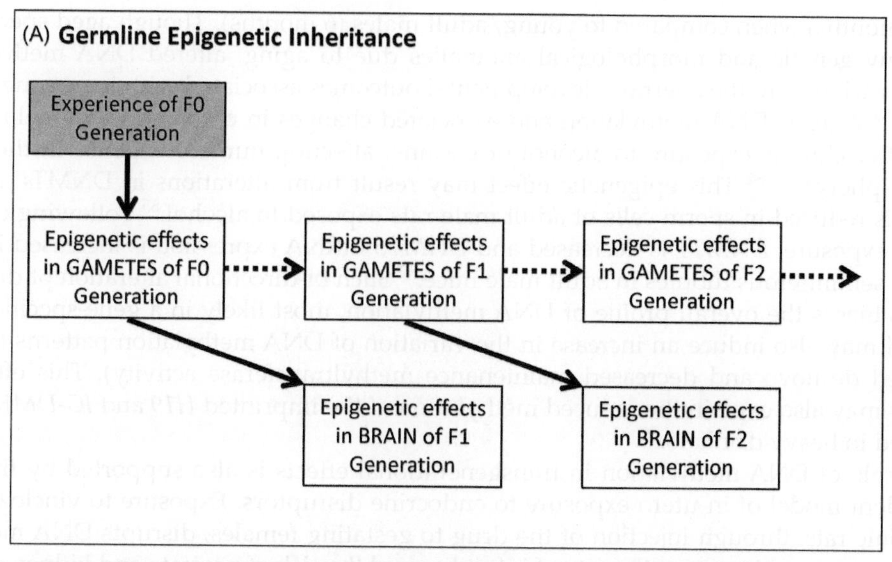

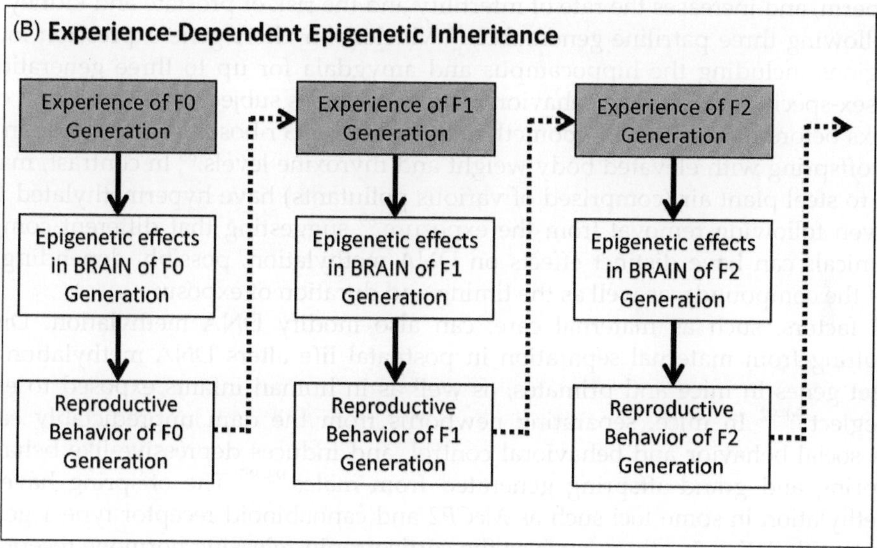

FIGURE 13.1 Illustration of the distinction between (A) germline epigenetic inheritance and (B) an experience-dependent inheritance of an epigenetic effect. (A) In germline inheritance of an environmental exposure (e.g. maternal separation, exposure to toxins, nutrition), the F0 generation is exposed to an environmental event which induces an epigenetic modification within the gametes (sperm or oocyte) in the F0 generation. This epigenetic variation is inherited by F1 and F2 generations via the germline, and transmitted across generations (dotted line) through gametic "epigenetic memory" that does not involve any continued exposure to the environmental event. (B) In experience-dependent epigenetic inheritance, such as the transmission of maternal behaviors across generations, an environmental exposure (e.g. stress, social experience, nutrition) experienced by the F0 generation induces an epigenetic change in the F0 brain with consequences for maternal/reproductive behavior. This behavioral change alters the developmental experience of the F1 generation leading to similar epigenetic, neurobiological, and behavioral changes with consequences for the F2 generation. Within this inheritance pathway, epigenetic variation does not persist unless the developmental experience of each generation re-establishes the epigenetic change.

(21–28 months) when compared to young/adult males (6 months). Though aged sperm carries many genetic and morphological anomalies due to aging, altered DNA methylation likely contributes to the aberrant developmental outcomes associated with increasing paternal age.[82] Altered DNA methylation and associated changes in chromatin remodeling also occur after chronic exposure to alcohol or cocaine, affecting numerous genes in the brain and periphery.[76,83,84] This epigenetic effect may result from alterations in DNMTs. *DNMT* mRNA is reduced in sperm cells of adult male rats exposed to alcohol.[85] Following chronic cocaine exposure, *DNMT1* is decreased and *DNMT3* mRNA expression is increased in cells of testis seminiferous tubules in adult male mice.[32] Such bi-directional alteration of different DNMTs biases the overall profile of DNA methylation, most likely in a gene-specific manner, and may also induce an increase in the variation of DNA methylation patterns (due to increased de novo and decreased maintenance methyltransferase activity). This effect on DNMTs may also explain the reduced methylation of the imprinted *H19* and *IG-DMR* genes observed in heavy drinkers.[76]

The role of DNA methylation in transgenerational effects is also supported by findings in a rodent model of in utero exposure to endocrine disruptors. Exposure to vinclozolin in embryonic rats, through injection of the drug to gestating females, disrupts DNA methylation in sperm, and increases the rate of infertility and the risk of prostate and kidney disease in the following three patriline generations.[86] Vinclozolin alters gene expression in several brain regions including the hippocampus and amygdala for up to three generations and induces sex-specific anxiety-like behaviors.[87] In mice, males subjected to chromium chloride two weeks before mating have hypomethylation of the 45S ribosomal RNA gene in sperm, and sire offspring with elevated body weight and thyroxine levels.[88] In contrast, male mice exposed to steel plant air (comprised of various pollutants) have hypermethylated DNA in sperm even following removal from the exposure,[89] suggesting that different compounds and chemicals can have distinct effects on DNA methylation, possibly depending on the nature of the compounds, as well as the timing and duration of exposure.

Social factors, such as maternal care, can also modify DNA methylation. Disrupted care resulting from maternal separation in postnatal life alters DNA methylation in several target genes in mice and primates, as well as in human infants exposed to early life social neglect.[90–92] In mice, separating newborns from the dam unpredictably each day perturbs social behavior and behavioral control, and induces depressive-like behaviors in the offspring and grand-offspring generated from males.[93–95] The offspring have higher DNA methylation in some loci such as *MeCP2* and cannabinoid receptor type 1 genes, but decreased methylation in others, such as the corticotropin-releasing hormone receptor gene, in sperm.[94] These changes are observed in the sperm of the separated males, but also in the brain and sperm of the offspring generation, strongly suggesting the involvement of germ cells in the inheritance of the effects of early-life adversity.

In addition to epigenetic marks on the DNA itself, germ cells also carry other non-genomic vectors that can modulate developmental processes and be transmitted. Evidence has accumulated that various non-coding RNAs including miRNAs, piRNAs and snoRNAs in sperm cells carry functional epigenetic information that can be inherited transgenerationally. One of the primary examples of RNA-mediated inheritance is that of paramutation, in which the interaction between two alleles of a single locus results in heritable variation. This phenomenon is well described in plants,[96] but was only recently reported to occur in the mouse. One example of paramutation involves the *Kit* gene that encodes a tyrosine kinase receptor involved in the synthesis of melanin.[97] In mice, individuals heterozygous for a mutation of the *Kit* gene have reduced *Kit* mRNA expression and distinctive white pigmentation in feet and tail. When either male or female heterozygotes are crossed with wild-type individuals, some of the wild-type offspring (Kit*) have reduced *Kit* mRNA levels and inherit the paramutation phenotype (white pigmentation). These individuals can then pass these alterations to their own offspring. The reduced level of *Kit* mRNA and RNAs in testes and mature sperm in mutant mice led to the hypothesis that RNAs may contribute to the inheritance. Injection of *Kit* mRNA from heterozygotes and miRNAs against *Kit* mRNAs into fertilized zygotes reproduces the white pigmentation phenotype in the resulting progeny.[97]

Other illustrative examples are that of the induced inheritance of particular traits following the injection of specific miRNAs in fertilized eggs in mice. Injection of the cardiac-specific miRNA, *miR-1*, induces anatomical and physiological signs of cardiac hypertrophy in the resulting offspring.[98] Likewise, injection of *miR-124*, a miRNA critical for brain development, results in offspring with increased growth rate.[99] Both manipulations modify the expression of the genes known to be targeted by the respective miRNAs during development and in adulthood and, in some cases, even alter chromatin structure. Further, similar to the *Kit* paramutation, miRNAs were found at detectable levels in sperm, and although not measured, may also be present in oocytes since transmission also occurred through maternal lineages for up to three generations. Taken together, these studies suggest that RNAs expressed in germ cells, in both males and females, likely carry functional epigenetic information that induces persistent transgenerational effects. Therefore, it is perhaps not surprising that environmental variables such as aging, smoking or exposure to chemicals perturbs RNA expression, including small RNAs, in germ cells.[100–103]

EXPERIENCE-DEPENDENT TRANSFER THROUGH FEMALES

In addition to germline-based epigenetic transmission, other non-germline modes of inheritance can underlie the transgenerational inheritance of phenotypes induced by environmental factors.[8] One example is the transmission of variability in maternal behaviors across generations, a process that involves epigenetic modifications, in particular changes in DNA methylation. In rats, female offspring of low LG mothers have increased methylation at the 1B promoter region of the estrogen receptor α (*ESR1* encoding ERα), a gene that regulates postpartum maternal behaviors. Higher methylation decreases *ESR1* mRNA expression by preventing the binding of signal transducer and activator of transcription (Stat5) to the *ESR1* promoter, and reduces the sensitivity to estrogen in the medial preoptic area of the hypothalamus.[104,105] High LG in infancy in contrast, is associated with reduced methylation of several sites within the *ESR1* promoter.[106] Thus, the transfer of variations in maternal LG from mother to daughter is associated with stable epigenetic modifications in specific regions of the brain that are established in infancy and influence behavior in adulthood.[107] Experience-dependent epigenetic changes are also induced by maternal abuse in rats. Exposure to periods of abusive maternal care (dragging, burying etc.) is associated with increased methylation of exon 1V of the *Bdnf* gene promoter resulting in decreased *Bdnf* mRNA expression in the prefrontal cortex.[21] Such altered methylation also affects the F1 offspring of abused females, suggesting that it can be re-instated in an experience-dependent manner. Such re-instatement implies that epigenetic alterations induced by environmental factors in the brain and which have a consequence on the expression of the gene essential for maternal behavior, likely underlie the mechanisms of behavioral transfer (see Figure 13.1).

DISSOCIATING PATERNAL AND MATERNAL INFLUENCES ON SUBSEQUENT GENERATIONS

Although epigenetic inheritance through the patriline is likely accounted for by germline factors, the interplay between paternal and maternal contributions to patrilineal inheritance is an important consideration. Empirical and theoretical studies of reproductive behavior suggest that females can dynamically adjust their maternal investment depending on the phenotypic quality of their mate with consequences for offspring development.[108,109] The relationship between maternal care (involving both prenatal and postnatal reproductive investment) and mate quality was first outlined by the differential allocation hypothesis (DAH). DAH states that females mating with high quality males (fit and attractive) should increase their investment in the offspring if the cost of reproduction is high.[110,111] Alternatively to DAH, the compensation hypothesis proposes that females paired with unattractive or non-preferred males should increase their parental investment to counteract the disadvantages that their offspring may inherit from their father.[112] These hypotheses have been tested in different taxa and were both supported by experimental data.[108,109,111,113] In wild mice, females mated with a "preferred" male give birth to larger litters, and to socially dominant offspring that are better nest-builders, respond rapidly to the presence of a predator, and have a decreased mortality rate when compared to the offspring of females mated with non-preferred males.[114] Similarly, Balb/c females mated with males exposed to social enrichment provide more nursing and LG to their offspring than females mated with socially isolated males.[115] This study, using genetically identical mice, provides strong support for the importance of social and environmental conditions (the only differing parameters) of males in driving differential allocation by female mates.[115] Indeed, maternal factors may also contribute to the paternal transmission of the effect of social stress to offspring. Some of the behavioral alterations resulting from male exposure to chronic social defeat in adulthood are not observed when the offspring are produced by in vitro fertilization, a manipulation that removes the potential effects of maternal investment.[116] In paternal vinclozolin exposure, mate preference analyses have indicated that females have lower preference for males derived from exposed lineages which has potential consequences for the offspring.[117]

The interaction between maternal and paternal contribution is further complicated by the epigenetic characteristics of the offspring. Epigenetic variation in offspring may induce differences in maternal investment via changes in placental function or through behavioral effects on offspring such as suckling ability or locomotor activity.[74,82] Such changes, in turn, may induce differential maternal investment towards the offspring. Thus, if the father influences fetal growth, this effect could potentially induce changes in maternal care. Consequent to these paternal–maternal interactions, the influence of indirect maternal effects and inherited germline epigenetic mechanisms can have complex phenotypic outcomes.

FUTURE DIRECTIONS IN THE STUDY OF EPIGENETICS AND INHERITANCE

The transmission of environmentally-induced traits across generations is a challenging concept that has gained great momentum in the past years.[118,119] Exploration of the role of epigenetic mechanisms in such transgenerational effects has generated interesting hypotheses regarding the "epigenetic memory" of gametes, and the possibility that epigenetic pathways provide a route for the inheritance of acquired traits. The current knowledge of these mechanisms is still limited, and further investigation of the nature of epigenetic modifications, of their plasticity and stability, and their modes of transmission within- and across-generations will be essential to elucidate how epigenetic variation can become heritable. Importantly, these studies should consider the type and timing of exposure, specificity of target genes, sex-specificity of transmission, and the interaction between different types of epigenetic modifications. The mechanisms of interaction between experience-dependent and germline epigenetic inheritance will also need to be better understood. Manipulations such as embryo transfer and cross-fostering, and the characterization of maternal investment in offspring should help to provide some insight into the origins of parental epigenetic effects. Overall, a transgenerational perspective that incorporates epigenetic mechanisms significantly complements our current understanding of the mechanisms of inheritance and the pathways linking genotype to phenotype.

Acknowledgments

The authors wish to acknowledge funding received from Grant Number DP2OD001674-01 from the Office of the Director, National Institutes of Health, from the University Zürich, the Swiss National Science Foundation and Roche.

References

1. Caspi A, Moffitt TE. Gene-environment interactions in psychiatry: joining forces with neuroscience. *Nat Rev.* 2006;7(7):583–590.
2. Meaney MJ. Epigenetics and the biological definition of gene x environment interactions. *Child Dev.* 2010;81(1):41–79.
3. Turner B. *Chromatin and Gene Regulation.* Oxford: Blackwell Science Ltd; 2001.
4. Jones PA, Taylor SM. Cellular differentiation, cytidine analogs and DNA methylation. *Cell.* 1980;20(1):85–93.
5. Santos F, Hendrich B, Reik W, Dean W. Dynamic reprogramming of DNA methylation in the early mouse embryo. *Dev Biol.* 2002;241(1):172–182.
6. Champagne FA. Epigenetic influence of social experiences across the lifespan. *Dev Psychobiol.* 2010;52(4):299–311.
7. Jirtle RL, Skinner MK. Environmental epigenomics and disease susceptibility. *Nat Rev.* 2007;8(4):253–262.
8. Danchin E, et al. Beyond DNA: integrating inclusive inheritance into an extended theory of evolution. *Nat Rev.* 2011;12(7):475–486.
9. Benoit D, Parker KC. Stability and transmission of attachment across three generations. *Child Dev.* 1994;65(5):1444–1456.
10. Sroufe LA. Attachment and development: a prospective, longitudinal study from birth to adulthood. *Attach Hum Dev.* 2005;7(4):349–367.
11. Miller L, Kramer R, Warner V, Wickramaratne P, Weissman M. Intergenerational transmission of parental bonding among women. *J Am Acad Child Adolesc Psychiatry.* 1997;36(8):1134–1139.

12. Berman C. Intergenerational transmission of maternal rejection rates among free-ranging rheus monkeys on Cayo Santiago. *Anim Behav*. 1990;44:247–258.
13. Maestripieri D. Fatal attraction: interest in infants and infant abuse in rhesus macaques. *Am J Phys Anthropol*. 1999;110(1):17–25.
14. Maestripieri D, Tomaszycki M, Carroll KA. Consistency and change in the behavior of rhesus macaque abusive mothers with successive infants. *Dev Psychobiol*. 1999;34(1):29–35.
15. Maestripieri D. Early experience affects the intergenerational transmission of infant abuse in rhesus monkeys. *Proc Natl Acad Sci USA*. 2005;102(27):9726–9729.
16. Champagne FA, Francis DD, Mar A, Meaney MJ. Variations in maternal care in the rat as a mediating influence for the effects of environment on development. *Physiol Behav*. 2003;79(3):359–371.
17. Champagne FA, Meaney MJ. Transgenerational effects of social environment on variations in maternal care and behavioral response to novelty. *Behav Neurosci*. 2007;121(6):1353–1363.
18. Francis D, Diorio J, Liu D, Meaney MJ. Nongenomic transmission across generations of maternal behavior and stress responses in the rat. *Science*. 1999;286(5442):1155–1158.
19. Champagne FA, Meaney MJ. Stress during gestation alters postpartum maternal care and the development of the offspring in a rodent model. *Biol Psychiatry*. 2006;59(12):1227–1235.
20. Connor KL, Vickers MH, Beltrand J, Meaney MJ, Sloboda DM. Nature, nurture or nutrition? Impact of maternal nutrition on maternal care, offspring development and reproductive function. *J Physiol*. 2012;590(9):2167–2180.
21. Roth TL, Lubin FD, Funk AJ, Sweatt JD. Lasting epigenetic influence of early-life adversity on the BDNF gene. *Biol Psychiatry*. 2009;65(9):760–769.
22. Crowcroft P, Rowe FP. Social organization and territorial behavior in the wild house mice (Mus musculus L.). *Proc Zool Soc London*. 1963;140:517–531.
23. Curley JP, Davidson S, Bateson P, Champagne FA. Social enrichment during postnatal development induces transgenerational effects on emotional and reproductive behavior in mice. *Front Behav Neurosci*. 2009;3:25.
24. Arai JA, Li S, Hartley DM, Feig LA. Transgenerational rescue of a genetic defect in long-term potentiation and memory formation by juvenile enrichment. *J Neurosci*. 2009;29(5):1496–1502.
25. Abel E. Paternal contribution to fetal alcohol syndrome. *Addiction Biol*. 2004;9(2):127–133. discussion 135-126.
26. Bielawski DM, Abel EL. Acute treatment of paternal alcohol exposure produces malformations in offspring. *Alcohol (Fayetteville, NY)*. 1997;14(4):397–401.
27. Cicero TJ, et al. Acute paternal alcohol exposure impairs fertility and fetal outcome. *Life Sci*. 1994;55(2):PL33–PL36.
28. Ledig M, et al. Paternal alcohol exposure: developmental and behavioral effects on the offspring of rats. *Neuropharmacology*. 1998;37(1):57–66.
29. Meek LR, Myren K, Sturm J, Burau D. Acute paternal alcohol use affects offspring development and adult behavior. *Physiol Behav*. 2007;91(1):154–160.
30. Wozniak DF, Cicero TJ, Kettinger 3rd L, Meyer ER. Paternal alcohol consumption in the rat impairs spatial learning performance in male offspring. *Psychopharmacology*. 1991;105(2):289–302.
31. Abel EL, Moore C, Waselewsky D, Zajac C, Russell LD. Effects of cocaine hydrochloride on reproductive function and sexual behavior of male rats and on the behavior of their offspring. *J Androl*. 1989;10(1):17–27.
32. He F, Lidow IA, Lidow MS. Consequences of paternal cocaine exposure in mice. *Neurotoxicol Teratol*. 2006;28(2):198–209.
33. Kahn AJ. Alteration of paternal environment prior to mating: effect on hemoglobin concentration in offspring of CF1 mice. *Growth*. 1970;34(2):215–220.
34. Drake AJ, Walker BR, Seckl JR. Intergenerational consequences of fetal programming by in utero exposure to glucocorticoids in rats. *Am J Physiol*. 2005;288(1):R34–R38.
35. Kaati G, Bygren LO, Edvinsson S. Cardiovascular and diabetes mortality determined by nutrition during parents' and grandparents' slow growth period. *Eur J Hum Genet*. 2002;10(11):682–688.
36. Kaati G, Bygren LO, Pembrey M, Sjostrom M. Transgenerational response to nutrition, early life circumstances and longevity. *Eur J Hum Genet*. 2007;15(7):784–790.
37. Pembrey ME, et al. Sex-specific, male-line transgenerational responses in humans. *Eur J Hum Genet*. 2006;14(2):159–166.
38. Anderson LM, et al. Preconceptional fasting of fathers alters serum glucose in offspring of mice. *Nutr*. 2006;22(3):327–331.
39. Zambrano E, et al. Sex differences in transgenerational alterations of growth and metabolism in progeny (F2) of female offspring (F1) of rats fed a low protein diet during pregnancy and lactation. *J Physiol*. 2005;566(1):225–236.

40. Carone BR, et al. Paternally induced transgenerational environmental reprogramming of metabolic gene expression in mammals. *Cell*. 2010;143(7):1084–1096.
41. Ng SF, et al. Chronic high-fat diet in fathers programs beta-cell dysfunction in female rat offspring. *Nature*. 2010;467(7318):963–966.
42. Dunn GA, Bale TL. Maternal high-fat diet promotes body length increases and insulin insensitivity in second-generation mice. *Endocrinology*. 2009;150(11):4999–5009.
43. Dunn GA, Bale TL. Maternal high-fat diet effects on third-generation female body size via the paternal lineage. *Endocrinology*. 2011;152(6):2228–2236.
44. Boucher BJ, Ewen SW, Stowers JM. Betel nut (Areca catechu) consumption and the induction of glucose intolerance in adult CD1 mice and in their F1 and F2 offspring. *Diabetologia*. 1994;37(1):49–55.
45. Chen TH, Chiu YH, Boucher BJ. Transgenerational effects of betel-quid chewing on the development of the metabolic syndrome in the Keelung Community-based Integrated Screening Program. *Am J Clin Nutrit*. 2006;83(3):688–692.
46. Skinner MK. What is an epigenetic transgenerational phenotype? F3 or F2. *Reproduc Toxicol 2*. 2008;5(1):2–6.
47. Alter MD, et al. Paternal transmission of complex phenotypes in inbred mice. *Biol Psychiatry*. 2009;66(11):1061–1066.
48. Brown AS, et al. Paternal age and risk of schizophrenia in adult offspring. *Am J Psychiatry*. 2002;159(9):1528–1533.
49. Malaspina D, et al. Advancing paternal age and the risk of schizophrenia. *Arch Gen Psychiatry*. 2001;58(4):361–367.
50. Reichenberg A, et al. Advancing paternal age and autism. *Arch Genl Psychiatry*. 2006;63(9):1026–1032.
51. Garcia-Palomares S, et al. Delayed fatherhood in mice decreases reproductive fitness and longevity of offspring. *Biol Reproduct*. 2009;80(2):343–349.
52. Garcia-Palomares S, et al. Long-term effects of delayed fatherhood in mice on postnatal development and behavioral traits of offspring. *Biol Reproduct*. 2009;80(2):337–342.
53. Feng J, Fouse S, Fan G. Epigenetic regulation of neural gene expression and neuronal function. *Pediatr Res*. 2007;61(5 Pt 2):58R–63R.
54. Razin A. CpG methylation, chromatin structure and gene silencing-a three-way connection. *Embo J*. 1998;17(17):4905–4908.
55. Fan G, Hutnick L. Methyl-CpG binding proteins in the nervous system. *Cell Res*. 2005;15(4):255–261.
56. Jenuwein T, Allis CD. Translating the histone code. *Science*. 2001;293(5532):1074–1080.
57. Peterson CL, Laniel MA. Histones and histone modifications. *Curr Biol*. 2004;14(14):R546–R551.
58. Barski A, et al. High-resolution profiling of histone methylations in the human genome. *Cell*. 2007;129(4):823–837.
59. Koch CM, et al. The landscape of histone modifications across 1% of the human genome in five human cell lines. *Genome Res*. 2007;17(6):691–707.
60. Sato F, Tsuchiya S, Meltzer SJ, Shimizu K. MicroRNAs and epigenetics. *FEBS J*. 2011;278(10):1598–1609.
61. Pelaez N, Carthew RW. Biological robustness and the role of microRNAs: a network perspective. *Curr Top Dev Biol*. 2012;99:237–255.
62. Banisch TU, Goudarzi M, Raz E. Small RNAs in germ cell development. *Curr Top Dev Biol*. 2012;99:79–113.
63. Hemberger M, Dean W, Reik W. Epigenetic dynamics of stem cells and cell lineage commitment: digging Waddington's canal. *Nat Rev Mol Cell Biol*. 2009;10(8):526–537.
64. Levenson JM, Sweatt JD. Epigenetic mechanisms in memory formation. *Nat Rev*. 2005;6(2):108–118
65. Shi L, Wu J. Epigenetic regulation in mammalian preimplantation embryo development. *Reproduct Biol Endocrinol*. 2009;7:59.
66. Allegrucci C, Thurston A, Lucas E, Young L. Epigenetics and the germline. *Reproduction*. 2005;129(2):137–149.
67. Lane N, et al. Resistance of IAPs to methylation reprogramming may provide a mechanism for epigenetic inheritance in the mouse. *Genesis*. 2003;35(2):88–93.
68. Morgan HD, Sutherland HG, Martin DI, Whitelaw E. Epigenetic inheritance at the agouti locus in the mouse. *Nat Genet*. 1999;23(3):314–318.
69. Rakyan VK, et al. Transgenerational inheritance of epigenetic states at the murine Axin(Fu) allele occurs after maternal and paternal transmission. *Proc Natl Acad Sci USA*. 2003;100(5):2538–2543.
70. Cropley JE, Suter CM, Beckman KB, Martin DI. Germ-line epigenetic modification of the murine avy allele by nutritional supplementation. *Proc Natl Acad Sci USA*. 2006;103(46):17308–17312.

71. Dolinoy DC, Huang D, Jirtle RL. Maternal nutrient supplementation counteracts bisphenol A-induced DNA hypomethylation in early development. *Proc Natl Acad Sci USA*. 2007;104(32):13056–13061.
72. Kaminen-Ahola N, et al. Maternal ethanol consumption alters the epigenotype and the phenotype of offspring in a mouse model. *PLoS Genet*. 2010;6(1):e1000811.
73. Katari S, et al. DNA methylation and gene expression differences in children conceived in vitro or in vivo. *Hum Mol Genet*. 2009;18(20):3769–3778.
74. Keverne EB, Curley JP. Epigenetics, brain evolution and behaviour. *Front Neuroendocrinol*. 2008;29(3):398–412.
75. Schalkwyk LC, et al. Allelic skewing of DNA methylation is widespread across the genome. *Am J Hum Genet*. 2010;86(2):196–212.
76. Ouko LA, et al. Effect of alcohol consumption on CpG methylation in the differentially methylated regions of H19 and IG-DMR in male gametes: implications for fetal alcohol spectrum disorders. *Alcohol Clin Exp Res*. 2009;33(9):1615–1627.
77. Stouder C, Deutsch S, Paoloni-Giacobino A. Superovulation in mice alters the methylation pattern of imprinted genes in the sperm of the offspring. *Reproduct Toxicol*. 2009;28(4):536–541.
78. Brykczynska U, et al. Repressive and active histone methylation mark distinct promoters in human and mouse spermatozoa. *Nat Struct Mol Biol*. 2010;17(6):679–687.
79. Hammoud SS, et al. Distinctive chromatin in human sperm packages genes for embryo development. *Nature*. 2009;460(7254):473–478.
80. Fraga MF, et al. Epigenetic differences arise during the lifetime of monozygotic twins. *Proc Natl Acad Sci USA*. 2005;102(30):10604–10609.
81. Oakes CC, Smiraglia DJ, Plass C, Trasler JM, Robaire B. Aging results in hypermethylation of ribosomal DNA in sperm and liver of male rats. *Proc Natl Acad Sci USA*. 2003;100(4):1775–1780.
82. Curley JP, Mashoodh R, Champagne FA. Epigenetics and the origins of paternal effects. *Hormones Behav*. 2011;59(3):306–314.
83. Novikova SI, et al. Maternal cocaine administration in mice alters DNA methylation and gene expression in hippocampal neurons of neonatal and prepubertal offspring. *PloS one*. 2008;3(4):e1919.
84. Pandey SC, Ugale R, Zhang H, Tang L, Prakash A. Brain chromatin remodeling: a novel mechanism of alcoholism. *J Neurosci*. 2008;28(14):3729–3737.
85. Bielawski DM, Zaher FM, Svinarich DM, Abel EL. Paternal alcohol exposure affects sperm cytosine methyltransferase messenger RNA levels. *Alcohol Clin Exp Res*. 2002;26(3):347–351.
86. Anway MD, Cupp AS, Uzumcu M, Skinner MK. Epigenetic transgenerational actions of endocrine disruptors and male fertility. *Science*. 2005;308(5727):1466–1469.
87. Skinner MK, Anway MD, Savenkova MI, Gore AC, Crews D. Transgenerational epigenetic programming of the brain transcriptome and anxiety behavior. *PloS one*. 2008;3(11):e3745.
88. Cheng RY, Hockman T, Crawford E, Anderson LM, Shiao YH. Epigenetic and gene expression changes related to transgenerational carcinogenesis. *Mol Carcinogen*. 2004;40(1):1–11.
89. Yauk C, et al. Germ-line mutations, DNA damage, and global hypermethylation in mice exposed to particulate air pollution in an urban/industrial location. *Proc Natl Acad Sci USA*. 2008;105(2):605–610.
90. Kinnally EL, et al. Epigenetic regulation of serotonin transporter expression and behavior in infant rhesus macaques. *Genes Brain Behav*. 2010;9:575–582.
91. Murgatroyd C, et al. Dynamic DNA methylation programs persistent adverse effects of early-life stress. *Nat Neurosci*. 2009;12(12):1559–1566.
92. Naumova OY, et al. Differential patterns of whole-genome DNA methylation in institutionalized children and children raised by their biological parents. *Dev Psychopathol*. 2012;24(1):143–155.
93. Franklin TB, Linder N, Russig H, Thony B, Mansuy IM. Influence of early stress on social abilities and serotonergic functions across generations in mice. *PloS one*. 2011;6(7):e21842.
94. Franklin TB, et al. Epigenetic transmission of the impact of early stress across generations. *Biol Psychiatry*. 2010;68(5):408–415.
95. Weiss IC, Franklin TB, Vizi S, Mansuy IM. Inheritable effect of unpredictable maternal separation on behavioral responses in mice. *Front Behav Neurosci*. 2011;5:3.
96. Erhard Jr KF, Hollick JB. Paramutation: a process for acquiring trans-generational regulatory states. *Curr Opin Plant Biol*. 2011;14(2):210–216.
97. Rassoulzadegan M, et al. RNA-mediated non-Mendelian inheritance of an epigenetic change in the mouse. *Nature*. 2006;441(7092):469–474.

98. Wagner KD, et al. RNA induction and inheritance of epigenetic cardiac hypertrophy in the mouse. *Dev Cell*. 2008;14(6):962–969.
99. Grandjean V, et al. The miR-124-Sox9 paramutation: RNA-mediated epigenetic control of embryonic and adult growth. *Development*. 2009;136(21):3647–3655.
100. Hamatani T, et al. Age-associated alteration of gene expression patterns in mouse oocytes. *Hum Mol Genet*. 2004;13(19):2263–2278.
101. Linschooten JO, et al. Use of spermatozoal mRNA profiles to study gene-environment interactions in human germ cells. *Mutat Res*. 2009;667(1-2):70–76.
102. Marczylo EL, Amoako AA, Konje JC, Gant TW, Marczylo TH. Smoking induces differential miRNA expression in human spermatozoa: A potential transgenerational epigenetic concern? *Epigenetics*. 2012;7:5.
103. Song R, et al. Male germ cells express abundant endogenous siRNAs. *Proc Natl Acad Sci USA*. 2011;108(32):13159–13164.
104. Champagne F, Diorio J, Sharma S, Meaney MJ. Naturally occurring variations in maternal behavior in the rat are associated with differences in estrogen-inducible central oxytocin receptors. *Proc Natl Acad Sci USA*. 2001;98(22):12736–12741.
105. Champagne FA, Weaver IC, Diorio J, Sharma S, Meaney MJ. Natural variations in maternal care are associated with estrogen receptor alpha expression and estrogen sensitivity in the medial preoptic area. *Endocrinology*. 2003;144(11):4720–4724.
106. Champagne FA, et al. Maternal care associated with methylation of the estrogen receptor-alpha1b promoter and estrogen receptor-alpha expression in the medial preoptic area of female offspring. *Endocrinology*. 2006;147(6):2909–2915.
107. Champagne FA. Epigenetic mechanisms and the transgenerational effects of maternal care. *Front Neuroendocrinol*. 2008;29(3):386–397.
108. Harris WE, Uller T. Reproductive investment when mate quality varies: differential allocation versus reproductive compensation. *Philosoph Transact Roy Soc Lond B, Biol Sci*. 2009;364(1520):1039–1048.
109. Ratikainen II, Kokko H. Differential allocation and compensation: who deserves the silver spoon? *Behav Ecol*. 2009;21(1):195–200.
110. Burley N. The differential-allocation hypothesis – An experimental test. *Am Nat*. 1988;132(5):611–628.
111. Sheldon BC. Differential allocation: tests, mechanisms and implications. *Trends Ecol Evol*. 2000;15(10):397–402.
112. Gowaty PA, et al. The hypothesis of reproductive compensation and its assumptions about mate preferences and offspring viability. *Proc Natl Acad Sci USA*. 2007;104(38):15023–15027.
113. Bluhm CK, Gowaty PA. Reproductive compensation for offspring viability deficits by female mallards, Anas platyrhynchos. *Anim Behav*. 2004;68(5):982–985.
114. Drickamer LC, Gowaty PA, Holmes CM. Free female mate choice in house mice affects reproductive success and offspring viability and performance. *Anim Behav*. 2000;59:371–378.
115. Mashoodh R, Franks B, Curley JP, Champagne FA. Paternal social enrichment effects on maternal behavior and offspring growth. *Proc Natl Acad Sci USA*. 2012;109(Suppl 2):17232–17238.
116. Dietz DM, et al. Paternal transmission of stress-induced pathologies. *Biol Psychiatry*. 2011;70(5):408–414.
117. Crews D, et al. Transgenerational epigenetic imprints on mate preference. *Proc Natl Acad Sci USA*. 2007;104(14):5942–5946.
118. Bohacek J, Mansuy IM. Epigenetic inheritance of disease and disease risk. *Neuropsychopharmacology*. 2012 In press.
119. Franklin TB, Mansuy IM. Epigenetic inheritance in mammals: evidence for the impact of adverse environmental effects. *Neurobiol Dis*. 2010;39(1):61–65.

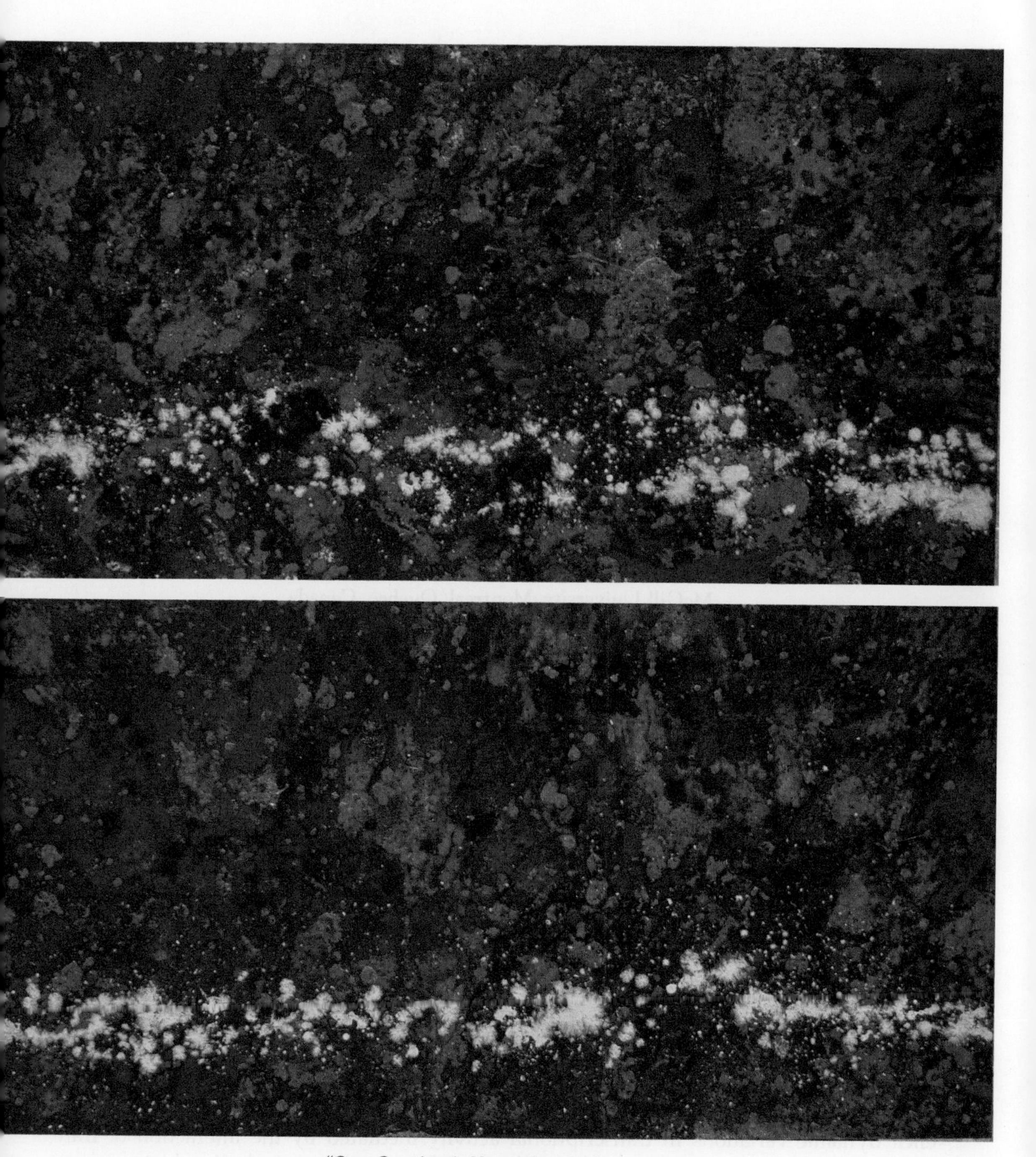

"Open Questions in Neuroepigenetics"

J. David Sweatt, acrylic on wood panel (diptych of two 24 x 48 panels), 2012

CHAPTER

14

Epigenetics: Defining the Frontiers of Genomic Function

Michael J. Meaney,[1] *Schahram Akbarian,*[2] *Eric J. Nestler*[3] *and J. David Sweatt*[4]

[1]Departments of Psychiatry, Neurology, and Neurosurgery, Douglas Institute, McGill University, Montreal, Quebec, Canada

[2]Department of Psychiatry, Friedman Brain Institute, Mount Sinai School of Medicine, New York, New York, USA

[3]Fishberg Department of Neuroscience, Friedman Brain Institute, Mount Sinai School of Medicine, New York, New York, USA

[4]McKnight Brain Institute, Department of Neurobiology, University of Alabama at Birmingham, Birmingham, Alabama, USA

The fusion of neuroscience and epigenetics offers remarkable opportunities for studies of the cellular and biochemical mechanisms underlying the variation in the structure and function of neurons and glia. Epigenetics provides a candidate mechanism for the stable changes in transcription that underlie the enduring effects on neuronal function of early experience, chronic environmental conditions, such as stress or repeated exposure to addictive substances, and which mediate the synaptic plasticity that permits learning and memory. Environmental stimuli activate neural circuits and consequently intracellular signals that elicit dynamic variations in transcriptional activity mediated by the interaction of transcription factors and co-factors with regulatory DNA sequences, including histone post-translational modifications (PTMs), that directly regulate the physicochemical properties of

Epigenetic Regulation in the Nervous System
DOI: http://dx.doi.org/10.1016/B978-0-12-391494-1.00014-8

chromatin. However, the activation of these pathways alone cannot explain the sustained "programming" of transcription that accompanies such conditions. While epigenetic modifications such as histone acetylation and phosphorylation are core features of the transient interactions of transcription factor complexes with the DNA, it is possible that modifications such as histone and DNA methylation could serve as more enduring signals that can exert a sustained influence over genome structure and function, and thus maintain an environmental imprint over the genome. Here is literally where the experience of the organism meets its DNA.

Environmental programming of genomic structure and function is of particular importance for the brain where replication-independent, enduring alterations in cellular function are essential for normal activity.[1] Indeed, the value of the energetically-costly brain is to guide the function of the organism in accordance with life history. The ability to mastermind such adaptation to circumstance relies upon plasticity at the level of genomic structure and function.[1,2]

Neuroepigenetics is not merely the application of the study of epigenetic mechanisms to neural function. As noted in Chapter 1, one definition of epigenetics includes the stipulation that the biochemical mark is heritable across cell division. This concern is fundamental for cell and tissue differentiation. The resistance to the notion of environmentally-induced epigenetic remodeling in fully differentiated cells derived from the apparent contradiction between the maintenance of a differentiated state and epigenetic "plasticity". How could a cell remain stably differentiated if the epigenome was in flux? However, heritable maintenance of differentiation is not a fundamental concern in brain. Rather the issue is that of generating variation in phenotype from a stable cell population as a function of development and experience. The business of the brain is plasticity; its loss is synonymous with states of dysfunction. Neuroepigenetics will thus have an emphasis that differs from the study of epigenetics in most other fields of biology (although immunology will be an interesting exception).

It is important to note that environmental regulation of the epigenome operates within limits. There is no suggestion that environmental signals will redefine the activity of densely packed heterochromatin: the genes that confer the ability to produce glucose and which were stably silenced during neuronal differentiation are not about to be suddenly rendered active as a function of maternal care, cocaine, or fear conditioning. Rather, environmentally-regulated genomic sites are more likely to lie within facultative euchromatin, where activity is conditional on cellular context or to regulate the level of activity within constitutively active euchromatin. This does not preclude more subtle changes in heterochromatic regions, as demonstrated recently in brain after cocaine exposure.[3] Indeed, one of the many challenges facing those who study epigenetics in the CNS is that of defining the features that render specific sites vulnerable to environmental modification. A recent study[4] suggests that about 1.5% of the CpG sites show alterations in methylation state as a function of seizure-induced activity in the hippocampus. This estimate may appear underwhelming. But only ≈1% of sequences in the human genome show variation across individuals. This figure has hardly dissuaded the study of genetics. The nature of these more volatile sites and their functional importance will become central to the study of epigenetics.

EPIGENETICS AND TRANSCRIPTION: CAUSE OR CONSEQUENCE?

A major theme in this book is that of positioning of epigenetics as an attractive candidate mechanism lying between environmental signals and stable alterations in transcriptional activity. Several chapters present evidence for the importance of transcription factor/co-factor complexes in initiating the remodeling of the epigenome in response to environmental conditions. For example, the activation of Gadd45b appears essential for the alterations in DNA methylation produced by intense hippocampal activation[4] and NGFI-A appears obligatory for similar maternal effects on the epigenome.[5] Recent studies from the Schubuler lab confirm the importance of transcription factor binding in defining DNA methylation profiles at specific genomic regions.[6,7] These findings suggest that the transcriptional machinery itself might associate with chromatin remodeling. This raises the obvious question of whether transcription is a consequence or a driver of epigenetic states. The Schubuler studies suggest that transcription factor binding, even in the absence of transcriptional activation, *can* remodel DNA methylation marks. Altered transcription factor binding, an event linked to environmental conditions, can initiate remodeling of the epigenome.

If the epigenome at specific sites across the genome remains vulnerable to environmentally-regulated remodeling, then what accounts for subsequent stability? How does an environmental effect on the epigenome endure? The notion of a critical threshold offers one possible explanation, particularly for alterations to states maintained by strong chemical bonds, such as DNA methylation. Thus, the remodeling of DNA methylation might require environmental signals of sufficient intensity to induce epigenetic remodeling. A threshold might ensure some measure of stability. While the strength of the chemical bonds underlying epigenetic states is likely a consideration, the complexities of epigenetic remodeling require a more sophisticated explanation. For example, mere threshold models could not account for the observation that specific environmental conditions, in this case fear conditioning, can a produce a transient alteration in DNA methylation in the hippocampus, but a far more persistent alteration in the prefrontal cortex.[8] Moreover, as noted above, seizure activity in the hippocampus, an undeniably intense stimulus, remodels only about 1–2% of the CpG sites.[4] A richer approach to the question of stability as well as the issue of regional specificity is to consider epigenetic remodeling within the context of the physiochemical context of transcriptional activation, which recruits perhaps a hundred proteins to a modified gene locus. This highlights future challenges, one of which is to establish the determinants of the relative susceptibility of specific genomic sites to epigenetic remodeling in response to environmental signals.

Transcriptional signaling complexes open chromatin, facilitating the access of the transcriptional machinery to the DNA. Transcription associates with nucleosome disruption or re-positioning (i.e. nucleosome sliding). Indeed, epigenetic mechanisms define the process of transcriptional activation. However, the termination of transcription is also an epigenetically-mediated process through the recruitment of histone deacetylases (HDACs)[9,10] (and see below). Activational complexes are recruited during initiation or elongation followed by recruitment of repressive complexes (JMJ2D and Sin3-HDAC) during the termination of transcription.[9] This process establishes the temporal limits for transcription and constrains cryptic transcriptional initiation. The termination of transcription also implies the repackaging of the nucleosome,[9] which is then critical in determining both the subsequent stability of the remodeled epigenome and the probability of subsequent transcriptional activation.[11,12] It is important to note that over the past 10 years the definition of epigenetics has migrated from a rather singular focus on cell differentiation that implies stable biochemical states that explain the fidelity of phenotype, to include an emphasis in dynamic variation linked to an altered probability of future transcriptional activity.

Nucleosome repacking provides the opportunity for the influence of the transcriptional signaling complex to remodel epigenetic states thus determining the subsequent probability of activation. Thus, in vitro studies show that exposure to HDAC inhibitors, which show no inherent capacity for demethylation, can result in the alteration of histone and DNA methylation states.[13–15] Presumably, HDAC inhibition enhances histone acetylation at numerous genes, including those that control other chromatin modifications. As another example, Weber and colleagues[16] suggest that an active chromatin state is involved in precluding DNA methyltransferases (DNMT) recruitment to CpG islands, indeed H3K4me3, which often with open chromatin, is inversely related to DNA methylation.[17] A corollary of this reasoning is that CpG methylation requires active maintenance, which would explain the high expression of DNMT1 in adult brain. In neuronal cell cultures, the knock-down of DNMT1 results in replication-independent demethylation.[18]

The introduction of the temporal component into epigenetic remodeling also reveals the limitations of current approaches that rely heavily on static images provided by chromatin immunoprecipitation (ChIP) assays. Approaches that respect the full dynamics of transcription will provide a more complete image of epigenetic remodeling, and this will include studies of nucleosome positioning and the inclusion of histone variants in newly packaged nucleosomes. Indeed, the study of histone variants may well offer novel insights into the stability of remodeled chromatin.[12] For example, H2A can be replaced by H2AZ, which associates with reduced nucleosome stability, or by H2AX, which associates with DNA repair activity, itself a potential source of demethylation. MacroH2A is associated with transcriptional inactivity, and thus enriched at inactive X chromosomal regions. H3.1/2 can be replaced by H3.3, which associates with greater activity. Importantly, such processes are subject to dynamic regulation. The Swr1 remodeling enzyme enhances the incorporation of the variant histone H2A.Z into nucleosomes.[19] The functional importance of processes such a nucleosome repositioning and histone variants has not yet been studied in detail in neuroepigenetics. Studies of DNA methylation and histone PTMs are only a starting point in explaining dynamic variation in transcriptional activity.

This perspective leads to a deeper level of analysis of the entire epigenetic landscape. One challenge here is that of abandoning our reliance on the notion of a simple histone code.[1,9] The metaphor of a histone "code" was coined within the context of studies in structural biology.[20] The code refers to the affinity of specific histone PTMs for the binding domains present in protein effectors that associate with chromatin and regulate transcriptional activity[21] such as bromodomain for lysine acetylation, and "chromo", "Tudor", "MBT", "WD40repeat", "PHD finger" domains targeting methylated lysines or arginines in a residue-specific manner.[21,22] The code thus describes the relation between histone PTMs and classes of transcriptional mediators; the code *does not refer* to a relation between the histone PTM and transcriptional activity. It is incorrect to assume there is a simple linear relation between the presence of a single histone PTM and ON/OFF transcriptional states. There are at least three reasons underlying this reasoning.

Context Counts

Earlier studies associated DNA methylation as well as specific histone PTMs, such as H3K9me3, with transcriptional silencing. However, this relation is apparent only when these marks occur at regions lying within regulatory elements. For example, DNA methylation lying within gene bodies is *positively* correlated to transcriptional activity, as seems to be the case for H3K9me3.[23] Moreover, the density of the underlying CpGs also moderates the relation between DNA methylation and transcriptional activity even within regulatory sites.[16] The relation between DNA methylation and transcription is closer within regions of low CpG density.

A similar relation between context and transcriptional activity is apparent in a detailed spatial analysis of lysine methylation across the human *PABPC1* gene in relation to active transcription.[16] Thus, H3K9me3 is virtually absent at the promoter region during active transcription, but highly enriched across the region lying 5 kb downstream of the transcriptional start site. This same region shows a depletion of H3K9ac. Stable H3K9me3 signals identified at pericentromeric regions associate with closed chromatin and transcriptional repression; dynamic H3K9me3 signals associate with transcriptionally active sites. Likewise, H3K27me1, a mark commonly associated with heterochromatin, while depleted at the transcriptional start site, was otherwise enriched across the *PABPC1* gene during transcription. Indeed, much of our knowledge of the epigenetic correlates of transcriptional activation is based on studies of modifications at promoter regions, which offers a limited vision of the scope of epigenetic remodeling involved in transcriptional processes. Another limitation is that most studies to date have characterized genome-wide epigenetic modifications in non-neural cells. Studies of the spatio-temporal control of transcription in neurons should reveal the contextual dependency of the relation between transcriptional states and any specific epigenetic modification, as well as control of alternative transcripts that might be critical to the region-specific effects that define neural function.

Bivalency

The assumption that epigenetic marks are uniformly associated with binary on or off transcriptional states contrasts with the results of sequential ChIP experiments showing that both "activating" and "repressive" marks, in particular H3K4me3 and H3K27me3, can co-localize even within common genomic domains.[24] Such bivalency is common in embryonic stem (ES) cells suggesting a relation to phenotypic plasticity. Interestingly, the adult brain is also enriched for these same bivalent marks, perhaps revealing an epigenetic state poised for transcriptional activation in response to appropriate environmental cues.

A common explanation of H3K4me3/H3K27me3 bivalency is that this state confers cellular plasticity. In general, it appears that genes that subsequently become active maintain H3K4me3 and lose H3K27me3, while the opposite occurs for silenced regions. Thus, in ES cells that are transitioned to neural precursor cells and then to neurons, there is a selective increase in H3K27me and its associated polycomb repressor complex-mediated repression among non-neuronal genes.[25] This alteration is especially prominent in the ES cell to neural precursor cell stage. Interestingly, a subsequent increase in DNA methylation appears to then follow a period of polycomb repressor complex-mediated repression during cell differentiation (and see[26]). Thus, DNA methylation follows a period of repression, at this stage an apparent consequence rather than cause of transcriptional silencing. DNA methylation may then further stabilize the repressed state and become statistically correlated with repression. Studies such as those of Mohn and colleagues[25] highlight the bi-directional relation between transcription and epigenetic state.

Promiscuity

The strongest case against the notion of a simple transcriptional code is the remarkable promiscuity of the epigenetic marks. The same epigenetic mark can recruit complexes that activate as well as others that repress transcription. This is true even for histone PTMs previously considered as classic signatures of transcriptional activation *or* repression. These marks are now characterized as potential partners for both activating and repressive effectors.[9] H3K4me3 localizes to the 5′ region of open reading frames and its absence seems a prerequisite for the recruitment of de novo DNA methyltransferases and DNA methylation.[27,28] H3K4me3 targets the nucleosome remodeling factor (NURF) and the PHD containing Yng1 protein in the NuA3 (nucleosomal acetyltransferase of histone H3) complex to genes increasing histone acetylation and transcriptional activation. These interactions are consistent with the idea that H3K4me3 recruits effector complexes that favor transcriptional activation. However, H3K4me3 also binds the Sin3–HDAC1 deacetylation complex through the PHD domain of the Ing2 protein. The Jumonji-D containing protein lysine demethylase, JMJD2A, contained within the N-CoR (nuclear hormone co-repressor complex) also targets H3K4me3; N-CoR represses transcription. Little imaginal discs 2 (Lid2) binding to H3K4me3 results in the methylation of H3K9 and the recruitment of an RNAi mediated-repression complex.[29] Interestingly, Lid2 contains both Jmj-c and PHD domains, and can thus function directly to mediate histone demethylation and indirectly, to enhance histone methylation, depending upon its partners. It may be that the apparent promiscuity of the H3K4me3 mark has a temporal basis, serving the orderly activation and then termination of transcription. Likewise, H3K36me3, which spatially associates with RNA polymerase II, actually serves to recruit HDACs through a chromodomain module that removes acetyl groups and terminates transcription.[10]

The situation is not unique to H3K4me or H3K36me3. H3K9me3 associates with heterochromatin protein 1 (HP1), heterochromatin formation and transcriptional silencing. Yet, both H3K9me3 and HP1 also accompany transcriptional induction.[23] Indeed, JMJD2A deletion, which demethylates H3K9me3, increases H3K9me3 and transcriptional activity.[21,30] Likewise, H4K20me1, initially associated with transcriptional repression[31,32] is now also associated with transcriptional activation, which was confirmed using genome-wide analyses.[33]

As noted above, such apparent contradictions may have a spatial resolution. That aside, it seems safe to assume that individual histone PTMs can mediate bi-directional effects on gene transcription. The challenge is that of identifying the relevant effectors, and then defining the causal relation between effector and transcriptional outcome. Studies of cocaine seeking identified a role for H3K9me2, which is catalyzed by G9a, an effector commonly associated with transcriptional repression.[3] Targeted G9a knock-down reversed the effects of cocaine. This study reveals the importance of moving beyond the mere identification of associated histone PTMs, to direct analysis of cause–effect relations by targeting relevant transcriptional mediators.

The study of DNA methylation is no less complex. DNA methylation of CpG islands within gene promoters is generally associated with transcriptional repression. DNA methylation represses transcription, in part, through the recruitment of methylated DNA binding proteins, in particular methylated-CpG-binding protein 2 (MeCP2) and MBD2. Both MeCP2 and MBD2 are mediators of DNA methylation-induced repression. However, this characterization ignores the multiple potential roles of these proteins in transcriptional regulation. MeCP2 binds at multiple sites, only some of which are methylated CpG regions. This observation might explain why MeCP2 overexpression associates with *both* increased as well as decreased transcriptional activity.[34]

There are also examples where MeCP2 or MBD2 can mediate transcriptional *activation*, even when bound to methylated CpG sites. For example, MBD2 complexes with transforming-acid-coiled–coil 3 (TACC3) and the pCAF histone acetyltransferase (p300/CBP associated factor) to enhance transcriptional activity.[35] The presence of MBD2 within the TACC3/pCAF complex increases histone acetyltransferase activity. Similarly, an MBD2/RNase helicaseA complex enhances CREB-dependent gene expression.[36] Studies of focal adhesion kinase (FAK)-induced muscle differentiation show that MBD2 mediates *both* transcriptional repression and activation from the same site depending upon the presence or absence of FAK.[37] These examples dovetail with studies noted above with neuronal tissues revealing that MeCP2 binding associates with both active and repressed transcriptional states depending upon the presence of CREB.[34] Indeed, MeCP2 may have a special role in the brain consistent with the idea that there are essential tissue-specific differences in the repertoires and roles played by these epigenetic mediators. For example, MeCP2 in the brain is suggested to have a genome-wide "policing" function in repressing transcriptional noise in a methylation dependent manner[38] (e.g. the FAK example noted above). This idea suggests a complexity well beyond simple notions of binary regulation of gene transcription.[1] Epigenetics is the study of structural adaptations of chromosomal regions that "register, signal or perpetuate altered transcriptional states".[11,12] The challenge for any model of neural function is that of describing the relevant mediators that couple the epigenetic signal and transcriptional activity.

IMPLICATIONS OF EPIGENETICS FOR HEALTH SCIENCES

One of the most obvious implications of epigenetics for the health sciences is the ability to explain the variation in genotype–phenotype relations apparent in studies of disease incidence of monozygotic twins.[39] Rates of even highly heritable, common mental disorders vary considerably among individuals with common genotypes. There is considerable evidence for inter-individual variation in DNA methylation across monozygotic twins.[40–45] Such differences are apparent in early childhood and appear to expand with age. One study compared DNA methylation across three genes commonly (though weakly) implicated in psychiatric disorder (i.e. DRD4, SLC6A4 and MAOA*)* in samples from mono- and dizygotic twins. The results suggested differential susceptibility of the different regions to environmental influences, with a pronounced effect on variation in *SLC6A4* DNA methylation.

Environmentally-induced alterations in epigenetic states offer an obvious explanation for both the variation in DNA methylation observed in samples obtained from "genetically-identical" twins and for the discordance in health status. Indeed, as described in several chapters in this volume, there is clear evidence for the idea that environmentally-induced alterations in cell signaling can remodel epigenetic marks. One implication is that of the use of epigenetic states as biomarkers for vulnerability as well as disease state. The rationale for such approaches is that epigenetic states are produced by the interaction of environmental signals and the underlying DNA sequence, and thus reflect gene × environment interactions. As described earlier in this volume, epigenetic marks are linked to mental disorders as well as to states of vulnerability.[46–49]

There is currently enthusiasm in psychiatric epidemiology for studies that examine epigenetic states as a candidate mechanism for the effects of environmental and gene × environmental influences on health status. Indeed, there is preliminary evidence that the quality of early childhood conditions statistically associate with epigenomic states, although to date such studies are limited to measures of DNA methylation.[50,51] However, the application of such approaches to brain-based disorders is compromised by the inaccessibility of CNS samples. Biological psychiatry will thus be limited to proxy samples such as blood or buccal epithelial cells, which raises concerns about validity: to what extent do samples of different cell types reflect the epigenetic marks in relevant brain regions? This is familiar territory for biological psychiatry where pioneering studies on neurotransmitter receptor levels and transmitter metabolism in relation to clinical states were commonly conducted on human samples obtained from blood (e.g. platelets or lymphocytes) or even urine. The question then, as now, is to what degree such measures reflect conditions in the relevant neuronal and glial populations? A comprehensive paper by Mill and colleagues[52] addressed this question comparing DNA methylation signals across blood, cerebellum, and multiple cortical regions. Not surprisingly, the DNA methylation signal corresponded to developmental origin, with blood and cerebellum revealing profiles distinct from cortical regions, particularly in genes associated with cell fate and differentiation. The significant correlation in the DNA methylation signals across blood and brain will no doubt serve as a source of encouragement. Differentially methylated regions across tissues were also associated with differences in gene expression. While these findings are certainly supportive of future research, it is important to note that we would expect DNA methylation levels to correlate to some degree between brain and almost any other tissue: CpG islands lying within promoters tend to be commonly unmethylated, while other regions, such as retrotransposons, should be highly methylated regardless of cell type. A more critical question concerns the correlation in the DNA methylation signal in those regions that show the greatest degree of inter-individual variation, such as the CpG island shores and low-density CpG islands. This concern highlights the need for approaches to the study of epigenetics in neuroscience that focus on genomic regions other than those most commonly associated with transcriptional regulation, such as CpG island promoters.

The relation between the epigenetic state of the genome in neural populations will also vary as a function of the gene. Specific regions of the genome may be influenced by common signals in brain and peripheral cells, such that, for certain genes, peripheral cells may indeed reflect the epigenetic state of the relevant brain cells. This will be difficult to evaluate for some time since we know little about the origins of the variation in the relevant epigenetic mark in the brain. However, we should remain sensitive to the possibility that for certain candidate genes, peripheral cells might well provide an adequate reflection of relevant epigenetic states in the brain. Tyrka and colleagues[53] reported increased DNA methylation of the exon 1F glucocorticoid receptor promoter in leukocytes obtained from adults as a function of childhood adversity. This same relation between childhood adversity and DNA methylation of the exon 1F promoter was previously reported in hippocampal post-mortem samples.[49] This will require a focused approach and consultation of the databases emerging from various international epigenome initiatives, such as the human epigenome roadmap project.[24]

The challenges of epidemiological epigenetic studies have been rendered more tractable with the introduction of more powerful tools, such as the Illumina 450K methylation array. This array surveys about 480000 CpGs across the human genome, or about 20-fold greater coverage than previous arrays. There is a very strong correlation between estimates of DNA methylation using the 450K array and traditional sequencing-based approaches techniques such as reduced representation bisulfite sequencing.[54,55] Likewise, the necessary informatics approaches for this and other tools are advancing to the point where complex data sets can be analyzed in a timely manner.

The field will benefit greatly from the availability of more powerful commercial arrays, data processing that eliminates common pitfalls, such as batch effects, appropriate and more informative informatics analyses and complete mapping of the human epigenome. However, we should not lose sight of the contribution of hypothesis-driven approaches. Epigenetics is the study of transcriptional regulation, and a valuable starting point for any program is that of detailed analysis of variation in transcription in relation to the phenotype of interest. This approach is evident in a number of the chapters presented here, such as that on the effects of drugs of abuse on neural function and behavior. Another example is that of the study of epigenetic regulation of the DNA methylation to schizophrenia. Cortical dysfunction in schizophrenia is associated with changes in GABAergic circuitry[56] and a decrease in the expression of the *GAD1* gene that encodes glutamic acid decarboxylase (GAD_{67}) and there is decreased expression of GAD_{67} in cortical tissues from schizophrenic patients[46,57] as well as reelin,[58] which associates with synaptic plasticity. GABAergic neurons in the schizophrenic brain that express reelin and GAD_{67} exhibit an increase in DNA methyltransferases 1 (DNMT1)[59] and the *reelin* promoter shows increased methylation in the brains of patients with schizophrenia compared with control subjects.[60,61] Inhibition of DNMT1 in neuronal cell lines decreased MeCP2 binding and increased expression of both reelin and GAD_{67}.[18] Interestingly, both in human and rodent cerebral cortex, Gad_{67} promoter-associated DNA methylation and histone PTMs are highly regulated during the course of normal development and aging,[47,48,62–64] and are sensitive to exposure of HDAC inhibitor drugs and even to the atypical antipsychotic, clozapine.[48,65] Therefore, GAD_{67} and other GABAergic gene promoters interconnect the molecular pathology of schizophrenia and related psychiatric disease with developmental mechanisms and environmental influences including drug exposure.

CONCLUSIONS

Such approaches will become critical in establishing cause–effect relations between epigenetic states and transcriptional activity. This is particularly true in light of evidence for sequence-dependent allele-specific methylation, where the epigenetic state varies as a function of underlying DNA sequence. Likewise, the resulting epigenetic state can moderate genotype–phenotype relations, and this influence will be critical for the study of genomic variants that associate with vulnerability for illness. The study of cause–effect relations will also require appropriate in vitro models that permit the study of transcriptional activity from constructs bearing the target sequence variation under varying epigenetic states. The study of recombinant mouse strains, many of which have been well studied with respect to phenotype, including gene expression (see genenetwork.org) would provide complementary in vivo approaches.

Since the transcriptional activity of the gene, in addition to its coding sequence per se, is relevant for health outcomes, the epigenetic state may serve to moderate the influence of sequence variation. This consideration cuts two ways. The epigenetic state might serve to enhance or dampen the functional consequences of a sequence-dependent effect on transcription, such that genotype–phenotype relations would depend upon the local epigenetic context. The same reasoning would suggest that the relation between the epigenetic state and transcriptional activity might be moderated by the underlying DNA sequence variant. The emerging data sets that include measures of genotype, epigenotype, and phenotype will be invaluable in addressing these issues. This will be especially interesting when such measures are derived from longitudinal assessments[66] that will allow for studies of the influence of sequence-based variation on the susceptibility of the epigenome to environmental influences over time.

There is clearly good reason for the excitement associated with studies of epigenetics and mental health. While the inaccessibility of the human brain is indeed a major limitation, it is worth noting that earlier studies of peripheral samples contributed to our understanding of the biochemical basis of psychiatric disease, and revealed the degree to which the underlying biological processes were dynamically regulated by environmental signals. Thus, perhaps the most important contribution will be the capacity to provide finally the biological basis for the integration of nature and nurture, and reveal the degree to which the study of one enriches our understanding of the other.

References

1. Meaney MJ, Ferguson-Smith AC. Epigenetic regulation of the neural transcriptome: the meaning of the marks. *Nat Neurosci*. 2010;13(11):1313–1318.
2. Gluckman PD, Hanson MA. Living with the past: evolution, development, and patterns of disease. *Science*. 2004;305:1733–1736.
3. Maze I, Feng J, Wilkinson MB, Sun HS, Shen L, Nestler EJ. Cocaine dynamically regulates heterochromatin and repetitive element unsilencing in nucleus accumbens. *Proc Natl Acad Sci USA*. 2011;108:3035–3040.
4. Guo JU, Ma DK, Mo H, et al. Neuronal activity modifies the DNA methylation landscape in the adult brain. *Nat Neurosci*. 2011;14:1345–1351.
5. Weaver IC, D'Alessio AC, Brown SE, et al. The transcription factor nerve growth factor-inducible protein a mediates epigenetic programming: altering epigenetic marks by immediate-early genes. *J Neurosci*. 2007;27(7):1756–1768.

6. Lienert F, Wirbelauer C, Som I, Dean A, Mohn F, Schübeler D. Identification of genetic elements that autonomously determine DNA methylation states. *Nat Genet*. 2011;43(11):1091–1097.
7. Stadler MB, Murr R, Burger L, et al. DNA binding factors shape the mouse methylome at distal regulatory regions. *Nature*. 2011;480:490–495.
8. Day JJ, Sweatt JD. Epigenetic mechanisms in cognition. *Neuron*. 2011;70:813–829.
9. Berger SL. The complex language of chromatin regulation during transcription. *Nature*. 2007;447:407–412.
10. Keogh MC, Kurdistani SK, Morris SA, et al. Cotranscriptional Set2 methylation of histone H3 lysine 36 recruits a repressive Rpd3 complex. *Cell*. 2005;123:593–605.
11. Bird AP. Perceptions of epigenetics. *Nature*. 2007;447:396–398.
12. Hake SB, Allis CD. Histone H3 variants and their potential role in indexing mammalian genomes: the "H3 barcode hypothesis". *Proc Natl Acad Sci USA*. 2006;103:6428–6435.
13. Renthal W, Maze I, Krishnan V, et al. Histone deacetylase 5 epigenetically controls behavioral adaptations to chronic emotional stimuli. *Neuron*. 2007;56:517–529.
14. Weaver IC, Cervoni N, Champagne FA, et al. Epigenetic programming through maternal behavior. *Nat Neurosci*. 2004;7:847–854.
15. Szyf M. Epigenetics DNA methylation, and chromatin modifying drugs. *Annu Rev Pharmacol Toxicol*. 2009;49:243–263.
16. Weber M, Hellmann I, Stadler MB, et al. Distribution, silencing potential and evolutionary impact of promoter DNA methylation in the human genome. *Nature Genet*. 2007;39:457–466.
17. Thomson JP, Skene PJ, Selfridge J, et al. CpG islands influence chromatin structure via the CpG-binding protein Cfp1. *Nature*. 2010;464:1082–1087.
18. Kundakovic M, Chen Y, Costa E, Grayson DR. DNA methyltransferase inhibitors coordinately induce expression of the human reelin and glutamic acid decarboxylase 67 genes. *Mol Pharmacol*. 2007;71:644–653.
19. Mizuguchi G, Shen X, Landry J, Wu WH, Sen S, Wu C. ATP-driven exchange of histone H2AZ variant catalyzed by SWR1 chromatin remodeling complex. *Science*. 2004;303(5656):343–348.
20. Jenuwein T, Allis CD. Translating the histone code. *Science*. 2001;293:1074–1080.
21. Taverna SD, Li H, Ruthenburg AJ, Allis CD, Patel DJ. How chromatin-binding modules interpret histone modifications: lessons from professional pocket pickers. *Nat Struct Mol Biol*. 2007;14:1025–1040.
22. Qin J, Whyte WA, Anderssen E, et al. The polycomb group protein L3mbtl2 assembles an atypical PRC1-family complex that is essential in pluripotent stem cells and early development. *Cell Stem Cell*. 2012;11(3):319–332.
23. Vakoc CR, Mandat SA, Olenchock BA, Blobel GA. Histone H3 lysine 9 methylation and HP1γ are associated with transcription elongation through mammalian chromatin. *Mol Cell*. 2005;19:381–391.
24. Bernstein BE, Stamatoyannopoulos JA, Costello JF, et al. The NIH roadmap epigenomics mapping consortium. *Nat Biotechnol*. 2010;28:1045–1048.
25. Mohn F, Weber M, Rebhan M, et al. Lineage-specific polycomb targets and de novo DNA methylation define restriction and potential of neuronal progenitors. *Mol Cell*. 2008;30:755–766.
26. Weiss A, Cedar H. The role of DNA demethylation during development. *Genes Cells*. 1997;2:481–486.
27. Ciccone DN, Su H, Hevi S, et al. KDM1B is a histone H3K4 demethylase required to establish maternal genomic imprints. *Nature*. 2009;461:415–418.
28. Ooi SK, Qiu C, Bernstein E, et al. DNMT3L connects unmethylated lysine 4 of histone H3 to de novo methylation of DNA. *Nature*. 2007;448:714–717.
29. Li F, Huarte M, Zaratiegui M, et al. Lid2 is required for coordinating H3K4 and H3K9 methylation of heterochromatin and euchromatin. *Cell*. 2008;135:272–283.
30. Whetstine JR, Nottke A, Lan F, et al. Reversal of histone lysine trimethylation by the JMJD2 family of histone demethylases. *Cell*. 2006;125:467–481.
31. Kalakonda N, Fischle W, Boccuni P, et al. Histone H4 lysine 20 monomethylation promotes transcriptional repression by L3MBTL1. *Oncogene*. 2008;27(31):4293–4304.
32. Karachentsev D, Sarma K, Reinberg D, Steward R. PR-Set7-dependent methylation of histone H4 Lys 20 functions in repression of gene expression and is essential for mitosis. *Genes Dev*. 2005;19(4):431–435.
33. Barski A, Cuddapah S, Cui K, et al. High-resolution profiling of histone methylations in the human genome. *Cell*. 2007;129(4):823–837.
34. Chahrour M, Jung SY, Shaw C, et al. MeCP2, a key contributor to neurological disease, activates and represses transcription. *Science*. 2008;320:1224–1229.
35. Angrisano T, Lembo F, Pero R, et al. TACC3 mediates the association of MBD2 with histone acetyltransferases and relieves transcriptional repression of methylated promoters. *Nuc Acids Res*. 2006;34:364–372.

36. Fujita H, Fujii R, Aratani S, Amano T, Fukamizu A, Nakajima T. Antithetic effects of MBD2a on gene regulation. *Molec Cell Biol*. 2003;23:2645–2657.
37. Luo SW, Zhang C, Zhang B, et al. Regulation of heterochromatin remodeling and myogenin expression during muscle differentiation by FAK interaction with MBD2. *EMBO J*. 2009;28:2568–2582.
38. Skene PJ, Illingworth RS, Webb S, et al. Neuronal MeCP2 is expressed at near histone-octamer levels and globally alters chromatin state. *Mol Cell*. 2010;37:457–468.
39. Petronis A. Epigenetics as a unifying principle in the aetiology of complex traits and diseases. *Nature*. 2010;465:721–727.
40. Boks MP, Derks EM, Weisenberger DJ, et al. The relationship of DNA methylation with age, gender and genotype in twins and healthy controls. *PLoS One*. 2009;4:6767.
41. Fraga MF, Ballestar E, Paz MF, et al. Epigenetic differences arise during the lifetime of monozygotic twins. *Proc Natl Acad Sci USA*. 2005;102:10604–10609.
42. Kaminsky ZA, Tang T, Wang SC, et al. DNA methylation profiles in monozygotic and dizygotic twins. *Nat Genet*. 2009;41:240–245.
43. Mill J, Dempster E, Caspi A, Williams B, Moffitt T, Craig I. Evidence for monozygotic twin (MZ) discordance in methylation level at two CpG sites in the promoter region of the catechol-O-methyltransferase (COMT) gene. *Am J Med Genet B Neuropsychiatr Genet*. 2006;141:421–425.
44. Petronis A, Gottesman II, Kan P, et al. Monozygotic twins exhibit numerous epigenetic differences: clues to twin discordance? *Schizophr Bull*. 2003;29:169–178.
45. Wong CC, Caspi A, Williams B, et al. A longitudinal study of epigenetic variation in twins. *Epigenetics*. 2010;5:516–526.
46. Costa E, Davis JM, Dong E, et al. A GABAergic cortical deficit dominates schizophrenia pathophysiology. *Crit Rev Neurobiol*. 2004;16:1–23.
47. Huang HS, Akbarian S. GAD1 mRNA expression and DNA methylation in prefrontal cortex of subjects with schizophrenia. *PLoS One*. 2007;2:e809.
48. Huang HS, Matevossian A, Whittle C, et al. Prefrontal dysfunction in schizophrenia involves mixed-lineage leukemia 1-regulated histone methylation at GABAergic gene promoters. *J Neurosci*. 2007;27:11254–11262.
49. McGowan PO, Sasaki A, D'Alessio AC, et al. Epigenetic regulation of the glucocorticoid receptor in human brain associates with childhood abuse. *Nat Neurosci*. 2009;12(3):342–348.
50. Borghol N, Suderman M, McArdle W, et al. Associations with early-life socio-economic position in adult DNA methylation. *Int J Epidemiol*. 2012;41:62–74.
51. Essex MJ, Thomas Boyce W, Hertzman C, et al. Epigenetic vestiges of early developmental adversity: childhood stress exposure and DNA methylation in adolescence. *Child Dev*. 2011;doi: 10.1111/j.1467-8624.2011.01641.x
52. Davies MN, Volta M, Pidsley R, et al. Functional annotation of the human brain methylome identifies tissue-specific epigenetic variation across brain and blood. *Genome Biol*. 2012;13:R58.
53. Tyrka AR, Prince LH, Marsit C, Walters OC, Carpenter LL. Childhood adversity and epigenetic modulation of the leukocyte glucocorticoid receptor: preliminary findings in healthy adults. *PloS One*. 2012;7:e30148.
54. Bock C, Tomazou EM, Brinkman AB, et al. Quantitative comparison of genome-wide DNA methylation mapping technologies. *Nat Biotechnol*. 2010;28:1106–1114.
55. Pan H, Chen L, Dogra S, et al. Measuring the methylome in clinical samples – improved processing of the Infinium Human Methylation450 BeadChip Array. *Epigenetics* 2012;7(10). [Epub ahead of print]
56. Benes FM, Berretta S. GABAergic interneurons: implications for understanding schizophrenia and bipolar disorder. *Neuropsychopharmacology*. 2001;25:1–27.
57. Akbarian S, Huang HS. Molecular and cellular mechanisms of altered GAD1/GAD67 expression in schizophrenia and related disorders. *Brain Res Brain Res Rev*. 2006;52:293–304.
58. Eastwood SL, Harrison PJ. Interstitial white matter neurons express less reelin and are abnormally distributed in schizophrenia: towards an integration of molecular and morphologic aspects of the neurodevelopmental hypothesis. *Mol Psychiatry*. 2003;769:821–831.
59. Veldic M, Caruncho HJ, Liu S, et al. DNA-methyltransferase 1 mRNA is selectively overexpressed in telencephalic GABAergic interneurons of schizophrenia brains. *Proc Natl Acad Sci USA*. 2004;101:348–353.
60. Abdolmaleky HM, Cheng K, Russo A, et al. Hypermethylation of the reelin (RELN) promoter in the brain of schizophrenic patients: a preliminary report. *Am J Med Genet B Neuropsychiatr Genet*. 2005;134:60–66.

61. Grayson DR, Jia X, Chen Y, et al. Reelin promoter hypermethylation in schizophrenia. *Proc Natl Acad Sci USA*. 2005;102:9341–9346.
62. Siegmund KD, Connor CM, Campan M, et al. DNA methylation in the human cerebral cortex is dynamically regulated throughout the life span and involves differentiated neurons. *PLoS One*. 2007;2:e895.
63. Tang B, Dean B, Thomas EA. Disease- and age-related changes in histone acetylation at gene promoters in psychiatric disorders. *Transl Psychiatry*. 2011:e64.
64. Zhang TY, Hellstrom IC, Bagot RC, Wen X, Diorio J, Meaney MJ. Maternal care and DNA methylation of a glutamic acid decarboxylase 1 promoter in rat hippocampus. *J Neurosci*. 2010;30:13130–13137.
65. Chen Y, Dong E, Grayson DR. Analysis of the GAD1 promoter: trans-acting factors and DNA methylation converge on the 5′ untranslated region. *Neuropharmacology*. 2011;60:1075–1087.
66. Ng JWY, Laura M, Barrett LM, et al. The role of longitudinal cohort studies in epigenetic epidemiology: challenges and opportunities. *Genome Biol*. 2012;13:246.

Index

Note: Page numbers followed by "*f*" and "*t*" refer to figures and tables respectively.

N

O

P

U

V

W

Printed and bound by CPI Group (UK) Ltd, Croydon, CR0 4YY
08/05/2025
01865018-0002